MERS-CoV

MERS-CoV

Special Issue Editors

Fang Li
Lanying Du

MDPI • Basel • Beijing • Wuhan • Barcelona • Belgrade

MDPI

Special Issue Editors

Fang Li
Department of Veterinary
and Biomedical Sciences,
University of Minnesota
USA

Lanying Du
Viral Immunology Laboratory,
Lindsley F. Kimball Research Institute,
New York Blood Center
USA

Editorial Office
MDPI
St. Alban-Anlage 66
4052 Basel, Switzerland

This is a reprint of articles from the Special Issue published online in the open access journal *Viruses* (ISSN 1999-4915) from 2018 to 2019 (available at: https://www.mdpi.com/journal/viruses/special_issues/MERS_CoV).

For citation purposes, cite each article independently as indicated on the article page online and as indicated below:

LastName, A.A.; LastName, B.B.; LastName, C.C. Article Title. *Journal Name* **Year**, *Article Number*, Page Range.

ISBN 978-3-03921-850-9 (Pbk)
ISBN 978-3-03921-851-6 (PDF)

Contents

About the Special Issue Editors

Fang Li is an Associate Professor in the Department of Veterinary and Biomedical Sciences at the University of Minnesota. His main line of research examines the invasion mechanisms of viruses. Specifically, his group investigates the structures and functions of virus-surface proteins that mediate receptor recognition and cell entry of viruses. His other line of research explores the structural and molecular basis for cancer and abnormal blood pressure. Specifically, his group investigates the structures and functions of mammalian-cell-surface enzymes that are critical for tumor cell growth and blood pressure regulation. Based on these structural and functional studies, his group further develops novel therapy strategies to treat human diseases. His research tools include X-ray crystallography, cryo-electron microscopy, protein biochemistry, molecular virology, and vaccine and drug designs.

Lanying Du is an Associate Member and Head of Viral Immunology Laboratory at Lindsley F. Kimball Research Institute of New York Blood Center, USA. Her research focuses are to: (1) design and develop effective and safe vaccines and therapeutic agents against coronaviruses (including MERS-CoV, SARS-CoV, and other coronaviruses with pandemic potential), influenza viruses, and flaviviruses (including the Zika virus and dengue virus); (2) understand protective mechanisms of the developed vaccines and therapeutics; and (3) study pathogenic mechanisms of these viruses, based on which to design novel vaccines and therapeutics. Her research tools include structure-based design of novel vaccines and therapeutics, mRNA technology, drug screening, and antibody production and evaluation.

Editorial

MERS Coronavirus: An Emerging Zoonotic Virus

Fang Li [1],* and Lanying Du [2],*

[1]   Department of Veterinary and Biomedical Sciences, College of Veterinary Medicine, University of Minnesota, Saint Paul, MN 55108, USA

[2]   Lindsley F. Kimball Research Institute, New York Blood Center, New York, NY 10065, USA

*   Correspondence: lifang@umn.edu (F.L.); ldu@nybc.org (L.D.);
      Tel.: +1-612-625-6149 (F.L.); +1-212-570-3459 (L.D.)

Received: 16 July 2019; Accepted: 17 July 2019; Published: 19 July 2019

Middle East respiratory syndrome coronavirus (MERS-CoV) is an emerging virus that was first reported in humans in June 2012 [1]. To date, MERS-CoV continues to infect humans with a fatality rate of ~35%. At least 27 countries have reported human infections with MERS-CoV (https://www.who.int/emergencies/mers-cov/en/). MERS-CoV is a zoonotic virus. Like severe acute respiratory syndrome coronavirus (SARS-CoV), MERS-CoV is believed to have originated from bats [2,3]. However, whereas the bat-to-human transmission of SARS-CoV was likely mediated by palm civets as intermediate hosts, humans likely acquired MERS-CoV from dromedary camels [4–6]. Human-to-human transmission of MERS-CoV does occur, but it is limited mostly to health care environments [7,8]. Moreover, whereas SARS-CoV recognizes angiotensin-converting enzyme 2 (ACE2) as a cellular receptor [9,10], MERS-CoV uses dipeptidyl peptidase 4 (DPP4) to enter target cells [11,12]. Currently, no vaccines or antiviral therapeutics have been approved for the prevention or treatment of MERS-CoV infection, although a number of them have been developed preclinically and/or tested clinically [13–16].

The articles in this special issue of *Viruses* were written by researchers working in the MERS-CoV field. The main aims of this issue are to (i) better understand MERS-CoV transmission, epidemiology, and pathogenesis; (ii) summarize current progress on MERS-CoV animal models, vaccines, and therapeutics; and (iii) discuss future prospects for MERS-CoV research. This issue includes seven review articles and nine original research papers, each providing detailed updates on current MERS-CoV studies.

Studies on the transmission, epidemiology, and pathogenesis of MERS-CoV form one of the foundations of MERS-CoV research. In this issue, Farag and colleagues summarize the possible drivers of the emergence of MERS-CoV and its spillover to humans in Qatar, explaining the potential reasons for the camel-to-human transmission of MERS-CoV [17]. The review article by Song and colleagues provides an overall description of the epidemiology, pathogenesis, and other important aspects of MERS-CoV [18]. Widagdo and colleagues review the host determinants of the transmission and pathogenesis of MERS-CoV, indicating that receptor DPP4 plays an important role in these processes [19]. A research article by Yan and colleagues characterizes the role of lipid profiles in the pathogenesis and infectivity of human coronaviruses, including MERS-CoV, suggesting that lipid metabolism may be involved in the propagations of these coronaviruses [20]. These reports provide insights into how MERS-CoV infects cells and spreads within and across host species. They have also laid the foundations for developing animal models.

Animal models are essential tools for the preclinical evaluation of anti-MERS-CoV countermeasures. Dromedary camels, alpacas, and non-human primates are susceptible to MERS-CoV infection [21–23]; however, the virus does not infect small animals such as mice, hamsters, and ferrets [24–26]. Several mouse models that express human DPP4 (hDPP4) have been established for MERS-CoV infection [27–29]. In this issue, Widagdo and colleagues examine rabbits as potential hosts for MERS-CoV, showing that MERS-CoV infects rabbits without causing symptoms; they also analyze the route of MERS-CoV transmission in rabbits [30]. Fan and colleagues report the development of an hDPP4-expressing mouse model through inserting hDPP4 gene into a constitutive and ubiquitous gene expression locus using CRISPR/Cas9 technology. This mouse model is susceptible to MERS-CoV infection [31]. These articles have established platforms for testing vaccines and therapeutic agents targeting MERS-CoV.

Effective vaccines are essential for preventing MERS-CoV infection. The MERS-CoV surface spike (S) protein is a key target for vaccine design [14]. The S protein comprises two subunits: the S1 subunit is responsible for binding to the DPP4 receptor via a receptor-binding domain (RBD), and the S2 subunit mediates virus–host membrane fusion [32–35]. Several MERS-CoV S protein-based vaccines have been developed; when tested in animal models, they showed protective efficacy against MERS-CoV [14]. In this issue, Schindewolf and Menachery summarize the progress of MERS-CoV S-protein-based vaccine development and also describe potential challenges [36]. Zhou and colleagues review current advances in RBD-based MERS-CoV vaccines [37]. A research paper by Adney and colleagues evaluates the efficacy of a MERS-CoV S1 subunit vaccine aided by adjuvants; the authors report reduced and delayed viral shedding in dromedary camels as well as the complete protection of alpacas from MERS-CoV infection [38]. This and other studies demonstrate that the protective efficacy of MERS vaccines positively correlates with neutralizing antibody titers in serum [38,39]. In addition to inducing neutralizing antibodies, some types of vaccines can induce cellular immune responses against MERS-CoV. Other than the S protein, structural proteins such as the nucleocapsid (N) protein may also serve as vaccine targets. Here, Veit and colleagues report that a MERS-CoV N protein-based vaccine, which is delivered through a modified Vaccinia virus, induces CD8$^+$ T cell responses in a mouse model; they further identify a MERS-CoV N protein-specific CD8$^+$ T cell epitope on the vaccine [40]. Overall, these reports demonstrate that a variety of promising vaccine tools are available to prevent MERS-CoV infection in humans and other animals.

Therapeutics are critical tools for treating MERS-CoV infection. Again, the MERS-CoV S protein is an important target for therapeutic development [16]. MERS-CoV S2 contains two heptad repeat regions, HR1 and HR2, that are critical for S protein-mediated membrane fusion [34]. Hence, peptides mimicking HR1 or HR2 may interfere with the viral membrane-fusion process [34]. Moreover, RBD-targeting neutralizing monoclonal antibodies (mAbs) can block the viral attachment step [37]. In addition to conventional mAbs, single-domain antibodies isolated from camelids, called nanobodies (Nbs), can also block RBD/receptor interactions; these Nbs have been gaining popularity as therapeutic agents due to their small size and high stability [41,42]. Thus, both the HR1/HR2 peptide mimics and RBD-targeting mAbs and Nbs may serve as MERS-CoV entry inhibitors. Furthermore, small molecules targeting the S protein or nonstructural proteins may serve as therapeutic alternatives to peptide mimics and antibodies [16,43]. In this issue, two review articles report the current advances in therapeutic neutralizing antibodies, one by Han and colleagues and the other by Zhou and colleagues [37,44]. The latter article also discusses potential strategies and challenges to improving the efficacy of therapeutic neutralizing antibodies. In a research article, He and colleagues describe the construction and expression of dimeric and trimeric Nbs that target MERS-CoV RBD and further demonstrate the strong stability and high neutralizing activity of these Nbs against multiple MERS-CoV strains [42]. In another research article, Xia and colleagues report that three peptides mimicking HR2 from HKU4 (which is a MERS-related coronavirus from bats) strongly inhibit MERS-CoV infection [45]. Interestingly, Wang and colleagues report that the combination of a MERS-CoV HR2 peptide mimic and an RBD-targeting neutralizing mAb demonstrate potent synergistic effects in inhibiting MERS-CoV S protein-mediated viral entry [46]. In another research article, Jiang and colleagues report that an antibody targeting complement receptor C5aR1 inhibits MERS-CoV infection, indicating that MERS-CoV infection elicits the over-activation of the complement system, and this process can be blocked by anti-C5aR1 antibodies [47]. Moreover, Liang and colleagues review advances in the development of small-molecular MERS-CoV inhibitors [48]. Overall, these articles confirm that anti-MERS-CoV therapeutics have great potentials in treating MERS-CoV infections in humans and other animals.

To summarize, significant progress has been made in MERS-CoV research in the past seven years since the virus was discovered. This progress includes, but is not limited to, the epidemiology, transmission, and pathogenesis of MERS-CoV, as well as animal models, vaccines, and antivirals for MERS-CoV. This special issue of *Viruses* provides updated reports on this progress. However, challenges remain. For example, we still do not understand how exactly MERS-CoV transmits from bats to camels or humans. Moreover, compared to HIV and influenza viruses, the potential market for MERS-CoV vaccines and therapeutics is much smaller, making commercialization of MERS-related products more challenging. Nevertheless, the past two decades have witnessed the emergence of two highly pathogenic coronaviruses, MERS-CoV and SARS-CoV. While these two viruses remain significant threats to global health, future novel coronaviruses with pandemic potential may emerge from their animal reservoirs and infect humans. Thus, research into MERS-CoV should remain a high priority for the virology community. In fact, the impressive progress in MERS-CoV research has benefitted tremendously from previous research into coronaviruses including SARS-CoV. Therefore, scientists' current efforts regarding MERS-CoV will prepare humans to battle any future novel coronaviruses with pandemic potential.

Acknowledgments: Our studies are supported by the NIH grants (R01AI139092, R01AI137472, R01AI089728, and R01AI110700). We would like to thank all authors and reviewers for their contributions to this special issue of *Viruses*.

Conflicts of Interest: We declare no competing interests.

References

1. Zaki, A.M.; van Boheemen, S.; Bestebroer, T.M.; Osterhaus, A.D.; Fouchier, R.A. Isolation of a novel coronavirus from a man with pneumonia in Saudi Arabia. *N. Engl. J. Med.* **2012**, *367*, 1814–1820. [CrossRef] [PubMed]
2. Yang, Y.; Du, L.; Liu, C.; Wang, L.; Ma, C.; Tang, J.; Baric, R.S.; Jiang, S.; Li, F. Receptor usage and cell entry of bat coronavirus HKU4 provide insight into bat-to-human transmission of MERS coronavirus. *Proc. Natl. Acad. Sci. USA* **2014**, *111*, 12516–12521. [CrossRef] [PubMed]
3. Wang, L.F.; Shi, Z.; Zhang, S.; Field, H.; Daszak, P.; Eaton, B.T. Review of bats and SARS. *Emerg. Infect. Dis.* **2006**, *12*, 1834–1840. [CrossRef]
4. Du, L.; He, Y.; Zhou, Y.; Liu, S.; Zheng, B.J.; Jiang, S. The spike protein of SARS-CoV–a target for vaccine and therapeutic development. *Nat. Rev. Microbiol.* **2009**, *7*, 226–236. [CrossRef] [PubMed]
5. Alshukairi, A.N.; Zheng, J.; Zhao, J.; Nehdi, A.; Baharoon, S.A.; Layqah, L.; Bokhari, A.; Al Johani, S.M.; Samman, N.; Boudjelal, M.; et al. High prevalence of MERS-CoV infection in camel workers in Saudi Arabia. *MBio* **2018**, *9*, e01985-18. [CrossRef] [PubMed]
6. Haagmans, B.L.; Al Dhahiry, S.H.; Reusken, C.B.; Raj, V.S.; Galiano, M.; Myers, R.; Godeke, G.J.; Jonges, M.; Farag, E.; Diab, A.; et al. Middle East respiratory syndrome coronavirus in dromedary camels: An outbreak investigation. *Lancet Infect. Dis.* **2014**, *14*, 140–145. [CrossRef]
7. Hunter, J.C.; Nguyen, D.; Aden, B.; Al, B.Z.; Al, D.W.; Abu, E.K.; Khudair, A.; Al, M.M.; El, S.F.; Imambaccus, H.; et al. Transmission of Middle East respiratory syndrome coronavirus infections in healthcare settings, Abu Dhabi. *Emerg. Infect. Dis.* **2016**, *22*, 647–656. [CrossRef] [PubMed]
8. Oboho, I.K.; Tomczyk, S.M.; Al-Asmari, A.M.; Banjar, A.A.; Al-Mugti, H.; Aloraini, M.S.; Alkhaldi, K.Z.; Almohammadi, E.L.; Alraddadi, B.M.; Gerber, S.I.; et al. 2014 MERS-CoV outbreak in Jeddah—A link to health care facilities. *N. Engl. J. Med.* **2015**, *372*, 846–854. [CrossRef]
9. Li, F.; Li, W.; Farzan, M.; Harrison, S.C. Structure of SARS coronavirus spike receptor-binding domain complexed with receptor. *Science* **2005**, *309*, 1864–1868. [CrossRef]
10. Li, W.; Moore, M.J.; Vasilieva, N.; Sui, J.; Wong, S.K.; Berne, M.A.; Somasundaran, M.; Sullivan, J.L.; Luzuriaga, K.; Greenough, T.C.; et al. Angiotensin-converting enzyme 2 is a functional receptor for the SARS coronavirus. *Nature* **2003**, *426*, 450–454. [CrossRef]
11. Raj, V.S.; Mou, H.; Smits, S.L.; Dekkers, D.H.; Muller, M.A.; Dijkman, R.; Muth, D.; Demmers, J.A.; Zaki, A.; Fouchier, R.A.; et al. Dipeptidyl peptidase 4 is a functional receptor for the emerging human coronavirus-EMC. *Nature* **2013**, *495*, 251–254. [CrossRef] [PubMed]

12. Li, F. Receptor recognition mechanisms of coronaviruses: A decade of structural studies. *J. Virol.* **2015**, *89*, 1954–1964. [CrossRef] [PubMed]

13. Haagmans, B.L.; van den Brand, J.M.; Raj, V.S.; Volz, A.; Wohlsein, P.; Smits, S.L.; Schipper, D.; Bestebroer, T.M.; Okba, N.; Fux, R.; et al. An orthopoxvirus-based vaccine reduces virus excretion after MERS-CoV infection in dromedary camels. *Science* **2016**, *351*, 77–81. [CrossRef] [PubMed]

14. Zhou, Y.; Jiang, S.; Du, L. Prospects for a MERS-CoV spike vaccine. *Expert Rev. Vaccines* **2018**, *17*, 677–686. [CrossRef] [PubMed]

15. Beigel, J.H.; Voell, J.; Kumar, P.; Raviprakash, K.; Wu, H.; Jiao, J.A.; Sullivan, E.; Luke, T.; Davey, R.T., Jr. Safety and tolerability of a novel, polyclonal human anti-MERS coronavirus antibody produced from transchromosomic cattle: A phase 1 randomised, double-blind, single-dose-escalation study. *Lancet Infect. Dis.* **2018**, *18*, 410–418. [CrossRef]

16. Du, L.; Yang, Y.; Zhou, Y.; Lu, L.; Li, F.; Jiang, S. MERS-CoV spike protein: A key target for antivirals. *Expert Opin. Ther. Targets* **2017**, *21*, 131–143. [CrossRef] [PubMed]

17. Farag, E.; Sikkema, R.S.; Vinks, T.; Islam, M.M.; Nour, M.; Al-Romaihi, H.; Al, T.M.; Atta, M.; Alhajri, F.H.; Al-Marri, S.; et al. Drivers of MERS-CoV emergence in Qatar. *Viruses* **2018**, *11*, 22. [CrossRef] [PubMed]

18. Song, Z.; Xu, Y.; Bao, L.; Zhang, L.; Yu, P.; Qu, Y.; Zhu, H.; Zhao, W.; Han, Y.; Qin, C. From SARS to MERS, thrusting coronaviruses into the spotlight. *Viruses* **2019**, *11*, 59. [CrossRef]

19. Widagdo, W.; Sooksawasdi Na Ayudhya, S.; Hundie, G.B.; Haagmans, B.L. Host determinants of MERS-CoV transmission and pathogenesis. *Viruses* **2019**, *11*, 280. [CrossRef]

20. Yan, B.; Chu, H.; Yang, D.; Sze, K.H.; Lai, P.M.; Yuan, S.; Shuai, H.; Wang, Y.; Kao, R.Y.; Chan, J.F.; et al. Characterization of the lipidomic profile of human coronavirus-infected cells: Implications for lipid metabolism remodeling upon coronavirus replication. *Viruses* **2019**, *11*, 73. [CrossRef]

21. Adney, D.R.; van Doremalen, N.; Brown, V.R.; Bushmaker, T.; Scott, D.; de Wit, E.; Bowen, R.A.; Munster, V.J. Replication and shedding of MERS-CoV in upper respiratory tract of inoculated dromedary camels. *Emerg. Infect. Dis.* **2014**, *20*, 1999–2005. [CrossRef] [PubMed]

22. Yao, Y.; Bao, L.; Deng, W.; Xu, L.; Li, F.; Lv, Q.; Yu, P.; Chen, T.; Xu, Y.; Zhu, H.; et al. An animal model of MERS produced by infection of rhesus macaques with MERS coronavirus. *J. Infect. Dis.* **2014**, *209*, 236–242. [CrossRef] [PubMed]

23. Adney, D.R.; Bielefeldt-Ohmann, H.; Hartwig, A.E.; Bowen, R.A. Infection, replication, and transmission of Middle East respiratory syndrome coronavirus in alpacas. *Emerg. Infect. Dis.* **2016**, *22*, 1031–1037. [CrossRef] [PubMed]

24. De Wit, E.; Prescott, J.; Baseler, L.; Bushmaker, T.; Thomas, T.; Lackemeyer, M.G.; Martellaro, C.; Milne-Price, S.; Haddock, E.; Haagmans, B.L.; et al. The Middle East respiratory syndrome coronavirus (MERS-CoV) does not replicate in Syrian hamsters. *PLoS ONE* **2013**, *8*, e69127. [CrossRef] [PubMed]

25. Raj, V.S.; Smits, S.L.; Provacia, L.B.; van den Brand, J.M.; Wiersma, L.; Ouwendijk, W.J.; Bestebroer, T.M.; Spronken, M.I.; van Amerongen, G.; Rottier, P.J.; et al. Adenosine deaminase acts as a natural antagonist for dipeptidyl peptidase 4-mediated entry of the Middle East respiratory syndrome coronavirus. *J. Virol.* **2014**, *88*, 1834–1838. [CrossRef] [PubMed]

26. Coleman, C.M.; Matthews, K.L.; Goicochea, L.; Frieman, M.B. Wild-type and innate immune-deficient mice are not susceptible to the Middle East respiratory syndrome coronavirus. *J. Gen. Virol.* **2014**, *95*, 408–412. [CrossRef]

27. Zhao, G.; Jiang, Y.; Qiu, H.; Gao, T.; Zeng, Y.; Guo, Y.; Yu, H.; Li, J.; Kou, Z.; Du, L.; et al. Multi-organ damage in human dipeptidyl peptidase 4 transgenic mice infected with Middle East respiratory syndrome-coronavirus. *PLoS ONE* **2015**, *10*, e0145561. [CrossRef]

28. Zhao, J.; Li, K.; Wohlford-Lenane, C.; Agnihothram, S.S.; Fett, C.; Zhao, J.; Gale, M.J., Jr.; Baric, R.S.; Enjuanes, L.; Gallagher, T.; et al. Rapid generation of a mouse model for Middle East respiratory syndrome. *Proc. Natl. Acad. Sci. USA* **2014**, *111*, 4970–4975. [CrossRef]

29. Li, K.; Wohlford-Lenane, C.L.; Channappanavar, R.; Park, J.E.; Earnest, J.T.; Bair, T.B.; Bates, A.M.; Brogden, K.A.; Flaherty, H.A.; Gallagher, T.; et al. Mouse-adapted MERS coronavirus causes lethal lung disease in human DPP4 knockin mice. *Proc. Natl. Acad. Sci. USA* **2017**, *114*, E3119–E3128. [CrossRef]

30. Widagdo, W.; Okba, N.M.A.; Richard, M.; de Meulder, D.; Bestebroer, T.M.; Lexmond, P.; Farag, E.A.B.A.; Al-Hajri, M.; Stittelaar, K.J.; de Waal, L.; et al. Lack of Middle East respiratory syndrome coronavirus transmission in rabbits. *Viruses* **2019**, *11*, 381. [CrossRef]

31. Fan, C.; Wu, X.; Liu, Q.; Li, Q.; Liu, S.; Lu, J.; Yang, Y.; Cao, Y.; Huang, W.; Liang, C.; et al. A human DPP4-knockin mouse's susceptibility to infection by authentic and pseudotyped MERS-CoV. *Viruses* **2018**, *10*, 448. [CrossRef] [PubMed]

32. Chen, Y.; Rajashankar, K.R.; Yang, Y.; Agnihothram, S.S.; Liu, C.; Lin, Y.L.; Baric, R.S.; Li, F. Crystal structure of the receptor-binding domain from newly emerged Middle East respiratory syndrome coronavirus. *J. Virol.* **2013**, *87*, 10777–10783. [CrossRef] [PubMed]

33. Wang, N.; Shi, X.; Jiang, L.; Zhang, S.; Wang, D.; Tong, P.; Guo, D.; Fu, L.; Cui, Y.; Liu, X.; et al. Structure of MERS-CoV spike receptor-binding domain complexed with human receptor DPP4. *Cell Res.* **2013**, *23*, 986–993. [CrossRef] [PubMed]

34. Lu, L.; Liu, Q.; Zhu, Y.; Chan, K.H.; Qin, L.; Li, Y.; Wang, Q.; Chan, J.F.; Du, L.; Yu, F.; et al. Structure-based discovery of Middle East respiratory syndrome coronavirus fusion inhibitor. *Nat. Commun.* **2014**, *5*, 3067. [CrossRef] [PubMed]

35. Li, F. Structure, function, and evolution of coronavirus spike proteins. *Annu. Rev. Virol.* **2016**, *3*, 237–261. [CrossRef] [PubMed]

36. Schindewolf, C.; Menachery, V.D. Middle East respiratory syndrome vaccine candidates: Cautious optimism. *Viruses* **2019**, *11*, 74. [CrossRef] [PubMed]

37. Zhou, Y.; Yang, Y.; Huang, J.; Jiang, S.; Du, L. Advances in MERS-CoV vaccines and therapeutics based on the receptor-binding domain. *Viruses* **2019**, *11*, 60. [CrossRef]

38. Adney, D.R.; Wang, L.; van Doremalen, N.; Shi, W.; Zhang, Y.; Kong, W.P.; Miller, M.R.; Bushmaker, T.; Scott, D.; de Wit, E.; et al. Efficacy of an adjuvanted Middle East respiratory syndrome coronavirus spike protein vaccine in dromedary camels and alpacas. *Viruses* **2019**, *11*, 212. [CrossRef]

39. Wang, Y.; Tai, W.; Yang, J.; Zhao, G.; Sun, S.; Tseng, C.K.; Jiang, S.; Zhou, Y.; Du, L.; Gao, J. Receptor-binding domain of MERS-CoV with optimal immunogen dosage and immunization interval protects human transgenic mice from MERS-CoV infection. *Hum. Vaccines Immunother.* **2017**, *13*, 1615–1624. [CrossRef]

40. Veit, S.; Jany, S.; Fux, R.; Sutter, G.; Volz, A. CD8+ T cells responding to the Middle East respiratory syndrome coronavirus nucleocapsid protein delivered by vaccinia virus MVA in mice. *Viruses* **2018**, *10*, 718. [CrossRef]

41. Zhao, G.; He, L.; Sun, S.; Qiu, H.; Tai, W.; Chen, J.; Li, J.; Chen, Y.; Guo, Y.; Wang, Y.; et al. A novel nanobody targeting Middle East respiratory syndrome coronavirus (MERS-CoV) receptor-binding domain has potent cross-neutralizing activity and protective efficacy against MERS-CoV. *J. Virol.* **2018**, *92*, e00837-18. [CrossRef] [PubMed]

42. He, L.; Tai, W.; Li, J.; Chen, Y.; Gao, Y.; Li, J.; Sun, S.; Zhou, Y.; Du, L.; Zhao, G. Enhanced ability of oligomeric nanobodies targeting MERS coronavirus receptor-binding domain. *Viruses* **2019**, *11*, 166. [CrossRef] [PubMed]

43. De Wilde, A.H.; Jochmans, D.; Posthuma, C.C.; Zevenhoven-Dobbe, J.C.; van Nieuwkoop, S.; Bestebroer, T.M.; van den Hoogen, B.G.; Neyts, J.; Snijder, E.J. Screening of an FDA-approved compound library identifies four small-molecule inhibitors of Middle East respiratory syndrome coronavirus replication in cell culture. *Antimicrob. Agents Chemother.* **2014**, *58*, 4875–4884. [CrossRef]

44. Han, H.J.; Liu, J.W.; Yu, H.; Yu, X.J. Neutralizing monoclonal antibodies as promising therapeutics against Middle East respiratory syndrome coronavirus infection. *Viruses* **2018**, *10*, 680. [CrossRef] [PubMed]

45. Xia, S.; Lan, Q.; Pu, J.; Wang, C.; Liu, Z.; Xu, W.; Wang, Q.; Liu, H.; Jiang, S.; Lu, L. Potent MERS-CoV fusion inhibitory peptides identified from HR2 domain in spike protein of bat coronavirus HKU4. *Viruses* **2019**, *11*, 56. [CrossRef] [PubMed]

46. Wang, C.; Hua, C.; Xia, S.; Li, W.; Lu, L.; Jiang, S. Combining a fusion inhibitory peptide targeting the MERS-CoV S2 protein HR1 domain and a neutralizing antibody specific for the S1 protein receptor-binding domain (RBD) showed potent synergism against pseudotyped MERS-CoV with or without mutations in RBD. *Viruses* **2019**, *11*, 31. [CrossRef]

47. Jiang, Y.; Li, J.; Teng, Y.; Sun, H.; Tian, G.; He, L.; Li, P.; Chen, Y.; Guo, Y.; Li, J.; et al. Complement receptor C5aR1 inhibition reduces pyroptosis in hDPP4-transgenic mice infected with MERS-CoV. *Viruses* **2019**, *11*, 39. [CrossRef] [PubMed]
48. Liang, R.; Wang, L.; Zhang, N.; Deng, X.; Su, M.; Su, Y.; Hu, L.; He, C.; Ying, T.; Jiang, S.; et al. Development of small-molecule MERS-CoV inhibitors. *Viruses* **2018**, *10*, 721. [CrossRef] [PubMed]

viruses

MDPI

Review

Drivers of MERS-CoV Emergence in Qatar

Elmoubasher Farag [1,†,*], Reina S. Sikkema [2,†,*], Tinka Vinks [3], Md Mazharul Islam [4], Mohamed Nour [1], Hamad Al-Romaihi [1], Mohammed Al Thani [1], Muzzamil Atta [4], Farhoud H. Alhajri [4], Salih Al-Marri [1], Mohd AlHajri [1], Chantal Reusken [2] and Marion Koopmans [2]

[1] Ministry of Public of Health, Doha 42, Qatar; mnour@moph.gov.qa (M.N.); halromaihi@moph.gov.qa (H.A.-R.); malthani@moph.gov.qa (M.A.T.); dralmarri@moph.gov.qa (S.A.-M.); malhajri1@moph.gov.qa (M.A.)

[2] Department of Viroscience, Erasmus University Medical Center, Wytemaweg 80, 3015 CN Rotterdam, The Netherlands; c.reusken@erasmusmc.nl (C.R.); m.koopmans@erasmusmc.nl (M.K.)

[3] Division Veterinary Public Health, Institute of Risk Assessment Sciences, Faculty of Veterinary Medicine, Yalelaan 2, 3584 CM Utrecht, The Netherlands; tinkavinks@gmail.com

[4] Department of Animal Resources, Ministry of Municipality and Environment, Doha 35081, Qatar; walidbdvet@gmail.com (M.M.I.); muzamilata@yahoo.com (M.A.); m6066@mme.gov.qa (F.H.A.)

*   Correspondence: eabdfarag@moph.gov.qa (E.F.); r.sikkema@erasmusmc.nl (R.S.S.)

†   Contributed equally to the manuscript.

Received: 10 October 2018; Accepted: 22 December 2018; Published: 31 December 2018

Abstract: MERS-CoV (Middle East respiratory syndrome corona virus) antibodies were detected in camels since 1983, but the first human case was only detected in 2012. This study sought to identify and quantify possible drivers for the MERS-CoV emergence and spillover to humans. A list of potential human, animal and environmental drivers for disease emergence were identified from literature. Trends in possible drivers were analyzed from national and international databases, and through structured interviews with experts in Qatar. The discovery and exploitation of oil and gas led to a 5-fold increase in Qatar GDP coupled with a 7-fold population growth in the past 30 years. The lifestyle gradually transformed from Bedouin life to urban sedentary life, along with a sharp increase in obesity and other comorbidities. Owing to substantial governmental support, camel husbandry and competitions flourished, exacerbating the already rapidly occurring desertification that forced banning of free grazing in 2005. Consequently, camels were housed in compact barns alongside their workers. The transition in husbandry leading to high density camel farming along with increased exposure to humans, combined with the increase of camel movement for the racing and breeding industry, have led to a convergence of factors driving spillover of MERS-CoV from camels to humans.

Keywords: Drivers; MERS-CoV; Qatar

1. Introduction

Emerging infectious diseases are a cause for increasing global concern, because of their impact on global health and economics [1]. The Ebola outbreak in West Africa during 2014-2015 showed that pathogens which previously caused small and easy to control outbreaks had the potential to infect thousands of people under the right circumstances [2]. This is also a concern for the Middle East Respiratory Syndrome coronavirus (MERS-CoV), which until now has been the cause of sporadic cases and hospital outbreaks [3]. To date, there have been 2220 confirmed laboratory cases worldwide, with 790 deaths [4]. All MERS index cases are linked to the Arabian Peninsula. Dromedary camels have been identified as a reservoir of MERS-CoV with occasional zoonotic transmission to humans [5,6]. Human-to-human transmission is also common, with around 30% of the MERS cases reported to

WHO being health care associated [7,8]. However, the source of infection of many index cases remains unclear [9,10].

Studies have shown that MERS-CoV, or related viruses have been circulating among camels at least since 1983 [11]. Since that period, massive changes have occurred in people's lives and in animal husbandry across the Arabian Peninsula. Understanding these changes may help to reconstruct the events that led to the emergence of MERS-CoV as a human disease. Past research identified several drivers of emerging zoonoses, such as urbanisation, population growth and demography, and environmental and agricultural changes [12–14]. The drivers which could have potentially influenced the MERS-CoV emergence in humans have only sporadically been investigated [15,16]. By reviewing changes involving humans and camels over the past 30 years in Qatar, this study sought to identify the key drivers of the emergence and spread of MERS-CoV.

2. Methods

Potential drivers for disease emergence were identified from literature and from discussions with national and international experts in MERS-CoV. The final list had the following categories: economic development; human demography and behavior; international travel, commerce, sports and leisure; political environment; agriculture and food industry change, including camel demography, husbandry and movement; changes in climate and land use. Data from 1980 onwards were collected from national and international databases. If multiple data sources were available, data from both sources were collected. All data were entered in an excel datasheet and reviewed and discussed with the project team (Supplementary 1).

Qualitative information and remaining data gaps were addressed by interviews with a group of 15 experts and stakeholders from Qatar. Criteria to select experts included 5 years or more experience in a camel-related business (farming, trading and racing) or professional services related to camels and being familiar with cultural aspects of the Qatari community. Using a structured interview guide (Supplementary 2) and a moderator, a series of 4 interviews were conducted in Arabic, each lasting approximately for 3 hours. The main themes that were covered during the interviews included: (changes in) people's living conditions; customs and purposes of camel ownership; cultural habits related to camels; educational level and personal behaviors of camel owners and workers; camel movement; demographic distribution of camels in Qatar; camel farming practices: feeding, grazing, and slaughter. A detailed transcript was shared with the experts for authentication. A literature search was done to complement findings from the quantitative and qualitative study, using PubMed, Google Scholar and the local sources of information including the Ministry of Public Health (MoPH), Ministry of Municipality and Environment (MME), Ministry of Development and Planning Statistics (MDPS), and Qatar Statistical Authority (QSA).

The funder had no role in study design, data analysis, data interpretation, or writing of the review.

3. Results

3.1. Changes in the Economic Situation

Historically, Qatari inhabitants were mostly Bedouins along with a few settled people [17,18]. The Bedouins owned limited numbers of camels, sheep, and goats [19]. Camels were used as a source of food (milk and meat) and means for transportation. In 1939, oil and natural gas resources were discovered. However, large-scale exploitation started in the 1950s [20]. From the 1950s onwards, Qatar's economy has been steadily growing. However, the year 2000 marked a significant turning point as Qatar's GDP almost increased by more than 5-fold during the period 2000–2006 (Figure 1A) [20,21]. Qatar is currently considered to be one of the wealthiest countries in the world [20].

3.2. Changes in Human Demography and Health

The thriving economy was paralleled by major demographic and life style changes. In the late 1950s, around 16,000 people lived in Qatar [22]. In response to demands for a larger workforce after the exploitation of oil and gas began, foreign laborers started to migrate to Qatar from countries in the region, like Palestine, Oman, Iran, and the Kingdom of Saudi Arabia (KSA). Later, immigrants from Pakistan, India, Nepal, Sri Lanka, Bangladesh, the Philippines, and Indonesia joined the older migrant populations, increasing the number of inhabitants to 369,079 by 1986 and recently to 2,617,634 (Figure 2A) [23]. In 2016, non-Qatari males made up 78% of the residents of working age (15-64) and non-Qatari made up more than 90% of the total number of Qatar inhabitants older than 15 years of age (Figure 2B,C) [24]. Most recent estimations of the origins of the non-Qatari population are that 25% is Indian, 11% Bangladeshi, 14% Nepali, 10% Filipinos, 9% Egyptian, 5% Pakistani, and 2% Iranian [25]. The total number of males in Qatar increased from 67.2% of the total population in 1986 to 75.5% in 2016 (Figure 2B). In 2004, almost 50% of residents were between 15 and 39 years old, and this has risen to more than 60% in 2015 (Figure 2A). Detailed accounts on age distribution were not available before 2004 [26].

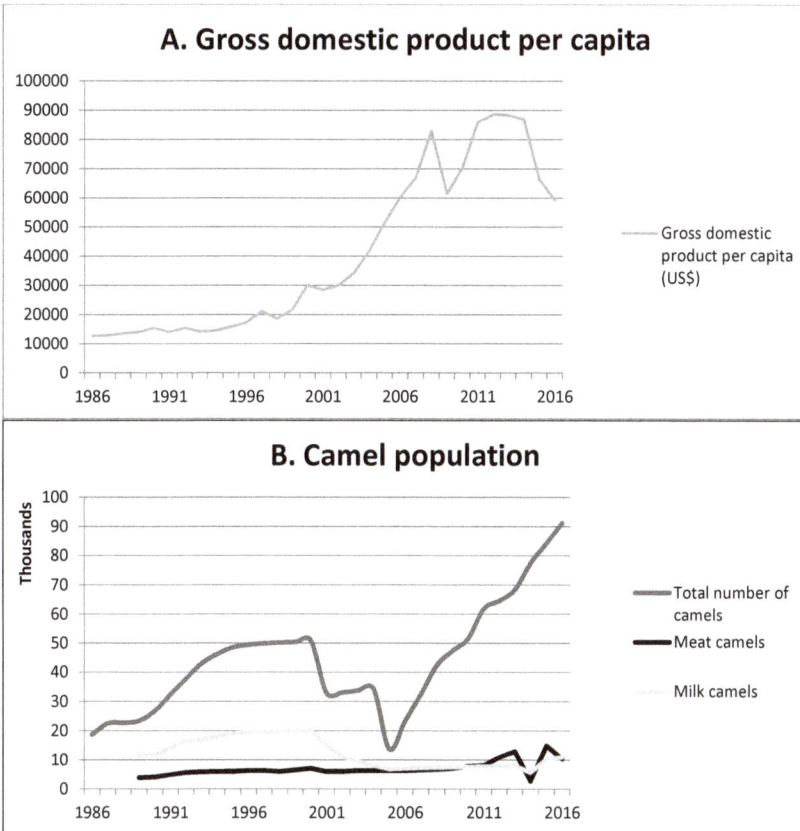

Figure 1. *Cont.*

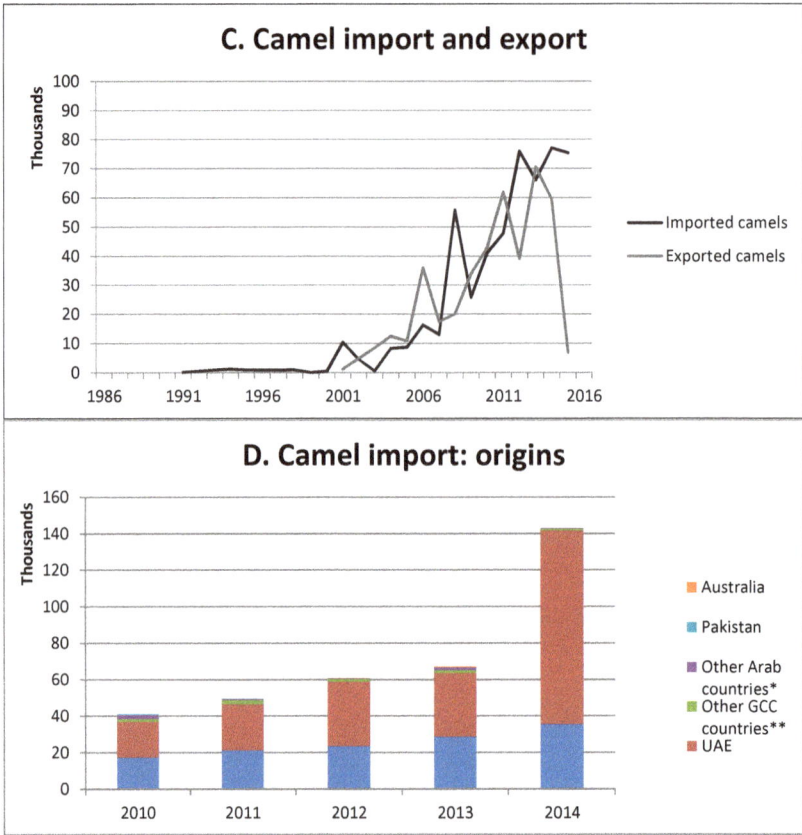

Figure 1. Developments in economy, camel demography, production, and trade. (**A**). Development over time of the gross domestic product per capita; (**B**) Development over time of the camel population; (**C**) Development over time of camel import and export; (**D**) Development over time of camel importation per country of origin. *Other Arab countries: Algeria, Comoros, Djibouti, Egypt, Iraq, Jordan, Lebanon, Libya, Mauritania, Morocco, Palestine, Somalia, Sudan, Syria, Tunisia, and Yemen. **Other GCC countries: Bahrain, Kuwait, Oman, Qatar.

Most people in Qatar live in urban areas. The percentage of residents living in cities increased from 85.3% in 1960, to 90.4% in 1986, and 99.3% in 2016 [20]. Doha, the capital and the biggest city of Qatar, hosts the greatest number of people. However, there has also been a large increase in number of people living in the Al-Rayan area, where most of the camel farms are located. The number of tourists visiting Qatar also increased, especially since 2000. Most tourists came from other GCC countries, but the number of visitors from Europe and America were also increasing (Figure 2D) [20,27].

According to experts, the economic development and population increase coincided with major changes in life style. The Bedouin nomadic lifestyle gradually decreased as most of the Qatari tribes shifted to an urban, settled lifestyle; cars and planes rapidly replaced camels as transportation means. This transformation to a more sedentary lifestyle is reflected in the profile of comorbidities. More than 70% of adults are overweight and almost half of them obese [28]. Male obesity increased from 17% in 1986 to 34% in 2014, which is extremely high compared to the current 11% prevalence in men worldwide [29]. In 1998, 7% of residents above 15 years were hypertensive, rising to 14% in 2006, and 33% in 2012 [30,31]. Prevalence of high blood sugar among adults in 2015 was 14%, compared to a worldwide prevalence of 9% [28]. The Qatar Stepwise Report reported in 2012 that 15% of adults were daily smokers. Yet, Qatar has a low death rate: 1.49/1000, compared to the worldwide death rate of 7.72/1000, and its healthcare system has developed rapidly over the past twenty years [28,31,32].

3.3. Changes in Camel Husbandry and Practices

The increase in the number of dromedary camels reflects the increasing popularity of camels as sports animals (Figure 1B). With the changing life style and increasing wealth, the purchase and breeding of (expensive) racing camels came within reach of an increasingly large segment of the Qatari national population. According to experts, although camel racing has traditionally been part of the Bedouin culture, the organized racing business went through major changes over the past decades. This was partly due to financial and regulatory support from the Qatari government. This support increased the social and economic value of camels in Qatar, further stimulating their popularity. The Al-Shehaniya camel-racing track, one of the biggest tracks in the Gulf, was opened in 1990 [33]. The camel farms that are located near the Al-Shehaniya camel racing area are mostly used for racing camels. There are about 1500 racing camel holdings at the Al-Shehaniya camel racing area. Some of the camels in Qatar are used to compete in camel beauty contests that are organized around the Arabian Peninsula. According to the FAO, in 1960 there were about 6,000 camels in Qatar. This rose to over 43,000 in 1992, 50,000 in 2000, and more than 90,000 in 2016 (Figure 1B). More than 83% of the animals are currently kept for racing [34]. Across the Gulf region, Qatar has the highest camel density, with 6.77 units/km^2, compared to 4.74 units/km^2 in United Arab Emirates (UAE) and 0.11 units/km^2 in the KSA [35]. In 2005, the total number of camel farms was 1300 and by 2014 it had increased to 9594 [34].

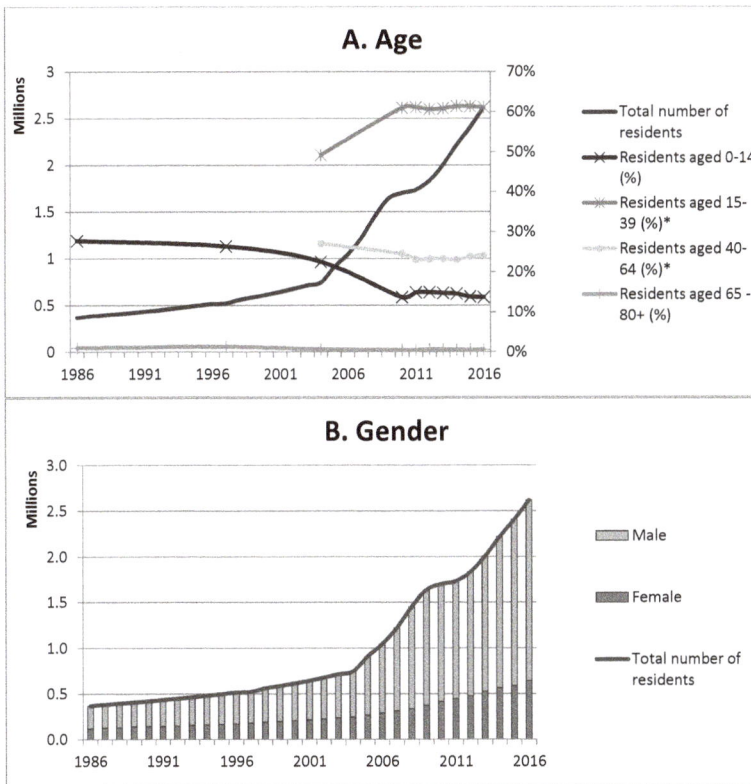

Figure 2. *Cont.*

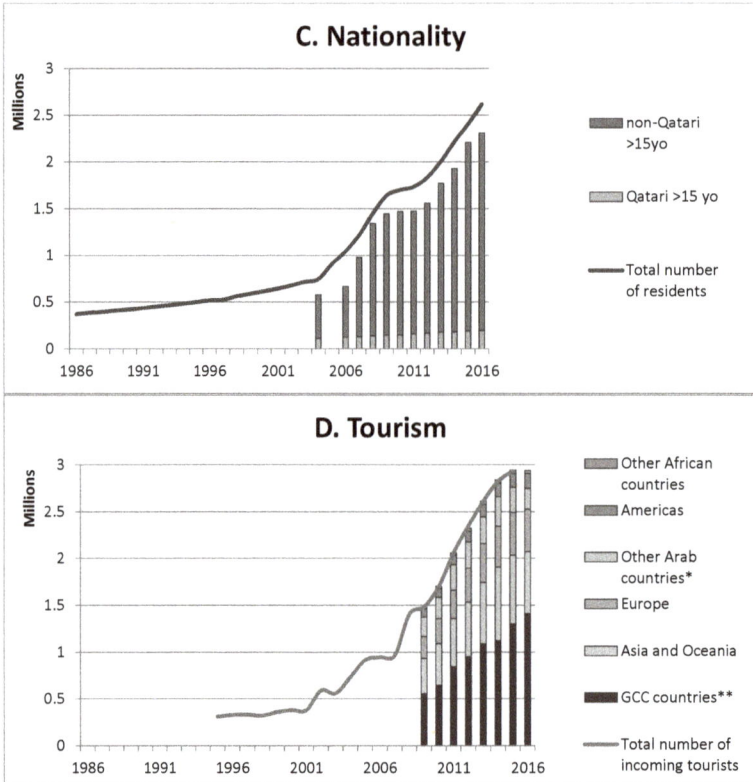

Figure 2. Developments in human demography in Qatar; (**A**) Development over time of the age structure of the population; (**B**) Development over time of the human sex ratio; (**C**) Development over time of the ration Qatari vs. non Qatari; (**D**) Development over time of tourism per country of origin *Other Arab countries: Algeria, Comoros, Djibouti, Egypt, Iraq, Jordan, Lebanon, Libya, Mauritania, Morocco, Palestine, Somalia, Sudan, Syria, Tunisia, and Yemen. **GCC countries: Bahrain, Kuwait, Oman, Qatar, KSA, and the United Arab Emirates.

As a result of the loss of traditional methods of rangeland management, the vegetation coverage decreased from 10% to only 1% of total land cover. Overgrazing of the green areas due to the increased population of camels and other livestock accelerated the desertification of Qatar [36]. Therefore, the government decided to assign natural protected areas in 2004 [37,38], and started to sanction the free grazing of livestock since 2005 [39]. By 2011, open grazing was completely banned [40]. According to the experts' opinions, this led to changes in farming practices, as herds were then moved outside of Qatar to areas where free grazing remained possible. Moreover, in Qatar, camels are now raised in closed systems and within 1 of 9 designated farming areas (camel complexes) in the residential districts. Camel workers also live on the premises of the camel complexes. Typically, a camel complex has a reception room (Majlis) for social activities of the camel owners. The Al-Rayyan municipality, where the Al-Shehaniya camel racing area is also located, currently holds about 83% of the total camel population and 61% of camel holdingss (Figure 3) [34,35]. According to the experts, this newly adopted closed farming system led to the increase of disease incidence, especially of parasitic diseases. However, we did not find any disease statistics to substantiate these findings.

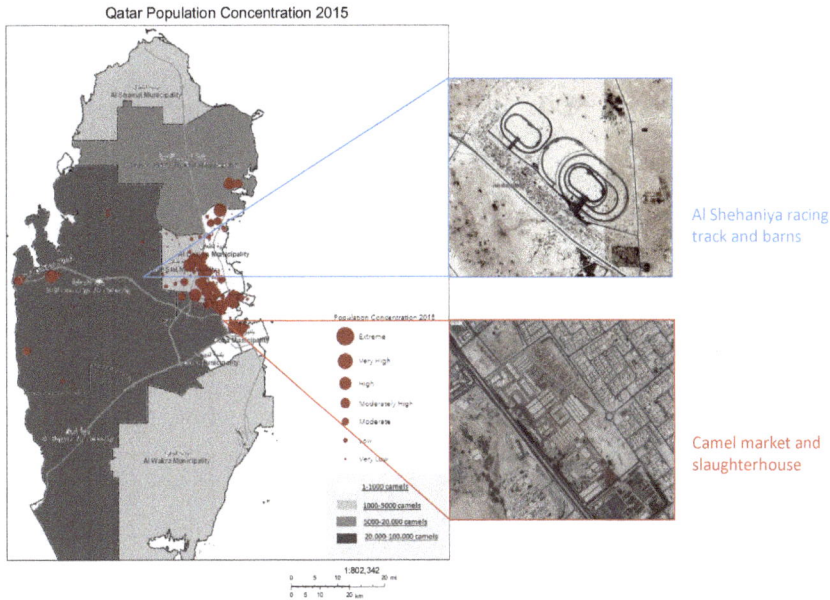

Figure 3. Human and camel density map of Qatar (adapted from Ministry of Development Planning and Statistics, Population Concentration map, 2015). The density map shows the density of camels (source: Ministry of Municipality and Environment) and humans in Qatar. Most people live in and around Doha, where the Doha animal market and slaughterhouse are also located. The highest camel density can be found in the Al-Rayyan area, where the Al-Shehaniya racing tracks are also located. A small, but growing, part of the Qatar population also lives in the Al-Rayyan area.

3.4. Changes in Race Camel Farming and Practices

The increasing focus on camel race competitions caused big changes in camel farming practices. Previously, the calves were weaned when the next calf was born. Currently, weaning occurs at around 7 months of age. After being weaned, young camels are directly taken for acclimatization (during the period mid-July through mid-August) from the general livestock farms (located across the region) to the racing farms, mostly located within the Al-Shehaniya area. This involves drastic changes in feeding systems, intense training for races, and mock races alongside camels from other farms and older training camels. The off-season for camel racing is during summer (mid-April to August) (Figure 4). During this time, most of the owners travel abroad, the frequency of visits to the farms substantially decreases, and workers are permitted to take annual vacations. From September onward, training intensifies, in preparation of the racing season, which lasts from mid-September through mid-April. During that time, 14,000 registered camels from different origins, ages, gender, nationalities, and breeds compete together at the Al-Shehaniya camel-racing track. During the racing season, up to 24 rounds take place, approximately five days per week.

3.5. Changes in International Camel Movements and Travel

An unprecedented, increasingly intensified mobility of camels inside and outside Qatar has been seen over the recent decades. The domestic and cross border mobility does not only involve camels, but also people who look after the camels to provide care along the journey. Import and export of camels have especially increased since the year 2000 (Figure 1C). The imported camels mainly come from the UAE and KSA (Figure 1D).

The dynamics and travel patterns of Qatari camels are complex (Figure 4). Camels are transported to and from different locations, for a variety of purposes, and with a noticeable seasonal pattern.

Mobility gets more intensive during the racing and trading season (September to April). Experts believe that the ban of open grazing in Qatar played a key role in the intensity and frequency of camel movements. They mention that there has been a remarkable increase after 2011 in numbers of camel workers and owners who cross the borders to and from KSA along with their animals, although this recently stopped with the KSA-Qatar political situation. The ban of open grazing stimulated camel owners to establish farms in KSA and UAE where open grazing is still permitted. Therefore, camels are moved through Gulf Countries, particularly during the winter season.

Camel races and beauty contests that are routinely organized in nearly all Gulf countries are another factor that boost the national and international movement of camels. Compared to other types of camels, racing camels dominate in terms of numbers and frequency of mobility both across borders and domestically, particularly between September and April. As per the records of the Camel Racing Committee, in the 2016 racing competitions, 14,000 camels from Qatar and camels from the other GCC countries contested [35]. However, owing to the lack of standardized identification system, it was difficult to determine the exact figures and the extent of these movements.

Camels are also being mobilized for reproduction purposes (Figure 4). Mating season (also known as camels' honeymoon) starts in the middle of August and continues through February of the next year, with the high season in the September-October period. Female camels are usually taken from their own location to other farms where selected males are kept particularly for reproduction purposes. About 14,000 female camels are annually being moved for mating. They spend around 1 week at a breeding farm with male camels before they are taken back to their original farms. Programmed mating is exclusively being practiced for race and show camels. The mating season is another seasonal activity that entails intensive movements of camels, camel owners, workers, car drivers and veterinarians.

3.6. Changes in Camel Trade

The Doha wholesale market constitutes the primary hub for camel trading. In parallel with the increased number of camel races, Al-Shehaniya City also grew as a market and has become a hub for trade of racing and beauty show camels in Qatar. The wholesale market in Doha hosts camels and other types of livestock from countries all over the Gulf region. The camels typically stay at the market until they are sold. Camel workers live at the market premises. Camels that are being sold (calves in particular) serve a variety of purposes. They are sold to be slaughtered at the Doha wholesale market abattoir, for breeding purposes, to be trained as racing camel, or to be prepared for camel show competitions.

In recent years, the Doha wholesale market has been surrounded by rapidly growing residential areas. Animals in the market are now in close proximity to the residents. As of 2005, slaughter practices were banned inside residential premises, and can only be performed in official slaughterhouses and exclusively by licensed persons.

3.7. Changes in Use of Camel Meat, Milk, and Urine

Camel meat and milk are no longer part of the daily diet of most Qatar inhabitants. Nonetheless, camel meat is a fundamental ingredient of Qatari social events and family celebrations. Production of camel meat and milk has remained stable in the past 30 years. Camel milk is generally kept for personal use, particularly for the perceived therapeutic merits of raw camel milk, as well as camel urine. Experts state that there is an unshakable belief that the regular consumption of camel milk helps to prevent and control diabetes. It is also widely believed in the Qatari community that camel urine and milk can heal skin lesions and other diseases. Camel urine is also regularly used to whiten the skin and face and lighten the hair. The majority of camel owners offer camel milk and urine for free, as a practice of generosity.

(A) Camel movements

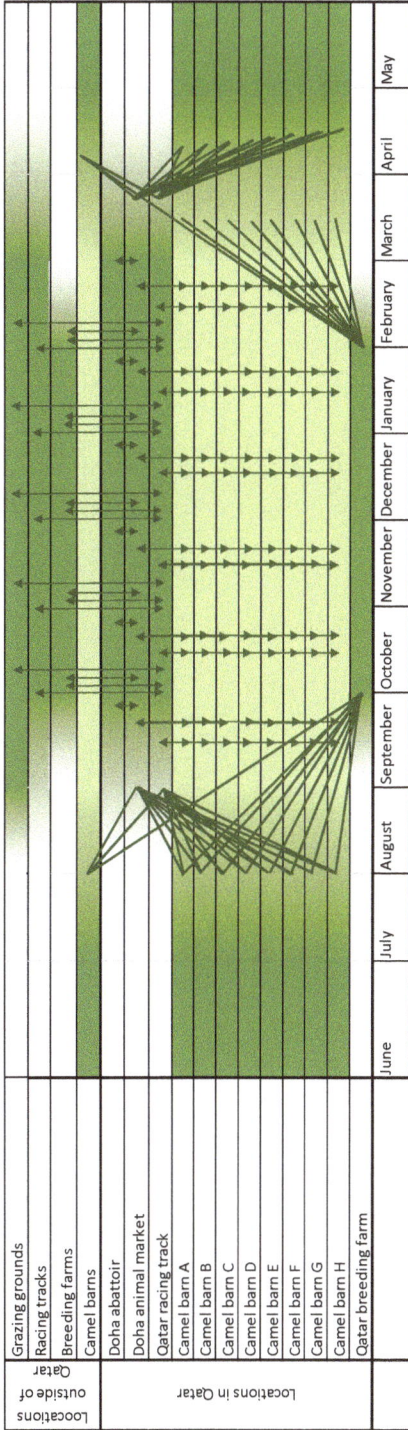

Figure 4. *Cont.*

(**B**) Activities involving camels

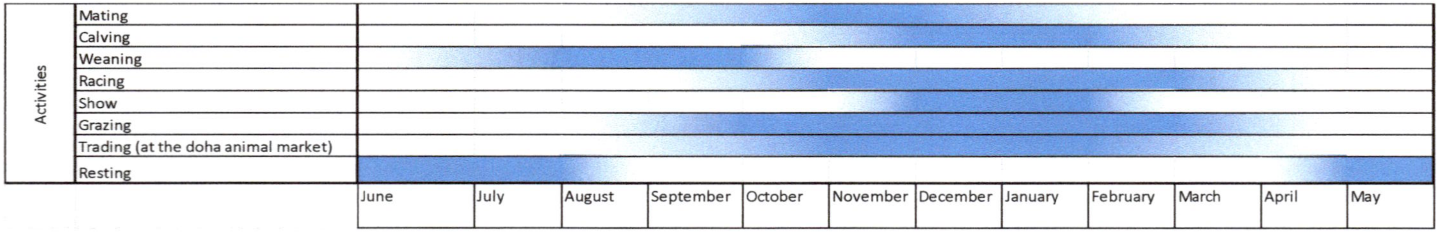

Activities		June	July	August	September	October	November	December	January	February	March	April	May
	Mating												
	Calving												
	Weaning												
	Racing												
	Show												
	Grazing												
	Trading (at the doha animal market)												
	Resting												

Figure 4. Seasonality and movements of camels and camel activities. (**A**) Dark green locations and periods of time indicate a high concentration of camels. The arrows show the direction of camel movements. It is shown that many camels gather and mix at the animal market, racing track, and breeding farms in Qatar in August, September and October. Moreover, there is constant movement to and from grazing grounds and racing tracks outside of Qatar. In March and April, most camels travel back to their barns. (**B**) shows the seasonality of camel related activities. Most activities take place in the "cold season" from September to April.

4. Discussion

The role of camels in the transmission of MERS-CoV is well documented [5]. Despite the fact that MERS antibodies have already been detected in camels since 1983 [11] and human contact with animals is not new, human MERS cases were only detected in 2012 [41]. Based on institutional and literature data and in-depth interviews with key professionals, this study sought to examine the changes involving the human, animal, and environmental drivers that may have contributed to the spread and virus spillover to humans.

Our reconstruction of events over the past decades, based on available literature, statistics, and expert opinions, lead to the conclusion that the discovery of oil and natural gas resources has been the starting point of a chain of events that ultimately led to conditions favoring the emergence of MERS-CoV (Figure 5). This discovery led to massive economic growth. Owning a camel represents the wealth and status of its owner in Arabic culture. Governmental sponsorship of camel ownership and camel racing further stimulated the camel industry, especially the camel-racing sector. This in turn lead to an accelerating increase of the camel population, a change in camel farming, and a concomitant increase in the number of camel workers [34]. The human population of Qatar has increased by 7-fold over the last decades [23]. This is unlike other high-income countries, that have a yearly overall population growth of only 0.6% [20]. Population growth and high population density have been shown previously to be important risk factors for disease emergence [42]. Moreover, consistent with the disease profile of wealthy countries where sedentary lifestyle prevails, the prevalence of chronic diseases increased in Qatar in accordance with the increasing GPD, ultimately rendering the Qatar population not only vulnerable to virus transmission, but also to its deadly complications [43].

The intimate nature and number of interactions between camels and humans has also increased significantly in the past 30 years, increasing the risk of any zoonotic spillover. At camel complexes, workers intimately reside, sleep, and eat with their camels. Camel owners, on the other hand, pay regular visits to their barns and stay there for considerable hours every day (even longer during weekends, holidays, and winter season) in the majlis built at the corner of their barns. Owners, who are often of advanced age with multiple comorbidities, enjoy drinking fresh camel milk and entertaining guests. Those who suffer certain diseases tend to visit the camel barns to use camel urine or drink fresh camel milk for its perceived curative properties.

Among the variety of changes that involved camel husbandry in Qatar, the shift from open grazing to close housing systems seems to be most significant. Opportunities for camel-to-camel and camel-to-human spread have greatly increased since then. It is possible that housing camels in barns, with poor biosecurity and hygienic standards, turned these barns into 'melting pots' for the virus that ultimately acquired the ability to cross the human-animal barrier. The increase of cross-border movement of camels increased chances and frequency of (international) virus spread. Camels are transported freely across borders for a variety of purposes through multiple routes and means of transportation. When camels and the humans that accompany them, arrive at the site of a race or beauty event in Qatar, they are housed with the local camels. Owners are welcomed in the majlis at the camel complexes. The mixing of camel and human of different origins further increase chances of virus transmission.

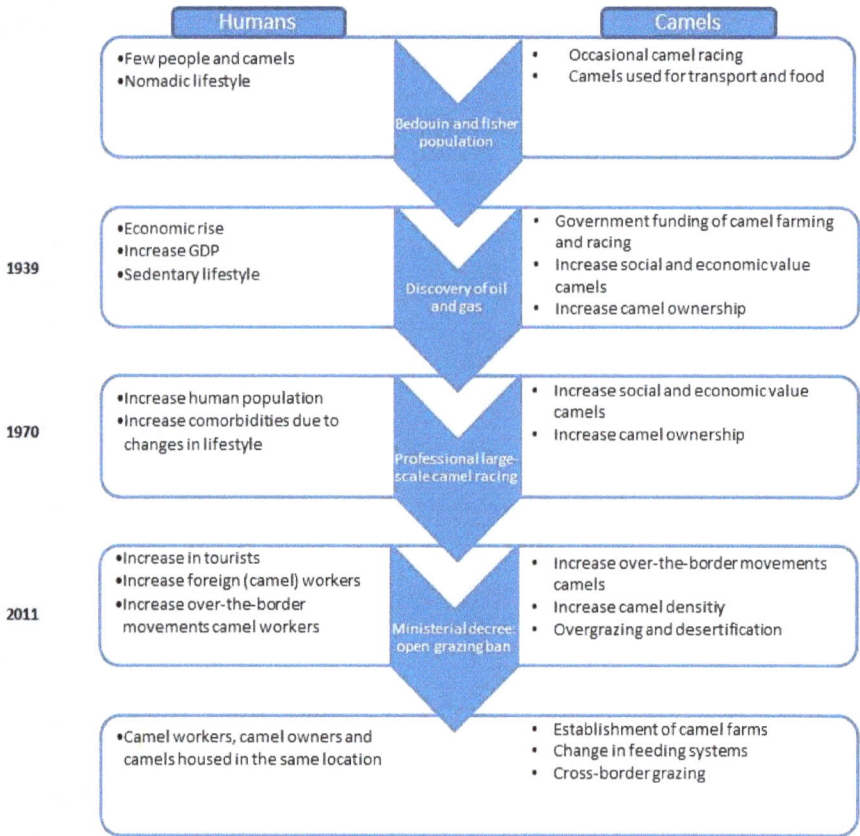

Figure 5. Visual summary/timeline of relevant events.

Although much effort was made to study MERS-CoV viral sequences and MERS-CoV transmission between dromedary camels and humans, it is still unknown which genetic mechanisms have caused the viral spillover of dromedary camels to humans. However, the most important determinant of host specificity seems to be the Spike S1 protein, that recognizes and binds to host-cell receptor DPP4 [44]. Recently it has been shown that the MERS-CoV spike can rapidly adapt to species variation in DPP4 [45]. As such, the increasing human-animal interface that is described in this paper may have facilitated the adaptation of the spike protein to human DDP4. However, much remains unknown, also in view of the findings that MERS-CoV from East Africa were not phenotypically different from the viruses from the Middle East, while human MERS patients have not been reported from the African continent [46].

Finally, the changes in animal husbandry practices, earlier weaning, frequent grouping and transportation of animals, and the introduction of an entirely new feeding system, may induce stress in the camels. These changes and movements often involve young weaned animals, at the same time as maternal antibodies are waning, which are linked to the shedding of the virus [42,47]. Most of the limitations of this study were related to the availability of data. Firstly, statistics on animals, import and export, animal workers, and land use were only found since 2000 onwards, limiting the chance to study the trends and changes prior to that year. Secondly, even the available national data on the animals, humans, and environment were found to be sometimes inconsistent, limiting the possibility to provide "hard evidence" of causality. Nevertheless, this is the first comprehensive quantitative overview of possible drivers of MERS-CoV in Qatar.

5. Conclusions

Several key changes were shown to involve camels, humans, the economy, and the environment in Qatar during the last 30 years. Our study indicates that the rapid increase in camel ownership, leading to the presence of camels from different origins in a high-density environment mixed together with human and other animal species may have offered the right circumstances for the virus to spread from camels to humans. The other key changes that were described collectively contributed to this situation. Further understanding of the drivers that led to the emergence of MERS-CoV can serve as input for MERS-CoV surveillance and control measures to prevent further spread of MERS-CoV and reduce transmission from camels to humans.

Supplementary Materials: The following are available online at http://www.mdpi.com/1999-4915/11/1/22/s1, Supplementary 1: Categories, subcategories, and data sources used for information gathering in this review; Supplementary 2: Questionnaires used for qualitative information gathering in this review.

Author Contributions: E.F., R.S.S., C.R., and M.K. conceived and developed the conceptual framework; T.V. and R.S.S. executed the literature search; E.F., M.M.I., M.N., and M.A. designed and executed expert interviews; E.F., R.S.S., T.V., and M.K. executed the data analysis; E.F., R.S.S., C.R., T.V., M.N., M.A.T., M.M.I., H.A.R., M.A., F.A., S.A.M., M.A., and M.K. contributed to the interpretation of the results; R.S.S., E.F., M.N., M.M.I., and M.K. wrote the manuscript. All authors provided critical feedback and helped shape the research, analysis and manuscript.

Funding: This study was financially supported by Ministry of Public of Health, Doha, Qatar and the European Commission's H2020 program under contract number 643476 (http://www.compare-europe.eu/).

Conflicts of Interest: All authors declare no conflicts of interest.

References

1. Jones, K.E.; Patel, N.G.; Levy, M.A.; Storeygard, A.; Balk, D.; Gittleman, J.L.; Daszak, P. Global trends in emerging infectious diseases. *Nature* **2008**, *451*, 990–994. [CrossRef]
2. Coltart, C.E.M.; Lindsey, B.; Ghinai, I.; Johnson, A.M.; Heymann, D.L. The Ebola outbreak, 2013–2016: Old lessions for new epidemics. *Philos. Trans. R. Soc. B* **2017**, *372*, 20160297. [CrossRef]
3. Zumla, A.; Hui, D.S.; Perlman, S. Middle East respiratory syndrome. *Lancet* **2015**, *386*, 995–1007. [CrossRef]
4. World Health Organization. Middle East Respiratory Syndrome Coronavirus. Available online: http://www.who.int/emergencies/mers-cov/en/ (accessed on 13 June 2018).
5. Reusken, C.B.; Raj, V.S.; Koopmans, M.P.; Haagmans, B.L. Cross host transmission in the emergency of MERS coronavirus. *Curr. Opin. Virol.* **2016**, *16*, 55–62. [CrossRef]
6. Azhar, E.I.; El-Kafrawy, S.A.; Farraj, S.A.; Hassan, A.M.; Al-Saeed, M.S.; Hashem, A.M.; Madani, T.A. Evidence for camel-to-human transmission of MERS coronavirus. *N. Engl. J. Med.* **2014**, *370*, 2499–2505. [CrossRef]
7. Arwady, M.A.; Alraddadi, B.; Basler, C.; Azhar, E.I.; Abuelzein, E.; Sindy, A.I.; Sadiq, B.M.B.; Althaqafi, A.O.; Shabouni, O.; Banjar, A.; et al. Middle East respiratory syndrome coronavirus transmission coronavirus transmission in extended family, Saudi Arabia, 2014. *Emerg. Infect. Dis.* **2016**, *22*, 1395–1402. [CrossRef]
8. Cho, S.Y.; Kang, J.; Ha, Y.E.; Park, G.E.; Lee, J.Y.; Ko, J.; Lee, J.Y.; Kim, J.M.; Kang, C.; Jo, I.J.; et al. MERS-CoV outbreak following a single patient exposure in an emergency room in South Korea: An epidemiological outbreak study. *Lancet* **2016**, *388*, 994–1001. [CrossRef]
9. Alagaili, A.N.; Briese, T.; Mishra, N.; Kapoor, V.; Sameroff, S.C.; Wit, E.; Munster, V.J.; Hensley, L.E.; Zalmout, I.S.; Kapoor, A.; et al. Middle East respiratory syndrome coronavirus infection in dromedary camels in Saudi Arabia. *mBio* **2014**, *5*, e00884-14. [CrossRef]
10. Lee, J.Y.; Kim, Y.; Chung, E.H.; Kim, D.; Jeong, I.; Lee, J.Y.; Kim, Y.; Chung, E.H.; Kim, D.; Jeong, I.; et al. The clinical and virological features of the first imported case causing MERS-CoV outbreak in South Korea, 2015. *BMC Infect. Dis.* **2017**, *17*, 498. [CrossRef]

11. Müller, M.A.; Corman, V.M.; Jores, J.; Meyer, B.; Younan, M.; Liljander, A.; Bosch, B.; Lattwein, E.; Hilali, M.; Musa, B.E.; et al. MERS coronavirus neutralizing antibodies in camels, Eastern Africa, 1983–1997. *Emerg. Infect. Dis.* **2014**, *20*, 2093–2095. [CrossRef]

12. Hassell, J.M.; Begon, M.; Ward, M.J.; Fèvre, E.M. Urbanization and disease emergence: Dynamics at the wildlife-livestock-human interface. *Trends Ecol. Evol.* **2017**, *32*, 55–67. [CrossRef]

13. Jones, B.A.; Grace, D.; Kock, R.; Alonso, S.; Rushton, J.; Said, M.Y.; KcKeever, D.; Mutua, F.; Young, J.; McDermott, J.; et al. Zoonosis emergence linked to agricultural intensification and environmental change. *Proc. Natl. Acad. Sci. USA* **2013**, *110*, 8399–8404. [CrossRef]

14. Hui, E.K. Reasons for the increase in emerging and re-emerging viral infectious diseases. *Microbes Infect.* **2006**, *8*, 905–916. [CrossRef]

15. Hemida, M.G.; Elmoslemany, A.; Al-Hizab, F.; Alnaeem, A.; Almathen, F.; Faye, B.; Chu, D.K.W.; Perera, R.A.P.M.; Peiris, M. Dromedary camels and the transmission of Middle East respiratory syndrome coronavirus (MERS-CoV). *Transbound. Emerg. Dis.* **2017**, *64*, 344–353. [CrossRef]

16. Cauchemez, S.; Nouvellet, P.; Cori, A.; Jombart, T.; Garske, T.; Clapham, H.; Moore, S.; Mills, H.L.; Salje, H.; Collins, C.; et al. Unraveling the drivers of MERS-CoV transmission. *Proc. Natl. Acad. Sci. USA* **2016**, *113*, 9081–9086. [CrossRef]

17. Hobbs, M. Divers Are a Pearl's Best Friend: Pearl Diving in the Gulf 1840S–1930S. Qatar Digital Library. Available online: https://www.qdl.qa/en/divers-are-pearl%E2%80%99s-best-friend-pearl-diving-gulf-1 840s%E2%80%931930s (accessed on 6 May 2017).

18. Al Janahi, B.M. National Identity Formation in Modern Qatar: New Perspective. Master's Thesis. Available online: http://hdl.handle.net/10576/3247 (accessed on 6 May 2017).

19. Crystal, J. Oil and Politics in the Gulf. In *Rulers and Merchants in Kuwait and Qatar*; CUP: Melbourne, Australia; Cambridge, UK, 1990. [CrossRef]

20. World Bank. Qatar Country Indicators. Available online: http://data.worldbank.org/country/qatar (accessed on 30 July 2017).

21. Embassy of the State of Qatar in Brussels. Qatar History. Available online: http://www.qatarembassy.be/ QatarEmbassy/English/History.html (accessed on 6 May 2017).

22. Dougherty, R.L. *Bedouins of Qatar, Klaus Ferdinand*; Thames and Hudson: London, UK, 1995. [CrossRef]

23. Ministry of Development Planning and Statistics. Quarterly Bulletin for Population and Social Statistics—Third Quarter 2016. Available online: https://www.mdps.gov.qa/en/statistics1/pages/lateststats/20170320.aspx (accessed on 6 May 2017).

24. Ministry of Development Planning and Statistics. Labor Force Survey 2016. Available online: http://www.md ps.gov.qa/en/statistics/StatisticalReleases/Social/LaborForce/2016/Labour_force_2016_AE.pdf (accessed on 28 June 2017).

25. Snoj, J. Population of Qatar by Nationality—2017 Report. Available online: http://priyadsouza.com/popula tion-of-qatar-by-nationality-in-2017 (accessed on 25 July 2017).

26. Ministry of Development Planning and Statistics. Population and Social Statistics 2016. Available online: http://www.mdps.gov.qa/en/statistics/StatisticalReleases/Population/Population/2016/Popul ation_social_1_2016_AE.pdf (accessed on 26 June 2017).

27. Qatar Tourism Authority. Annual Tourism Performance Report. Available online: https://www.visitqatar.q a/corporate/planning/data-and-statistics.html (accessed on 26 June 2017).

28. Bakri, A.H. *Chronic Disease Risk Factor Surveillance: Qatar Stepwise Report 2012*; Supreme Council of Health: Doha, Qatar, 2013.

29. World Health Organization. Global Health Observatory (GHO) Data: Overweight and Obesity. 2017. Available online: http://www.who.int/gho/ncd/risk_factors/overweight/en/ (accessed on 26 June 2017).

30. National Health Authority Qatar. World Health Survey Qatar. Available online: https://static-content.sprin ger.com/esm/art%3A10.1186%2F1478-7954-12-18/MediaObjects/12963_2013_244_MOESM1_ESM.pdf (accessed on 26 June 2017).

31. World Health Organization. Noncommunicable Diseases and Their Risk Factors; STEPwise Approach to Surveillance (STEPS). Available online: http://www.who.int/ncds/surveillance/steps/en/ (accessed on 26 June 2017).

32. World Bank. Death Rate, Crude Death. Available online: https://data.worldbank.org/indicator/SP.DYN. CDRT.IN (accessed on 26 June 2017).

33. Camel Racing Committee. *The Reports of The Camel Racing Organizing Committee*; Ministry of Culture and Sport: Doha, Qatar, 2016.

34. Department of Animal Resources. *Yearbook of Animal Statistics 2015*; Ministry of Municipality and Environment: Doha, Qatar, 2016.

35. World Organisation for Animal Health (OIE). Animal Population; World Animal Health Information Database (WAHIS Interface)—Version 1. Available online: https://www.oie.int/wahis_2/public/wahid.php/Countryinformation/Animalpopulation (accessed on 26 June 2017).

36. Ministry of Municipality and Environment. Doha, Qatar. Available online: http://www.mme.gov.qa/cui/index.dox?siteID=2,2014 (accessed on 26 June 2017).

37. Ministry of Development Planning and Statics. *Environment Statistics Annual Report. 2013*; Ministry of Development Planning and Statics: Doha, Qatar, 2014.

38. Ministry of Development Planning and Statistics. Environment Statistics in the State of Qatar. Doha-Qatar 2015. Available online: https://www.mdps.gov.qa/en/statistics/Statistical%20Releases/Environmental/EnvironmentalStatistics/Environment_QSA_EN_2015.pdf (accessed on 26 June 2017).

39. Supreme Council for Environment and Natural Reserves. Protected Area Action Plan 2008–2013, Conversion of Biological Diversity (CBD). Available online: https://www.cbd.int/doc/world/qa/qa-nbsap-oth-en.pdf (accessed on 26 June 2017).

40. Elford, C.J. Opportunities for the Sustainable Use of the Camel in Qatar. Master's Thesis, Virginia Commonwealth University, Richmond, VA, USA, 2013. Available online: https://core.ac.uk/download/pdf/51293060.pdf (accessed on 17 July 2017).

41. Zaki, A.M.; Boheemen, S.V.; Bestebroer, T.M.; Osterhaus, A.D.; Fouchier, R.A. Isolation of a novel coronavirus from a man with pneumonia in Saudi Arabia. *N. Engl. J. Med.* **2012**, *367*, 1814–1820. [CrossRef]

42. Khalfallah, A.I.; Lu, X.; Mubarak, A.I.A.; Dalab, A.H.S.; Al-Busadah, K.A.S.; Erdman, D.D. MERS-CoV in upper respiratory tract and lungs of dromedary camels, Saudi Arabia, 2013–2014. *Emerg. Infect. Dis.* **2015**, *21*, 1153–1158. [CrossRef]

43. Badawi, A.; Ryo, S.G. Prevalence of comorbidities in the Middle East respiratory syndrome coronavirus (MERS-CoV): A systematic review and meta-analysis. *Int. J. Infect. Dis.* **2016**, *49*, 129–133. [CrossRef]

44. Lu, G.; Wang, Q.; Gao, G.F. Bat-to-human: Spike features determining 'host jump' of coronaviruses SARS-CoV, MERS-CoV, and beyond. *Trends Microbiol.* **2015**, *23*, 468–478. [CrossRef]

45. Letko, M.; Miazgowicz, K.; McMinn, R.; Seifert, S.N.; Sola, I.; Enjuanes, L.; Carmody, A.; van Doremalen, N.; Munster, V. Adaptive Evolution of MERS-CoV to Species Variation in DPP4. *Cell Rep.* **2018**, *24*, 1730–1737. [CrossRef]

46. Adney, D.R.; Doremalen, N.V.; Brown, V.R.; Bushmaker, T.; Scott, D.; Wit, E.D.; Bowen, R.A.; Munster, V.J. Replication and shedding of MERS-CoV in upper respiratory tract of inoculated dromedary camels. *Emerg. Infect. Dis.* **2014**, *20*, 1999–2005. [CrossRef]

47. Chu, D.K.W.; Hui, K.P.Y.; Perera, R.A.P.M.; Miguel, E.; Niemeyer, D.; Zhao, J.; Channappanavar, R.; Dudas, G.; Oladipo, J.O.; Traoré, A.; et al. MERS coronaviruses from camels in Africa exhibit region-dependent genetic diversity. *Proc. Natl. Acad. Sci. USA* **2018**, *115*, 3144–3149. [CrossRef]

Review

From SARS to MERS, Thrusting Coronaviruses into the Spotlight

Zhiqi Song [1,2,3], **Yanfeng Xu** [1,2,3], **Linlin Bao** [1,2,3], **Ling Zhang** [1,2,3], **Pin Yu** [1,2,3], **Yajin Qu** [1,2,3], **Hua Zhu** [1,2,3], **Wenjie Zhao** [1,2,3], **Yunlin Han** [1,2,3] and **Chuan Qin** [1,2,3,*]

[1] Institute of Laboratory Animal Science, Chinese Academy of Medical Sciences (CAMS) & Comparative Medicine Centre, Peking Union Medical Collage (PUMC), Beijing 100021, China; songzhiqi1989@foxmail.com (Z.S.); xuyanf2009@163.com (Y.X.); bllmsl@aliyun.com (L.B.); zhangling@cnilas.org (L.Z.); pinyucau@gmail.com (P.Y.); quyj@cnilas.org (Y.Q.); zhuh@cnilas.org (H.Z.); hnndwenjiezhao@163.com (W.Z.); 18510165683@163.com (Y.H.)

[2] NHC Key Laboratory of Human Disease Comparative Medicine, the Institute of Laboratory Animal Sciences, CAMS&PUMC, Beijing 100021, China

[3] Beijing Key Laboratory for Animal Models of Emerging and Reemerging Infectious, Beijing 100021, China

* Correspondence: qinchuan@pumc.edu.cn

Received: 16 December 2018; Accepted: 9 January 2019; Published: 14 January 2019

Abstract: Coronaviruses (CoVs) have formerly been regarded as relatively harmless respiratory pathogens to humans. However, two outbreaks of severe respiratory tract infection, caused by the severe acute respiratory syndrome coronavirus (SARS-CoV) and the Middle East respiratory syndrome coronavirus (MERS-CoV), as a result of zoonotic CoVs crossing the species barrier, caused high pathogenicity and mortality rates in human populations. This brought CoVs global attention and highlighted the importance of controlling infectious pathogens at international borders. In this review, we focus on our current understanding of the epidemiology, pathogenesis, prevention, and treatment of SARS-CoV and MERS-CoV, as well as provides details on the pivotal structure and function of the spike proteins (S proteins) on the surface of each of these viruses. For building up more suitable animal models, we compare the current animal models recapitulating pathogenesis and summarize the potential role of host receptors contributing to diverse host affinity in various species. We outline the research still needed to fully elucidate the pathogenic mechanism of these viruses, to construct reproducible animal models, and ultimately develop countermeasures to conquer not only SARS-CoV and MERS-CoV, but also these emerging coronaviral diseases.

Keywords: coronaviruses; SARS-CoV; MERS-CoV; spike proteins; animal model; prevention and treatment

---

1. Introduction

Before the first outbreak of severe acute respiratory syndrome (SARS), a limited number of coronaviruses were known to be circulating in humans, causing only mild illnesses, such as the common cold [1]. Following the 2003 SARS pandemic [2,3], it became apparent that coronaviruses could cross the species barrier and cause life-threatening infections in humans. Therefore, further attention needs to be paid to these new coronaviruses.

The 21st century has seen the worldwide spread of two previously unrecognized coronaviruses, the severe acute respiratory syndrome coronavirus (SARS-CoV) [4] and Middle East respiratory syndrome coronavirus (MERS-CoV), both of which are highly pathogenic. Starting from November 2002 in China [5], there have been unprecedented nosocomial transmissions from person to person of SARS-CoV, accompanied by high fatality rates. A united global effort led to the rapid identification of the SARS coronavirus and remarkable scientific advancements in epidemic prevention. Additionally, the zoonotic transmission of SARS from December 2003 to January 2004 [6] provided insight for researchers into the origin of this novel coronavirus. Notably, the SARS pandemic was declared to be over in 2004 when no more infections in patients were being detected. Subsequently, certain SARS-CoV-like viruses found in bats demonstrated the ability to infect human cells without prior adaptation [7,8] which indicates the possibility of the re-emergence of SARS-CoV or SARS-CoV-like viruses.

A decade later in June 2012, another highly pathogenic and novel coronavirus, MERS-CoV, was isolated from the sputum of a male patient who died from acute pneumonia and renal failure in Saudi Arabia [9]. Nosocomial infections were reported, and international travel led to the transmission of MERS-CoV to countries outside of the Arabian Peninsula, causing it to become a global pathophoresis. In May 2015, an outbreak of MERS occurred in South Korea due to an individual returning from the Middle East [10]. Based on the lessons learned from managing SARS-CoV prevalence over the last decade, tremendous progress toward unraveling the biological characteristics of MERS-CoV has been achieved at an unprecedented speed. Scientific advancements have allowed for rapid and systemic progress in our understanding of the epidemiology and pathogenesis of MERS-CoV.

SARS-CoV and MERS-CoV share several important common features that contribute to nosocomial transmission, preferential viral replication in the lower respiratory tract, and viral immunopathology. This review highlights the epidemiology and pathogenesis of these viruses, including our current understanding of their biological characteristics, their transmission, and their replication in the host. The spike proteins (S proteins) of CoVs play pivotal roles in viral infection and pathogenesis. As critical surface-located trimeric glycoproteins of CoVs, they guide entry into host cells. In this review, we summarize the structure and function of the S proteins and therapeutics designed to target them. Moreover, we will explore how CoV–host interactions cause pathogenic outcomes and discuss potential treatment options, as well as describe recent mammalian models that closely recapitulate the pathogenic process and have contributed to the development of prevention and treatment strategies for SARS-CoV and MERS-CoV. Although several potential therapies have been identified with SARS and MERS in animal and in vitro models, human clinical trials remain lacking, hindering the advancement of these potential countermeasures.

2. Epidemiology of SARS-CoV and MERS-CoV

Prior to the outbreaks of SARS and MERS [2,9], the clinical importance and epidemic possibility of CoVs had been recognized by researchers, (Table 1). In 2002, a SARS epidemic that originated in Guangdong Province in China resulted in 916 deaths among more than 8098 patients in 29 countries [11], identifying SARS as the first new infectious disease of the 21st century. Ten years later, the World Health Organization (WHO) published 2254 laboratory-confirmed cases of MERS-CoV that occurred from 2012 to 16 September 2018, with at least 800 deaths in 27 countries. Remarkably, more than 80% of recent research into the virology and genetics of this infection indicated that bats could be the possible natural reservoirs of both SARS and MERS-CoV. Palm civets [12] and dromedary camels [13] are also possible intermediary hosts of SARS and MERS, respectively, before dissemination to humans [14].

The transmission mechanism of SARS-CoV and MERS-CoV has yet to be fully understood. For transmission from animals to humans, direct contact with the intermediary host might be one route. Recent reports demonstrated that camel workers in Saudi Arabia with high prevalence of MERS-CoV infection may contribute to the transmission of MERS [15]. Some customs and habits may also be conducive to transmission, such as the consumption of milk, urine, or uncooked meat. In this way, MERS-CoV was transmitted from dromedary camels directly to humans, principally in the Arabian Peninsula, and this is considered to be the main route of transmission from animals to humans, causing significant morbidity and mortality [16,17]. Human-to-human spread has also been detected, especially through nosocomial transmission. Delays in diagnosis in hospitals might lead to secondary cases among healthcare workers, family members, or other patients sharing rooms [18–22].

Among the reported cases of SARS, 22% were healthcare workers in China and more than 40% were healthcare workers in Canada [23]. Nosocomial transmission for MERS has similarly been seen in the Middle East [16] and in the Republic of Korea [22]. Outbreaks in other countries all resulted from the reported cases in the Middle East or North Africa, and transmission was the result of international travel. Both SARS and MERS caused large outbreaks with significant public health and economic consequences.

Table 1. Epidemiology and biological characteristics of the severe acute respiratory syndrome coronavirus (SARS-CoV) and the Middle East respiratory syndrome coronavirus (MERS-CoV).

		SARS-CoV	MERS-CoV
	Genus	Beta-CoVs, lineage B	Beta-CoVs, lineage C
	Possible Natural Reservoir	Bat	Bat
	Possible Intermediary Host	Palm civet	Dromedary camel
	Origin	Guangdong province, China	Arabian Peninsula
Clinical Epidemiology	Total global number reported to WHO	More than 8098 people	2254 (from 2012 through 16 September 2018)
	Affected countries	29	27
	Number of deaths	916	800
	Mortality	More than 10%	More than 35%
	Transmission region	Globally	Regionally
	Transmission patterns	From animal to human; from human to human	
	The predominant receptor	Human angiotensin-converting enzyme 2 (ACE2)	Human dipeptidyl peptidase 4 (DPP4 or CD26)
	Receptor distribution	Arterial and venous endothelium; arterial smooth muscle; small intestine; respiratory tract epithelium; alveolar monocytes and macrophages	Respiratory tract epithelium; kidney; small intestine; liver and prostate; activated leukocytes
	Cell line susceptibility	Respiratory tract; kidney; liver	Respiratory tract; intestinal tract; genitourinary tract; liver, kidney, neurons; monocyte; Tlymphocyte; and histiocytic cell lines
	Viral replication efficiency	High	Higher
	Ability to inhibit IFN production	Delayed recognition and proinflammatory response	Delayed recognition and proinflammatory response

3. Pathogenesis of SARS-CoV and MERS-CoV

Although our current understanding of the pathogenesis of the SARS-CoV and MERS-CoV infection remains unclear, we summarize what is presently known (Table 2).

Table 2. The genomic characterization of SARS-CoV and MERS-CoV.

		SARS-CoV	MERS-CoV
Length of nucleotides		29,727	30,119
Open reading frames (ORFs)		11	11
Structural protein		4	4
Spike protein (length of amino acids)		1255	1353
S1 subunit	Receptor-binding domain (RBD)	318–510	367–588
	Receptor-binding motif (RBM)	424–494	484–567
S2 subunit	Heptad repeat 1 (HR1) domains	892–1013	984–1104
	Heptad repeat 2 (HR2) domains	1145–1195	1246–1295
Non-structural proteins (NSPs)		At least 5	16
Accessory proteins		8	5
A characteristic gene order			5′-replicase ORF1ab, spike (S), envelope (E), membrane (M), and nucleocapsid (N)-3′

Coronaviruses are the largest kind of positive-strand RNA viruses (26–32 kb) as they are about 125 nm in diameter [24], and comprise four genera (alpha-, beta-, gamma-, and delta-coronavirus) [25]. Currently, six human CoVs (HCoVs) have been confirmed: HCoV-NL63 and HCoV-229E, which belong to the alpha-coronavirus genus; and HCoV-OC43, HCoV-HKU1, SARS-CoV, and MERS-CoV, which belong to the beta-coronavirus genus. SARS-CoV and MERS-CoV are the two major causes of severe pneumonia in humans and share some common coronavirus structural characteristics. Similarly, their genomic organization is typical of coronaviruses, having an enveloped, single, positive-stranded RNA genome that encodes four major viral structural proteins, namely spike (S), envelope (E), membrane (M), and nucleocapsid (N) proteins 3–5, that follow the characteristic gene order [5′-replicase (*rep* gene), spike (S), envelope (E), membrane (M), nucleocapsid (N)-3′] with short untranslated regions at both termini (Figure 1). The viral membrane contains S, E, and M proteins, and the spike protein plays a vital functional role in viral entry. The *rep* gene encodes the non-structural protein and constitutes approximately two-thirds of the genome at the 5′ end. In detail, the S protein is in charge of receptor-binding and subsequent viral entry into host cells, and is therefore a major therapeutic target [26,27]. The M and E proteins play important roles in viral assembly, and the N protein is necessary for RNA synthesis.

The SARS-CoV genome has 29,727 nucleotides in length, including 11 open reading frames (ORFs). The SARS-CoV *rep* gene, containing about two-thirds of the genome, encodes at least two polyproteins (encoded by ORF1a and ORF1b) that undergo the process of cotranslational proteolysis. Between ORF1b and S of group 2 and some group 3 coronaviruses, there is a gene that encodes hemagglutinin-esterase [4], while this was not detected in SARS-CoV. This virus is significantly different from previously reported coronaviruses for many reasons, such as the short anchor of the S protein, the specific number and location of small ORFs, and the presence of only one copy of PLP^{pro}.

The MERS-CoV genome is larger than that of SARS-CoV at 30,119 nucleotides in length, and comprises a 5′ terminal cap structure, along with a poly (A) tail at the 3′ end, as well as the *rep* gene containing 16 non-structural proteins (nsp1–16) at the 5′ end of the genome. Four structural proteins (S, E, M, and N) and five accessory proteins (ORF3, ORF4a, ORF4b, ORF5, and ORF8) constitute about 10 kb at the 3′ end of the genome. Unlike some other beta-coronaviruses, the MERS-CoV genome does not encode a hemagglutinin-esterase (HE) protein [1]. Genomic analysis of MERS-CoV implies the potential for genetic recombination during a MERS-CoV outbreak [9]. MERS-CoV and SARS-CoV possess five and eight accessory proteins, respectively, which might help the virus evade the immune system by being harmful to the innate immune response. These differences might lead to greater sensitivity to the effects of induction and signaling of type 1 interferons (IFNs) in MERS-CoV than SARS-CoV.

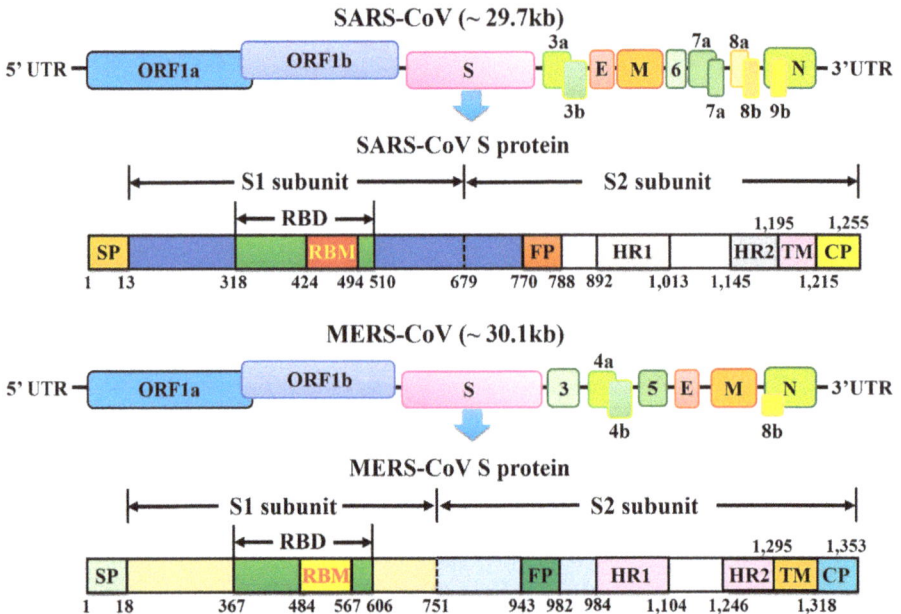

Figure 1. Schematic representation of the genome organization and functional domains of S protein for SARS-CoV and MERS-CoV. The single-stranded RNA genomes of SARS-CoV and MERS-CoV encode two large genes, the ORF1a and ORF1b genes, which encode 16 non-structural proteins (nsp1–nsp16) that are highly conserved throughout coronaviruses. The structural genes encode the structural proteins, spike (S), envelope (E), membrane (M), and nucleocapsid (N), which are common features to all coronaviruses. The accessory genes (shades of green) are unique to different coronaviruses in terms of number, genomic organization, sequence, and function. The structure of each S protein is shown beneath the genome organization. The S protein mainly contains the S1 and S2 subunits. The residue numbers in each region represent their positions in the S protein of SARS and MERS, respectively. The S1/S2 cleavage sites are highlighted by dotted lines. SARS-CoV, severe acute respiratory syndrome coronavirus; MERS-CoV, Middle East respiratory syndrome coronavirus; CP, cytoplasm domain; FP, fusion peptide; HR, heptad repeat; RBD, receptor-binding domain; RBM, receptor-binding motif; SP, signal peptide; TM, transmembrane domain.

4. Comparative Pathology and Life Cycles of SARS-CoV and MERS-CoV

Both SARS and MERS cause severe pneumonia resulting from these novel coronaviruses, sharing some similarities in their pathogenesis (Figure 2) [28].

SARS is an emerging infectious viral disease characterized by severe clinical manifestations of the lower respiratory tract, resulting in diffuse alveolar damage. SARS-CoV spreads through respiratory secretions, such as droplets, via direct person-to-person contact. Upon exposure of the host to the virus, the virus binds to cells expressing the virus receptors, of which the angiotensin-converting enzyme 2 (ACE2) is one of the main receptors, and CD209L is an alternative receptor with a much lower affinity [29]. In the respiratory tract, ACE2 is widely expressed on the epithelial cells of alveoli, trachea, bronchi, bronchial serous glands [30], and alveolar monocytes and macrophages [31]. The virus enters and replicates in these target cells. The mature virions are then released from primary cells and infect new target cells [32]. Furthermore, as a surface molecule, ACE2 is also diffusely localized on the endothelial cells of arteries and veins, the mucosal cells of the intestines, tubular epithelial cells of the kidneys, epithelial cells of the renal tubules, and cerebral neurons and immune cells, providing a variety of susceptible cells to SARS-CoV [33,34]. Respiratory secretions, urine, stools, and sweat from patients with SARS contain infective viral particles, which may be excreted into and contaminate the environment. Atypical pneumonia with rapid respiratory deterioration and failure can be induced by SARS-CoV infection because of increased levels of activated proinflammatory chemokines and cytokines [35].

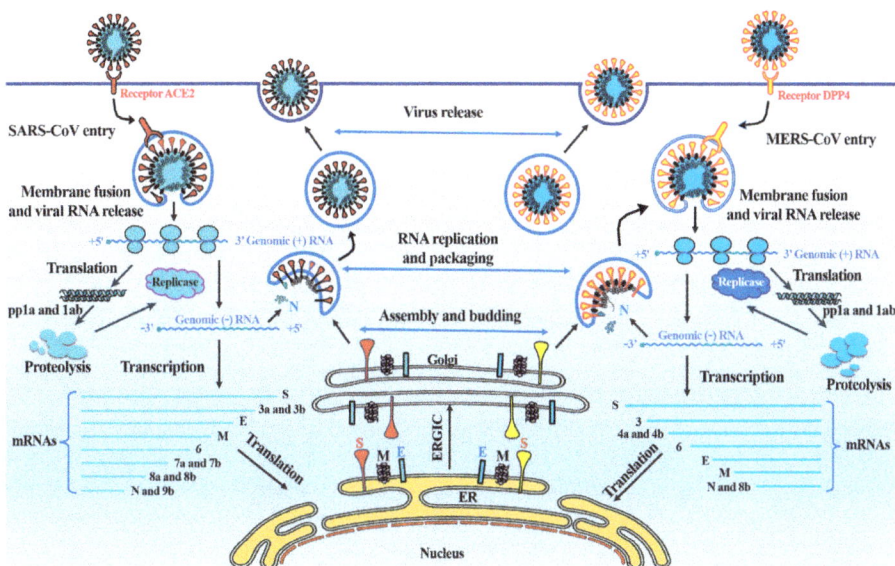

Figure 2. The life cycle of SARS-CoV and MERS-CoV in host cells. SARS-CoV and MERS-CoV enter target cells through an endosomal pathway. The S proteins of SARS and MERS bind to cellular receptor angiotensin-converting enzyme 2 (ACE2) and cellular receptor dipeptidyl peptidase 4 (DPP4), respectively. Following entry of the virus into the host cell, the viral RNA is unveiled in the cytoplasm. ORF1a and ORF1ab are translated to produce pp1a and pp1ab polyproteins, which are cleaved by the proteases that are encoded by ORF1a to yield 16 non-structural proteins that form the RNA replicase–transcriptase complex. This complex drives the production of negative-sense RNAs [(−) RNA] through both replication and transcription. During replication, full-length (−) RNA copies of the genome are produced and used as templates for full-length (+) RNA genomes. During transcription, a subset of 7–9 sub-genomic RNAs, including those encoding all structural proteins, is produced through discontinuous transcription. Although the different sub-genomic mRNAs may contain several open reading frames (ORFs), only the first ORF (that closest to the 5′ end) is translated. Viral nucleocapsids are assembled from genomic RNA and N protein in the cytoplasm, followed by budding into the lumen of the ERGIC (endoplasmic reticulum (ER)–Golgi intermediate compartment). Virions are then released from the infected cell through exocytosis. SARS-CoV, severe acute respiratory syndrome coronavirus; MERS-CoV, Middle East respiratory syndrome coronavirus; S, spike; E, envelope; M, membrane; N, nucleocapsid.

For MERS-CoV infection of humans, the primary receptor is a multifunctional cell surface protein, dipeptidyl peptidase 4 (DPP4, also known as CD26) [36], which is widely expressed on epithelial cells in the kidney, alveoli, small intestine, liver, and prostate, and on activated leukocytes [37]. Consistent with this, MERS-CoV can infect several human cell lines, including lower respiratory, kidney, intestinal, and liver cells, as well as histiocytes, as shown by a cell-line susceptibility study [38], indicating that the range of MERS-CoV tissue tropism in vitro was broader than that of any other CoV. MERS-CoV causes acute, highly lethal pneumonia and renal dysfunction with various clinical symptoms, including—but not restricted to—fever, cough, sore throat, myalgia, chest pain, diarrhea, vomiting, and abdominal pain [39,40]. Lung infection in the MERS animal model demonstrated infiltration of neutrophils and macrophages and alveolar edema [41]. The entry receptor (DPP4) for MERS-CoV is also highly expressed in the kidney, causing renal dysfunctions by either hypoxic damage or direct infection of the epithelia [42]. Remarkably, unlike SARS-CoV, MERS-CoV has the ability to infect human dendritic cells [43] and macrophages [44] in vitro, thus helping the virus to disrupt the immune system. T cells are another target for MERS-CoV because of their high amounts of CD26 [45]. This virus might deregulate antiviral T-cell responses due to the stimulation of T-cell apoptosis [45,46]. MERS-CoV might also lead to immune dysregulation [47] by stimulating attenuated innate immune responses, with delayed proinflammatory cytokine induction in vitro and in vivo [44,48,49].

5. SARS and MERS-CoV Spike Protein: A Key Target for Antivirals

5.1. Structure of the SARS-CoV and MERS-CoV Spike Protein

Trimers of the S protein make up the spikes of SARS-CoV and provide the formation of a 1255-amino-acids-length surface glycoprotein precursor. Most of the protein and the amino terminus are situated on the outside of the virus particle or the cell surface [50]. The expected structure of the S protein comprises four parts: a signal peptide located at the N terminus from amino acids 1 to 12, an extracellular domain from amino acids 13 to 1195, a transmembrane domain from amino acids 1196 to 1215, and an intracellular domain from amino acids 1216 to 1255. Proteases such as factor Xa, trypsin, and cathepsin L cleave the SARS-CoV S protein into two subunits, the S1 and S2 subunits. A minimal receptor-binding domain (RBD) located in the S1 subunit (amino acids 318–510) can combine with the host cell receptor, ACE2. The RBD displays a concave surface during interaction with the receptor. The entire receptor-binding loop, known as the receptor-binding motif (RBM) (amino acids 424–494), is located on the RBD and is responsible for complete contact with ACE2. Importantly, two residues in the RBM at positions 479 and 487 determine the progression of the SARS disease and the tropism of SARS-CoV [51,52]. Recent studies using civets, mice, and rats demonstrated that any change in these two residues might improve animal-to-human or human-to-human transmission and facilitate efficient cross-species infection [53]. The S2 subunit mediates the fusion between SARS-CoV and target cells, and includes the heptad repeat 1 (HR1) and HR2 domains, whose HR1 region is longer than the HR2 region.

Similar to SARS-CoV, during the infection process, the S protein of MERS-CoV is cleaved into a receptor-binding subunit S1 and a membrane-fusion subunit S2 [54–57]. The MERS-CoV S1 subunit also includes an RBD, mediating the attachment between virus and target cells [54,55,58,59]. Unlike SARS-CoV, MERS-CoV requires DPP4 (also known as CD26) as its cellular receptor [60,61] but not ACE2. The RBDs of MERS-CoV and SARS-CoV differ, although they share a high degree of structural similarity in their core subdomains, explaining the different critical receptors noted above [57,62]. The core subdomain of RBD is stabilized by three disulfide bonds, and includes a five-stranded antiparallel β-sheet and several connecting helices. The RBM comprises a four-stranded antiparallel β-sheet for connecting to the core via loops [57,62]. Two N-linked glycans, N410 and N487, are seated in the core and RBM, respectively. Particularly, the residues 484–567 of RBM take charge of interacting with the extracellular β-propeller domain of DPP4. The fusion core formation of MERS-CoV resembles that of SARS-CoV; however, it is different from that of other coronaviruses, such as the mouse hepatitis virus (MHV) and HCoV-NL63 [63–66].

5.2. Functions of the SARS-CoV and MERS-CoV S Protein

The SARS-CoV S protein plays pivotal roles in viral infection and pathogenesis [67,68]. The S1 subunit recognizes and binds to host receptors, and the subsequent conformational changes in the S2 subunit mediate fusion between the viral envelope and the host cell membrane [69,70]. The RBD in the S1 subunit is responsible for virus binding to host cell receptors [61,70,71]. ACE2 is a functional receptor for SARS-CoV that makes contact with 14 amino acids in the RBD of SARS-CoV among its 18 residues [53]. The RBD in the S1 subunit is responsible for virus binding to host cell receptors [61,70,71]. Position R453 in the RBD and position K341 in ACE2 play indispensable roles in complex formation [72]. Furthermore, the N479 and T487 in the RBD of the S protein are pivotal positions for the affinity with ACE2 [52], and R441 or D454 in the RBD influences the antigenic structure and binding activity between RBD and ACE2 [73]. From a pre-fusion structure to a post-fusion structure, binding of the RBD in the S1 subunit to the receptor ACE2 stimulates a conformational change in S2. Accordingly, the supposed fusion peptide (amino acids 770–788) [74] builds in the target cell membrane of the host. Meanwhile, a six-helix bundle fusion core structure is made up by the HR1 and HR2 domains for bringing the viral envelope and the target cell membrane into close proximity and contributing to fusion [74]. Resembling the S2 subunit of SARS-CoV, the MERS-CoV S2 subunit is in charge of membrane fusion. The HR1 and HR2 regions in S2 play essential and complementary roles [56,63]. Furthermore, SARS-CoV displays an alternative method of binding to the host cell via other potential receptors. Dendritic cell-specific intercellular adhesion molecule-3-grabbing non-integrin (DC-SIGN) and/or liver/lymph node-SIGN (L-SIGN) are two examples of such receptors [29,75]. Seven residue sites, at positions 109, 118, 119, 158, 227, 589, and 699 of the S protein displaying asparagine-linked glycosylation are crucial for DC-SIGN or L-SIGN-mediated virus entry. These residues, unlike those of the ACE2-binding domain, function independently of ACE2 [76].

5.3. Vaccines Based on the SARS-CoV and MERS-CoV S Protein

In order to control the outbreak of viruses, vaccinations were developed against SARS-CoV and MERS-CoV. There are various approaches of different vaccines, and the development and advantages/disadvantages of these are listed in Table 3 (this table includes updates about SARS-CoV and MERS-CoV since 2013; SARS-CoV-related parts were modified by Graham et al. in Nature Reviews Microbiology, 2013 [77]).

Importantly, among all the functional/non-functional structural proteins of SARS-CoV and MERS-CoV, the S protein is the principal antigenic component that induces antibodies to block virus-binding, stimulate host immune responses, fuse or neutralize antibodies and/or protect the immune system against virus infection. Therefore, the S protein has been selected as a significant target for the development of vaccines. It has been noted that antibodies raised against subunit S1 (amino acids 485–625) or S2 (amino acids 1029–1192) neutralize infection by SARS-CoV strains in Vero E6 cells [78,79]. Researchers have constructed an attenuated parainfluenza virus encoding the full-length S protein of the SARS-CoV Urbani strain for the vaccination of African green monkeys. This vaccine could protect monkeys from subsequent homologous SARS-CoV infection, demonstrating highly effective immunization with the S protein [80]. Other studies in a mouse model structured a DNA vaccine encoding the full-length S protein of the SARS-CoV Urbani strain that not only induced T-cell and neutralizing-antibody responses, but also stimulated protective immunity [81]. Furthermore, monkeys or mice were vaccinated with a highly attenuated, modified vaccine virus, Ankara, encoding the full-length S protein of the SARS-CoV strain HKU39849 or Urbani [82]. However, full-length S protein-based SARS vaccines may induce harmful immune responses, causing liver damage in the vaccinated animals or enhancing infection after being challenged with homologous SARS-CoV [83,84]. Researchers are thus concerned about the safety and ultimate protective efficacy of vaccines that include the full-length SARS-CoV S protein.

There are still no commercial vaccines available against MERS-CoV [26]. Multiple vaccine candidates targeting the S protein, which is responsible for viral entry, have been developed, including subunit vaccines [85,86], recombinant vector vaccines [87,88], and DNA vaccines [89,90]. Importantly, compared with other regions of the S protein, the RBD fragment induced the highest-titer IgG antibodies in mice [85]. Modified vaccines, including recombinant vectors of Ankara and adenoviruses expressing the MERS-CoV S glycoprotein, showed immunogenicity in mice [25]. Attenuated live vaccines also showed a protective function, but there were concerns regarding the degree of attenuation [91]. After intranasal vaccination with the CoV N protein, airway memory CD4 T cells were generated and mediated the protection following a CoV challenge [92]. These cells could induce anti-viral innate responses at an early stage of infection, and facilitated CD8 T-cell responses by stimulating recombinant dendritic cell migration and CD8 T-cell mobilization [92]. The stimulation of airway memory CD4 T cells should be regarded as an essential part of any HCoV vaccine strategy, because these CD4 T cells target a conserved epitope within the N protein that cross-reacts with several other CoVs [92]. Furthermore, DNA vaccines expressing the MERS-CoV S1 gene produced antigen-specific humoral and cellular immune responses in mice [89].

Table 3. Vaccine strategies of SARS-CoV and MERS-CoV.

Vaccine Strategy	Process of Production	References		Advantages	Disadvantages
		SARS	MERS		
Inactivated virus vaccines	Virus particles are inactivated by heat, chemicals, or radiation	Whole virus, with or without adjuvant (promote an effective immune response against the inactivated pathogen) [93,94]	Whole virus, with or without adjuvant (promote an effective immune response against the inactivated pathogen) [91,95]	Maintained virus particles structure; rapidly develop; easy to prepare; safety; high-titer neutralizing antibodies [93]; protection with adjuvant [96,97].	Potential inappropriate for highly immunosuppressed individuals; possible T$_H$2 cell-distortive immune response [98,99].
Live-attenuated virus vaccines	Attenuated the virulence, but still keeping it viable by mutagenesis or targeted deletions	Envelope protein deletion [100]; non-structural protein 14 (nsp14) and exonuclease (ExoN) inactivation [101]	Full-length infectious cDNA clone or mutant viruses [102]	Inexpensive; quick immunity; less adverse effect; activates all phases of the immune system [103]; more durable immunity; more targeted [77].	Phenotypic or genotypic reversion possible; need sufficient viral replication [77].
Viral vector vaccines	Genetically engineered unrelated viral genome with deficient packaging elements for encoding targeted gene	Spike and nucleocapsid proteins [100,104]	Spike and nucleocapsid proteins [87,88]	Safety; stronger and specific cellular and humoral immune responses [77].	Varies inoculation routes may produce different immune responses [96]; possibly incomplete protection; may fail in aged vaccinees; possible T$_H$2 cell-distortive immune response [105].
Subunit vaccines	Antigenic components inducing the immune system without introducing viral particles, whole or otherwise.	Spike and nucleocapsid proteins [53,59,106]	Spike and nucleocapsid proteins [85,86,107,108]	High safety; consistent production; can induce cellular and humoral immune responses; high-titer neutralizing antibodies [109].	Uncertain cost-effectiveness; relatively lower immunogenicity; need appropriate adjuvants [77].
DNA vaccines	Genetically engineered DNA for directly producing an antigen	Spike and nucleocapsid proteins [110,111]	Spike and nucleocapsid proteins [89,90]	Easier to design; high safety; high-titer neutralizing antibodies [110].	Lower immune responses; potential T$_H$2 cell-distortive immune response results; potential ineffective; possibly delayed-type hypersensitivity [112].

5.4. S Protein-Based Therapeutics for SARS-CoV and MERS-CoV

Despite the presence of extensive research reporting on SARS-CoV and MERS-CoV therapies, it was not possible to establish whether treatments benefited patients during their outbreak. In the absence of fundamental, clinically proven, effective antiviral therapy against SARS-CoV and MERS-CoV, patients mainly receive supportive care supplemented by diverse combinations of drugs. Several approaches are being considered to treat infections of SARS-CoV [113] and MERS-CoV (Table 4, MERS-CoV-related table previously reviewed by de Wit et al. in Nature Reviews Microbiology, 2016 [10]), including the use of antibodies, IFNs, and inhibitors of viral and host proteases.

The vital role of the S protein of SARS-CoV makes this protein an important therapeutic target, and numerous studies have explored potential therapeutics. Firstly, peptides that block RBD–ACE2-binding derived from both RBD [114] and ACE2 [76] could be developed as novel therapeutics against SARS-CoV infection. Secondly, peptides binding to the S protein interfere with the cleavage of S1 and S2. This step inhibits the production of functional S1 and S2 subunits and the consequent fusion of the viral envelope with the host cell membrane. Thirdly, anti-SARS-CoV peptides blocking the HR1–HR2 interaction by forming a fusion-active core have viral fusion inhibitory activity at the micromolar level [115–117]. However, the potential selection of escape mutants with altered host range phenotypes is one of the disadvantages of this strategy that needs further modification [118]. Furthermore, mouse monoclonal antibodies (mAbs) targeting assorted fragments of the SARS-CoV S protein have effectively inhibited SARS-CoV infection [79,119–122]. A series of neutralizing human mAbs were generated from the B cells of patients infected with SARS-CoV [123,124]. Another strategy used human immunoglobulin transgenic mice immunized with full-length SARS-CoV S proteins [125–127]. 80R and CR3014 binding to the ACE2 receptor are examples of S-specific mAbs [128,129].

Similarly, the therapeutic agents that have been developed against MERS-CoV are based on the S protein and basically restrain the binding of receptors or the fusion of membrane proteins, thereby leading to the inhibition of MERS-CoV infection. These methods mainly involve peptidic fusion inhibitors [56,63,116,130], anti-MERS-CoV neutralizing mAbs [86,131], anti-DPP4 mAbs [86,132,133], DPP4 antagonists [134], and protease inhibitors [135–137]. However, none of these anti-MERS-CoV curative agents are approved for commercial use in humans.

Table 4. Potential therapeutics for severe acute respiratory syndrome (SARS) and MERS.

Treatment	Stage of Development	
	SARS (Notes)	MERS (Notes)
Host protease inhibitors	Effective in mouse models [138]	In vitro inhibition [138]
Viral protease inhibitors	In vitro inhibition [139]	In vitro inhibition [140]
Monoclonal and polyclonal antibodies	Effective in mouse, ferrets, golden Syrian hamster [124,141,142] and non-human primate models [143,144]	Effective in mouse, rabbit, and non-human primate models [10,145]
Convalescent plasma	Off-label use in patients [146,147]	Effective in a mouse model; clinical trial approved [10]
Interferons	Off-label use in patients (often in combination with immunoglobulins or thymosins) [146,147]	Effective in non-human primate models; off-label use in patients (often in combination with a broad-spectrum antibiotic and oxygen) [10]
Ribavirin	Off-label use in patients (often in combination with corticosteroids) [146,147]	Effective in a non-human primate model; off-label use in patients (often in combination with a broad-spectrum antibiotic and oxygen) [10]
Lopinavir and ritonavir	Off-label use in patients (improved the outcome in combination with ribavirin) [146,147]	Effective in a non-human primate model; off-label use in patients [10,148]
Common Feature	None of these therapeutic agents are approved for commercial use in humans	

6. The Animal Models of SARS and MERS-CoV

International coordination and cooperation led to the rapid identification of SARS-CoV and MERS-CoV. Emergency control measures and laboratory detection systems which were put in place in response to SARS-CoV and MERS-CoV outbreaks were both exemplary. To establish optimal prevention and control strategies for SARS and MERS, numerous efforts to develop animal models were undertaken in several laboratories, despite the fact that some conflicting results have been reported. It is therefore necessary to compare and document the features and disadvantages of different animal models to better understand viral replication, transmission, pathogenesis, prevention, and treatment. Notably, several animal species were suggested as suitable disease models of SARS-CoV, but most laboratory animals are refractory or only semi-permissive to MERS-CoV infection.

6.1. Animal Models of SARS-CoV

SARS-CoV replication has been studied in mice, Syrian golden and Chinese hamsters, civet cats, and non-human primates. The most severe symptoms of SARS were observed in aged animals. To develop epidemiological symptoms that advanced age resulted in increased mortality, aged mouse model of SARS-CoV has been generated. Transgenic mice expressing human ACE2 were also developed to closely mimic SARS-CoV infection in humans. Some animal models have been tested and analyzed on the genomic and proteomics level to study the pathogenesis of SARS.

6.1.1. Mouse Models

Mouse species that have been used as SARS-CoV-infected animal models include BALB/c [149,150], C57BL6 (B6) [151], and 129SvEv-lineage mice. The most relevant transgenic and knockout lines are accessible based on these susceptible animals [152]. Signal transducers and activators of transcription 1 (STAT1)-knockout and myeloid differentiation primary response 88 (MYD88)-knockout mice [149,151,153,154] are examples of mouse models with innate immune deficiency, and such animals display severe effects of the disease, such as pneumonitis, bronchiolitis, and weight loss, and often die within 9 days of infection. Notably, young mice require more mutations and passages than aged mice to produce SARS-CoV mouse-adapted strains. More severe pathological lesions and increased mortality were observed in one-year-old animals, along with fewer mutations at miscellaneous locations throughout the genome [98,155–159]. Intranasal inoculation of four- to eight-week-old BALB/c or B6 mice with SARS-CoV resulted in nasal turbinate in the upper respiratory tract and a high titer of virus replication in the lungs of the lower respiratory tract, and this model was highly reproducible without any signs of morbidity or mortality [149,151]. Neutralizing antibody responses could be generated in sub-lethally infected mice protecting recipients from subsequent lethal challenges, which probably reflected the situation in infected humans during an epidemic [160]. However, on day 2–3 post-infection (pi), virus replication in the respiratory tract peaked but was not accompanied by massive pulmonary inflammation or pneumonitis. By day 5–7 pi, the virus had been eliminated from the lungs [149,151]. It was obvious that viremia is common and long-lasting in patients, while it is rare and transient in mouse models [161]. Mice could therefore be used as a stable and reproducible animal model for the evaluation of vaccines, immune-prophylaxis, and antiviral drugs against SARS-CoV [81,96,109,124,149,162–166].

6.1.2. Hamster Model

Golden Syrian and Chinese hamsters have also been evaluated and shown to be excellent models of SARS-CoV infection, owing to their high titer of virus replication in the respiratory tract, associated with diffuse alveolar damage, interstitial pneumonitis, and pulmonary consolidation [104,167–169]. On day 2 pi, peak levels of viral replication were detected in the lower respiratory tract, and the virus was cleared without obvious clinical illness 7–10 days after infection. Similarly to mice, infected hamsters also produced a protective neutralizing-antibody response to subsequent SARS-CoV challenges [104,170]. Resulting from the extremely high titers and reproducible pulmonary pathological lesions in SARS-CoV-infected hamsters, this animal model is ideal for studies on the immunoprophylaxis and treatment of SARS [104,170]. However, there are still limited resources in terms of genetically established animal lines and accurate immunological and cellular biomarkers for hamster models.

6.1.3. Ferret Model

Ferrets were found to be susceptible to SARS-CoV infection [171] but could also transmit the virus at low levels by direct contact [84,172–174]. They showed diverse clinical symptoms in different studies [171,174]. Importantly, ferrets could develop fever, which is a characteristic clinical symptom of SARS-CoV-infected patients [93,175]. Similar to rodent models, infection of ferrets with SARS-CoV did not result in significant mortality. However, there are still some conflicting reports regarding the histopathological lesions and severity of clinical observations in the ferret model that require further investigation.

6.1.4. Non-Human Primate Models

Several species of non-human primates (NHP) were evaluated as animal models for SARS. At least six NHP species were tested including three Old World monkeys: rhesus macaques [176–180], cynomolgus macaques [177,181,182], and African green monkeys [177]; and three New World monkeys: the common marmoset [183], squirrel monkeys, and mustached tamarins [176–178,181–184]. Except for squirrel monkeys and mustached tamarins [185], all of the evaluated NHP species facilitated the replication of SARS-CoV [186]. Virus replication was detected in the respiratory tract of rhesus macaques, cynomolgus macaques, and African green monkeys. Pneumonitis was observed in each of these species in different studies [176–178,182]. SARS-infected common marmosets displayed a fever, watery diarrhea, pneumonitis, and hepatitis [183]. Unfortunately, research into the clinical signs of disease in cynomolgus and rhesus macaques gave conflicting results and therefore needs further investigation. The main reason for the lack of reproducibility in such studies may be the limited sample size.

6.2. Animal Models of MERS-CoV

Small animal models of MERS infection are urgently needed to elucidate MERS pathogenesis and explore potential vaccines and antiviral drugs. Previous studies have demonstrated the difficulties in developing such a model, such as that mice [187,188], ferrets [134], guinea pigs [189], and hamsters [189] are not susceptible to experimental MERS-CoV infection mainly because their homologous DPP4 molecules do not function as receptors for MERS-CoV entry. After administering a high dose of MERS-CoV, no viral replication could be detected in these animals [190]. In an animal model using New Zealand white rabbits, regardless of the fact that detectable viral RNA existed in the respiratory tract and moderate necrosis was observed in nasal turbinates, the animals showed no clinical symptoms of disease [191]. In another study, attempts to infect hamsters with MERS-CoV were not successful [192]. However, despite this, MERS-CoV is a broad host-range virus in vitro [25], and there is hope that a reproducible and stable animal model for human MERS-CoV infection can be improved in the near future.

6.2.1. Mouse Model for MERS Infection

Despite the fact that wild-type rodents are not susceptible to MERS-CoV infection [188], researchers have developed several models in which mice are susceptible to MERS-CoV infection [193–195]. The first mouse model of MERS infection reported in 2014 involved transducing animals with recombinant adenovirus 5 encoding human DPP4 (hDPP4) molecules intranasally, and this resulted in replication of MERS-CoV in the lungs. This mouse model also showed clinical symptoms of interstitial pneumonia, including inflammatory cell infiltration, and thickened alveolar and mild edema [195]. However, there are certain limitations to this model, such as the uncontrolled expression and distribution of hDPP4. In 2015, the establishment of hDPP4 transgenic mice was reported [194]. MERS-CoV could infect this mouse model effectively. However, similarly to SARS-CoV-infected ACE2 transgenic mice [196], systemic expressions led to multiple organ lesions [194], resulting in the death of the animals. Most recently, the homologous hDPP4 gene was used in several MERS transgenic mouse models [193,197]. Remarkably, hDPP4 knockin (KI) mice, where mouse DPP4 gene fragments had been displaced by homologous human DPP4 fragments, showed effective receptor binding. Furthermore, a mouse-adapted MERS-CoV strain (MERS$_{MA}$) including 13–22 mutations was produced in the lungs of hDPP4-KI mice after 30 serial passages, causing effective weight loss and mortality in this mouse model [193]. Both this hDPP4-KI mouse and the MERS$_{MA}$ strain provide better tools to explore the pathogenesis of MERS and potential novel treatments.

6.2.2. Camelidae

As a reservoir of MERS-CoV, dromedary camels showed mild upper respiratory infections after the administration of MERS-CoV [198]. Oronasal infection of MERS-CoV in alpacas, a close relative within the Camelidae family, resulted in an asymptomatic infection with no signs of upper or lower respiratory tract disease [199,200]. Additionally, owing to their high cost and relatively large size, these animal models are not available for high-throughput studies of MERS.

6.2.3. Non-Human Primates

NHPs, such as the rhesus macaques [201] and common marmosets [202], are useful models for studying the pathogenesis of mild MERS-CoV infection and evaluating novel therapies for humans, although the degree of replication and disease severity vary [192,201,203,204]. MERS-CoV caused transient lower respiratory tract infection in rhesus macaques, with associated pneumonia. Clinical signs were observed by day 1 pi and resolved as early as day 4 pi [201]. Relatively mild clinical symptoms were observed early on in infection without fatalities, indicating that rhesus macaques do not recapitulate the severe infections observed in human cases; however, treatment of MERS-CoV-infected rhesus macaques with IFN-α and ribavirin decreased virus replication, alleviated the host response, and improved the clinical outcome [205]. Infection of MERS-CoV in common marmosets demonstrated various extents of damage depending on the study, but successfully reproduced several features of MERS-CoV infection in humans. Importantly, one study indicated that the infection became progressive severe pneumonia [203], while other groups found that MERS-CoV-infected common marmosets only developed mild to moderate nonlethal respiratory diseases by intratracheal administration [206].

7. Role of Host Receptors in Animal Models of SARS-CoV and MERS-CoV

The reasons for host restriction, none or limited clinical symptoms observed in varies animal models are complexity. The interaction between the host receptor and functional proteins of SARS and MERS, respectively, plays an important and predominant role. In the context of animal models of SARS-CoV infection, researchers have compared the ACE2 amino acids that interact with the S protein RBD from several species. In agreement with the permissive nature of these species, the ACE2 residues of marmoset and hamster are similar to those of hACE2 [53]. By comparison, many residues of mouse ACE2 are different from those of hACE2, and this meets with decreased replication of SARS-CoV in mouse cells [207] and the lungs of young mice [149]. The changes at positions 353 (histidine) and 82 (asparagine) of rat ACE2 relative to hACE2 partially disrupt the S protein-DPP4 interaction and contribute to abrogation of binding. Interestingly, ferrets are permissive to SARS-CoV infection, but most of their ACE2 interaction residues are different from those of hACE2 [53], while many of the ACE2 residues between civet and ferret are the same, which may result in similar affinity [208]. For MERS-CoV, 14 residues of the S protein RBD have direct contact with 15 residues of hDPP4 [57].

Comparisons of human DPP4 binding affinity to that of other species indicated that human DPP4 had the highest affinity to the S protein of MERS-CoV, where the decreasing order of affinity is as follows: human > horses > camels > goats > bats [209]. Further evidence demonstrates that the host restriction of MERS-CoV remarkably depends on the sequence of DPP4, such as the characterization of amino acid residues at the connector of DPP4 with the RBD of S proteins in mice [187,210], hamsters, and cotton rats [210]. However, the multiplicity in severity of disease between rhesus macaques and common marmosets indicate that other host factors can perhaps affect the infection and replication of the virus, such as the presence of S-cleaving proteases [187]. In general, although the structural analysis of receptors-S protein interactions cannot fully explain all the observations for host restriction, they agree with the improved replication in several animal models and that it should be the premier and remarkable focus of small-animal model development. These special residues for host affinity are important to build up transgenic animal models enhancing the permissiveness and infection of SARS and MERS.

8. Outlook and Summary

Unlike SARS-CoV, which resolved without more reported cases, continued outbreaks of MERS-CoV present an ongoing threat to public health. It should be noted that no specific treatment is currently available for HCoVs, and further research into the pathogenesis of HCoV infection is therefore imperative to identify appropriate therapeutic targets. Accordingly, at present, the prevention of viral transmission is of utmost importance to limit the spread of MERS. The enormous ratio of nosocomial infections indicates that preventive measures in hospitals have not been sufficiently implemented. Additionally, as an emerging zoonotic virus, prevention of transmission from dromedary camels is another possibility to reduce the quantity of MERS cases. Regarding clinical therapies, a combination of treatment administered as early as possible and aimed at synchronously disrupting viral replication, inhibiting viral dissemination, and restraining the host response is likely to be most suitable, due to the acute clinical features of MERS with diffuse lung damage and the important role of immunopathology.

Potential treatments must undergo in vitro and in vivo studies to select the most promising options. The development of stable and reproducible animal models of MERS, especially in NHPs, is therefore a decisive step forward. The next step in the development of standardized and controllable therapies against SARS and MERS will be clinical trials in humans, validating a standard protocol for dosage and timing, and accruing data in real time during future outbreaks to monitor specific adverse effects and help inform treatment.

The comprehensive lessons and experiences that have resulted from the outbreaks of SARS and MERS provide valuable insight and advancements in how to react to future emerging and re-emerging infectious agents. Rapid identification of the pathogen via effective diagnostic assays is the first step, followed by the implementation of preventive measures, including raising awareness of the new agent, reporting and recording (suspected) cases, and infection control management in medical facilities. Studies are currently needed that focus on the epidemiology of these organisms, especially in terms of pathogen transmission and potential reservoirs and/or intermediate hosts. Animal models and prophylactic and therapeutic approaches should be promoted, followed by fast-tracked clinical trials.

Our increasing understanding of novel emerging coronaviruses will be accompanied by increasing opportunities for the reasonable design of therapeutics. Importantly, understanding this basic information will not only aid our public health preparedness against SARS-CoV and MERS-CoV, but also help prepare for novel coronaviruses that may emerge.

Funding: This research was funded by CAMS Innovation Fund for Medical Science (CIFMS), grant number 2016-12M-2-006.

Acknowledgments: We thank Kate Fox, DPhil, from Liwen Bianji, Edanz Group China (www.liwenbianji.cn/ac), for editing the English text of a draft of this manuscript.

Conflicts of Interest: The authors declare no conflict of interest.

References

1. Yin, Y.; Wunderink, R.G. MERS, SARS and other coronaviruses as causes of pneumonia. *Respirology* **2018**, *23*, 130–137. [CrossRef] [PubMed]
2. Drosten, C.; Gunther, S.; Preiser, W.; van der Werf, S.; Brodt, H.R.; Becker, S.; Rabenau, H.; Panning, M.; Kolesnikova, L.; Fouchier, R.A.; et al. Identification of a novel coronavirus in patients with severe acute respiratory syndrome. *N. Engl. J. Med.* **2003**, *348*, 1967–1976. [CrossRef] [PubMed]
3. Ksiazek, T.G.; Erdman, D.; Goldsmith, C.S.; Zaki, S.R.; Peret, T.; Emery, S.; Tong, S.; Urbani, C.; Comer, J.A.; Lim, W.; et al. A novel coronavirus associated with severe acute respiratory syndrome. *N. Engl. J. Med.* **2003**, *348*, 1953–1966. [CrossRef] [PubMed]
4. Rota, P.A. Characterization of a Novel Coronavirus Associated with Severe Acute Respiratory Syndrome. *Science* **2003**, *300*, 1394–1399. [CrossRef] [PubMed]
5. Zhong, N.S.; Zheng, B.J.; Li, Y.M.; Poon, L.L.; Xie, Z.H.; Chan, K.H.; Li, P.H.; Tan, S.Y.; Chang, Q.; Xie, J.P.; et al. Epidemiology and cause of severe acute respiratory syndrome (SARS) in Guangdong, People's Republic of China, in February, 2003. *Lancet* **2003**, *362*, 1353–1358. [CrossRef]
6. Lee, N.; Hui, D.; Wu, A.; Chan, P.; Cameron, P.; Joynt, G.M.; Ahuja, A.; Yung, M.Y.; Leung, C.B.; To, K.F.; et al. A major outbreak of severe acute respiratory syndrome in Hong Kong. *N. Engl. J. Med.* **2003**, *348*, 1986–1994. [CrossRef] [PubMed]
7. Ge, X.Y.; Li, J.L.; Yang, X.L.; Chmura, A.A.; Zhu, G.; Epstein, J.H.; Mazet, J.K.; Hu, B.; Zhang, W.; Peng, C.; et al. Isolation and characterization of a bat SARS-like coronavirus that uses the ACE2 receptor. *Nature* **2013**, *503*, 535–538. [CrossRef]
8. Menachery, V.D.; Yount, B.J.; Debbink, K.; Agnihothram, S.; Gralinski, L.E.; Plante, J.A.; Graham, R.L.; Scobey, T.; Ge, X.Y.; Donaldson, E.F.; et al. A SARS-like cluster of circulating bat coronaviruses shows potential for human emergence. *Nat. Med.* **2015**, *21*, 1508–1513. [CrossRef]
9. Zaki, A.M.; van Boheemen, S.; Bestebroer, T.M.; Osterhaus, A.D.; Fouchier, R.A. Isolation of a novel coronavirus from a man with pneumonia in Saudi Arabia. *N. Engl. J. Med.* **2012**, *367*, 1814–1820. [CrossRef]
10. de Wit, E.; van Doremalen, N.; Falzarano, D.; Munster, V.J. SARS and MERS: Recent insights into emerging coronaviruses. *Nat. Rev. Microbiol.* **2016**, *14*, 523–534. [CrossRef]
11. World Health Organization. WHO Guidelines for the Global Surveillance of Severe Acute Respiratory Syndrome (SARS). Updated Recommendations. October 2004. Available online: http://www.who.int/mediacentre/factsheets/mers-cov/en/ (accessed on 8 October 2018).
12. Guan, Y.; Zheng, B.J.; He, Y.Q.; Liu, X.L.; Zhuang, Z.X.; Cheung, C.L.; Luo, S.W.; Li, P.H.; Zhang, L.J.; Guan, Y.J.; et al. Isolation and characterization of viruses related to the SARS coronavirus from animals in southern China. *Science* **2003**, *302*, 276–278. [CrossRef]
13. Azhar, E.I.; El-Kafrawy, S.A.; Farraj, S.A.; Hassan, A.M.; Al-Saeed, M.S.; Hashem, A.M.; Madani, T.A. Evidence for camel-to-human transmission of MERS coronavirus. *N. Engl. J. Med.* **2014**, *370*, 2499–2505. [CrossRef] [PubMed]
14. Cui, J.; Li, F.; Shi, Z. Origin and evolution of pathogenic coronaviruses. *Nat. Rev. Microbiol.* **2018**. [CrossRef] [PubMed]
15. Alshukairi, A.N.; Zheng, J.; Zhao, J.; Nehdi, A.; Baharoon, S.A.; Layqah, L.; Bokhari, A.; Al, J.S.; Samman, N.; Boudjelal, M.; et al. High Prevalence of MERS-CoV Infection in Camel Workers in Saudi Arabia. *mBio* **2018**, *9*, e01985-18. [CrossRef] [PubMed]
16. Assiri, A.; Al-Tawfiq, J.A.; Al-Rabeeah, A.A.; Al-Rabiah, F.A.; Al-Hajjar, S.; Al-Barrak, A.; Flemban, H.; Al-Nassir, W.N.; Balkhy, H.H.; Al-Hakeem, R.F.; et al. Epidemiological, demographic, and clinical characteristics of 47 cases of Middle East respiratory syndrome coronavirus disease from Saudi Arabia: A descriptive study. *Lancet Infect. Dis.* **2013**, *13*, 752–761. [CrossRef]
17. Arabi, Y.M.; Arifi, A.A.; Balkhy, H.H.; Najm, H.; Aldawood, A.S.; Ghabashi, A.; Hawa, H.; Alothman, A.; Khaldi, A.; Al, R.B. Clinical course and outcomes of critically ill patients with Middle East respiratory syndrome coronavirus infection. *Ann. Intern. Med.* **2014**, *160*, 389–397. [CrossRef] [PubMed]
18. Al-Abdallat, M.M.; Payne, D.C.; Alqasrawi, S.; Rha, B.; Tohme, R.A.; Abedi, G.R.; Al, N.M.; Iblan, I.; Jarour, N.; Farag, N.H.; et al. Hospital-associated outbreak of Middle East respiratory syndrome coronavirus: A serologic, epidemiologic, and clinical description. *Clin. Infect. Dis.* **2014**, *59*, 1225–1233. [CrossRef] [PubMed]

19. Al, H.F.; Pringle, K.; Al, M.M.; Kim, L.; Pham, H.; Alami, N.N.; Khudhair, A.; Hall, A.J.; Aden, B.; El, S.F.; et al. Response to Emergence of Middle East Respiratory Syndrome Coronavirus, Abu Dhabi, United Arab Emirates, 2013–2014. *Emerg. Infect. Dis.* **2016**, *22*, 1162–1168.

20. Assiri, A.; McGeer, A.; Perl, T.M.; Price, C.S.; Al, R.A.; Cummings, D.A.; Alabdullatif, Z.N.; Assad, M.; Almulhim, A.; Makhdoom, H.; et al. Hospital outbreak of Middle East respiratory syndrome coronavirus. *N. Engl. J. Med.* **2013**, *369*, 407–416. [CrossRef]

21. Drosten, C.; Muth, D.; Corman, V.M.; Hussain, R.; Al, M.M.; HajOmar, W.; Landt, O.; Assiri, A.; Eckerle, I.; Al, S.A.; et al. An observational, laboratory-based study of outbreaks of middle East respiratory syndrome coronavirus in Jeddah and Riyadh, kingdom of Saudi Arabia, 2014. *Clin. Infect. Dis.* **2015**, *60*, 369–377. [CrossRef]

22. Park, H.Y.; Lee, E.J.; Ryu, Y.W.; Kim, Y.; Kim, H.; Lee, H.; Yi, S.J. Epidemiological investigation of MERS-CoV spread in a single hospital in South Korea, May to June 2015. *Eurosurveillance* **2015**, *20*, 21169. [CrossRef] [PubMed]

23. Skowronski, D.M.; Astell, C.; Brunham, R.C.; Low, D.E.; Petric, M.; Roper, R.L.; Talbot, P.J.; Tam, T.; Babiuk, L. Severe acute respiratory syndrome (SARS): A year in review. *Annu. Rev. Med.* **2005**, *56*, 357–381. [CrossRef] [PubMed]

24. Fehr, A.R.; Perlman, S. Coronaviruses: An overview of their replication and pathogenesis. *Methods Mol. Biol.* **2015**, *1282*, 1–23.

25. Wang, Y.; Sun, J.; Zhu, A.; Zhao, J.; Zhao, J. Current understanding of middle east respiratory syndrome coronavirus infection in human and animal models. *J. Thorac. Dis.* **2018**, *10*, S2260–S2271. [CrossRef] [PubMed]

26. Du, L.; Yang, Y.; Zhou, Y.; Lu, L.; Li, F.; Jiang, S. MERS-CoV spike protein: A key target for antivirals. *Expert Opin. Ther. Targets* **2017**, *21*, 131–143. [CrossRef] [PubMed]

27. Du, L.; He, Y.; Zhou, Y.; Liu, S.; Zheng, B.; Jiang, S. The spike protein of SARS-CoV—A target for vaccine and therapeutic development. *Nat. Rev. Microbiol.* **2009**, *7*, 226–236. [CrossRef]

28. Zumla, A.; Hui, D.S.; Perlman, S. Middle East respiratory syndrome. *Lancet* **2015**, *386*, 995–1007. [CrossRef]

29. Jeffers, S.A.; Tusell, S.M.; Gillim-Ross, L.; Hemmila, E.M.; Achenbach, J.E.; Babcock, G.J.; Thomas, W.J.; Thackray, L.B.; Young, M.D.; Mason, R.J.; et al. CD209L (L-SIGN) is a receptor for severe acute respiratory syndrome coronavirus. *Proc. Natl. Acad. Sci. USA* **2004**, *101*, 15748–15753. [CrossRef] [PubMed]

30. Liu, L.; Wei, Q.; Alvarez, X.; Wang, H.; Du, Y.; Zhu, H.; Jiang, H.; Zhou, J.; Lam, P.; Zhang, L.; et al. Epithelial Cells Lining Salivary Gland Ducts Are Early Target Cells of Severe Acute Respiratory Syndrome Coronavirus Infection in the Upper Respiratory Tracts of Rhesus Macaques. *J. Virol.* **2011**, *85*, 4025–4030. [CrossRef] [PubMed]

31. Kuba, K.; Imai, Y.; Rao, S.; Gao, H.; Guo, F.; Guan, B.; Huan, Y.; Yang, P.; Zhang, Y.; Deng, W.; et al. A crucial role of angiotensin converting enzyme 2 (ACE2) in SARS coronavirus-induced lung injury. *Nat. Med.* **2005**, *11*, 875–879. [CrossRef] [PubMed]

32. Qinfen, Z.; Jinming, C.; Xiaojun, H.; Huanying, Z.; Jicheng, H.; Ling, F.; Kunpeng, L.; Jingqiang, Z. The life cycle of SARS coronavirus in Vero E6 cells. *J. Med. Virol.* **2004**, *73*, 332–337. [CrossRef] [PubMed]

33. Guo, Y.; Korteweg, C.; McNutt, M.A.; Gu, J. Pathogenetic mechanisms of severe acute respiratory syndrome. *Virus Res.* **2008**, *133*, 4–12. [CrossRef] [PubMed]

34. Gu, J.; Korteweg, C. Pathology and pathogenesis of severe acute respiratory syndrome. *Am. J. Pathol.* **2007**, *170*, 1136–1147. [CrossRef] [PubMed]

35. Ding, Y.; He, L.; Zhang, Q.; Huang, Z.; Che, X.; Hou, J.; Wang, H.; Shen, H.; Qiu, L.; Li, Z.; et al. Organ distribution of severe acute respiratory syndrome (SARS) associated coronavirus (SARS-CoV) in SARS patients: Implications for pathogenesis and virus transmission pathways. *J. Pathol.* **2004**, *203*, 622–630. [CrossRef] [PubMed]

36. Meyerholz, D.K.; Lambertz, A.M.; McCray, P.J. Dipeptidyl Peptidase 4 Distribution in the Human Respiratory Tract: Implications for the Middle East Respiratory Syndrome. *Am. J. Pathol.* **2016**, *186*, 78–86. [CrossRef] [PubMed]

37. Widagdo, W.; Raj, V.S.; Schipper, D.; Kolijn, K.; van Leenders, G.; Bosch, B.J.; Bensaid, A.; Segales, J.; Baumgartner, W.; Osterhaus, A.; et al. Differential Expression of the Middle East Respiratory Syndrome Coronavirus Receptor in the Upper Respiratory Tracts of Humans and Dromedary Camels. *J. Virol.* **2016**, *90*, 4838–4842. [CrossRef]

38. Oboho, I.K.; Tomczyk, S.M.; Al-Asmari, A.M.; Banjar, A.A.; Al-Mugti, H.; Aloraini, M.S.; Alkhaldi, K.Z.; Almohammadi, E.L.; Alraddadi, B.M.; Gerber, S.I.; et al. 2014 MERS-CoV outbreak in Jeddah—A link to health care facilities. *N. Engl. J. Med.* **2015**, *372*, 846–854. [CrossRef]

39. Chu, K.H.; Tsang, W.K.; Tang, C.S.; Lam, M.F.; Lai, F.M.; To, K.F.; Fung, K.S.; Tang, H.L.; Yan, W.W.; Chan, H.W.; et al. Acute renal impairment in coronavirus-associated severe acute respiratory syndrome. *Kidney Int.* **2005**, *67*, 698–705. [CrossRef]

40. Saad, M.; Omrani, A.S.; Baig, K.; Bahloul, A.; Elzein, F.; Matin, M.A.; Selim, M.A.; Al, M.M.; Al, N.D.; Al, A.A.; et al. Clinical aspects and outcomes of 70 patients with Middle East respiratory syndrome coronavirus infection: A single-center experience in Saudi Arabia. *Int. J. Infect. Dis.* **2014**, *29*, 301–306. [CrossRef]

41. Alagaili, A.N.; Briese, T.; Mishra, N.; Kapoor, V.; Sameroff, S.C.; Burbelo, P.D.; de Wit, E.; Munster, V.J.; Hensley, L.E.; Zalmout, I.S.; et al. Middle East respiratory syndrome coronavirus infection in dromedary camels in Saudi Arabia. *mBio* **2014**, *5*, e814–e884. [CrossRef]

42. Lambeir, A.M.; Durinx, C.; Scharpe, S.; De Meester, I. Dipeptidyl-peptidase IV from bench to bedside: An update on structural properties, functions, and clinical aspects of the enzyme DPP IV. *Crit. Rev. Clin. Lab. Sci.* **2003**, *40*, 209–294. [CrossRef] [PubMed]

43. Chu, H.; Zhou, J.; Wong, B.H.; Li, C.; Cheng, Z.S.; Lin, X.; Poon, V.K.; Sun, T.; Lau, C.C.; Chan, J.F.; et al. Productive replication of Middle East respiratory syndrome coronavirus in monocyte-derived dendritic cells modulates innate immune response. *Virology* **2014**, *454–455*, 197–205. [CrossRef]

44. Zhou, J.; Chu, H.; Li, C.; Wong, B.H.; Cheng, Z.S.; Poon, V.K.; Sun, T.; Lau, C.C.; Wong, K.K.; Chan, J.Y.; et al. Active replication of Middle East respiratory syndrome coronavirus and aberrant induction of inflammatory cytokines and chemokines in human macrophages: Implications for pathogenesis. *J. Infect. Dis.* **2014**, *209*, 1331–1342. [CrossRef]

45. Chu, H.; Zhou, J.; Wong, B.H.; Li, C.; Chan, J.F.; Cheng, Z.S.; Yang, D.; Wang, D.; Lee, A.C.; Li, C.; et al. Middle East Respiratory Syndrome Coronavirus Efficiently Infects Human Primary T Lymphocytes and Activates the Extrinsic and Intrinsic Apoptosis Pathways. *J. Infect. Dis.* **2016**, *213*, 904–914. [CrossRef]

46. Yeung, M.; Yao, Y.; Jia, L.; Chan, J.F.W.; Chan, K.; Cheung, K.; Chen, H.; Poon, V.K.M.; Tsang, A.K.L.; To, K.K.W.; et al. MERS coronavirus induces apoptosis in kidney and lung by upregulating Smad7 and FGF2. *Nat. Microbiol.* **2016**, *1*, 16004. [CrossRef]

47. Liu, W.J.; Lan, J.; Liu, K.; Deng, Y.; Yao, Y.; Wu, S.; Chen, H.; Bao, L.; Zhang, H.; Zhao, M.; et al. Protective T Cell Res.ponses Featured by Concordant Recognition of Middle East Respiratory Syndrome Coronavirus-Derived CD8+ T Cell Epitopes and Host MHC. *J. Immunol.* **2017**, *198*, 873–882. [CrossRef] [PubMed]

48. Chan, R.W.; Chan, M.C.; Agnihothram, S.; Chan, L.L.; Kuok, D.I.; Fong, J.H.; Guan, Y.; Poon, L.L.; Baric, R.S.; Nicholls, J.M.; et al. Tropism of and innate immune responses to the novel human betacoronavirus lineage C virus in human ex vivo respiratory organ cultures. *J. Virol.* **2013**, *87*, 6604–6614. [CrossRef] [PubMed]

49. Lau, S.K.; Lau, C.C.; Chan, K.H.; Li, C.P.; Chen, H.; Jin, D.Y.; Chan, J.F.; Woo, P.C.; Yuen, K.Y. Delayed induction of proinflammatory cytokines and suppression of innate antiviral response by the novel Middle East respiratory syndrome coronavirus: Implications for pathogenesis and treatment. *J. Gen. Virol.* **2013**, *94*, 2679–2690. [CrossRef] [PubMed]

50. Marra, M.A.; Jones, S.J.; Astell, C.R.; Holt, R.A.; Brooks-Wilson, A.; Butterfield, Y.S.; Khattra, J.; Asano, J.K.; Barber, S.A.; Chan, S.Y.; et al. The Genome sequence of the SARS-associated coronavirus. *Science* **2003**, *300*, 1399–1404. [CrossRef] [PubMed]

51. Qu, X.X.; Hao, P.; Song, X.J.; Jiang, S.M.; Liu, Y.X.; Wang, P.G.; Rao, X.; Song, H.D.; Wang, S.Y.; Zuo, Y.; et al. Identification of two critical amino acid residues of the severe acute respiratory syndrome coronavirus spike protein for its variation in zoonotic tropism transition via a double substitution strategy. *J. Biol. Chem.* **2005**, *280*, 29588–29595. [CrossRef] [PubMed]

52. Li, W.; Zhang, C.; Sui, J.; Kuhn, J.H.; Moore, M.J.; Luo, S.; Wong, S.K.; Huang, I.C.; Xu, K.; Vasilieva, N.; et al. Receptor and viral determinants of SARS-coronavirus adaptation to human ACE2. *EMBO J.* **2005**, *24*, 1634–1643. [CrossRef] [PubMed]

53. Li, F.; Li, W.; Farzan, M.; Harrison, S.C. Structure of SARS coronavirus spike receptor-binding domain complexed with receptor. *Science* **2005**, *309*, 1864–1868. [CrossRef] [PubMed]

54. Li, F. Receptor recognition mechanisms of coronaviruses: A decade of structural studies. *J. Virol.* **2015**, *89*, 1954–1964. [CrossRef] [PubMed]

55. Lu, G.; Hu, Y.; Wang, Q.; Qi, J.; Gao, F.; Li, Y.; Zhang, Y.; Zhang, W.; Yuan, Y.; Bao, J.; et al. Molecular basis of binding between novel human coronavirus MERS-CoV and its receptor CD26. *Nature* **2013**, *500*, 227–231. [CrossRef] [PubMed]

56. Gao, J.; Lu, G.; Qi, J.; Li, Y.; Wu, Y.; Deng, Y.; Geng, H.; Li, H.; Wang, Q.; Xiao, H.; et al. Structure of the fusion core and inhibition of fusion by a heptad repeat peptide derived from the S protein of Middle East respiratory syndrome coronavirus. *J. Virol.* **2013**, *87*, 13134–13140. [CrossRef]

57. Wang, N.; Shi, X.; Jiang, L.; Zhang, S.; Wang, D.; Tong, P.; Guo, D.; Fu, L.; Cui, Y.; Liu, X.; et al. Structure of MERS-CoV spike receptor-binding domain complexed with human receptor DPP4. *Cell Res.* **2013**, *23*, 986–993. [CrossRef]

58. Zhang, N.; Jiang, S.; Du, L. Current advancements and potential strategies in the development of MERS-CoV vaccines. *Expert Rev. Vaccines* **2014**, *13*, 761–774. [CrossRef] [PubMed]

59. Du, L.; Zhao, G.; Kou, Z.; Ma, C.; Sun, S.; Poon, V.K.; Lu, L.; Wang, L.; Debnath, A.K.; Zheng, B.J.; et al. Identification of a receptor-binding domain in the S protein of the novel human coronavirus Middle East respiratory syndrome coronavirus as an essential target for vaccine development. *J. Virol.* **2013**, *87*, 9939–9942. [CrossRef] [PubMed]

60. Raj, V.S.; Mou, H.; Smits, S.L.; Dekkers, D.H.; Muller, M.A.; Dijkman, R.; Muth, D.; Demmers, J.A.; Zaki, A.; Fouchier, R.A.; et al. Dipeptidyl peptidase 4 is a functional receptor for the emerging human coronavirus-EMC. *Nature* **2013**, *495*, 251–254. [CrossRef]

61. Li, W.; Moore, M.J.; Vasilieva, N.; Sui, J.; Wong, S.K.; Berne, M.A.; Somasundaran, M.; Sullivan, J.L.; Luzuriaga, K.; Greenough, T.C.; et al. Angiotensin-converting enzyme 2 is a functional receptor for the SARS coronavirus. *Nature* **2003**, *426*, 450–454. [CrossRef]

62. Chen, Y.; Rajashankar, K.R.; Yang, Y.; Agnihothram, S.S.; Liu, C.; Lin, Y.L.; Baric, R.S.; Li, F. Crystal structure of the receptor-binding domain from newly emerged Middle East respiratory syndrome coronavirus. *J. Virol.* **2013**, *87*, 10777–10783. [CrossRef]

63. Lu, L.; Liu, Q.; Zhu, Y.; Chan, K.H.; Qin, L.; Li, Y.; Wang, Q.; Chan, J.F.; Du, L.; Yu, F.; et al. Structure-based discovery of Middle East respiratory syndrome coronavirus fusion inhibitor. *Nat. Commun.* **2014**, *5*, 3067. [CrossRef]

64. Zheng, Q.; Deng, Y.; Liu, J.; van der Hoek, L.; Berkhout, B.; Lu, M. Core structure of S2 from the human coronavirus NL63 spike glycoprotein. *Biochemistry* **2006**, *45*, 15205–15215. [CrossRef] [PubMed]

65. Xu, Y.; Lou, Z.; Liu, Y.; Pang, H.; Tien, P.; Gao, G.F.; Rao, Z. Crystal structure of severe acute respiratory syndrome coronavirus spike protein fusion core. *J. Biol. Chem.* **2004**, *279*, 49414–49419. [CrossRef] [PubMed]

66. Xu, Y.; Liu, Y.; Lou, Z.; Qin, L.; Li, X.; Bai, Z.; Pang, H.; Tien, P.; Gao, G.F.; Rao, Z. Structural basis for coronavirus-mediated membrane fusion. Crystal structure of mouse hepatitis virus spike protein fusion core. *J. Biol. Chem.* **2004**, *279*, 30514–30522. [CrossRef]

67. Hofmann, H.; Hattermann, K.; Marzi, A.; Gramberg, T.; Geier, M.; Krumbiegel, M.; Kuate, S.; Uberla, K.; Niedrig, M.; Pohlmann, S. S protein of severe acute respiratory syndrome-associated coronavirus mediates entry into hepatoma cell lines and is targeted by neutralizing antibodies in infected patients. *J. Virol.* **2004**, *78*, 6134–6142. [CrossRef]

68. Holmes, K.V. SARS-associated coronavirus. *N. Engl. J. Med.* **2003**, *348*, 1948–1951. [CrossRef]

69. Li, F.; Berardi, M.; Li, W.; Farzan, M.; Dormitzer, P.R.; Harrison, S.C. Conformational states of the severe acute respiratory syndrome coronavirus spike protein ectodomain. *J. Virol.* **2006**, *80*, 6794–6800. [CrossRef] [PubMed]

70. Wong, S.K.; Li, W.; Moore, M.J.; Choe, H.; Farzan, M. A 193-amino acid fragment of the SARS coronavirus S protein efficiently binds angiotensin-converting enzyme 2. *J. Biol. Chem.* **2004**, *279*, 3197–3201. [CrossRef] [PubMed]

71. Xiao, X.; Chakraborti, S.; Dimitrov, A.S.; Gramatikoff, K.; Dimitrov, D.S. The SARS-CoV S glycoprotein: Expression and functional characterization. *Biochem. Biophys. Res. Commun.* **2003**, *312*, 1159–1164. [CrossRef] [PubMed]

72. Zhang, Y.; Zheng, N.; Hao, P.; Cao, Y.; Zhong, Y. A molecular docking model of SARS-CoV S1 protein in complex with its receptor, human ACE2. *Comput. Biol. Chem.* **2005**, *29*, 254–257. [CrossRef] [PubMed]

73. He, Y.; Li, J.; Jiang, S. A single amino acid substitution (R441A) in the receptor-binding domain of SARS coronavirus spike protein disrupts the antigenic structure and binding activity. *Biochem. Biophys. Res. Commun.* **2006**, *344*, 106–113. [CrossRef] [PubMed]

74. Sainz, B.J.; Rausch, J.M.; Gallaher, W.R.; Garry, R.F.; Wimley, W.C. Identification and characterization of the putative fusion peptide of the severe acute respiratory syndrome-associated coronavirus spike protein. *J. Virol.* **2005**, *79*, 7195–7206. [CrossRef] [PubMed]

75. Yang, Z.Y.; Huang, Y.; Ganesh, L.; Leung, K.; Kong, W.P.; Schwartz, O.; Subbarao, K.; Nabel, G.J. pH-dependent entry of severe acute respiratory syndrome coronavirus is mediated by the spike glycoprotein and enhanced by dendritic cell transfer through DC-SIGN. *J. Virol.* **2004**, *78*, 5642–5650. [CrossRef] [PubMed]

76. Han, D.P.; Lohani, M.; Cho, M.W. Specific asparagine-linked glycosylation sites are critical for DC-SIGN- and L-SIGN-mediated severe acute respiratory syndrome coronavirus entry. *J. Virol.* **2007**, *81*, 12029–12039. [CrossRef]

77. Graham, R.L.; Donaldson, E.F.; Baric, R.S. A decade after SARS: Strategies for controlling emerging coronaviruses. *Nat. Rev. Microbiol.* **2013**, *11*, 836–848. [CrossRef]

78. Keng, C.T.; Zhang, A.; Shen, S.; Lip, K.M.; Fielding, B.C.; Tan, T.H.; Chou, C.F.; Loh, C.B.; Wang, S.; Fu, J.; et al. Amino acids 1055 to 1192 in the S2 region of severe acute respiratory syndrome coronavirus S protein induce neutralizing antibodies: Implications for the development of vaccines and antiviral agents. *J. Virol.* **2005**, *79*, 3289–3296. [CrossRef]

79. Zhou, T.; Wang, H.; Luo, D.; Rowe, T.; Wang, Z.; Hogan, R.J.; Qiu, S.; Bunzel, R.J.; Huang, G.; Mishra, V.; et al. An exposed domain in the severe acute respiratory syndrome coronavirus spike protein induces neutralizing antibodies. *J. Virol.* **2004**, *78*, 7217–7226. [CrossRef]

80. Bukreyev, A.; Lamirande, E.W.; Buchholz, U.J.; Vogel, L.N.; Elkins, W.R.; St, C.M.; Murphy, B.R.; Subbarao, K.; Collins, P.L. Mucosal immunisation of African green monkeys (*Cercopithecus aethiops*) with an attenuated parainfluenza virus expressing the SARS coronavirus spike protein for the prevention of SARS. *Lancet* **2004**, *363*, 2122–2127. [CrossRef]

81. Yang, Z.Y.; Kong, W.P.; Huang, Y.; Roberts, A.; Murphy, B.R.; Subbarao, K.; Nabel, G.J. A DNA vaccine induces SARS coronavirus neutralization and protective immunity in mice. *Nature* **2004**, *428*, 561–564. [CrossRef]

82. Kam, Y.W.; Kien, F.; Roberts, A.; Cheung, Y.C.; Lamirande, E.W.; Vogel, L.; Chu, S.L.; Tse, J.; Guarner, J.; Zaki, S.R.; et al. Antibodies against trimeric S glycoprotein protect hamsters against SARS-CoV challenge despite their capacity to mediate FcgammaRII-dependent entry into B cells in vitro. *Vaccine* **2007**, *25*, 729–740. [CrossRef] [PubMed]

83. Czub, M.; Weingartl, H.; Czub, S.; He, R.; Cao, J. Evaluation of modified vaccinia virus Ankara based recombinant SARS vaccine in ferrets. *Vaccine* **2005**, *23*, 2273–2279. [CrossRef]

84. Weingartl, H.; Czub, M.; Czub, S.; Neufeld, J.; Marszal, P.; Gren, J.; Smith, G.; Jones, S.; Proulx, R.; Deschambault, Y.; et al. Immunization with modified vaccinia virus Ankara-based recombinant vaccine against severe acute respiratory syndrome is associated with enhanced hepatitis in ferrets. *J. Virol.* **2004**, *78*, 12672–12676. [CrossRef] [PubMed]

85. Tai, W.; Wang, Y.; Fett, C.A.; Zhao, G.; Li, F.; Perlman, S.; Jiang, S.; Zhou, Y.; Du, L. Recombinant Receptor-Binding Domains of Multiple Middle East Respiratory Syndrome Coronaviruses (MERS-CoVs) Induce Cross-Neutralizing Antibodies against Divergent Human and Camel MERS-CoVs and Antibody Escape Mutants. *J. Virol.* **2017**, *91*, e01651-16. [CrossRef] [PubMed]

86. Wang, L.; Shi, W.; Joyce, M.G.; Modjarrad, K.; Zhang, Y.; Leung, K.; Lees, C.R.; Zhou, T.; Yassine, H.M.; Kanekiyo, M.; et al. Evaluation of candidate vaccine approaches for MERS-CoV. *Nat. Commun.* **2015**, *6*, 7712. [CrossRef] [PubMed]

87. Gilbert, S.C.; Warimwe, G.M. Rapid development of vaccines against emerging pathogens: The replication-deficient simian adenovirus platform technology. *Vaccine* **2017**, *35*, 4461–4464. [CrossRef]

88. Kim, E.; Okada, K.; Kenniston, T.; Raj, V.S.; AlHajri, M.M.; Farag, E.A.; AlHajri, F.; Osterhaus, A.D.; Haagmans, B.L.; Gambotto, A. Immunogenicity of an adenoviral-based Middle East Respiratory Syndrome coronavirus vaccine in BALB/c mice. *Vaccine* **2014**, *32*, 5975–5982. [CrossRef]

89. Chi, H.; Zheng, X.; Wang, X.; Wang, C.; Wang, H.; Gai, W.; Perlman, S.; Yang, S.; Zhao, J.; Xia, X. DNA vaccine encoding Middle East respiratory syndrome coronavirus S1 protein induces protective immune responses in mice. *Vaccine* **2017**, *35*, 2069–2075. [CrossRef]

90. Al-Amri, S.S.; Abbas, A.T.; Siddiq, L.A.; Alghamdi, A.; Sanki, M.A.; Al-Muhanna, M.K.; Alhabbab, R.Y.; Azhar, E.I.; Li, X.; Hashem, A.M. Immunogenicity of Candidate MERS-CoV DNA Vaccines Based on the Spike Protein. *Sci. Rep.* **2017**, *7*, 44875. [CrossRef]

91. Modjarrad, K. MERS-CoV vaccine candidates in development: The current landscape. *Vaccine* **2016**, *34*, 2982–2987. [CrossRef]

92. Zhao, J.; Zhao, J.; Mangalam, A.K.; Channappanavar, R.; Fett, C.; Meyerholz, D.K.; Agnihothram, S.; Baric, R.S.; David, C.S.; Perlman, S. Airway Memory CD4+ T Cells Mediate Protective Immunity against Emerging Respiratory Coronaviruses. *Immunity* **2016**, *44*, 1379–1391. [CrossRef]

93. Roper, R.L.; Rehm, K.E. SARS vaccines: Where are we? *Expert Rev. Vaccines* **2009**, *8*, 887–898. [CrossRef] [PubMed]

94. Lin, J.; Zhang, J.; Su, N.; Xu, J.; Wang, N.; Chen, J.; Chen, X.; Liu, Y.; Gao, H.; Jia, Y.; et al. Safety and immunogenicity from a phase I trial of inactivated severe acute respiratory syndrome coronavirus vaccine. *Antivir. Ther.* **2007**, *12*, 1107–1113. [PubMed]

95. Deng, Y.; Lan, J.; Bao, L.; Huang, B.; Ye, F.; Chen, Y.; Yao, Y.; Wang, W.; Qin, C.; Tan, W. Enhanced protection in mice induced by immunization with inactivated whole viruses compare to spike protein of middle east respiratory syndrome coronavirus. *Emerg. Microbes Infect.* **2018**, *7*, 60. [CrossRef] [PubMed]

96. See, R.H.; Zakhartchouk, A.N.; Petric, M.; Lawrence, D.J.; Mok, C.P.; Hogan, R.J.; Rowe, T.; Zitzow, L.A.; Karunakaran, K.P.; Hitt, M.M.; et al. Comparative evaluation of two severe acute respiratory syndrome (SARS) vaccine candidates in mice challenged with SARS coronavirus. *J. Gen. Virol.* **2006**, *87*, 641–650. [CrossRef] [PubMed]

97. Enjuanes, L.; Dediego, M.L.; Alvarez, E.; Deming, D.; Sheahan, T.; Baric, R. Vaccines to prevent severe acute respiratory syndrome coronavirus-induced disease. *Virus Res.* **2008**, *133*, 45–62. [CrossRef] [PubMed]

98. Bolles, M.; Deming, D.; Long, K.; Agnihothram, S.; Whitmore, A.; Ferris, M.; Funkhouser, W.; Gralinski, L.; Totura, A.; Heise, M.; et al. A double-inactivated severe acute respiratory syndrome coronavirus vaccine provides incomplete protection in mice and induces increased eosinophilic proinflammatory pulmonary response upon challenge. *J. Virol.* **2011**, *85*, 12201–12215. [CrossRef] [PubMed]

99. Ishioka, T.; Kimura, H.; Kita, H.; Obuchi, M.; Hoshino, H.; Noda, M.; Nishina, A.; Kozawa, K.; Kato, M. Effects of respiratory syncytial virus infection and major basic protein derived from eosinophils in pulmonary alveolar epithelial cells (A549). *Cell Biol. Int.* **2011**, *35*, 467–474. [CrossRef] [PubMed]

100. Fett, C.; DeDiego, M.L.; Regla-Nava, J.A.; Enjuanes, L.; Perlman, S. Complete protection against severe acute respiratory syndrome coronavirus-mediated lethal respiratory disease in aged mice by immunization with a mouse-adapted virus lacking E protein. *J. Virol.* **2013**, *87*, 6551–6559. [CrossRef] [PubMed]

101. Graham, R.L.; Becker, M.M.; Eckerle, L.D.; Bolles, M.; Denison, M.R.; Baric, R.S. A live, impaired-fidelity coronavirus vaccine protects in an aged, immunocompromised mouse model of lethal disease. *Nat. Med.* **2012**, *18*, 1820–1826. [CrossRef]

102. Almazan, F.; DeDiego, M.L.; Sola, I.; Zuniga, S.; Nieto-Torres, J.L.; Marquez-Jurado, S.; Andres, G.; Enjuanes, L. Engineering a replication-competent, propagation-defective Middle East respiratory syndrome coronavirus as a vaccine candidate. *mBio* **2013**, *4*, e613–e650. [CrossRef] [PubMed]

103. Vignuzzi, M.; Wendt, E.; Andino, R. Engineering attenuated virus vaccines by controlling replication fidelity. *Nat. Med.* **2008**, *14*, 154–161. [CrossRef] [PubMed]

104. Roberts, A.; Vogel, L.; Guarner, J.; Hayes, N.; Murphy, B.; Zaki, S.; Subbarao, K. Severe acute respiratory syndrome coronavirus infection of golden Syrian hamsters. *J. Virol.* **2005**, *79*, 503–511. [CrossRef] [PubMed]

105. Deming, D.; Sheahan, T.; Heise, M.; Yount, B.; Davis, N.; Sims, A.; Suthar, M.; Harkema, J.; Whitmore, A.; Pickles, R.; et al. Vaccine efficacy in senescent mice challenged with recombinant SARS-CoV bearing epidemic and zoonotic spike variants. *PLoS Med.* **2006**, *3*, e525. [CrossRef] [PubMed]

106. Chen, Z.; Zhang, L.; Qin, C.; Ba, L.; Yi, C.E.; Zhang, F.; Wei, Q.; He, T.; Yu, W.; Yu, J.; et al. Recombinant modified vaccinia virus Ankara expressing the spike glycoprotein of severe acute respiratory syndrome coronavirus induces protective neutralizing antibodies primarily targeting the receptor binding region. *J. Virol.* **2005**, *79*, 2678–2688. [CrossRef] [PubMed]

107. Lan, J.; Yao, Y.; Deng, Y.; Chen, H.; Lu, G.; Wang, W.; Bao, L.; Deng, W.; Wei, Q.; Gao, G.F.; et al. Recombinant Receptor Binding Domain Protein Induces Partial Protective Immunity in Rhesus Macaques Against Middle East Respiratory Syndrome Coronavirus Challenge. *EBioMedicine* **2015**, *2*, 1438–1446. [CrossRef]

108. Jiaming, L.; Yanfeng, Y.; Yao, D.; Yawei, H.; Linlin, B.; Baoying, H.; Jinghua, Y.; Gao, G.F.; Chuan, Q.; Wenjie, T. The recombinant N-terminal domain of spike proteins is a potential vaccine against Middle East respiratory syndrome coronavirus (MERS-CoV) infection. *Vaccine* **2017**, *35*, 10–18. [CrossRef]

109. Bisht, H.; Roberts, A.; Vogel, L.; Subbarao, K.; Moss, B. Neutralizing antibody and protective immunity to SARS coronavirus infection of mice induced by a soluble recombinant polypeptide containing an N-terminal segment of the spike glycoprotein. *Virology* **2005**, *334*, 160–165. [CrossRef]

110. Woo, P.C.; Lau, S.K.; Tsoi, H.W.; Chen, Z.W.; Wong, B.H.; Zhang, L.; Chan, J.K.; Wong, L.P.; He, W.; Ma, C.; et al. SARS coronavirus spike polypeptide DNA vaccine priming with recombinant spike polypeptide from Escherichia coli as booster induces high titer of neutralizing antibody against SARS coronavirus. *Vaccine* **2005**, *23*, 4959–4968. [CrossRef]

111. Shi, S.; Peng, J.; Li, Y.; Qin, C.; Liang, G.; Xu, L.; Yang, Y.; Wang, J.; Sun, Q. The expression of membrane protein augments the specific responses induced by SARS-CoV nucleocapsid DNA immunization. *Mol. Immunol.* **2006**, *43*, 1791–1798. [CrossRef]

112. Gupta, V.; Tabiin, T.M.; Sun, K.; Chandrasekaran, A.; Anwar, A.; Yang, K.; Chikhlikar, P.; Salmon, J.; Brusic, V.; Marques, E.T.; et al. SARS coronavirus nucleocapsid immunodominant T-cell epitope cluster is common to both exogenous recombinant and endogenous DNA-encoded immunogens. *Virology* **2006**, *347*, 127–139. [CrossRef] [PubMed]

113. Zhuang, M.; Jiang, H.; Suzuki, Y.; Li, X.; Xiao, P.; Tanaka, T.; Ling, H.; Yang, B.; Saitoh, H.; Zhang, L.; et al. Procyanidins and butanol extract of Cinnamomi Cortex inhibit SARS-CoV infection. *Antivir. Res.* **2009**, *82*, 73–81. [CrossRef]

114. Hu, H.; Li, L.; Kao, R.Y.; Kou, B.; Wang, Z.; Zhang, L.; Zhang, H.; Hao, Z.; Tsui, W.H.; Ni, A.; et al. Screening and identification of linear B-cell epitopes and entry-blocking peptide of severe acute respiratory syndrome (SARS)-associated coronavirus using synthetic overlapping peptide library. *J. Comb. Chem.* **2005**, *7*, 648–656. [CrossRef]

115. Zheng, B.J.; Guan, Y.; Hez, M.L.; Sun, H.; Du, L.; Zheng, Y.; Wong, K.L.; Chen, H.; Chen, Y.; Lu, L.; et al. Synthetic peptides outside the spike protein heptad repeat regions as potent inhibitors of SARS-associated coronavirus. *Antivir. Ther.* **2005**, *10*, 393–403. [PubMed]

116. Bosch, B.J.; Martina, B.E.; Van Der Zee, R.; Lepault, J.; Haijema, B.J.; Versluis, C.; Heck, A.J.; De Groot, R.; Osterhaus, A.D.; Rottier, P.J. Severe acute respiratory syndrome coronavirus (SARS-CoV) infection inhibition using spike protein heptad repeat-derived peptides. *Proc. Natl. Acad. Sci. USA* **2004**, *101*, 8455–8460. [CrossRef]

117. Yuan, K.; Yi, L.; Chen, J.; Qu, X.; Qing, T.; Rao, X.; Jiang, P.; Hu, J.; Xiong, Z.; Nie, Y.; et al. Suppression of SARS-CoV entry by peptides corresponding to heptad regions on spike glycoprotein. *Biochem. Biophys. Res. Commun.* **2004**, *319*, 746–752. [CrossRef]

118. McRoy, W.C.; Baric, R.S. Amino acid substitutions in the S2 subunit of mouse hepatitis virus variant V51 encode determinants of host range expansion. *J. Virol.* **2008**, *82*, 1414–1424. [CrossRef]

119. He, Y.; Li, J.; Heck, S.; Lustigman, S.; Jiang, S. Antigenic and immunogenic characterization of recombinant baculovirus-expressed severe acute respiratory syndrome coronavirus spike protein: Implication for vaccine design. *J. Virol.* **2006**, *80*, 5757–5767. [CrossRef]

120. He, Y.; Li, J.; Li, W.; Lustigman, S.; Farzan, M.; Jiang, S. Cross-neutralization of human and palm civet severe acute respiratory syndrome coronaviruses by antibodies targeting the receptor-binding domain of spike protein. *J. Immunol.* **2006**, *176*, 6085–6092. [CrossRef] [PubMed]

121. He, Y.; Zhu, Q.; Liu, S.; Zhou, Y.; Yang, B.; Li, J.; Jiang, S. Identification of a critical neutralization determinant of severe acute respiratory syndrome (SARS)-associated coronavirus: Importance for designing SARS vaccines. *Virology* **2005**, *334*, 74–82. [CrossRef]

122. Lai, S.C.; Chong, P.C.; Yeh, C.T.; Liu, L.S.; Jan, J.T.; Chi, H.Y.; Liu, H.W.; Chen, A.; Wang, Y.C. Characterization of neutralizing monoclonal antibodies recognizing a 15-residues epitope on the spike protein HR2 region of severe acute respiratory syndrome coronavirus (SARS-CoV). *J. Biomed. Sci.* **2005**, *12*, 711–727. [CrossRef] [PubMed]

123. Nie, Y.; Wang, G.; Shi, X.; Zhang, H.; Qiu, Y.; He, Z.; Wang, W.; Lian, G.; Yin, X.; Du, L.; et al. Neutralizing antibodies in patients with severe acute respiratory syndrome-associated coronavirus infection. *J. Infect. Dis.* **2004**, *190*, 1119–1126. [CrossRef]

124. Traggiai, E.; Becker, S.; Subbarao, K.; Kolesnikova, L.; Uematsu, Y.; Gismondo, M.R.; Murphy, B.R.; Rappuoli, R.; Lanzavecchia, A. An efficient method to make human monoclonal antibodies from memory B cells: Potent neutralization of SARS coronavirus. *Nat. Med.* **2004**, *10*, 871–875. [CrossRef]

125. Rockx, B.; Corti, D.; Donaldson, E.; Sheahan, T.; Stadler, K.; Lanzavecchia, A.; Baric, R. Structural basis for potent cross-neutralizing human monoclonal antibody protection against lethal human and zoonotic severe acute respiratory syndrome coronavirus challenge. *J. Virol.* **2008**, *82*, 3220–3235. [CrossRef] [PubMed]

126. Zhu, Z.; Chakraborti, S.; He, Y.; Roberts, A.; Sheahan, T.; Xiao, X.; Hensley, L.E.; Prabakaran, P.; Rockx, B.; Sidorov, I.A.; et al. Potent cross-reactive neutralization of SARS coronavirus isolates by human monoclonal antibodies. *Proc. Natl. Acad. Sci. USA* **2007**, *104*, 12123–12128. [CrossRef] [PubMed]

127. Coughlin, M.; Lou, G.; Martinez, O.; Masterman, S.K.; Olsen, O.A.; Moksa, A.A.; Farzan, M.; Babcook, J.S.; Prabhakar, B.S. Generation and characterization of human monoclonal neutralizing antibodies with distinct binding and sequence features against SARS coronavirus using XenoMouse. *Virology* **2007**, *361*, 93–102. [CrossRef]

128. van den Brink, E.N.; Ter Meulen, J.; Cox, F.; Jongeneelen, M.A.; Thijsse, A.; Throsby, M.; Marissen, W.E.; Rood, P.M.; Bakker, A.B.; Gelderblom, H.R.; et al. Molecular and biological characterization of human monoclonal antibodies binding to the spike and nucleocapsid proteins of severe acute respiratory syndrome coronavirus. *J. Virol.* **2005**, *79*, 1635–1644. [CrossRef] [PubMed]

129. Sui, J.; Li, W.; Murakami, A.; Tamin, A.; Matthews, L.J.; Wong, S.K.; Moore, M.J.; Tallarico, A.S.; Olurinde, M.; Choe, H.; et al. Potent neutralization of severe acute respiratory syndrome (SARS) coronavirus by a human mAb to S1 protein that blocks receptor association. *Proc. Natl. Acad. Sci. USA* **2004**, *101*, 2536–2541. [CrossRef]

130. Liu, S.; Xiao, G.; Chen, Y.; He, Y.; Niu, J.; Escalante, C.R.; Xiong, H.; Farmar, J.; Debnath, A.K.; Tien, P.; et al. Interaction between heptad repeat 1 and 2 regions in spike protein of SARS-associated coronavirus: Implications for virus fusogenic mechanism and identification of fusion inhibitors. *Lancet* **2004**, *363*, 938–947. [CrossRef]

131. Corti, D.; Zhao, J.; Pedotti, M.; Simonelli, L.; Agnihothram, S.; Fett, C.; Fernandez-Rodriguez, B.; Foglierini, M.; Agatic, G.; Vanzetta, F.; et al. Prophylactic and postexposure efficacy of a potent human monoclonal antibody against MERS coronavirus. *Proc. Natl. Acad. Sci. USA* **2015**, *112*, 10473–10478. [CrossRef]

132. Li, Y.; Wan, Y.; Liu, P.; Zhao, J.; Lu, G.; Qi, J.; Wang, Q.; Lu, X.; Wu, Y.; Liu, W.; et al. A humanized neutralizing antibody against MERS-CoV targeting the receptor-binding domain of the spike protein. *Cell Res.* **2015**, *25*, 1237–1249. [CrossRef] [PubMed]

133. Du, L.; Zhao, G.; Yang, Y.; Qiu, H.; Wang, L.; Kou, Z.; Tao, X.; Yu, H.; Sun, S.; Tseng, C.T.; et al. A conformation-dependent neutralizing monoclonal antibody specifically targeting receptor-binding domain in Middle East respiratory syndrome coronavirus spike protein. *J. Virol.* **2014**, *88*, 7045–7053. [CrossRef] [PubMed]

134. Raj, V.S.; Smits, S.L.; Provacia, L.B.; van den Brand, J.M.; Wiersma, L.; Ouwendijk, W.J.; Bestebroer, T.M.; Spronken, M.I.; van Amerongen, G.; Rottier, P.J.; et al. Adenosine deaminase acts as a natural antagonist for dipeptidyl peptidase 4-mediated entry of the Middle East respiratory syndrome coronavirus. *J. Virol.* **2014**, *88*, 1834–1838. [CrossRef] [PubMed]

135. Zhou, N.; Pan, T.; Zhang, J.; Li, Q.; Zhang, X.; Bai, C.; Huang, F.; Peng, T.; Zhang, J.; Liu, C.; et al. Glycopeptide Antibiotics Potently Inhibit Cathepsin L in the Late Endosome/Lysosome and Block the Entry of Ebola Virus, Middle East Respiratory Syndrome Coronavirus (MERS-CoV), and Severe Acute Respiratory Syndrome Coronavirus (SARS-CoV). *J. Biol. Chem.* **2016**, *291*, 9218–9232. [CrossRef]

136. Gierer, S.; Bertram, S.; Kaup, F.; Wrensch, F.; Heurich, A.; Kramer-Kuhl, A.; Welsch, K.; Winkler, M.; Meyer, B.; Drosten, C.; et al. The spike protein of the emerging betacoronavirus EMC uses a novel coronavirus receptor for entry, can be activated by TMPRSS2, and is targeted by neutralizing antibodies. *J. Virol.* **2013**, *87*, 5502–5511. [CrossRef] [PubMed]

137. Shirato, K.; Kawase, M.; Matsuyama, S. Middle East respiratory syndrome coronavirus infection mediated by the transmembrane serine protease TMPRSS2. *J. Virol.* **2013**, *87*, 12552–12561. [CrossRef]

138. Zhou, Y.; Vedantham, P.; Lu, K.; Agudelo, J.; Carrion, R.; Nunneley, J.W.; Barnard, D.; Pöhlmann, S.; McKerrow, J.H.; Renslo, A.R.; Simmons, G. Protease inhibitors targeting coronavirus and filovirus entry. *Antivir. Res.* **2015**, *116*, 76–84. [CrossRef] [PubMed]

139. Momattin, H.; Mohammed, K.; Zumla, A.; Memish, Z.A.; Al-Tawfiq, J.A. Therapeutic Options for Middle East Respiratory Syndrome Coronavirus (MERS-CoV)–possible lessons from a systematic review of SARS-CoV therapy. *Int. J. Infect. Dis.* **2013**, *17*, e792–e798. [CrossRef]

140. Hart, B.J.; Dyall, J.; Postnikova, E.; Zhou, H.; Kindrachuk, J.; Johnson, R.F.; Olinger, G.J.; Frieman, M.B.; Holbrook, M.R.; Jahrling, P.B.; et al. Interferon-beta and mycophenolic acid are potent inhibitors of Middle East respiratory syndrome coronavirus in cell-based assays. *J. Gen. Virol.* **2014**, *95*, 571–577. [CrossRef]

141. ter Meulen, J.; van den Brink, E.N.; Poon, L.L.; Marissen, W.E.; Leung, C.S.; Cox, F.; Cheung, C.Y.; Bakker, A.Q.; Bogaards, J.A.; van Deventer, E.; et al. Human monoclonal antibody combination against SARS coronavirus: Synergy and coverage of escape mutants. *PLoS Med.* **2006**, *3*, e237. [CrossRef]

142. Zhang, J.S.; Chen, J.T.; Liu, Y.X.; Zhang, Z.S.; Gao, H.; Liu, Y.; Wang, X.; Ning, Y.; Liu, Y.F.; Gao, Q.; et al. A serological survey on neutralizing antibody titer of SARS convalescent sera. *J. Med. Virol.* **2005**, *77*, 147–150. [CrossRef]

143. Miyoshi-Akiyama, T.; Ishida, I.; Fukushi, M.; Yamaguchi, K.; Matsuoka, Y.; Ishihara, T.; Tsukahara, M.; Hatakeyama, S.; Itoh, N.; Morisawa, A.; et al. Fully Human Monoclonal Antibody Directed to Proteolytic Cleavage Site in Severe Acute Respiratory Syndrome (SARS) Coronavirus S Protein Neutralizes the Virus in a Rhesus Macaque SARS Model. *J. Infect. Dis.* **2011**, *203*, 1574–1581. [CrossRef]

144. Wang, Q.; Zhang, L.; Kuwahara, K.; Li, L.; Liu, Z.; Li, T.; Zhu, H.; Liu, J.; Xu, Y.; Xie, J.; et al. Immunodominant SARS Coronavirus Epitopes in Humans Elicited both Enhancing and Neutralizing Effects on Infection in Non-human Primates. *ACS Infect. Dis.* **2016**, *2*, 361–376. [CrossRef]

145. Chen, Z.; Bao, L.; Chen, C.; Zou, T.; Xue, Y.; Li, F.; Lv, Q.; Gu, S.; Gao, X.; Cui, S.; et al. Human Neutralizing Monoclonal Antibody Inhibition of Middle East Respiratory Syndrome Coronavirus Replication in the Common Marmoset. *J. Infect. Dis.* **2017**, *215*, 1807–1815. [CrossRef]

146. Stockman, L.J.; Bellamy, R.; Garner, P.; Low, D. SARS: Systematic Review of Treatment Effects. *PLoS Med.* **2006**, *3*, e343. [CrossRef]

147. Chan, P.K.S.; Tang, J.W.; Hui, D.S.C. SARS: Clinical presentation, transmission, pathogenesis and treatment options. *Clin. Sci.* **2006**, *110*, 193–204. [CrossRef]

148. Chan, J.F.; Yao, Y.; Yeung, M.; Deng, W.; Bao, L.; Jia, L.; Li, F.; Xiao, C.; Gao, H.; Yu, P.; et al. Treatment with Lopinavir/Ritonavir or Interferon-β1b Improves Outcome of MERS-CoV Infection in a Nonhuman Primate Model of Common Marmoset. *J. Infect. Dis.* **2015**, *212*, 1904–1913. [CrossRef]

149. Subbarao, K.; McAuliffe, J.; Vogel, L.; Fahle, G.; Fischer, S.; Tatti, K.; Packard, M.; Shieh, W.J.; Zaki, S.; Murphy, B. Prior infection and passive transfer of neutralizing antibody prevent replication of severe acute respiratory syndrome coronavirus in the respiratory tract of mice. *J. Virol.* **2004**, *78*, 3572–3577. [CrossRef]

150. Wentworth, D.E.; Gillim-Ross, L.; Espina, N.; Bernard, K.A. Mice susceptible to SARS coronavirus. *Emerg. Infect. Dis.* **2004**, *10*, 1293–1296. [CrossRef]

151. Glass, W.G.; Subbarao, K.; Murphy, B.; Murphy, P.M. Mechanisms of host defense following severe acute respiratory syndrome-coronavirus (SARS-CoV) pulmonary infection of mice. *J. Immunol.* **2004**, *173*, 4030–4039. [CrossRef]

152. Yang, X.H.; Deng, W.; Tong, Z.; Liu, Y.X.; Zhang, L.F.; Zhu, H.; Gao, H.; Huang, L.; Liu, Y.L.; Ma, C.M.; et al. Mice transgenic for human angiotensin-converting enzyme 2 provide a model for SARS coronavirus infection. *Comp. Med.* **2007**, *57*, 450–459. [PubMed]

153. Roberts, A.; Paddock, C.; Vogel, L.; Butler, E.; Zaki, S.; Subbarao, K. Aged BALB/c mice as a model for increased severity of severe acute respiratory syndrome in elderly humans. *J. Virol.* **2005**, *79*, 5833–5838. [CrossRef] [PubMed]

154. Hogan, R.J.; Gao, G.; Rowe, T.; Bell, P.; Flieder, D.; Paragas, J.; Kobinger, G.P.; Wivel, N.A.; Crystal, R.G.; Boyer, J.; et al. Resolution of primary severe acute respiratory syndrome-associated coronavirus infection requires Stat1. *J. Virol.* **2004**, *78*, 11416–11421. [CrossRef]

155. Frieman, M.; Yount, B.; Agnihothram, S.; Page, C.; Donaldson, E.; Roberts, A.; Vogel, L.; Woodruff, B.; Scorpio, D.; Subbarao, K.; et al. Molecular determinants of severe acute respiratory syndrome coronavirus pathogenesis and virulence in young and aged mouse models of human disease. *J. Virol.* **2012**, *86*, 884–897. [CrossRef] [PubMed]

156. Zhao, J.; Zhao, J.; Legge, K.; Perlman, S. Age-related increases in PGD(2) expression impair respiratory DC migration, resulting in diminished T cell responses upon respiratory virus infection in mice. *J. Clin. Investig.* **2011**, *121*, 4921–4930. [CrossRef]

157. Zhao, J.; Zhao, J.; Perlman, S. T cell responses are required for protection from clinical disease and for virus clearance in severe acute respiratory syndrome coronavirus-infected mice. *J. Virol.* **2010**, *84*, 9318–9325. [CrossRef] [PubMed]

158. Zhao, J.; Zhao, J.; Van Rooijen, N.; Perlman, S. Evasion by stealth: Inefficient immune activation underlies poor T cell response and severe disease in SARS-CoV-infected mice. *PLoS Pathog.* **2009**, *5*, e1000636. [CrossRef] [PubMed]

159. Sheahan, T.; Morrison, T.E.; Funkhouser, W.; Uematsu, S.; Akira, S.; Baric, R.S.; Heise, M.T. MyD88 is required for protection from lethal infection with a mouse-adapted SARS-CoV. *PLoS Pathog.* **2008**, *4*, e1000240. [CrossRef]

160. Li, C.K.; Wu, H.; Yan, H.; Ma, S.; Wang, L.; Zhang, M.; Tang, X.; Temperton, N.J.; Weiss, R.A.; Brenchley, J.M.; et al. T cell responses to whole SARS coronavirus in humans. *J. Immunol.* **2008**, *181*, 5490–5500. [CrossRef]

161. Chen, W.; Xu, Z.; Mu, J.; Yang, L.; Gan, H.; Mu, F.; Fan, B.; He, B.; Huang, S.; You, B.; et al. Antibody response and viraemia during the course of severe acute respiratory syndrome (SARS)-associated coronavirus infection. *J. Med. Microbiol.* **2004**, *53*, 435–438. [CrossRef]

162. Spruth, M.; Kistner, O.; Savidis-Dacho, H.; Hitter, E.; Crowe, B.; Gerencer, M.; Bruhl, P.; Grillberger, L.; Reiter, M.; Tauer, C.; et al. A double-inactivated whole virus candidate SARS coronavirus vaccine stimulates neutralising and protective antibody responses. *Vaccine* **2006**, *24*, 652–661. [CrossRef]

163. Greenough, T.C.; Babcock, G.J.; Roberts, A.; Hernandez, H.J.; Thomas, W.J.; Coccia, J.A.; Graziano, R.F.; Srinivasan, M.; Lowy, I.; Finberg, R.W.; et al. Development and characterization of a severe acute respiratory syndrome-associated coronavirus-neutralizing human monoclonal antibody that provides effective immunoprophylaxis in mice. *J. Infect. Dis.* **2005**, *191*, 507–514. [CrossRef]

164. Kapadia, S.U.; Rose, J.K.; Lamirande, E.; Vogel, L.; Subbarao, K.; Roberts, A. Long-term protection from SARS coronavirus infection conferred by a single immunization with an attenuated VSV-based vaccine. *Virology* **2005**, *340*, 174–182. [CrossRef]

165. Stadler, K.; Roberts, A.; Becker, S.; Vogel, L.; Eickmann, M.; Kolesnikova, L.; Klenk, H.D.; Murphy, B.; Rappuoli, R.; Abrignani, S.; et al. SARS vaccine protective in mice. *Emerg. Infect. Dis.* **2005**, *11*, 1312–1314. [CrossRef]

166. Bisht, H.; Roberts, A.; Vogel, L.; Bukreyev, A.; Collins, P.L.; Murphy, B.R.; Subbarao, K.; Moss, B. Severe acute respiratory syndrome coronavirus spike protein expressed by attenuated vaccinia virus protectively immunizes mice. *Proc. Natl. Acad. Sci. USA* **2004**, *101*, 6641–6646. [CrossRef]

167. Schaecher, S.R.; Stabenow, J.; Oberle, C.; Schriewer, J.; Buller, R.M.; Sagartz, J.E.; Pekosz, A. An immunosuppressed Syrian golden hamster model for SARS-CoV infection. *Virology* **2008**, *380*, 312–321. [CrossRef]

168. Luo, D.; Ni, B.; Zhao, G.; Jia, Z.; Zhou, L.; Pacal, M.; Zhang, L.; Zhang, S.; Xing, L.; Lin, Z.; et al. Protection from infection with severe acute respiratory syndrome coronavirus in a Chinese hamster model by equine neutralizing F(ab')2. *Viral Immunol.* **2007**, *20*, 495–502. [CrossRef]

169. Subbarao, K.; Roberts, A. Is there an ideal animal model for SARS? *Trends Microbiol.* **2006**, *14*, 299–303. [CrossRef]

170. Roberts, A.; Thomas, W.D.; Guarner, J.; Lamirande, E.W.; Babcock, G.J.; Greenough, T.C.; Vogel, L.; Hayes, N.; Sullivan, J.L.; Zaki, S.; et al. Therapy with a severe acute respiratory syndrome-associated coronavirus-neutralizing human monoclonal antibody reduces disease severity and viral burden in golden Syrian hamsters. *J. Infect. Dis.* **2006**, *193*, 685–692. [CrossRef]

171. Martina, B.E.; Haagmans, B.L.; Kuiken, T.; Fouchier, R.A.; Rimmelzwaan, G.F.; Van Amerongen, G.; Peiris, J.S.; Lim, W.; Osterhaus, A.D. Virology: SARS virus infection of cats and ferrets. *Nature* **2003**, *425*, 915. [CrossRef]

172. van den Brand, J.M.; Haagmans, B.L.; Leijten, L.; van Riel, D.; Martina, B.E.; Osterhaus, A.D.; Kuiken, T. Pathology of experimental SARS coronavirus infection in cats and ferrets. *Vet. Pathol.* **2008**, *45*, 551–562. [CrossRef]

173. Darnell, M.E.; Plant, E.P.; Watanabe, H.; Byrum, R.; St, C.M.; Ward, J.M.; Taylor, D.R. Severe acute respiratory syndrome coronavirus infection in vaccinated ferrets. *J. Infect. Dis.* **2007**, *196*, 1329–1338. [CrossRef]

174. ter Meulen, J.; Bakker, A.B.; van den Brink, E.N.; Weverling, G.J.; Martina, B.E.; Haagmans, B.L.; Kuiken, T.; de Kruif, J.; Preiser, W.; Spaan, W.; et al. Human monoclonal antibody as prophylaxis for SARS coronavirus infection in ferrets. *Lancet* **2004**, *363*, 2139–2141. [CrossRef]

175. See, R.H.; Petric, M.; Lawrence, D.J.; Mok, C.P.; Rowe, T.; Zitzow, L.A.; Karunakaran, K.P.; Voss, T.G.; Brunham, R.C.; Gauldie, J.; et al. Severe acute respiratory syndrome vaccine efficacy in ferrets: Whole killed virus and adenovirus-vectored vaccines. *J. Gen. Virol.* **2008**, *89*, 2136–2146. [CrossRef]

176. Qin, C.; Wang, J.; Wei, Q.; She, M.; Marasco, W.A.; Jiang, H.; Tu, X.; Zhu, H.; Ren, L.; Gao, H.; et al. An animal model of SARS produced by infection of Macaca mulatta with SARS coronavirus. *J. Pathol.* **2005**, *206*, 251–259. [CrossRef]

177. McAuliffe, J.; Vogel, L.; Roberts, A.; Fahle, G.; Fischer, S.; Shieh, W.J.; Butler, E.; Zaki, S.; St, C.M.; Murphy, B.; et al. Replication of SARS coronavirus administered into the respiratory tract of African Green, rhesus and cynomolgus monkeys. *Virology* **2004**, *330*, 8–15. [CrossRef]

178. Rowe, T.; Gao, G.; Hogan, R.J.; Crystal, R.G.; Voss, T.G.; Grant, R.L.; Bell, P.; Kobinger, G.P.; Wivel, N.A.; Wilson, J.M. Macaque model for severe acute respiratory syndrome. *J. Virol.* **2004**, *78*, 11401–11404. [CrossRef]

179. Li, B.; Tang, Q.; Cheng, D.; Qin, C.; Xie, F.Y.; Wei, Q.; Xu, J.; Liu, Y.; Zheng, B.; Woodle, M.C.; et al. Using siRNA in prophylactic and therapeutic regimens against SARS coronavirus in Rhesus macaque. *Nat. Med.* **2005**, *11*, 944–951. [CrossRef]

180. Liu, L.; Wei, Q.; Nishiura, K.; Peng, J.; Wang, H.; Midkiff, C.; Alvarez, X.; Qin, C.; Lackner, A.; Chen, Z. Spatiotemporal interplay of severe acute respiratory syndrome coronavirus and respiratory mucosal cells drives viral dissemination in rhesus macaques. *Mucosal Immunol.* **2016**, *9*, 1089–1101. [CrossRef]

181. Fouchier, R.A.; Kuiken, T.; Schutten, M.; van Amerongen, G.; van Doornum, G.J.; van den Hoogen, B.G.; Peiris, M.; Lim, W.; Stohr, K.; Osterhaus, A.D. Aetiology: Koch's postulates fulfilled for SARS virus. *Nature* **2003**, *423*, 240. [CrossRef]

182. Kuiken, T.; Fouchier, R.A.; Schutten, M.; Rimmelzwaan, G.F.; van Amerongen, G.; van Riel, D.; Laman, J.D.; de Jong, T.; van Doornum, G.; Lim, W.; et al. Newly discovered coronavirus as the primary cause of severe acute respiratory syndrome. *Lancet* **2003**, *362*, 263–270. [CrossRef]

183. Greenough, T.C.; Carville, A.; Coderre, J.; Somasundaran, M.; Sullivan, J.L.; Luzuriaga, K.; Mansfield, K. Pneumonitis and multi-organ system disease in common marmosets (*Callithrix jacchus*) infected with the severe acute respiratory syndrome-associated coronavirus. *Am. J. Pathol.* **2005**, *167*, 455–463. [CrossRef]

184. Lawler, J.V.; Endy, T.P.; Hensley, L.E.; Garrison, A.; Fritz, E.A.; Lesar, M.; Baric, R.S.; Kulesh, D.A.; Norwood, D.A.; Wasieloski, L.P.; et al. Cynomolgus macaque as an animal model for severe acute respiratory syndrome. *PLoS Med.* **2006**, *3*, e149. [CrossRef]

185. Roberts, A.; Subbarao, K. Animal models for SARS. *Adv. Exp. Med. Biol.* **2006**, *581*, 463–471.

186. Chen, Y.; Liu, L.; Wei, Q.; Zhu, H.; Jiang, H.; Tu, X.; Qin, C.; Chen, Z. Rhesus angiotensin converting enzyme 2 supports entry of severe acute respiratory syndrome coronavirus in Chinese macaques. *Virology* **2008**, *381*, 89–97. [CrossRef]

187. Cockrell, A.S.; Peck, K.M.; Yount, B.L.; Agnihothram, S.S.; Scobey, T.; Curnes, N.R.; Baric, R.S.; Heise, M.T. Mouse dipeptidyl peptidase 4 is not a functional receptor for Middle East respiratory syndrome coronavirus infection. *J. Virol.* **2014**, *88*, 5195–5199. [CrossRef]

188. Coleman, C.M.; Matthews, K.L.; Goicochea, L.; Frieman, M.B. Wild-type and innate immune-deficient mice are not susceptible to the Middle East respiratory syndrome coronavirus. *J. Gen. Virol.* **2014**, *95*, 408–412. [CrossRef]

189. de Wit, E.; Feldmann, F.; Horne, E.; Martellaro, C.; Haddock, E.; Bushmaker, T.; Rosenke, K.; Okumura, A.; Rosenke, R.; Saturday, G.; et al. Domestic Pig Unlikely Reservoir for MERS-CoV. *Emerg. Infect. Dis.* **2017**, *23*, 985–988. [CrossRef]

190. Peck, K.M.; Scobey, T.; Swanstrom, J.; Jensen, K.L.; Burch, C.L.; Baric, R.S.; Heise, M.T. Permissivity of Dipeptidyl Peptidase 4 Orthologs to Middle East Respiratory Syndrome Coronavirus Is Governed by Glycosylation and Other Complex Determinants. *J. Virol.* **2017**, *91*, e00534-17. [CrossRef]

191. Haagmans, B.L.; van den Brand, J.M.; Provacia, L.B.; Raj, V.S.; Stittelaar, K.J.; Getu, S.; de Waal, L.; Bestebroer, T.M.; van Amerongen, G.; Verjans, G.M.; et al. Asymptomatic Middle East respiratory syndrome coronavirus infection in rabbits. *J. Virol.* **2015**, *89*, 6131–6135. [CrossRef]

192. de Wit, E.; Prescott, J.; Baseler, L.; Bushmaker, T.; Thomas, T.; Lackemeyer, M.G.; Martellaro, C.; Milne-Price, S.; Haddock, E.; Haagmans, B.L.; et al. The Middle East respiratory syndrome coronavirus (MERS-CoV) does not replicate in Syrian hamsters. *PLoS ONE* **2013**, *8*, e69127. [CrossRef]

193. Li, K.; Wohlford-Lenane, C.L.; Channappanavar, R.; Park, J.E.; Earnest, J.T.; Bair, T.B.; Bates, A.M.; Brogden, K.A.; Flaherty, H.A.; Gallagher, T.; et al. Mouse-adapted MERS coronavirus causes lethal lung disease in human DPP4 knockin mice. *Proc. Natl. Acad. Sci. USA* **2017**, *114*, E3119–E3128. [CrossRef]

194. Zhao, G.; Jiang, Y.; Qiu, H.; Gao, T.; Zeng, Y.; Guo, Y.; Yu, H.; Li, J.; Kou, Z.; Du, L.; et al. Multi-Organ Damage in Human Dipeptidyl Peptidase 4 Transgenic Mice Infected with Middle East Respiratory Syndrome-Coronavirus. *PLoS ONE* **2015**, *10*, e145561. [CrossRef]

195. Zhao, J.; Li, K.; Wohlford-Lenane, C.; Agnihothram, S.S.; Fett, C.; Zhao, J.; Gale, M.J.; Baric, R.S.; Enjuanes, L.; Gallagher, T.; et al. Rapid generation of a mouse model for Middle East respiratory syndrome. *Proc. Natl. Acad. Sci. USA* **2014**, *111*, 4970–4975. [CrossRef]

196. Tseng, C.T.; Huang, C.; Newman, P.; Wang, N.; Narayanan, K.; Watts, D.M.; Makino, S.; Packard, M.M.; Zaki, S.R.; Chan, T.S.; et al. Severe acute respiratory syndrome coronavirus infection of mice transgenic for the human Angiotensin-converting enzyme 2 virus receptor. *J. Virol.* **2007**, *81*, 1162–1173. [CrossRef]

197. Cockrell, A.S.; Yount, B.L.; Scobey, T.; Jensen, K.; Douglas, M.; Beall, A.; Tang, X.C.; Marasco, W.A.; Heise, M.T.; Baric, R.S. A mouse model for MERS coronavirus-induced acute respiratory distress syndrome. *Nat. Microbiol.* **2016**, *2*, 16226. [CrossRef]

198. Haagmans, B.L.; van den Brand, J.M.; Raj, V.S.; Volz, A.; Wohlsein, P.; Smits, S.L.; Schipper, D.; Bestebroer, T.M.; Okba, N.; Fux, R.; et al. An orthopoxvirus-based vaccine reduces virus excretion after MERS-CoV infection in dromedary camels. *Science* **2016**, *351*, 77–81. [CrossRef]

199. Adney, D.R.; Bielefeldt-Ohmann, H.; Hartwig, A.E.; Bowen, R.A. Infection, Replication, and Transmission of Middle East Respiratory Syndrome Coronavirus in Alpacas. *Emerg. Infect. Dis.* **2016**, *22*, 1031–1037. [CrossRef]

200. Crameri, G.; Durr, P.A.; Klein, R.; Foord, A.; Yu, M.; Riddell, S.; Haining, J.; Johnson, D.; Hemida, M.G.; Barr, J.; et al. Experimental Infection and Response to Rechallenge of Alpacas with Middle East Respiratory Syndrome Coronavirus. *Emerg. Infect. Dis.* **2016**, *22*, 1071–1074. [CrossRef]

201. Yao, Y.; Bao, L.; Deng, W.; Xu, L.; Li, F.; Lv, Q.; Yu, P.; Chen, T.; Xu, Y.; Zhu, H.; et al. An Animal Model of MERS Produced by Infection of Rhesus Macaques with MERS Coronavirus. *J. Infect. Dis.* **2013**, *209*, 236–242. [CrossRef]

202. Yu, P.; Xu, Y.; Deng, W.; Bao, L.; Huang, L.; Xu, Y.; Yao, Y.; Qin, C. Comparative pathology of rhesus macaque and common marmoset animal models with Middle East respiratory syndrome coronavirus. *PLoS ONE* **2017**, *12*, e172093. [CrossRef] [PubMed]

203. Falzarano, D.; de Wit, E.; Feldmann, F.; Rasmussen, A.L.; Okumura, A.; Peng, X.; Thomas, M.J.; van Doremalen, N.; Haddock, E.; Nagy, L.; et al. Infection with MERS-CoV causes lethal pneumonia in the common marmoset. *PLoS Pathog.* **2014**, *10*, e1004250. [CrossRef] [PubMed]

204. Munster, V.J.; de Wit, E.; Feldmann, H. Pneumonia from human coronavirus in a macaque model. *N. Engl. J. Med.* **2013**, *368*, 1560–1562. [CrossRef] [PubMed]

205. Falzarano, D.; de Wit, E.; Rasmussen, A.L.; Feldmann, F.; Okumura, A.; Scott, D.P.; Brining, D.; Bushmaker, T.; Martellaro, C.; Baseler, L.; et al. Treatment with interferon-alpha2b and ribavirin improves outcome in MERS-CoV-infected rhesus macaques. *Nat. Med.* **2013**, *19*, 1313–1317. [CrossRef]

206. Johnson, R.F.; Via, L.E.; Kumar, M.R.; Cornish, J.P.; Yellayi, S.; Huzella, L.; Postnikova, E.; Oberlander, N.; Bartos, C.; Ork, B.L.; et al. Intratracheal exposure of common marmosets to MERS-CoV Jordan-n3/2012 or MERS-CoV EMC/2012 isolates does not result in lethal disease. *Virology* **2015**, *485*, 422–430. [CrossRef]

207. Li, W.; Greenough, T.C.; Moore, M.J.; Vasilieva, N.; Somasundaran, M.; Sullivan, J.L.; Farzan, M.; Choe, H. Efficient replication of severe acute respiratory syndrome coronavirus in mouse cells is limited by murine angiotensin-converting enzyme 2. *J. Virol.* **2004**, *78*, 11429–11433. [CrossRef] [PubMed]

208. Wu, D.; Tu, C.; Xin, C.; Xuan, H.; Meng, Q.; Liu, Y.; Yu, Y.; Guan, Y.; Jiang, Y.; Yin, X.; et al. Civets are equally susceptible to experimental infection by two different severe acute respiratory syndrome coronavirus isolates. *J. Virol.* **2005**, *79*, 2620–2625. [CrossRef]

209. Barlan, A.; Zhao, J.; Sarkar, M.K.; Li, K.; McCray, P.J.; Perlman, S.; Gallagher, T. Receptor variation and susceptibility to Middle East respiratory syndrome coronavirus infection. *J. Virol.* **2014**, *88*, 4953–4961. [CrossRef]

210. van Doremalen, N.; Miazgowicz, K.L.; Milne-Price, S.; Bushmaker, T.; Robertson, S.; Scott, D.; Kinne, J.; McLellan, J.S.; Zhu, J.; Munster, V.J. Host species restriction of Middle East respiratory syndrome coronavirus through its receptor, dipeptidyl peptidase 4. *J. Virol.* **2014**, *88*, 9220–9232. [CrossRef]

viruses

MDPI

Review

Host Determinants of MERS-CoV Transmission and Pathogenesis

W. Widagdo, Syriam Sooksawasdi Na Ayudhya, Gadissa B. Hundie and Bart L. Haagmans *

Department of Viroscience, Erasmus Medical Center, 3025 Rotterdam, The Netherlands;
w.widagdo@erasmusmc.nl (W.W.); s.sooksawasdi@erasmusmc.nl (S.S.N.A.); g.hundie@erasmusmc.nl (G.B.H.)
* Correspondence: b.haagmans@erasmusmc.nl

Received: 1 March 2019; Accepted: 13 March 2019; Published: 19 March 2019

Abstract: Middle East respiratory syndrome coronavirus (MERS-CoV) is a zoonotic pathogen that causes respiratory infection in humans, ranging from asymptomatic to severe pneumonia. In dromedary camels, the virus only causes a mild infection but it spreads efficiently between animals. Differences in the behavior of the virus observed between individuals, as well as between humans and dromedary camels, highlight the role of host factors in MERS-CoV pathogenesis and transmission. One of these host factors, the MERS-CoV receptor dipeptidyl peptidase-4 (DPP4), may be a critical determinant because it is variably expressed in MERS-CoV-susceptible species as well as in humans. This could partially explain inter- and intraspecies differences in the tropism, pathogenesis, and transmissibility of MERS-CoV. In this review, we explore the role of DPP4 and other host factors in MERS-CoV transmission and pathogenesis—such as sialic acids, host proteases, and interferons. Further characterization of these host determinants may potentially offer novel insights to develop intervention strategies to tackle ongoing outbreaks.

Keywords: MERS-CoV; transmission; pathogenesis; host factors; DPP4

1. Introduction

Middle East respiratory syndrome coronavirus (MERS-CoV) is a novel pathogen that was isolated in late 2012 [1]. Since then, the virus has caused multiple outbreaks and infected more than 2000 individuals, [2] who then develop a respiratory infection ranging in severity from asymptomatic to fatal [3,4]. Severe-to-fatal MERS-CoV patients have a higher chance of transmitting this virus since they shed a higher amount of virus progeny in comparison to the asymptomatic-to-mild ones [5–8]. Identifying and quarantining these patients in healthcare facilities where outbreaks have occurred, together with implementing proper infection control, has been effective in reducing transmission and containing these outbreaks [9,10]. However, new MERS-CoV cases are still being reported, especially in the Arabian Peninsula [2,11]. This is partly due to the continuous zoonotic introduction of this virus to the human population in this region by dromedaries [12]. The dromedary camel is the only animal species that has been reported to transmit this virus to humans [13–16]. MERS-CoV infection in these animals merely causes mild upper respiratory tract infection [17,18], but seroepidemiological studies showed that this virus has been circulating in dromedary camels for decades, suggesting the efficient transmission of MERS-CoV in this species [19–22].

Although the clinical manifestations, as well as transmission, are remarkably different in MERS-CoV-infected humans and dromedary camels, the viruses isolated from these two species are highly similar, if not indistinguishable [12,16]. This indicates that host factors play a significant role in MERS-CoV pathogenesis and transmission. However, the identity of these host factors and how they affect the pathogenesis and transmission of MERS-CoV are generally not well understood. Dipeptidyl peptidase-4 (DPP4)—the MERS-CoV receptor, sialic acids, proteases, and interferons are all examples of potentially critical host factors that have been shown to affect MERS-CoV infection in vitro [23–26]. This review highlights the role of some MERS-CoV-interacting host factors—especially DPP4—in MERS-CoV pathogenesis and transmission.

2. MERS-CoV-Interacting Host Factors

MERS-CoV infection of a target cell is initiated by the virus attachment to the cell surface [23,27]. MERS-CoV uses the N-terminal part of its spike (S)—the so called S1 protein (Figure 1A)—to bind to two host cell surface molecules, dipeptidyl peptidase-4 (DPP4) and α2,3-sialic acids [23,24]. DPP4 is the functional receptor of MERS-CoV; its absence renders cells resistant to this virus, while its transient expression in non-susceptible cells permits viral replication [23]. DPP4 is a serine exopeptidase, which is either expressed at the cell surface or shed in a soluble form. It has the capacity to cleave-off dipeptides from polypeptides with either L-proline or L-alanine at the penultimate position. Accordingly, DPP4 is capable of cutting various substrates, such as hormones, cytokines, chemokines, and neuropeptides, allowing it to be involved in multiple physiological functions as well as pathophysiological conditions [28]. This enzymatic activity is mediated by the α/β hydrolase domain of DPP4, while MERS-CoV infection is mediated by the binding of S1 protein to the β-propeller domain of this exopeptidase (Figure 1B) [28–31]. There are 11 critical residues within the β-propeller domain that directly interact with the S1 protein [29–31]. These residues are quite conserved in camelids, primates, and rabbits—species shown to be susceptible to MERS-CoV [17,31–33]. In contrast, ferrets, rats, and mice resist MERS-CoV infection due to differences in some critical DPP4 residues [31,34–36]. These data illustrate that DPP4 has the capacity to determine the host range of MERS-CoV.

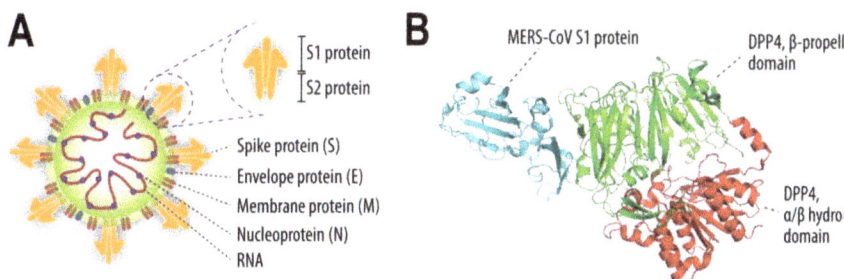

Figure 1. Schematic figure depicting four structural proteins of Middle East respiratory syndrome coronavirus (MERS-CoV), i.e., S, E, M, and N proteins (**A**); a cartoon representation of MERS-CoV S1 protein binding to DPP4 (PDB code 4L72) (**B**). The S protein consists of the S1 and S2 subunits. The α/β hydrolase domain of DPP4 is indicated in red, β-propeller domain in green, while part of the MERS-CoV S1 protein is shown in blue.

Other MERS-CoV-interacting host factors besides DPP4 are less extensively studied and have mostly been investigated in vitro. Glycotopes of α2,3-sialic acids coupled with 5-N-acetylated neuraminic acid are recognized by the S1 protein of MERS-CoV during attachment [24]. In the absence of these glycotopes, MERS-CoV entry is reduced but not abolished, indicating their function as an attachment factor rather than a receptor [24]. Besides α2,3-sialic acids, CEACAM5 and GRP78 have also been suggested to be attachment factors for MERS-CoV, but their roles in vivo during MERS-CoV infection are not clear at this moment [37,38]. Post attachment, MERS-CoV uses the C-terminal part of its S protein—known as S2 (Figure 1A)—to interact with host proteases, such as furin, TMPRSS2, and cathepsins [39–42]. These proteases cleave the S protein and induce conformational changes, allowing fusion between viral and host cellular membranes, resulting in the release of viral RNA into the cell cytoplasm [27]. TMPRSS2 and DPP4 are held in one complex at the cell surface by a scaffolding protein, the tetraspanin CD9, leading to a rapid and efficient entry of MERS-CoV into the susceptible cells [43]. Once fusion with host cell membranes has occurred, MERS-CoV subsequently replicates its genetic material and produces viral proteins in the cell cytoplasm to generate new virus progeny. During this stage, MERS-CoV uses its nsp3–4 polyproteins to build its replication organelles as well as its accessory proteins such as the 4a and 4b proteins to inhibit host anti-viral defense mechanisms [44–54]. However, the capacity of MERS-CoV accessory proteins to impede several pathways of host immune response in the lungs may be limited. MERS-CoV inoculation of macaques and genetically modified mice generally results in limited clinical manifestations; thus, adapting this virus through serial passaging or defecting the type I interferon pathway may be needed to enhance viral replication and pathogenesis in these animals [32,55–58]. These observations, together with studies showing type I interferon capacity to inhibit MERS-CoV infection in vitro [25,59], highlight the importance of the innate immune response, especially type I interferon, as an inhibiting factor for MERS-CoV.

3. Host Factors in MERS-CoV Transmission

So far MERS-CoV has been isolated from dromedary camels and humans [1,60]. Both species are not only susceptible to MERS-CoV infection, but also capable of transmitting this virus [7,12–18,22]. However, current data indicate that virus spread is more efficient in dromedary camels than in humans [5,7,19–21,61]. This difference in transmissibility could be partially due to the different tropism of MERS-CoV in these two species. In dromedaries, MERS-CoV has been shown to replicate in the nasal epithelium upon experimental in vivo infection [17], while in humans, MERS-CoV mainly replicates in the lower respiratory tract, particularly in the bronchiolar and alveolar epithelia [23,62–65]. Higher viral RNA levels in the sputum and lavage samples of MERS-CoV patients compared to nasal and throat swabs are consistent with the tropism of MERS-CoV in humans [66–68]. This different MERS-CoV tropism in dromedary camels and humans is in line with the localization of DPP4 in the respiratory tract tissues of these two species. In humans, DPP4 is absent in the nasal epithelium but present in the lower respiratory tract epithelium, mainly in type II pneumocytes [69,70]. In contrast, DPP4 is expressed in the nasal epithelium of dromedary camels [69]. This difference in DPP4 localization between humans and dromedary camels therefore explains MERS-CoV tropism in these two species and highlights DPP4 as an essential determinant of MERS-CoV tropism.

DPP4 localization has also been investigated in many other MERS-CoV-susceptible species. In Gambian and Egyptian fruit bats, DPP4 is expressed in the respiratory tract and intestinal epithelium, suggesting that MERS-CoV can target both tissues [71]. In line with this finding, MERS-CoV inoculation via intranasal and intraperitoneal routes in the Jamaican fruit bat led to viral RNA shedding both in the respiratory tract and the intestinal tract [72]. In contrast to frugivorous bats, DPP4 is limitedly expressed in the respiratory tract epithelium of two insectivorous bats, i.e., common pipistrelle and common serotine bats, but abundant in their intestinal epithelium [71]. Accordingly, sequences of MERS-like-CoVs were mainly obtained from rectal swabs and fecal samples of insectivorous bats [73–80]. These findings not only support insectivorous bats as the origin host of MERS-CoV [73–80], but also indicate the importance of intestinal tropism and fecal–oral transmission of MERS-like-CoV in these insectivorous bats.

Besides bats, humans, and dromedary camels, other animal species have also been proposed as potential hosts of MERS-CoV. Remarkably, DPP4 of horses, llamas, alpacas, pigs, bovines, goats, sheep, and rabbits has been demonstrated to recognize the S protein of MERS-CoV [81,82]. In most of these species, there is a preferential upper respiratory tract expression of DPP4 observed. Rabbits express DPP4 in the upper and lower respiratory tract epithelium, and thus may allow MERS-CoV to replicate in both compartments [33,83]. Horses, llamas, and pigs mainly express DPP4 in the upper respiratory tract—particularly the nasal epithelium [84]. Upon intranasal MERS-CoV inoculation, llamas, alpacas, and pigs developed upper respiratory tract infection, while horses did not seroconvert and only shed infectious virus in a limited amount [84–88]. The reason why horses seem to be less permissive to MERS-CoV remains to be investigated, but a chronic co-infection in the guttural pouch, a common disease among horses, might be one of the explanations. This guttural pouch infection results in excessive mucus production that might hinder MERS-CoV from attaching and entering the nasal epithelium [84,89,90]. Sheep, on the other hand, did not seem to express significant levels of DPP4 in their respiratory tract, and thus did not seroconvert nor shed infectious virus upon experimental MERS-CoV inoculation [84,88]. Comparable to sheep, goats limitedly shed infectious virus upon experimental infection and did not transmit this virus to other naïve goats upon direct contact [88]. The results of experimental MERS-CoV infection in livestock animals are in line with data from epidemiological studies. MERS-CoV seropositive llamas and alpacas are present in the field, while horses, goats, and sheep are generally found to be seronegative [22,86,87,91–98].

Given the fact that experimental in vivo infection studies and DPP4 expression analysis in different animal species revealed that dromedary camels are not the only animals in which MERS-CoV has an upper respiratory tract tropism [17,18,83,84], it is then relevant to question whether other animals can potentially spread MERS-CoV as well. New World camelids, i.e., alpacas and llamas, are able to transmit the virus to respective naïve animals upon contact [86]. Pigs and rabbits, on the other hand, hardly transmit the virus—neither by contact nor airborne routes [83,99]. Most likely, this is caused by the fact that pigs and rabbits, unlike dromedary camels, shed low levels of infectious virus upon MERS-CoV inoculation (Figure 2). This difference indicates that other host factors besides DPP4 could cause interspecies variation in MERS-CoV infection. Indeed, several glycotopes of α2,3-sialic acids that function as attachment factors of MERS-CoV are present in the nasal epithelium of dromedary camels but absent in that of rabbits and pigs (Figure 3) [24,100]. The lack of these glycotopes in pigs and rabbits might limit the susceptibility and transmission of MERS-CoV in these animals. Although the role of these glycotopes in MERS-CoV transmission still requires further investigation, it remains plausible that an efficient transmission of this virus might require the presence of both DPP4 and MERS-CoV-recognized glycotopes of α2,3-sialic acids (Figure 3).

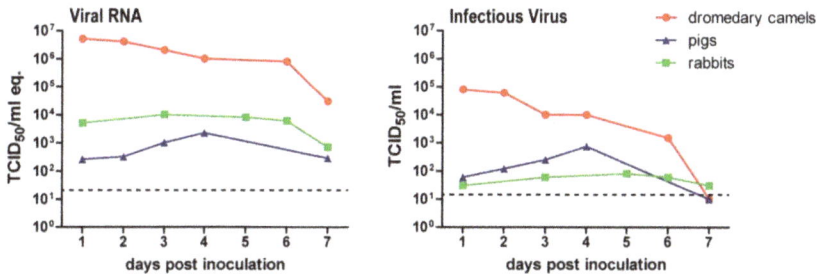

Figure 2. Schematic overview of viral RNA and infectious virus shedding of MERS-CoV-inoculated dromedary camels, pigs, and rabbits. Each data point represents the average data from previous experiments [17,33,84]. Viral RNA is measured in $TCID_{50}$/mL genome equivalents, while infectious virus is expressed in $TCID_{50}$/mL.

Besides entry and attachment receptors, MERS-CoV has been demonstrated to use both cell surface and lysosomal proteases to enter its target cells [39,40,43,101]. The preference of MERS-CoV to use certain host proteases is influenced by the type of target cell and the cleavage stage of their S protein prior to infection [40]. It has also been reported that the lysosomal proteases from bat cells support coronavirus spike-mediated virus entry more efficiently than their counterparts from human cells [39]. These observations suggest that host proteases from different host species may determine the species and tissue tropism of MERS-CoV.

Because MERS-CoV has been circulating in dromedary camels for decades before emerging in the human population [19–22], it is plausible that this virus inhibits the immune response of dromedary camels more efficiently than that of other species, including pigs and rabbits. The difference in immune response among MERS-CoV-susceptible species is therefore another factor that might yield interspecies variation in permissiveness to MERS-CoV. Characterizing the difference in host proteases and immune responses among MERS-CoV-susceptible species, as performed for DPP4 and MERS-CoV-recognized α2,3-sialic acid glycotopes (Figure 3), has not yet been investigated. These data, however, may further explain interspecies variation in MERS-CoV infection and transmission.

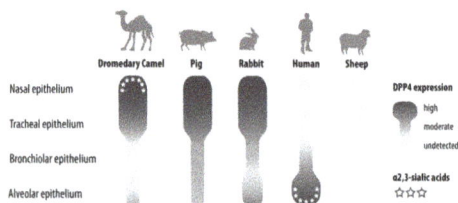

Figure 3. Schematic representation of DPP4 expression and MERS-CoV-recognized α2,3-sialic acid glycotopes in the respiratory tract of dromedary camel, pig, rabbit, human, and sheep.

4. Host Factors in MERS-CoV Pathogenesis

MERS-CoV causes respiratory infection in humans ranging from asymptomatic to severe pneumonia [3,4]. However, it is currently unclear what causes this intraspecies variation. Epidemiology data indicate that individuals with certain risk factors are at higher risk of developing severe MERS-CoV infection [4,102]. This implies that some host factors may dictate the outcome of MERS-CoV infection, thus rendering intraspecies variation. Two of the risk factors, i.e., smoking and chronic obstructive pulmonary disease (COPD), have been shown to upregulate DPP4 expression in the lungs [70,102–104], suggesting DPP4 as a possible reason for intraspecies variation observed among MERS-CoV patients. In healthy human lungs, DPP4 is almost exclusively expressed in type II pneumocytes [69,70]. Type II pneumocytes are small cuboidal cells that can regenerate alveolar epithelium upon injury, and roughly cover 2% of the alveolar surface area. Meanwhile, around 95% of the surface area of the alveolus is occupied by type I pneumocytes that are morphologically flat and responsible for gas exchange [105,106]. In the lungs of smokers and COPD patients, unlike in healthy human lungs, DPP4 is prominently expressed in both type I and II pneumocytes, indicating upregulated expression on type I pneumocytes [104]. Autopsy reports from fatal MERS-CoV patients showed that both type I and II pneumocytes expressed DPP4 and became infected by MERS-CoV, proposing a role of DPP4-expressing type I pneumocytes in MERS-CoV pathogenesis [64,107]. Damage to type I cells in the lung alveoli during viral infection may lead to diffuse alveolar damage [108]. In line with observations made in human MERS cases, common marmosets that express DPP4 in both type I and II pneumocytes have been reported to produce more infectious virus upon experimental MERS-CoV infection, compared to rhesus and cynomolgus macaques that merely expressed DPP4 in type II pneumocytes [58,109–112]. Accordingly, these common marmosets developed moderate-to-severe infection, while macaques generally developed mild transient pneumonia [32,58,109–112]. Similarly, in genetically modified mice that displayed MERS-CoV tropism for type II pneumocytes, only mild clinical manifestations were observed upon MERS-CoV infection [56,113]. Adapting MERS-CoV through serial passaging or upregulating DPP4 expression throughout the airway epithelium in mice, however, will induce severe clinical disease [55,56]. These data altogether support the role of DPP4-expressing type I pneumocytes in the pathogenesis of severe MERS-CoV infection.

The differential expression of host factors that limits the infection should also be taken into account. DPP4 in soluble form has been demonstrated to protect against MERS-CoV infection in vitro and in a mouse model [23,114]; however, its presence in the lungs and role in MERS-CoV pathogenesis remain to be investigated. The host immune response also has the capacity to inhibit MERS-CoV infection. MERS-CoV has been shown to replicate to higher levels in immunocompromised rhesus macaques [115], consistent with the observation that immunocompromised individuals have difficulties clearing MERS-CoV upon infection [68,107,116]. The survivors of MERS-CoV infection have been shown to develop virus-specific CD4+ and CD8+ T cell responses, implying the role of T cells in virus clearance [117]. However, the depletion of T cells in mice can either lead to failure in MERS-CoV clearance or improvement in clinical outcome, depending on the type of mouse model used [57,118]. Therefore, the role of adaptive immune response in MERS-CoV pathogenesis is currently unclear. On the other hand, one of the main components of the host innate immune response, type I interferon, inhibits MERS-CoV replication in susceptible cells, partly by inhibiting double membrane vesicles (DMV) formation [25,57,59,119,120]. The absence of type I interferon signaling in mice also resulted in more severe clinical manifestations and histopathological lesions upon MERS-CoV infection [57]. Advance age, which can cause delayed type I interferon response upon viral infection, is a well-known risk factor for fatal MERS-CoV infection [4,102,121–123]. Collectively, these data highlight the role of host innate immune response as a potent inhibitor for MERS-CoV infection.

It is indubitable that severe MERS-CoV infection is not solely driven by the pathogen. Additional underlying conditions increase MERS-CoV replication and induce severe-to-fatal clinical manifestations [4,11,103,124,125]. It is plausible that more than one underlying condition is needed to yield a fatal outcome. DPP4 upregulation in type I pneumocytes and insufficient type I interferon response might be crucial determinants for severe MERS-CoV infection (Figure 4). Further investigation of the host determinants of MERS-CoV pathogenesis may offer insights for developing novel therapeutic measures.

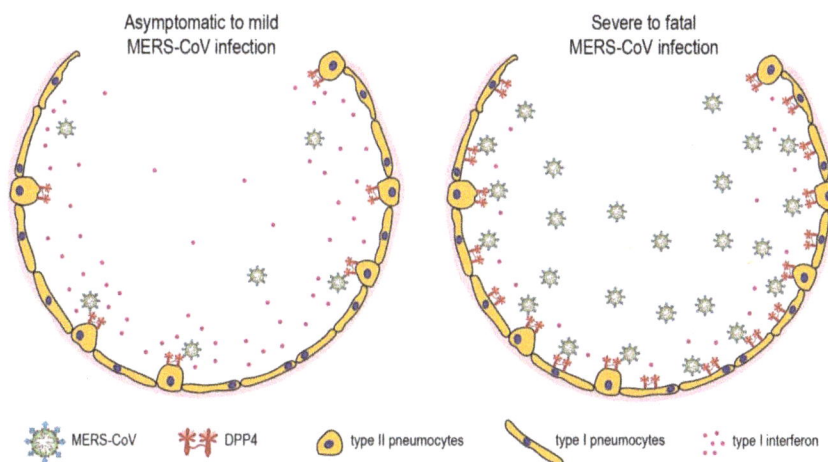

Figure 4. MERS-CoV infection in the lungs of asymptomatic-to-mild (left panel) and severe-to-fatal cases (right panel). Shown is a hypothetical model with two critical host determinants, DPP4 and interferon, differentially expressed in asymptomatic-to-mild and severe-to-fatal MERS-CoV infection.

5. Concluding Remarks and Future Perspectives

Although MERS-CoV has been reported to undergo some genotypic changes since it emerged in the human population [12,126–129], this has not resulted in distinct phenotypic changes so far [63,126]. Therefore, host factors remain the most significant determinant in explaining inter- and intraspecies variations observed in MERS-CoV pathogenesis and transmission. DPP4 and MERS-CoV-recognized α2,3-sialic acids might partially explain these variations, since their localization has been demonstrated to be variable between MERS-CoV-susceptible species [69,71,84,100]. DPP4 expression in human lungs has also been shown to vary due to certain comorbidities [70,96,104]. Nevertheless, it is undoubtable that the inter- and intraspecies variation in MERS-CoV pathogenesis and transmission is a complex phenomenon influenced by more than one host factor. Current data suggest proteases and interferons as other critical host factors, but how they instigate inter- and intraspecies variations, as well as their role in MERS-CoV pathogenesis and transmission, still remain to be further elucidated. Characterization of the host determinants of MERS-CoV pathogenesis and transmission could potentially offer insight into this virus epidemiology and guide novel therapeutic development. It may also help to identify the most vulnerable individuals to protect against MERS-CoV infection—for example, by using vaccination.

Author Contributions: All authors contributed to the writing of the manuscript and carefully evaluated the manuscript before submission.

Funding: Our work is supported by Zoonotic Anticipation and Preparedness Initiative (Innovative Medicines Initiative grant 115760), with assistance and financial support from Innovative Medicines Initiative and the European Commission, and contributions from European Federation of Pharmaceutical Industries and Associations partners. Syriam Sooksawasdi Na Ayudhya received the Royal Thai Government Scholarship supported by the Ministry of Science and Technology of Thailand to perform her doctoral study.

Conflicts of Interest: The authors declare no conflict of interest.

References

1. Zaki, A.M.; van Boheemen, S.; Bestebroer, T.M.; Osterhaus, A.D.; Fouchier, R.A. Isolation of a novel coronavirus from a man with pneumonia in Saudi Arabia. *N. Engl. J. Med.* **2012**, *367*, 1814–1820. [CrossRef]
2. World Health Organization MERS Situation Update, March 2018. Available online: http://www.emro.who. int/pandemic-epidemic-diseases/mers-cov/mers-situation-update-march-2018.html (accessed on 1 March 2019).
3. Widagdo, W.; Okba, N.M.A.; Stalin Raj, V.; Haagmans, B.L. MERS-coronavirus: From discovery to intervention. *One Health* **2017**, *3*, 11–16. [CrossRef]
4. The WHO Mers-CoV Research Group. State of Knowledge and Data Gaps of Middle East Respiratory Syndrome Coronavirus (MERS-CoV) in Humans. *PLoS Curr.* **2013**, *5*. [CrossRef]
5. Moon, S.Y.; Son, J.S. Infectivity of an Asymptomatic Patient With Middle East Respiratory Syndrome Coronavirus Infection. *Clin. Infect. Dis.* **2017**, *64*, 1457–1458. [CrossRef] [PubMed]
6. Memish, Z.A.; Al-Tawfiq, J.A.; Makhdoom, H.Q.; Al-Rabeeah, A.A.; Assiri, A.; Alhakeem, R.F.; AlRabiah, F.A.; Al Hajjar, S.; Albarrak, A.; Flemban, H.; et al. Screening for Middle East respiratory syndrome coronavirus infection in hospital patients and their healthcare worker and family contacts: A prospective descriptive study. *Clin. Microbiol. Infect.* **2014**, *20*, 469–474. [CrossRef]
7. Drosten, C.; Meyer, B.; Muller, M.A.; Corman, V.M.; Al-Masri, M.; Hossain, R.; Madani, H.; Sieberg, A.; Bosch, B.J.; Lattwein, E.; et al. Transmission of MERS-coronavirus in household contacts. *N. Engl. J. Med.* **2014**, *371*, 828–835. [CrossRef]
8. Kim, S.W.; Park, J.W.; Jung, H.D.; Yang, J.S.; Park, Y.S.; Lee, C.; Kim, K.M.; Lee, K.J.; Kwon, D.; Hur, Y.J.; et al. Risk factors for transmission of Middle East respiratory syndrome coronavirus infection during the 2015 outbreak in South Korea. *Clin. Infect. Dis.* **2017**, *64*, 551–557. [CrossRef] [PubMed]
9. Normile, D. South Korea finally MERS-free. *Science* **2015**. [CrossRef]
10. World Health Organization Middle East Respiratory Syndrome Coronavirus (MERS-CoV), Republic of Korea—Disease Outbreak News 25 October 2015. Available online: http://www.who.int/csr/don/25-october-2015-mers-korea/en/ (accessed on 1 March 2019).
11. World Health Organization MERS-CoV Global Summary and Assessment of Risk. July 2017. Available online: http://www.who.int/emergencies/mers-cov/risk-assessment-july-2017.pdf?ua=1 (accessed on 1 March 2019).
12. Dudas, G.; Carvalho, L.M.; Rambaut, A.; Bedford, T. MERS-CoV spillover at the camel-human interface. *Elife* **2018**, *7*, eLife.31257.
13. Haagmans, B.L.; Al Dhahiry, S.H.; Reusken, C.B.; Raj, V.S.; Galiano, M.; Myers, R.; Godeke, G.J.; Jonges, M.; Farag, E.; Diab, A.; et al. Middle East respiratory syndrome coronavirus in dromedary camels: An outbreak investigation. *Lancet Infect. Dis.* **2014**, *14*, 140–145. [CrossRef]
14. Memish, Z.A.; Cotten, M.; Meyer, B.; Watson, S.J.; Alsahafi, A.J.; Al Rabeeah, A.A.; Corman, V.M.; Sieberg, A.; Makhdoom, H.Q.; Assiri, A.; et al. Human infection with MERS coronavirus after exposure to infected camels, Saudi Arabia, 2013. *Emerg. Infect. Dis.* **2014**, *20*, 1012–1015. [CrossRef] [PubMed]

15. Paden, C.R.; Yusof, M.; Al Hammadi, Z.M.; Queen, K.; Tao, Y.; Eltahir, Y.M.; Elsayed, E.A.; Marzoug, B.A.; Bensalah, O.K.A.; Khalafalla, A.I.; et al. Zoonotic origin and transmission of Middle East respiratory syndrome coronavirus in the UAE. *Zoonoses. Public Health* **2018**, *65*, 322–333. [CrossRef]

16. Briese, T.; Mishra, N.; Jain, K.; Zalmout, I.S.; Jabado, O.J.; Karesh, W.B.; Daszak, P.; Mohammed, O.B.; Alagaili, A.N.; Lipkin, W.I. Middle East respiratory syndrome coronavirus quasispecies that include homologues of human isolates revealed through whole-genome analysis and virus cultured from dromedary camels in Saudi Arabia. *MBio* **2014**, *5*, e01146-14.

17. Haagmans, B.L.; van den Brand, J.M.; Raj, V.S.; Volz, A.; Wohlsein, P.; Smits, S.L.; Schipper, D.; Bestebroer, T.M.; Okba, N.; Fux, R.; et al. An orthopoxvirus-based vaccine reduces virus excretion after MERS-CoV infection in dromedary camels. *Science* **2016**, *351*, 77–81. [CrossRef]

18. Adney, D.R.; van Doremalen, N.; Brown, V.R.; Bushmaker, T.; Scott, D.; de Wit, E.; Bowen, R.A.; Munster, V.J. Replication and shedding of MERS-CoV in upper respiratory tract of inoculated dromedary camels. *Emerg. Infect. Dis.* **2014**, *20*, 1999–2005.

19. Meyer, B.; Muller, M.A.; Corman, V.M.; Reusken, C.B.; Ritz, D.; Godeke, G.J.; Lattwein, E.; Kallies, S.; Siemens, A.; van Beek, J.; et al. Antibodies against MERS coronavirus in dromedary camels, United Arab Emirates, 2003 and 2013. *Emerg. Infect. Dis.* **2014**, *20*, 552–559. [CrossRef] [PubMed]

20. Muller, M.A.; Corman, V.M.; Jores, J.; Meyer, B.; Younan, M.; Liljander, A.; Bosch, B.J.; Lattwein, E.; Hilali, M.; Musa, B.E.; et al. MERS coronavirus neutralizing antibodies in camels, Eastern Africa, 1983–1997. *Emerg. Infect. Dis.* **2014**, *20*, 2093–2095.

21. Reusken, C.B.; Messadi, L.; Feyisa, A.; Ularamu, H.; Godeke, G.J.; Danmarwa, A.; Dawo, F.; Jemli, M.; Melaku, S.; Shamaki, D.; et al. Geographic distribution of MERS coronavirus among dromedary camels, Africa. *Emerg. Infect. Dis.* **2014**, *20*, 1370–1374.

22. Alagaili, A.N.; Briese, T.; Mishra, N.; Kapoor, V.; Sameroff, S.C.; Burbelo, P.D.; de Wit, E.; Munster, V.J.; Hensley, L.E.; Zalmout, I.S.; et al. Middle East respiratory syndrome coronavirus infection in dromedary camels in Saudi Arabia. *MBio* **2014**, *5*, e00884-14. [CrossRef]

23. Raj, V.S.; Mou, H.; Smits, S.L.; Dekkers, D.H.; Muller, M.A.; Dijkman, R.; Muth, D.; Demmers, J.A.; Zaki, A.; Fouchier, R.A.; et al. Dipeptidyl peptidase 4 is a functional receptor for the emerging human coronavirus-EMC. *Nature* **2013**, *495*, 251–254.

24. Li, W.; Hulswit, R.J.G.; Widjaja, I.; Raj, V.S.; McBride, R.; Peng, W.; Widagdo, W.; Tortorici, M.A.; van Dieren, B.; Lang, Y.; et al. Identification of sialic acid-binding function for the Middle East respiratory syndrome coronavirus spike glycoprotein. *Proc. Natl. Acad. Sci. USA* **2017**, *114*, E8508–E8517. [CrossRef]

25. De Wilde, A.H.; Raj, V.S.; Oudshoorn, D.; Bestebroer, T.M.; van Nieuwkoop, S.; Limpens, R.W.; Posthuma, C.C.; van der Meer, Y.; Barcena, M.; Haagmans, B.L.; et al. MERS-coronavirus replication induces severe in vitro cytopathology and is strongly inhibited by cyclosporin A or interferon-alpha treatment. *J. Gen. Virol.* **2013**, *94 Pt 8*, 1749–1760. [CrossRef]

26. Iwata-Yoshikawa, N.; Okamura, T.; Shimizu, Y.; Hasegawa, H.; Takeda, M.; Nagata, N. TMPRSS2 contributes to virus spread and immunopathology in the airways of murine models after coronavirus infection. *J. Virol.* **2019**. [CrossRef]

27. Fehr, A.R.; Perlman, S. Coronaviruses: An overview of their replication and pathogenesis. *Methods. Mol. Biol.* **2015**, *1282*, 1–23. [PubMed]

28. Boonacker, E.; Van Noorden, C.J. The multifunctional or moonlighting protein CD26/DPPIV. *Eur. J. Cell. Biol.* **2003**, *82*, 53–73. [CrossRef] [PubMed]

29. Wang, N.; Shi, X.; Jiang, L.; Zhang, S.; Wang, D.; Tong, P.; Guo, D.; Fu, L.; Cui, Y.; Liu, X.; et al. Structure of MERS-CoV spike receptor-binding domain complexed with human receptor DPP4. *Cell Res.* **2013**, *23*, 986–993. [CrossRef]

30. Lu, G.; Hu, Y.; Wang, Q.; Qi, J.; Gao, F.; Li, Y.; Zhang, Y.; Zhang, W.; Yuan, Y.; Bao, J.; et al. Molecular basis of binding between novel human coronavirus MERS-CoV and its receptor CD26. *Nature* **2013**, *500*, 227–231. [CrossRef]

31. Bosch, B.J.; Raj, V.S.; Haagmans, B.L. Spiking the MERS-coronavirus receptor. *Cell Res.* **2013**, *23*, 1069–1070. [CrossRef] [PubMed]

32. De Wit, E.; Rasmussen, A.L.; Falzarano, D.; Bushmaker, T.; Feldmann, F.; Brining, D.L.; Fischer, E.R.; Martellaro, C.; Okumura, A.; Chang, J.; et al. Middle East respiratory syndrome coronavirus (MERS-CoV) causes transient lower respiratory tract infection in rhesus macaques. *Proc. Natl. Acad. Sci. USA* **2013**, *110*, 16598–16603. [CrossRef] [PubMed]

33. Haagmans, B.L.; van den Brand, J.M.; Provacia, L.B.; Raj, V.S.; Stittelaar, K.J.; Getu, S.; de Waal, L.; Bestebroer, T.M.; van Amerongen, G.; Verjans, G.M.; et al. Asymptomatic Middle East respiratory syndrome coronavirus infection in rabbits. *J. Virol.* **2015**, *89*, 6131–6135. [CrossRef]

34. Raj, V.S.; Smits, S.L.; Provacia, L.B.; van den Brand, J.M.; Wiersma, L.; Ouwendijk, W.J.; Bestebroer, T.M.; Spronken, M.I.; van Amerongen, G.; Rottier, P.J.; et al. Adenosine deaminase acts as a natural antagonist for dipeptidyl peptidase 4-mediated entry of the Middle East respiratory syndrome coronavirus. *J. Virol.* **2014**, *88*, 1834–1838. [CrossRef] [PubMed]

35. Coleman, C.M.; Matthews, K.L.; Goicochea, L.; Frieman, M.B. Wild-type and innate immune-deficient mice are not susceptible to the Middle East respiratory syndrome coronavirus. *J. Gen. Virol.* **2014**, *95 Pt 2*, 408–412. [CrossRef]

36. Iwata-Yoshikawa, N.; Fukushi, S.; Fukuma, A.; Suzuki, T.; Takeda, M.; Tashiro, M.; Hasegawa, H.; Nagata, N. Non Susceptibility of Neonatal and Adult Rats against the Middle East Respiratory Syndrome Coronavirus. *Jpn. J. Infect. Dis.* **2016**, *69*, 510–516. [CrossRef] [PubMed]

37. Chu, H.; Chan, C.M.; Zhang, X.; Wang, Y.; Yuan, S.; Zhou, J.; Au-Yeung, R.K.; Sze, K.H.; Yang, D.; Shuai, H.; et al. Middle East respiratory syndrome coronavirus and bat coronavirus HKU9 both can utilize GRP78 for attachment onto host cells. *J. Biol. Chem.* **2018**, *293*, 11709–11726. [CrossRef] [PubMed]

38. Chan, C.M.; Chu, H.; Wang, Y.; Wong, B.H.; Zhao, X.; Zhou, J.; Yang, D.; Leung, S.P.; Chan, J.F.; Yeung, M.L.; et al. Carcinoembryonic Antigen-Related Cell Adhesion Molecule 5 Is an Important Surface Attachment Factor That Facilitates Entry of Middle East Respiratory Syndrome Coronavirus. *J. Virol.* **2016**, *90*, 9114–9127. [CrossRef]

39. Zheng, Y.; Shang, J.; Yang, Y.; Liu, C.; Wan, Y.; Geng, Q.; Wang, M.; Baric, R.; Li, F. Lysosomal proteases are a determinant of coronavirus tropism. *J. Virol.* **2018**, *92*, e01504. [CrossRef]

40. Park, J.E.; Li, K.; Barlan, A.; Fehr, A.R.; Perlman, S.; McCray, P.B., Jr.; Gallagher, T. Proteolytic processing of Middle East respiratory syndrome coronavirus spikes expands virus tropism. *Proc. Natl. Acad. Sci. USA* **2016**, *113*, 12262–12267. [CrossRef]

41. Shirato, K.; Kawase, M.; Matsuyama, S. Middle East respiratory syndrome coronavirus infection mediated by the transmembrane serine protease TMPRSS2. *J. Virol.* **2013**, *87*, 12552–12561. [CrossRef] [PubMed]

42. Millet, J.K.; Whittaker, G.R. Host cell entry of Middle East respiratory syndrome coronavirus after two-step, furin-mediated activation of the spike protein. *Proc. Natl. Acad. Sci. USA* **2014**, *111*, 15214–15219. [CrossRef] [PubMed]

43. Earnest, J.T.; Hantak, M.P.; Li, K.; McCray, P.B., Jr.; Perlman, S.; Gallagher, T. The tetraspanin CD9 facilitates MERS-coronavirus entry by scaffolding host cell receptors and proteases. *PLoS Pathog.* **2017**, *13*, e1006546. [CrossRef] [PubMed]

44. Muller, C.; Hardt, M.; Schwudke, D.; Neuman, B.W.; Pleschka, S.; Ziebuhr, J. Inhibition of Cytosolic Phospholipase A2alpha Impairs an Early Step of Coronavirus Replication in Cell Culture. *J. Virol.* **2018**, *92*, e01463-17. [PubMed]

45. Belov, G.A.; van Kuppeveld, F.J. (+)RNA viruses rewire cellular pathways to build replication organelles. *Curr. Opin. Virol.* **2012**, *2*, 740–747. [CrossRef]

46. Belov, G.A.; Nair, V.; Hansen, B.T.; Hoyt, F.H.; Fischer, E.R.; Ehrenfeld, E. Complex dynamic development of poliovirus membranous replication complexes. *J. Virol.* **2012**, *86*, 302–312. [CrossRef]

47. Rabouw, H.H.; Langereis, M.A.; Knaap, R.C.; Dalebout, T.J.; Canton, J.; Sola, I.; Enjuanes, L.; Bredenbeek, P.J.; Kikkert, M.; de Groot, R.J.; et al. Middle East Respiratory Coronavirus Accessory Protein 4a Inhibits PKR-Mediated Antiviral Stress Responses. *PLoS Pathog.* **2016**, *12*, e1005982. [CrossRef]

48. Canton, J.; Fehr, A.R.; Fernandez-Delgado, R.; Gutierrez-Alvarez, F.J.; Sanchez-Aparicio, M.T.; Garcia-Sastre, A.; Perlman, S.; Enjuanes, L.; Sola, I. MERS-CoV 4b protein interferes with the NF-kappaB-dependent innate immune response during infection. *PLoS Pathog.* **2018**, *14*, e1006838. [CrossRef]

49. Siu, K.L.; Yeung, M.L.; Kok, K.H.; Yuen, K.S.; Kew, C.; Lui, P.Y.; Chan, C.P.; Tse, H.; Woo, P.C.; Yuen, K.Y.; et al. Middle east respiratory syndrome coronavirus 4a protein is a double-stranded RNA-binding protein that suppresses PACT-induced activation of RIG-I and MDA5 in the innate antiviral response. *J. Virol.* **2014**, *88*, 4866–4876. [CrossRef]

50. Nakagawa, K.; Narayanan, K.; Wada, M.; Popov, V.L.; Cajimat, M.; Baric, R.S.; Makino, S. The endonucleolytic RNA cleavage function of nsp1 of Middle East respiratory syndrome coronavirus promotes the production of infectious virus particles in specific human cell lines. *J. Virol.* **2018**, *92*, e01157-18. [CrossRef]

51. Knoops, K.; Kikkert, M.; Worm, S.H.; Zevenhoven-Dobbe, J.C.; van der Meer, Y.; Koster, A.J.; Mommaas, A.M.; Snijder, E.J. SARS-coronavirus replication is supported by a reticulovesicular network of modified endoplasmic reticulum. *PLoS Biol.* **2008**, *6*, e226. [CrossRef] [PubMed]

52. Thornbrough, J.M.; Jha, B.K.; Yount, B.; Goldstein, S.A.; Li, Y.; Elliott, R.; Sims, A.C.; Baric, R.S.; Silverman, R.H.; Weiss, S.R. Middle East Respiratory Syndrome Coronavirus NS4b Protein Inhibits Host RNase L Activation. *MBio* **2016**, *7*, e00258-16. [CrossRef] [PubMed]

53. Menachery, V.D.; Mitchell, H.D.; Cockrell, A.S.; Gralinski, L.E.; Yount, B.L., Jr.; Graham, R.L.; McAnarney, E.T.; Douglas, M.G.; Scobey, T.; Beall, A.; et al. MERS-CoV Accessory ORFs Play Key Role for Infection and Pathogenesis. *MBio* **2017**, *8*, e00665-17. [CrossRef] [PubMed]

54. Oudshoorn, D.; Rijs, K.; Limpens, R.; Groen, K.; Koster, A.J.; Snijder, E.J.; Kikkert, M.; Barcena, M. Expression and Cleavage of Middle East Respiratory Syndrome Coronavirus nsp3-4 Polyprotein Induce the Formation of Double-Membrane Vesicles That Mimic Those Associated with Coronaviral RNA Replication. *MBio* **2017**, *8*, e01658-17. [CrossRef]

55. Li, K.; Wohlford-Lenane, C.L.; Channappanavar, R.; Park, J.E.; Earnest, J.T.; Bair, T.B.; Bates, A.M.; Brogden, K.A.; Flaherty, H.A.; Gallagher, T.; et al. Mouse-adapted MERS coronavirus causes lethal lung disease in human DPP4 knockin mice. *Proc. Natl. Acad. Sci. USA* **2017**, *114*, E3119–E3128. [CrossRef]

56. Cockrell, A.S.; Yount, B.L.; Scobey, T.; Jensen, K.; Douglas, M.; Beall, A.; Tang, X.C.; Marasco, W.A.; Heise, M.T.; Baric, R.S. A mouse model for MERS coronavirus-induced acute respiratory distress syndrome. *Nat. Microbiol.* **2016**, *2*, 16226. [CrossRef]

57. Zhao, J.; Li, K.; Wohlford-Lenane, C.; Agnihothram, S.S.; Fett, C.; Zhao, J.; Gale, M.J., Jr.; Baric, R.S.; Enjuanes, L.; Gallagher, T.; et al. Rapid generation of a mouse model for Middle East respiratory syndrome. *Proc. Natl. Acad. Sci. USA* **2014**, *111*, 4970–4975. [CrossRef]

58. Yao, Y.; Bao, L.; Deng, W.; Xu, L.; Li, F.; Lv, Q.; Yu, P.; Chen, T.; Xu, Y.; Zhu, H.; et al. An animal model of MERS produced by infection of rhesus macaques with MERS coronavirus. *J. Infect. Dis.* **2014**, *209*, 236–242. [CrossRef] [PubMed]

59. Falzarano, D.; de Wit, E.; Martellaro, C.; Callison, J.; Munster, V.J.; Feldmann, H. Inhibition of novel beta coronavirus replication by a combination of interferon-alpha2b and ribavirin. *Sci. Rep.* **2013**, *3*, 1686. [CrossRef]

60. Raj, V.S.; Farag, E.A.; Reusken, C.B.; Lamers, M.M.; Pas, S.D.; Voermans, J.; Smits, S.L.; Osterhaus, A.D.; Al-Mawlawi, N.; Al-Romaihi, H.E.; et al. Isolation of MERS coronavirus from a dromedary camel, Qatar, 2014. *Emerg. Infect. Dis.* **2014**, *20*, 1339–1342. [CrossRef]

61. Cho, S.Y.; Kang, J.M.; Ha, Y.E.; Park, G.E.; Lee, J.Y.; Ko, J.H.; Lee, J.Y.; Kim, J.M.; Kang, C.I.; Jo, I.J.; et al. MERS-CoV outbreak following a single patient exposure in an emergency room in South Korea: An epidemiological outbreak study. *Lancet* **2016**, *388*, 994–1001. [CrossRef]

62. Chan, R.W.; Chan, M.C.; Agnihothram, S.; Chan, L.L.; Kuok, D.I.; Fong, J.H.; Guan, Y.; Poon, L.L.; Baric, R.S.; Nicholls, J.M.; et al. Tropism of and innate immune responses to the novel human betacoronavirus lineage C virus in human ex vivo respiratory organ cultures. *J. Virol.* **2013**, *87*, 6604–6614. [CrossRef] [PubMed]

63. Chan, R.W.; Hemida, M.G.; Kayali, G.; Chu, D.K.; Poon, L.L.; Alnaeem, A.; Ali, M.A.; Tao, K.P.; Ng, H.Y.; Chan, M.C.; et al. Tropism and replication of Middle East respiratory syndrome coronavirus from dromedary camels in the human respiratory tract: An in-vitro and ex-vivo study. *Lancet Respir. Med.* **2014**, *2*, 813–822. [CrossRef]

64. Ng, D.L.; Al Hosani, F.; Keating, M.K.; Gerber, S.I.; Jones, T.L.; Metcalfe, M.G.; Tong, S.; Tao, Y.; Alami, N.N.; Haynes, L.M.; et al. Clinicopathologic, Immunohistochemical, and Ultrastructural Findings of a Fatal Case of Middle East Respiratory Syndrome Coronavirus Infection in United Arab Emirates, April 2014. *Am. J. Pathol.* **2016**, *186*, 652–658. [CrossRef]

65. Hocke, A.C.; Becher, A.; Knepper, J.; Peter, A.; Holland, G.; Tonnies, M.; Bauer, T.T.; Schneider, P.; Neudecker, J.; Muth, D.; et al. Emerging human middle East respiratory syndrome coronavirus causes widespread infection and alveolar damage in human lungs. *Am. J. Respir. Crit. Care Med.* **2013**, *188*, 882–886. [CrossRef]

66. Bermingham, A.; Chand, M.A.; Brown, C.S.; Aarons, E.; Tong, C.; Langrish, C.; Hoschler, K.; Brown, K.; Galiano, M.; Myers, R.; et al. Severe respiratory illness caused by a novel coronavirus, in a patient transferred to the United Kingdom from the Middle East, September 2012. *Eurosurveillance* **2012**, *17*, 20290.

67. Corman, V.M.; Albarrak, A.M.; Omrani, A.S.; Albarrak, M.M.; Farah, M.E.; Almasri, M.; Muth, D.; Sieberg, A.; Meyer, B.; Assiri, A.M.; et al. Viral Shedding and Antibody Response in 37 Patients With Middle East Respiratory Syndrome Coronavirus Infection. *Clin. Infect. Dis.* **2016**, *62*, 477–483. [CrossRef]

68. Drosten, C.; Seilmaier, M.; Corman, V.M.; Hartmann, W.; Scheible, G.; Sack, S.; Guggemos, W.; Kallies, R.; Muth, D.; Junglen, S.; et al. Clinical features and virological analysis of a case of Middle East respiratory syndrome coronavirus infection. *Lancet Infect. Dis.* **2013**, *13*, 745–751. [CrossRef]

69. Widagdo, W.; Raj, V.S.; Schipper, D.; Kolijn, K.; van Leenders, G.J.; Bosch, B.J.; Bensaid, A.; Segales, J.; Baumgartner, W.; Osterhaus, A.D.; et al. Differential Expression of the Middle East Respiratory Syndrome Coronavirus Receptor in the Upper Respiratory Tracts of Humans and Dromedary Camels. *J. Virol.* **2016**, *90*, 4838–4842. [CrossRef]

70. Meyerholz, D.K.; Lambertz, A.M.; McCray, P.B., Jr. Dipeptidyl Peptidase 4 Distribution in the Human Respiratory Tract: Implications for the Middle East Respiratory Syndrome. *Am. J. Pathol.* **2016**, *186*, 78–86. [CrossRef] [PubMed]

71. Widagdo, W.; Begeman, L.; Schipper, D.; Run, P.R.V.; Cunningham, A.A.; Kley, N.; Reusken, C.B.; Haagmans, B.L.; van den Brand, J.M.A. Tissue Distribution of the MERS-Coronavirus Receptor in Bats. *Sci. Rep.* **2017**, *7*, 1193. [CrossRef]

72. Munster, V.J.; Adney, D.R.; van Doremalen, N.; Brown, V.R.; Miazgowicz, K.L.; Milne-Price, S.; Bushmaker, T.; Rosenke, R.; Scott, D.; Hawkinson, A.; et al. Replication and shedding of MERS-CoV in Jamaican fruit bats (*Artibeus jamaicensis*). *Sci. Rep.* **2016**, *6*, 21878. [CrossRef] [PubMed]

73. Annan, A.; Baldwin, H.J.; Corman, V.M.; Klose, S.M.; Owusu, M.; Nkrumah, E.E.; Badu, E.K.; Anti, P.; Agbenyega, O.; Meyer, B.; et al. Human betacoronavirus 2c EMC/2012-related viruses in bats, Ghana and Europe. *Emerg. Infect. Dis.* **2013**, *19*, 456–459. [CrossRef]

74. Yang, L.; Wu, Z.; Ren, X.; Yang, F.; Zhang, J.; He, G.; Dong, J.; Sun, L.; Zhu, Y.; Zhang, S.; et al. MERS-related betacoronavirus in Vespertilio superans bats, China. *Emerg. Infect. Dis.* **2014**, *20*, 1260–1262. [CrossRef]

75. Wacharapluesadee, S.; Sintunawa, C.; Kaewpom, T.; Khongnomnan, K.; Olival, K.J.; Epstein, J.H.; Rodpan, A.; Sangsri, P.; Intarut, N.; Chindamporn, A.; et al. Group C betacoronavirus in bat guano fertilizer, Thailand. *Emerg. Infect. Dis.* **2013**, *19*, 1349–1351. [CrossRef]

76. Kim, H.K.; Yoon, S.W.; Kim, D.J.; Koo, B.S.; Noh, J.Y.; Kim, J.H.; Choi, Y.G.; Na, W.; Chang, K.T.; Song, D.; et al. Detection of Severe Acute Respiratory Syndrome-Like, Middle East Respiratory Syndrome-Like Bat Coronaviruses and Group H Rotavirus in Faeces of Korean Bats. *Transbound. Emerg. Dis.* **2016**, *63*, 365–372. [CrossRef] [PubMed]

77. Corman, V.M.; Ithete, N.L.; Richards, L.R.; Schoeman, M.C.; Preiser, W.; Drosten, C.; Drexler, J.F. Rooting the phylogenetic tree of middle East respiratory syndrome coronavirus by characterization of a conspecific virus from an African bat. *J. Virol.* **2014**, *88*, 11297–11303. [CrossRef] [PubMed]

78. Ithete, N.L.; Stoffberg, S.; Corman, V.M.; Cottontail, V.M.; Richards, L.R.; Schoeman, M.C.; Drosten, C.; Drexler, J.F.; Preiser, W. Close relative of human Middle East respiratory syndrome coronavirus in bat, South Africa. *Emerg. Infect. Dis.* **2013**, *19*, 1697–1699. [CrossRef] [PubMed]

79. Anthony, S.J.; Gilardi, K.; Menachery, V.D.; Goldstein, T.; Ssebide, B.; Mbabazi, R.; Navarrete-Macias, I.; Liang, E.; Wells, H.; Hicks, A.; et al. Further Evidence for Bats as the Evolutionary Source of Middle East Respiratory Syndrome Coronavirus. *MBio* **2017**, *8*, e00373-17. [CrossRef]

80. Luo, C.M.; Wang, N.; Yang, X.L.; Liu, H.Z.; Zhang, W.; Li, B.; Hu, B.; Peng, C.; Geng, Q.B.; Zhu, G.J.; et al. Discovery of Novel Bat Coronaviruses in South China That Use the Same Receptor as Middle East Respiratory Syndrome Coronavirus. *J. Virol.* **2018**, *92*, e00116-18. [CrossRef]

81. Barlan, A.; Zhao, J.; Sarkar, M.K.; Li, K.; McCray, P.B., Jr.; Perlman, S.; Gallagher, T. Receptor variation and susceptibility to Middle East respiratory syndrome coronavirus infection. *J. Virol.* **2014**, *88*, 4953–4961. [CrossRef] [PubMed]

82. Van Doremalen, N.; Miazgowicz, K.L.; Milne-Price, S.; Bushmaker, T.; Robertson, S.; Scott, D.; Kinne, J.; McLellan, J.S.; Zhu, J.; Munster, V.J. Host species restriction of Middle East respiratory syndrome coronavirus through its receptor, dipeptidyl peptidase 4. *J. Virol.* **2014**, *88*, 9220–9232. [CrossRef] [PubMed]

83. Widagdo, W.; Okba, N.M.A.; Richard, M.; de Muelder, D.; Bestebroer, T.M.; Lexmond, P.; J.M.A. v. d., B.; Haagmans, B.L.; Herfst, S. Middle East respiratory syndrome coronavirus transmission in rabbits. Unpublished work. 2019.

84. Vergara-Alert, J.; van den Brand, J.M.; Widagdo, W.; Munoz, M.t.; Raj, S.; Schipper, D.; Solanes, D.; Cordon, I.; Bensaid, A.; Haagmans, B.L.; et al. Livestock Susceptibility to Infection with Middle East Respiratory Syndrome Coronavirus. *Emerg. Infect. Dis.* **2017**, *23*, 232–240. [CrossRef]

85. De Wit, E.; Feldmann, F.; Horne, E.; Martellaro, C.; Haddock, E.; Bushmaker, T.; Rosenke, K.; Okumura, A.; Rosenke, R.; Saturday, G.; et al. Domestic Pig Unlikely Reservoir for MERS-CoV. *Emerg. Infect. Dis.* **2017**, *23*, 985–988. [CrossRef]

86. Adney, D.R.; Bielefeldt-Ohmann, H.; Hartwig, A.E.; Bowen, R.A. Infection, Replication, and Transmission of Middle East Respiratory Syndrome Coronavirus in Alpacas. *Emerg. Infect. Dis.* **2016**, *22*, 1031–1037. [CrossRef]

87. Crameri, G.; Durr, P.A.; Klein, R.; Foord, A.; Yu, M.; Riddell, S.; Haining, J.; Johnson, D.; Hemida, M.G.; Barr, J.; et al. Experimental Infection and Response to Rechallenge of Alpacas with Middle East Respiratory Syndrome Coronavirus. *Emerg. Infect. Dis.* **2016**, *22*, 1071–1074. [CrossRef]

88. Adney, D.R.; Brown, V.R.; Porter, S.M.; Bielefeldt-Ohmann, H.; Hartwig, A.E.; Bowen, R.A. Inoculation of Goats, Sheep, and Horses with MERS-CoV Does Not Result in Productive Viral Shedding. *Viruses* **2016**, *8*, 230. [CrossRef] [PubMed]

89. Harris, S.R.; Robinson, C.; Steward, K.F.; Webb, K.S.; Paillot, R.; Parkhill, J.; Holden, M.T.; Waller, A.S. Genome specialization and decay of the strangles pathogen, Streptococcus equi, is driven by persistent infection. *Genome Res.* **2015**, *25*, 1360–1371. [CrossRef]

90. Waller, A.S. Strangles: A pathogenic legacy of the war horse. *Vet. Rec.* **2016**, *178*, 91–92. [CrossRef] [PubMed]

91. Reusken, C.B.; Haagmans, B.L.; Muller, M.A.; Gutierrez, C.; Godeke, G.J.; Meyer, B.; Muth, D.; Raj, V.S.; Smits-De Vries, L.; Corman, V.M.; et al. Middle East respiratory syndrome coronavirus neutralising serum antibodies in dromedary camels: A comparative serological study. *Lancet Infect. Dis.* **2013**, *13*, 859–866. [CrossRef]

92. Hemida, M.G.; Perera, R.A.; Wang, P.; Alhammadi, M.A.; Siu, L.Y.; Li, M.; Poon, L.L.; Saif, L.; Alnaeem, A.; Peiris, M. Middle East Respiratory Syndrome (MERS) coronavirus seroprevalence in domestic livestock in Saudi Arabia, 2010 to 2013. *Eurosurveillance* **2013**, *18*, 20659. [CrossRef]

93. Van Doremalen, N.; Hijazeen, Z.S.; Holloway, P.; Al Omari, B.; McDowell, C.; Adney, D.; Talafha, H.A.; Guitian, J.; Steel, J.; Amarin, N.; et al. High Prevalence of Middle East Respiratory Coronavirus in Young Dromedary Camels in Jordan. *Vector Borne Zoonotic Dis.* **2017**, *17*, 155–159. [CrossRef] [PubMed]

94. Meyer, B.; Garcia-Bocangra, I.; Wernery, U.; Wernery, R.; Sieberg, A.; Muller, M.A.; Drexler, J.F.; Drosten, C.; Eckerle, I. Serologic assessment of possibility for MERS-CoV infection in equids. *Emerg. Infect. Dis.* **2015**, *21*, 181–182. [CrossRef]

95. Hemida, M.G.; Chu, D.K.W.; Perera, R.; Ko, R.L.W.; So, R.T.Y.; Ng, B.C.Y.; Chan, S.M.S.; Chu, S.; Alnaeem, A.A.; Alhammadi, M.A.; et al. Coronavirus infections in horses in Saudi Arabia and Oman. *Transbound. Emerg. Dis.* **2017**, *64*, 2093–2103. [CrossRef]

96. Ali, M.; El-Shesheny, R.; Kandeil, A.; Shehata, M.; Elsokary, B.; Gomaa, M.; Hassan, N.; El Sayed, A.; El-Taweel, A.; Sobhy, H.; et al. Cross-sectional surveillance of Middle East respiratory syndrome coronavirus (MERS-CoV) in dromedary camels and other mammals in Egypt, August 2015 to January 2016. *Eurosurveillance* **2017**, *22*, 30487. [CrossRef]

97. David, D.; Rotenberg, D.; Khinich, E.; Erster, O.; Bardenstein, S.; van Straten, M.; Okba, N.M.A.; Raj, S.V.; Haagmans, B.L.; Miculitzki, M.; et al. Middle East respiratory syndrome coronavirus specific antibodies in naturally exposed Israeli llamas, alpacas and camels. *One Health* **2018**, *5*, 65–68. [CrossRef]

98. Reusken, C.B.; Schilp, C.; Raj, V.S.; De Bruin, E.; Kohl, R.H.; Farag, E.A.; Haagmans, B.L.; Al-Romaihi, H.; Le Grange, F.; Bosch, B.J.; et al. MERS-CoV Infection of Alpaca in a Region Where MERS-CoV is Endemic. *Emerg. Infect. Dis.* **2016**, *22*, 1129. [CrossRef] [PubMed]

99. Vergara-Alert, J.; Raj, V.S.; Munoz, M.; Abad, F.X.; Cordon, I.; Haagmans, B.L.; Bensaid, A.; Segales, J. Middle East respiratory syndrome coronavirus experimental transmission using a pig model. *Transbound. Emerg. Dis.* **2017**, *64*, 1342–1345. [CrossRef]

100. Widagdo, W.; Okba, N.M.A.; Li, W.; de Jong, A.; de Swart, R.; Begeman, L.; Cunningham, A.A.; van Riel, D.; van den Brand, J.M.A.; Segales, J.; et al. Species specific binding of the MERS-coronavirus S1^A protein. Unpublished work. 2019.

101. Yang, Y.; Du, L.; Liu, C.; Wang, L.; Ma, C.; Tang, J.; Baric, R.S.; Jiang, S.; Li, F. Receptor usage and cell entry of bat coronavirus HKU4 provide insight into bat-to-human transmission of MERS coronavirus. *Proc. Natl. Acad. Sci. USA* **2014**, *111*, 12516–12521. [CrossRef] [PubMed]

102. Nam, H.S.; Park, J.W.; Ki, M.; Yeon, M.Y.; Kim, J.; Kim, S.W. High fatality rates and associated factors in two hospital outbreaks of MERS in Daejeon, the Republic of Korea. *Int. J. Infect. Dis.* **2017**, *58*, 37–42. [CrossRef]

103. Alraddadi, B.M.; Watson, J.T.; Almarashi, A.; Abedi, G.R.; Turkistani, A.; Sadran, M.; Housa, A.; Almazroa, M.A.; Alraihan, N.; Banjar, A.; et al. Risk Factors for Primary Middle East Respiratory Syndrome Coronavirus Illness in Humans, Saudi Arabia, 2014. *Emerg. Infect. Dis.* **2016**, *22*, 49–55. [CrossRef]

104. Seys, L.J.M.; Widagdo, W.; Verhamme, F.M.; Kleinjan, A.; Janssens, W.; Joos, G.F.; Bracke, K.R.; Haagmans, B.L.; Brusselle, G.G. DPP4, the Middle East Respiratory Syndrome Coronavirus Receptor, is Upregulated in Lungs of Smokers and Chronic Obstructive Pulmonary Disease Patients. *Clin. Infect. Dis.* **2018**, *66*, 45–53. [CrossRef]

105. Dahlin, K.; Mager, E.M.; Allen, L.; Tigue, Z.; Goodglick, L.; Wadehra, M.; Dobbs, L. Identification of genes differentially expressed in rat alveolar type I cells. *Am. J. Respir. Cell. Mol. Biol.* **2004**, *31*, 309–316. [CrossRef] [PubMed]

106. Evans, M.J.; Cabral, L.J.; Stephens, R.J.; Freeman, G. Renewal of alveolar epithelium in the rat following exposure to NO$_2$. *Am. J. Pathol.* **1973**, *70*, 175–198. [PubMed]

107. Alsaad, K.O.; Hajeer, A.H.; Al Balwi, M.; Al Moaiqel, M.; Al Oudah, N.; Al Ajlan, A.; AlJohani, S.; Alsolamy, S.; Gmati, G.E.; Balkhy, H.; et al. Histopathology of Middle East respiratory syndrome coronovirus (MERS-CoV) infection-clinicopathological and ultrastructural study. *Histopathology* **2018**, *72*, 516–524. [CrossRef]

108. Haagmans, B.L.; Kuiken, T.; Martina, B.E.; Fouchier, R.A.; Rimmelzwaan, G.F.; van Amerongen, G.; van Riel, D.; de Jong, T.; Itamura, S.; Chan, K.H.; et al. Pegylated interferon-alpha protects type 1 pneumocytes against SARS coronavirus infection in macaques. *Nat. Med.* **2004**, *10*, 290–293. [CrossRef]

109. Chen, Z.; Bao, L.; Chen, C.; Zou, T.; Xue, Y.; Li, F.; Lv, Q.; Gu, S.; Gao, X.; Cui, S.; et al. Human Neutralizing Monoclonal Antibody Inhibition of Middle East Respiratory Syndrome Coronavirus Replication in the Common Marmoset. *J. Infect. Dis.* **2017**, *215*, 1807–1815. [CrossRef]

110. Chan, J.F.; Yao, Y.; Yeung, M.L.; Deng, W.; Bao, L.; Jia, L.; Li, F.; Xiao, C.; Gao, H.; Yu, P.; et al. Treatment With Lopinavir/Ritonavir or Interferon-beta1b Improves Outcome of MERS-CoV Infection in a Nonhuman Primate Model of Common Marmoset. *J. Infect. Dis.* **2015**, *212*, 1904–1913. [CrossRef]

111. Falzarano, D.; de Wit, E.; Feldmann, F.; Rasmussen, A.L.; Okumura, A.; Peng, X.; Thomas, M.J.; van Doremalen, N.; Haddock, E.; Nagy, L.; et al. Infection with MERS-CoV causes lethal pneumonia in the common marmoset. *PLoS. Pathog.* **2014**, *10*, e1004250. [CrossRef]

112. Widagdo, W.; Wiersma, L.C.M.; Smits, S.L.; de Vries, R.D.; Schipper, D.; Raj, V.S.; van den Ham, H.J.; Brown, R.; Zambon, M.; Kondova, I.; et al. DPP4-expressing type I pneumocytes in a fatal human MERS-coronavirus case. Unpublished work. 2019.

113. Li, K.; Wohlford-Lenane, C.; Perlman, S.; Zhao, J.; Jewell, A.K.; Reznikov, L.R.; Gibson-Corley, K.N.; Meyerholz, D.K.; McCray, P.B., Jr. Middle East Respiratory Syndrome Coronavirus Causes Multiple Organ Damage and Lethal Disease in Mice Transgenic for Human Dipeptidyl Peptidase 4. *J. Infect. Dis.* **2016**, *213*, 712–722. [CrossRef]

114. Algaissi, A.; Agrawal, A.S.; Han, S.; Peng, B.H.; Luo, C.; Li, F.; Chan, T.S.; Couch, R.B.; Tseng, C.K. Elevated Human Dipeptidyl Peptidase 4 Expression Reduces the Susceptibility of hDPP4 Transgenic Mice to Middle East Respiratory Syndrome Coronavirus Infection and Disease. *J. Infect. Dis.* **2018**, *219*, 829–835. [CrossRef] [PubMed]

115. Prescott, J.; Falzarano, D.; de Wit, E.; Hardcastle, K.; Feldmann, F.; Haddock, E.; Scott, D.; Feldmann, H.; Munster, V.J. Pathogenicity and Viral Shedding of MERS-CoV in Immunocompromised Rhesus Macaques. *Front. Immunol.* **2018**, *9*, 205. [CrossRef]

116. Kim, S.H.; Ko, J.H.; Park, G.E.; Cho, S.Y.; Ha, Y.E.; Kang, J.M.; Kim, Y.J.; Huh, H.J.; Ki, C.S.; Jeong, B.H.; et al. Atypical presentations of MERS-CoV infection in immunocompromised hosts. *J. Infect. Chemother.* **2017**, *23*, 769–773. [CrossRef]

117. Zhao, J.; Alshukairi, A.N.; Baharoon, S.A.; Ahmed, W.A.; Bokhari, A.A.; Nehdi, A.M.; Layqah, L.A.; Alghamdi, M.G.; Al Gethamy, M.M.; Dada, A.M.; et al. Recovery from the Middle East respiratory syndrome is associated with antibody and T-cell responses. *Sci. Immunol.* **2017**, *2*, eaan5393. [CrossRef] [PubMed]

118. Coleman, C.M.; Sisk, J.M.; Halasz, G.; Zhong, J.; Beck, S.E.; Matthews, K.L.; Venkataraman, T.; Rajagopalan, S.; Kyratsous, C.A.; Frieman, M.B. CD8+ T Cells and Macrophages Regulate Pathogenesis in a Mouse Model of Middle East Respiratory Syndrome. *J. Virol.* **2017**, *91*, e01825-16. [CrossRef]

119. Hart, B.J.; Dyall, J.; Postnikova, E.; Zhou, H.; Kindrachuk, J.; Johnson, R.F.; Olinger, G.G., Jr.; Frieman, M.B.; Holbrook, M.R.; Jahrling, P.B.; et al. Interferon-beta and mycophenolic acid are potent inhibitors of Middle East respiratory syndrome coronavirus in cell-based assays. *J. Gen. Virol.* **2014**, *95 Pt 3*, 571–577. [CrossRef]

120. Oudshoorn, D.; van der Hoeven, B.; Limpens, R.W.; Beugeling, C.; Snijder, E.J.; Barcena, M.; Kikkert, M. Antiviral Innate Immune Response Interferes with the Formation of Replication-Associated Membrane Structures Induced by a Positive-Strand RNA Virus. *MBio* **2016**, *7*, e01991-16. [CrossRef]

121. Uno, K.; Yagi, K.; Yoshimori, M.; Tanigawa, M.; Yoshikawa, T.; Fujita, S. IFN production ability and healthy ageing: Mixed model analysis of a 24 year longitudinal study in Japan. *BMJ Open* **2013**, *3*, e002113. [CrossRef]

122. Li, G.; Ju, J.; Weyand, C.M.; Goronzy, J.J. Age-Associated Failure To Adjust Type I IFN Receptor Signaling Thresholds after T Cell Activation. *J. Immunol.* **2015**, *195*, 865–874. [CrossRef] [PubMed]

123. Channappanavar, R.; Fehr, A.R.; Vijay, R.; Mack, M.; Zhao, J.; Meyerholz, D.K.; Perlman, S. Dysregulated Type I Interferon and Inflammatory Monocyte-Macrophage Responses Cause Lethal Pneumonia in SARS-CoV-Infected Mice. *Cell Host Microbe* **2016**, *19*, 181–193. [CrossRef]

124. Zumla, A.; Hui, D.S.; Perlman, S. Middle East respiratory syndrome. *Lancet* **2015**, *386*, 995–1007. [CrossRef]

125. Alfaraj, S.H.; Al-Tawfiq, J.A.; Alzahrani, N.A.; Altwaijri, T.A.; Memish, Z.A. The impact of co-infection of influenza A virus on the severity of Middle East Respiratory Syndrome Coronavirus. *J. Infect.* **2017**, *74*, 521–523. [CrossRef]

126. Chu, D.K.W.; Hui, K.P.Y.; Perera, R.; Miguel, E.; Niemeyer, D.; Zhao, J.; Channappanavar, R.; Dudas, G.; Oladipo, J.O.; Traore, A.; et al. MERS coronaviruses from camels in Africa exhibit region-dependent genetic diversity. *Proc. Natl. Acad. Sci. USA* **2018**, *115*, 3144–3149. [CrossRef]

127. Assiri, A.M.; Biggs, H.M.; Abedi, G.R.; Lu, X.; Bin Saeed, A.; Abdalla, O.; Mohammed, M.; Al-Abdely, H.M.; Algarni, H.S.; Alhakeem, R.F.; et al. Increase in Middle East Respiratory Syndrome-Coronavirus Cases in Saudi Arabia Linked to Hospital Outbreak With Continued Circulation of Recombinant Virus, July 1-August 31, 2015. *Open Forum Infect. Dis.* **2016**, *3*, ofw165. [CrossRef] [PubMed]

128. Payne, D.C.; Biggs, H.M.; Al-Abdallat, M.M.; Alqasrawi, S.; Lu, X.; Abedi, G.R.; Haddadin, A.; Iblan, I.; Alsanouri, T.; Al Nsour, M.; et al. Multihospital Outbreak of a Middle East Respiratory Syndrome Coronavirus Deletion Variant, Jordan: A Molecular, Serologic, and Epidemiologic Investigation. *Open Forum Infect. Dis.* **2018**, *5*, ofy095. [CrossRef] [PubMed]

129. Lamers, M.M.; Raj, V.S.; Shafei, M.; Ali, S.S.; Abdallh, S.M.; Gazo, M.; Nofal, S.; Lu, X.; Erdman, D.D.; Koopmans, M.P.; et al. Deletion Variants of Middle East Respiratory Syndrome Coronavirus from Humans, Jordan, 2015. *Emerg. Infect. Dis.* **2016**, *22*, 716–719. [CrossRef] [PubMed]

viruses

MDPI

Article

Characterization of the Lipidomic Profile of Human Coronavirus-Infected Cells: Implications for Lipid Metabolism Remodeling upon Coronavirus Replication

Bingpeng Yan [1,2,†]**, Hin Chu** [1,2,†]**, Dong Yang** [2,†]**, Kong-Hung Sze** [1,2,†]**, Pok-Man Lai** [2]**,**
Shuofeng Yuan [1,2]**, Huiping Shuai** [2]**, Yixin Wang** [2]**, Richard Yi-Tsun Kao** [1,2]**,**
Jasper Fuk-Woo Chan [1,2,3,4,5,*]**and Kwok-Yung Yuen** [1,2,3,4,5,6,*]

[1] State Key Laboratory of Emerging Infectious Diseases, The University of Hong Kong, Pokfulam,
    Hong Kong Special Administrative Region, China; ybp1205@hku.hk (B.Y.); hinchu@hku.hk (H.C.);
    khsze@hku.hk (K.-H.S.); yuansf@hku.hk (S.Y.); rytkao@hku.hk (R.Y.-T.K.)
[2] Department of Microbiology, Li Ka Shing Faculty of Medicine, The University of Hong Kong, Pokfulam,
    Hong Kong Special Administrative Region, China; u3005140@connect.hku.hk (D.Y.);
    vangor@hku.hk (P.-M.L.); shuaihp@connect.hku.hk (H.S.); jasyx@connect.hku.hk (Y.W.)
[3] Carol Yu Centre for Infection, Li Ka Shing Faculty of Medicine, The University of Hong Kong, Pokfulam,
    Hong Kong Special Administrative Region, China
[4] Hainan-Medical University-The University of Hong Kong Joint Laboratory of Tropical Infectious Diseases,
    Hainan Medical University, Haikou 96708, China
[5] Hainan-Medical University-The University of Hong Kong Joint Laboratory of Tropical Infectious Diseases,
    The University of Hong Kong, Pokfulam, Hong Kong Special Administrative Region, China
[6] The Collaborative Innovation Center for Diagnosis and Treatment of Infectious Diseases, The University of
    Hong Kong, Pokfulam, Hong Kong Special Administrative Region, China
* Correspondence: jfwchan@hku.hk (J.F.-W.C.); kyyuen@hku.hk (K.-Y.Y.);
    Tel.: +852-2255-2413 (J.F.-W.C. & K.-Y.Y.); Fax: +852-2855-1241 (J.F.-W.C. & K.-Y.Y.)
† These authors contributed equally to this work.

Received: 14 December 2018; Accepted: 15 January 2019; Published: 16 January 2019

Abstract: Lipids play numerous indispensable cellular functions and are involved in multiple steps in the replication cycle of viruses. Infections by human-pathogenic coronaviruses result in diverse clinical outcomes, ranging from self-limiting flu-like symptoms to severe pneumonia with extrapulmonary manifestations. Understanding how cellular lipids may modulate the pathogenicity of human-pathogenic coronaviruses remains poor. To this end, we utilized the human coronavirus 229E (HCoV-229E) as a model coronavirus to comprehensively characterize the host cell lipid response upon coronavirus infection with an ultra-high performance liquid chromatography-mass spectrometry (UPLC–MS)-based lipidomics approach. Our results revealed that glycerophospholipids and fatty acids (FAs) were significantly elevated in the HCoV-229E-infected cells and the linoleic acid (LA) to arachidonic acid (AA) metabolism axis was markedly perturbed upon HCoV-229E infection. Interestingly, exogenous supplement of LA or AA in HCoV-229E-infected cells significantly suppressed HCoV-229E virus replication. Importantly, the inhibitory effect of LA and AA on virus replication was also conserved for the highly pathogenic Middle East respiratory syndrome coronavirus (MERS-CoV). Taken together, our study demonstrated that host lipid metabolic remodeling was significantly associated with human-pathogenic coronavirus propagation. Our data further suggested that lipid metabolism regulation would be a common and druggable target for coronavirus infections.

Keywords: lipidomics; UHPLC–MS; HCoV-229E; MERS-CoV

1. Introduction

Coronaviruses are enveloped viruses with a large single-strand, positive-sense RNA genome [1,2]. As of today, there are a total of six coronaviruses that are known to infect humans, including human coronavirus OC43 (HCoV-OC43), human coronavirus 229E (HCoV-229E), severe acute respiratory syndrome coronavirus (SARS-CoV), human coronavirus HKU1 (HCoV-HKU1), human coronavirus NL63 (HCoV-NL63), and the Middle East respiratory syndrome coronavirus (MERS-CoV) [3]. These human-pathogenic coronaviruses cause a broad range of clinical manifestations. HCoV-OC43, HCoV-229E, HCoV-HKU1, and HCoV-NL63 cause mild, self-limiting upper respiratory tract infections. In contrast, SARS-CoV and the recently emerged MERS-CoV may cause severe pneumonia with acute respiratory distress syndrome, multi-organ failure, and death in both immunocompetent and immunocompromised hosts [4–7].

Lipids play crucial roles at various stages in the virus life cycle. First, lipids can serve as the direct receptors or entry co-factors for enveloped and non-enveloped viruses at the cell surface or the endosomes [8,9]. Second, lipids and lipid synthesis play important roles in the formation and function of the viral replication complex [10,11]. Third, lipid metabolism can generate the required energy for efficient viral replication [12]. Moreover, lipids can dictate the proper cellular distribution of viral proteins, as well as the trafficking, assembly, and release of virus particles [13,14]. In this regard, the host lipid biogenesis pathways play indispensable roles in modulating virus propagation.

As in other viruses, lipids play key roles in the life cycle of coronaviruses. Coronaviruses confiscate intracellular membranes of the host cells to generate new compartments known as double membrane vesicles (DMVs) for the amplification of the viral genome. DMVs are membranous structures that not only harbor viral proteins but also contain a specific array of hijacked host factors, which collectively orchestrate a unique lipid micro-environment optimal for coronavirus replication [15]. A recent study indicated that a key lipid processing enzyme, cytosolic phospholipase A2α enzyme (cPLA2α) that belongs to the phospholipase A2 (PLA2) superfamily, was closely associated with DMVs' formation and coronaviruses' replication [16]. The viral protein and RNA accumulation, as well as the production of infectious virus progeny, were significantly diminished in the presence of cPLA2α inhibitor [16]. At the same time, phospholipase A2 group IID (PLA2G2D), an enzyme that predominantly contributes to anti-inflammatory/pro-resolving lipid mediator expression, contributed to worsened outcomes in mice infected with SARS-CoV by modulating the immune response [17]. However, to date, the change and modulating effects of the specific lipids involved in lipid rearrangement upon coronavirus infection remains largely unexplored.

To obtain a comprehensive and unbiased profile of perturbed lipids upon coronavirus infection, we performed mass spectrometry (MS)-based lipidomics profiling on coronavirus-infected cells using HCoV-229E as a model virus. Specific lipids including glycerophospholipids and fatty acids (FAs) upon virus infection were identified, which represented the lipid species that were rearranged by HCoV-229E infection. Further pathway analysis revealed that the linoleic acid (LA) and arachidonic acid (AA) metabolism axis was the most perturbed pathway upon HCoV-229E infection. Importantly, supplement of additional LA and AA to coronavirus-infected cells significantly inhibited virus replication of both HCoV-229E and the highly virulent MERS-CoV, suggesting that the LA–AA metabolism axis is a common and essential pathway that could modulate coronavirus replication. In this regard, temporal modulation of the host lipid profile is a potential novel strategy to combat emerging human coronaviruses.

2. Materials and Methods

2.1. Materials

High performance liquid chromatography (HPLC)-grade methanol, acetonitrile, chloroform and 2-propanol were purchased from Merck (Darmstadt, Germany). HPLC-grade water was prepared using a Milli-Q water purification system (Millipore, Burlington, MA, USA). Analytical grade acetic acid and commercial standards used for biomarker identification were purchased from Sigma-Aldrich (St. Louis, MO, USA). Internal standards (IS) including Arachidonic acid-d8, 15(S)-HETE-d8, Leukotriene-B4-d4 and Platelet-activating factor C-16-d4 (PAF C-16-d4) were purchased from Cayman Chemical (Ann Arbor, MI, USA) [18].

2.2. Viruses and Cells

Huh-7 and VeroE6 cells were maintained in Dulbecco's modified Eagle medium (DMEM) supplemented with 10% heat-inactivated fetal bovine serum (FBS), 100 U/mL penicillin, and 100 g/mL streptomycin (5% CO_2 at 37 °C). MERS-CoV (EMC/2012 strain) was kindly provided by Professor Ron Fouchier (Erasmus Medical Center, Rotterdam, The Netherlands). MERS-CoV and HCoV-229E were cultured in VeroE6 cells in serum-free DMEM supplemented with 100 U/mL penicillin and 100 g/mL streptomycin as we described previously [19–21]. The supernatants were harvested when cytopathic effects (CPE) were observed and centrifuged to generate the viral stocks. The viral stocks were titrated by plaque assay on VeroE6 cells and stored at −80 °C as previously described [22,23]. Briefly, confluent VeroE6 cells were infected with 10-fold serial viral dilutions. The cells were incubated with diluted viruses at 37 °C for 1 h and subsequently overlaid with 1% low-melting-point agarose (Promega, Madison, WI, USA). The cells were fixed with 4% formaldehyde as the plaques were observed and then stained with 0.2% crystal violet. All experiments involving live MERS-CoV followed the approved standard operating procedures of the biosafety level 3 facility as previously described [24–27].

2.3. Lipid Treatment of Middle East Respiratory Syndrome Coronavirus (MERS-CoV)-Infected and Human Coronavirus (HCoV-229E)-Infected Huh-7 Cells

Huh-7 cells were seeded into 24-well plate to reach 90% confluency and infected with MERS-CoV or HCoV-229E at multiplicity of infection (MOI) of 0.005 or 1, respectively. After 1 h of inoculation, the cells were washed with phosphate-buffered saline (PBS) and maintained in lipids-supplemented medium at the indicated concentrations for 24 h. AA, LA, oleic acid (OA), and palmitic acid (PA) were dissolved in ethanol and ethanol was used as a negative control. The lipids were purchased from Cayman Chemical (Ann Arbor, MI, USA). The supernatants and cell lysates were collected at 24 h post-infection. The viral genome copy numbers were determined by reverse-transcription quantitative polymerase chain reaction (RT-qPCR) as previously described [28–30].

2.4. Lipid Extraction for Lipidomics Profiling

Confluent Huh-7 cells were mock infected or infected with HCoV-229E at MOI of 1 and incubated in DMEM medium. At 24 hpi, cells were collected for cellular lipid extraction. The lipid extraction was performed for liquid chromatography-mass spectrometry (LC-MS) analysis according to a previously described protocol with slight modifications [31,32]. Inactivation of virus infectivity was confirmed before further processing as we previously described with some modifications [33]. Briefly, 500 µL of ice-cold 150 mM ammonium bicarbonate solution was added to dissociate cells. Two millilitres of chloroform/methanol (v/v 2:1) containing IS were added, followed by vortexing and centrifugation at 4500 rpm for 10 min at 4 °C. The bottom phase was transferred to glass vials and dried using a vacuum concentrator for storage at −80 °C. The dried samples were reconstituted in 250 µL solvent mixture containing methanol/2-propanol/water ($v/v/v$ 5:4:1) for LC-MS analysis. After centrifugation at 14,000 rpm for 10 min at 4 °C, supernatants were transferred to LC vials for LC-MS analysis.

2.5. Ultra-High Performance Liquid Chromatography-Electrospray Ionization-Quadrupole-Time of Flight-Mass Spectrometry (UPLC-ESI-Q-TOF-MS) Analysis

The lipid extract was analyzed using an Acquity UPLC system coupled to a Synapt G2-Si High Definition Mass Spectrometry (HDMS) system (Waters Corp., Milford, MA, USA). The chromatography was performed on a Waters ACQUITY BEH C18 column (1.7 µm, 2.1 × 100 mm, I.D., 1.7 mm, Waters Corp., Milford, MA, USA). The mobile phase consisted of (A) 0.1% acetic acid in water and

(B) acetonitrile. Gradient elution applied for ultra-high performance liquid chromatography-mass spectrometry (UPLC-MS) analysis was described in Table S1. The column and autosampler temperature were maintained at 45 °C and 4 °C, respectively. The injection volume was 5 μL [34].

The mass spectral data were acquired in both positive and negative modes. The capillary voltage, sampling cone voltage and source offset were maintained at 2.5 kV, 60 V, and 60 V, respectively. Nitrogen was used as desolvation gas at a flow rate of 800 L/h. The source and desolvation temperatures were maintained at 120 °C and 400 °C, respectively. Mass spectra were acquired over the m/z range of 50 to 1200. The SYNAPT G2-Si HDMS system was calibrated using sodium formate clusters and operated in sensitivity mode. Leucine enkephalin was used as a lock mass for all experiments. MS/MS acquisition was operated in the same parameters as MS acquisition. Collision energy was applied at the range from 20 to 40 eV for fragmentation to allow putative identification and structural elucidation of the significant lipids.

2.6. Data Processing and Statistical Data Analysis

Acquisition of the raw data was performed using MassLynx software version 4.1 (Waters Corp., Milford, MA, USA) and raw data were converted to the common data format (NetCDF) files using conversion software Databridge (Waters Corp., Milford, MA, USA). The NetCDF data were subsequently deconvolved into a usable data matrix using the XCMS software (http://metlin.scripps.edu/download/) [35] and the grouping of features was performed using the CAMERA R package [36]. Preprocessed data were then exported as a .csv file for further data statistical analysis. MetaboAnalyst 3.0 (http://www.metaboanalyst.ca) and SIMCA-P V12.0 (Umetrics, Umeå, Sweden) were used for univariate and multivariate statistical analysis, respectively [37]. For univariate analysis, statistical significance of features was determined between the mock and HCoV-229E infected group using the Student's *t*-test and fold change. The *p*-value < 0.05 and fold change > 2 were used as criteria for significant features selection. For multivariate analysis, the features were subjected to Pareto scaling firstly then orthogonal partial least squares discriminant analysis (OPLS-DA) was performed as a supervised method to find important variables with discriminative power. The OPLS-DA model was evaluated with the relevant R2 and Q2. The variable importance in projection (VIP), which reflects both the loading weights for each component and the variability of the response explained by this component, was used for feature selection [38].

2.7. Lipids Identification

MS/MS fragmentation was performed on the significant features with high abundances. The significant features identification were carried out by searching accurate MS and MS/MS fragmentation pattern data in the METLIN database (Metabolomics Database, http://metlin.scripps.edu/), Human Metabolome Database (http://www.hmdb.ca/), and LIPD MAPS (Lipidomics Gateway, http://www.lipidmaps.org/). For confirmation of lipid identity using authentic chemical standard, MS/MS fragmentation pattern of the chemical standard was compared with that of candidate lipid under the same LC-MS condition to reveal any matching [18,39].

3. Results

3.1. Omics-Based Statistical Analysis for Significant Features

To investigate how coronavirus perturbs host lipid metabolism, we performed lipidomics analysis on HCoV-229E-infected Huh7 cells and compared the results with those of the mock-infected cells. The preliminary features list included precursor ions, adducts and isotope ions, which were imported into the MetaboAnalyst and SIMCA-P software for further analysis. The R2X/ R2Y, represented the X/Y variables explanation rate of the OPLS-DA model, were 83.0% and 98.8%, respectively. The predicted component, as estimated by cross-validation, was 0.97 (Q2). These cross-validated parameters were satisfactory for OPLS-DA mode (Supplementary Figure S1a). At the same time, the permutation test (100 times) also indicated that the validated model was satisfied (Supplementary Figure S1b). Overall, our results demonstrated that these significant lipid features could be selected by the validated statistical model for subsequent identification.

3.2. Identification of Lipids Specific to HCoV-229E

A total of 206 (positive mode) and 100 (negative mode) ion features were selected according to the omics-based statistical analysis method. These ion features were significantly discriminative between HCoV-229E-infected and mock-infected cells. To observe the discrimination trend in more detail, a hierarchical clustering analysis was performed based on the degree of similarity of lipid abundance profiles to show the overview trend of all significant ion features. As indicated in Figure 1, most of the significant features from both negative mode (Figure 1A) and positive mode (Figure 1B) expressed an up-regulation trend after HCoV-229E infection compared with the mock infection controls. Furthermore, to identify lipids specific to HCoV-229E infection, these significant features were grouped and annotated using the CAMERA software, and the potential precursor ions were used to perform further MS/MS experiments for obtaining their fragmentation patterns. Finally, a total of 24 lipids were identified, which could be classified into three lipid classes, including lysophosphatidylcholine (lysoPC), lysophosphatidylethanolamine (lysoPE) and fatty acid (FA). The chromatogram peak heights of these identified lipids were generated by LC-MS raw data and the ratio between infected and mock-infected cells was determined. As demonstrated in Figure 2, we found a consistent up-regulation trend of the identified lipids in HCoV-229E-infected cells. In particular, lysoPC was the predominant lipid class of all identified, accounting for approximately 60% of all identified lipids with significant elevation (Figure 2A). At the same time, arachidonic acid (AA), which belongs to the FA class, showed the highest increase in fold-change among all identified lipids with a maximum of 7.1-fold increase (Figure 2B). In addition, the level of lysoPEs (Figure 2C) was also up-regulated with a maximum fold change of 2.93, which was comparatively less than that of the lysoPCs and FAs. The identities of lysoPC (16:0/0:0), platelet-activating factor C-16 (PAF C-16), lysoPE (16:0/0:0), AA, LA, PA and OA were confirmed by matching the retention time (RT) and MS/MS fragmentation patterns of the authentic chemical standards that distinguish between HCoV-229E-infected cases and non-infected controls (Figure 3). The detailed information of the 24 identified lipids was listed in Table 1. MS/MS fragmentation patterns of five representative lipids and corresponding standards are also demonstrated in Supplementary Figure S2.

3.3. Pathway Analysis of HCoV-229E-Infected Huh7 Cells

Based on the list of significantly up-regulated lipids after HCoV-229E-infection, MetaboAnalyst (http://www.metaboanalyst.ca) was applied to investigate which pathway might be markedly perturbed. The result of the pathway analysis was graphically presented in Figure 4. From the enrichment analysis results, the LA metabolism pathway and FA biosynthesis pathway had a statistically significant raw p-value (raw $p < 0.05$, as shown in the Y-axis). Pathway impact results indicated that the LA metabolism and AA metabolism pathways presented higher impact than the other pathways, as indicated in the X-axis value. Combining the above two analysis results, we postulated that the LA metabolism pathway to be a markedly perturbed pathway that correlated with the lipid rearrangement process induced by HCoV-229E infection.

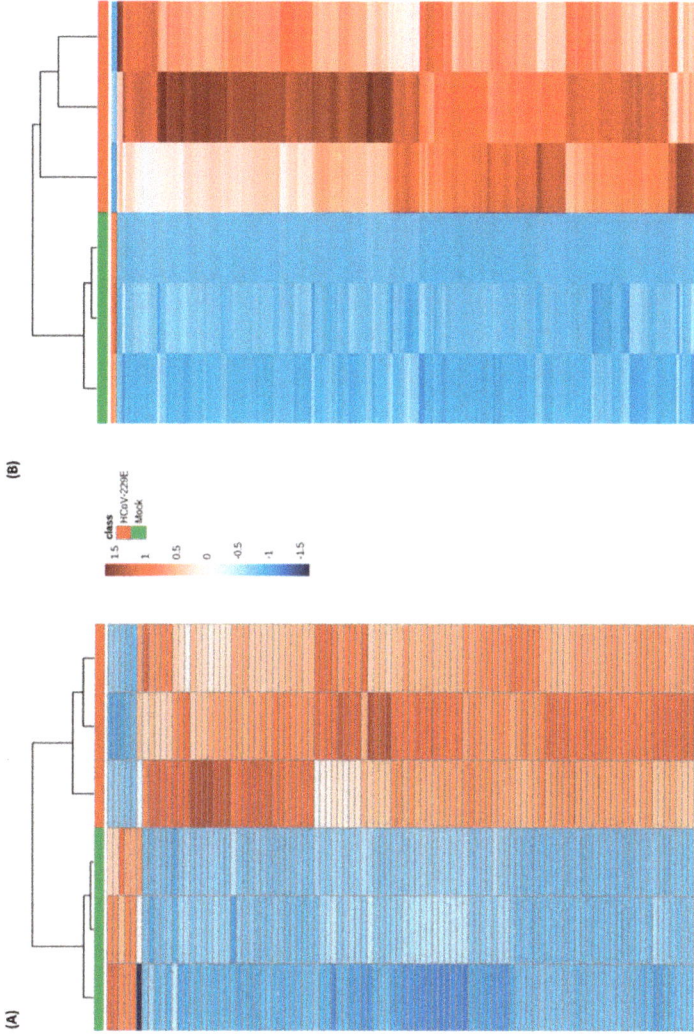

Figure 1. Heatmap showing the lipidomic analysis of human coronavirus 229E (HCoV-229E)-infected versus non-infected Huh-7 cells. Each rectangle represents an ion feature colored by its normalized intensity scale from blue (decreased level) to red (increased level). The dendrogram on the top was constructed based on the lipid intensity (similarity measure using Euclidean, and the Ward clustering algorithm). HCoV-229E, HCoV-229E-infected cells; Mock, non-infected cells. (**A**) Significant ion features in negative detection mode; (**B**) significant ion features in positive detection mode.

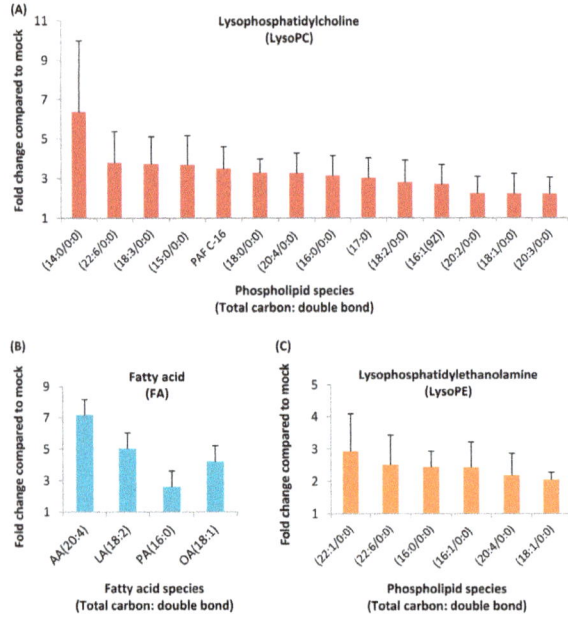

Figure 2. Liquid chromatography-mass spectrometry (LC/MS) analysis of HCoV-229E-infected cells revealed a homeostatic change in lipid levels. Huh-7 cells were mock- or HCoV-229E-infected and harvested at 24 hpi. The peak heights of these lipids were calculated and the fold change plotted with GraphPad Prism 5. (**A**) Lysophosphatidylcholine (LysoPC), (**B**) fatty acid (FA), (**C**) lysophosphatidylethanolamine (LysoPE). AA, arachidonic acid; LA, linoleic acid; PA, palmitic acid; OA, oleic acid.

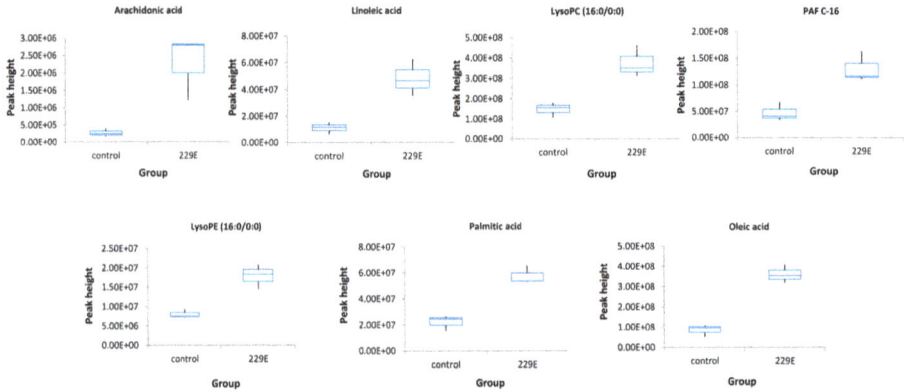

Figure 3. Box-whisker plots of the 7 standard confirmed lipids that were distinguished between the HCoV-229E-infected samples and the non-infected controls. The peak height was generated by LC-MS raw data. Control, non-infected cells; 229E, HCoV-229E-infected cells.

Table 1. The 24 lipids that were significantly different between HCoV-229E-infected and mock-infected samples.

Significant Lipids	Trend in HCoV-229E vs. Control	Molecular Formula	Detection Mode	Retention Time	Accurate Mass in Detection Mode	Fold Change	p-Value	VIP
lysoPC(16:0/0:0)[S]	up-regulation	C24H50NO7P	pos	12.75	496.34	3.14	0.0027	1.40
PAF C-16[S]	up-regulation	C26H54NO7P	pos	14.38	524.37	3.49	0.0214	1.60
LysoPC(18:1/0:0)[P]	up-regulation	C26H52NO7P	neg/pos	13.22	580.3611/522.3582	2.23	0.0049	4.71
LysoPC(18:0/0:0)[P]	up-regulation	C26H54NO7P	neg/pos	14.77	582.4761/524.3715	3.29	0.0051	8.61
LysoPC(16:1(9Z))[P]	up-regulation	C24H48NO7P	pos	11.48	494.32	2.7	0.0086	4.12
LysoPC(18:2/0:0)[P]	up-regulation	C26H50NO7P	pos	11.98	520.34	2.81	0.0158	3.33
LysoPC(18:3/0:0)[P]	up-regulation	C26H48NO7P	pos	12.74	518.32	3.73	0.0046	3.68
LysoPC(14:0/0:0)[P]	up-regulation	C22H46NO7P	pos	10.73	468.31	6.36	0.0081	3.35
LysoPC(20:2/0:0)[P]	up-regulation	C28H54NO7P	pos	13.74	548.37	2.24	0.0124	1.52
LysoPC(20:3/0:0)[P]	up-regulation	C28H52NO7P	pos	13.14	546.35	2.23	0.0146	2.26
LysoPC(20:4/0:0)[P]	up-regulation	C26H52NO7P	pos	11.93	544.34	3.27	0.0073	2.89
LysoPC(22:6/0:0)[P]	up-regulation	C30H50NO7P	pos	11.86	568.34	3.78	0.0079	1.97
LysoPC(15:0)[P]	up-regulation	C23H48NO7P	pos	11.72	482.32	3.68	0.0083	2.88
LysoPC(17:0)[P]	up-regulation	C25H52NO7P	pos	13.36	510.36	3.01	0.0101	5.02
LysoPE(16:0/0:0)[S]	up-regulation	C21H44NO7P	pos	12.65	454.29	2.44	0.0047	2.90
LysoPE(20:4/0:0)[P]	up-regulation	C25H44NO7P	pos	11.87	502.29	2.18	0.0147	3.98
LysoPE(22:6/0:0)[P]	up-regulation	C27H44NO7P	pos	11.80	526.29	2.52	0.0223	2.02
LysoPE(16:1/0:0)[P]	up-regulation	C21H42NO7P	pos	11.20	452.28	2.43	0.0165	1.75
LysoPE(18:1/0:0)[P]	up-regulation	C23H46NO7P	neg	13.07	478.48	2.04	0.0000	2.67
LysoPE(22:1/0:0)[P]	up-regulation	C27H54NO7P	pos	14.11	536.37	2.93	0.0142	2.34
Arachidonic acid[S]	up-regulation	C20H32O2	neg	17.15	303.20	7.16	0.0200	1.05
Linoleic acid[S]	up-regulation	C18H32O2	neg	16.55	279.23	5.03	0.0085	6.57
Palmitic acid[S]	up-regulation	C16H32O2	neg	18.02	281.18	2.61	0.0028	2.20
Oleic acid[S]	up-regulation	C18H34O2	neg	17.72	255.45	4.21	0.0009	3.28

pos and neg represented positive mode and negative mode respectively; s Lipids that were confirmed with authentic standards; p Lipids that putatively annotated and matched the fragmentation pattern with the database.

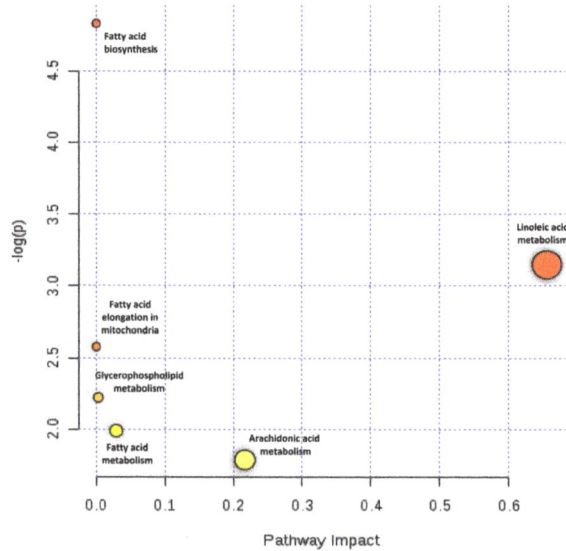

Figure 4. Pathway analysis associate with HCoV-229E infection was carried out by MetaboAnalyst. The *Y*-axis, "log(p)", represented the transformation of the original *p*-value calculated from the enrichment analysis. The *X*-axis, "Pathway Impact", represented the value calculated from the pathway topology analysis.

To better understand the current pathway analysis results and the cellular lipid signaling response upon HCoV-229E infection, we constructed a global LA pathway map based on the pathway information in the Kyoto Encyclopedia of Genes and Genomes (KEGG) database (https://www.genome.jp/kegg/) and literature mining (Figure 5). Upon HCoV-229E infection, the glycerophospholipids, as main components of the cell membrane, were metabolized to lysophospholipids and FAs after cPLA2 enzyme activation. Lysophospholipids such as lysoPCs and lysoPEs were correspondingly increased after HCoV-229E infection. Moreover, lysoPCs could be further metabolized to platelet-activating factor. FAs were also released from glycerophospholipids but only LA and AA could initiate downstream pathways to generate corresponding metabolites. The up-regulation of both lysophospholipids and FAs were partially confirmed by authentic standards. Furthermore, to investigate the downstream pathways trend of FAs, the authentic standards were also applied in LC-MS method to confirm whether these downstream lipids were changed correspondingly. As illustrated in Figure 5, AA is a downstream lipid of LA and the origin lipid of AA metabolism pathway. The identity of AA was confirmed by authentic standard (Supplementary Figure S2D), which was found to be significantly up-regulated. Therefore, combining pathway analysis and the authentic standards verification results, our data suggested that the LA–AA metabolism axis was the most significantly perturbed pathway and might be associated with lipids rearrangement or other processes in HCoV-229E infection.

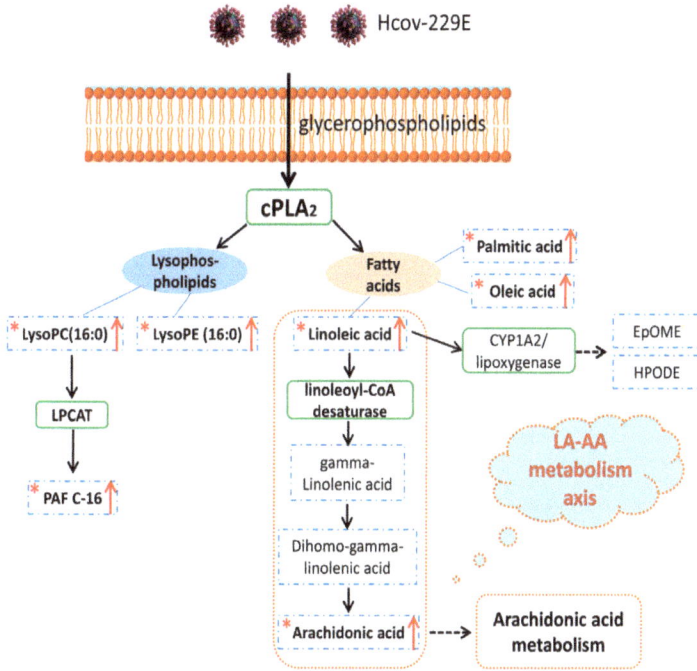

Figure 5. The pathway map based on identified lipids and linoleic acid metabolism recorded in the Kyoto Encyclopedia of Genes and Genomes (KEGG) PATHWAY Database. The star mark "*" indicates the lipids could be matched with commercial standards and have an up-regulation trend. The red arrow represents the up-regulation trend. The blue dashed rectangle and green solid rectangles represent lipids and corresponding enzyme in this pathway, respectively. The orange dashed line represents the LA–AA metabolism axis.

3.4. Lipids Treatment of Virus-Infected Cells

To investigate the potential implication of the perturbed LA-AA metabolism axis in HCoV-229E infection, we treated HCoV-229E-infected Huh7 cells with LA and AA and included PA and OA for comparison. The LA and AA were mapped and played a vital role in the perturbed LA-AA metabolism axis (Figure 5). In contrast, PA and OA were not mapped in the perturbed pathway and may only be produced from glycerophospholipids due to cPLA2 enzyme activation. Huh-7 cells were infected with HCoV-229E and treated with AA, LA, PA, or OA. The cell lysates and culture supernatants were harvested at 24 h post-infection to determine the viral genome copy number by RT-qPCR. As shown in Figure 6, LA and AA consistently inhibited the replication of HCoV-229E as evidenced by the decrease in virus genome copies in both cell lysate (Figure 6A,C) and supernatant samples (Figure 6B–D). In contrast, PA inhibited HCoV-229E replication only when supplied at high concentration while HCoV-229E replication was largely independent of OA (Figure 6A–D).

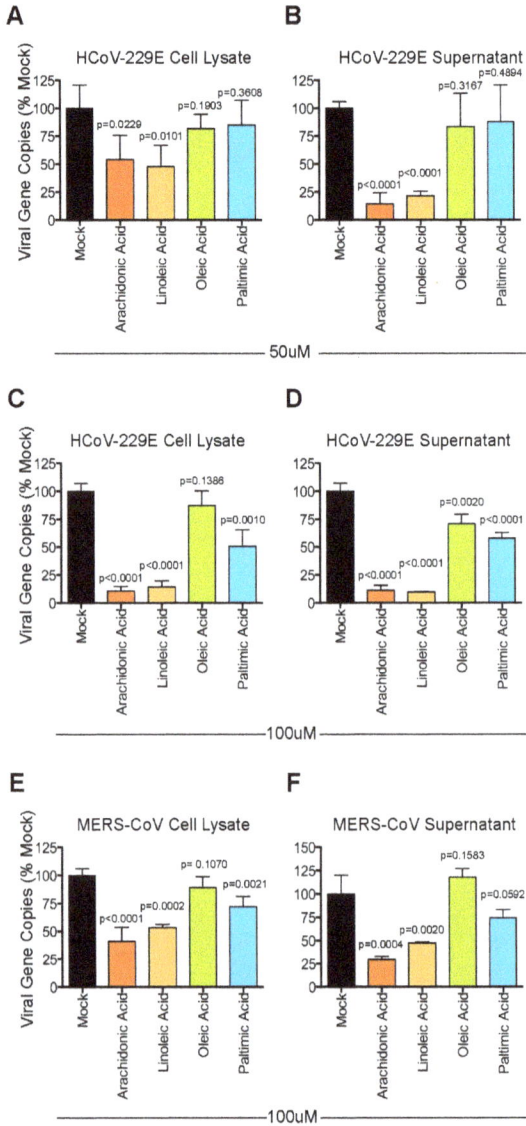

Figure 6. Modulatory effect of lipids on HCoV-229E and Middle East respiratory syndrome coronavirus (MERS-CoV). Huh-7 cells were infected with HCoV-229E. After 1 h of inoculation, the virus inoculum was replaced with medium containing 50 μM (**A,B**) or 100 μM (**C,D**) of lipids and incubated for 24 h. The supernatants and cell lysates were collected for reverse-transcription quantitative polymerase chain reaction (RT-qPCR) analysis. In parallel, Huh-7 cells were infected with MERS-CoV. After 1 h of inoculation, the virus inoculum was replaced with medium containing 100 μM (**E,F**) of lipids and incubated for 24 h. The supernatants and cell lysates were collected for RT-qPCR analysis. Statistical significance was determined by Student's t-test by comparing the individual lipid-treated group with the mock-treated group ($n = 4$). The difference was considered significant when $p < 0.05$.

To further investigate if the modulatory effects of LA and AA were conserved among other human-pathogenic coronaviruses, we evaluated the effects of these lipids on the replication of the recently emerged and highly virulent MERS-CoV. Our data demonstrated that LA and AA potently suppressed MERS-CoV replication in a similar manner as HCoV-229E (Figure 6E,F). Overall, our results demonstrated that exogenously supplied LA and AA could interfere with the optimal replication of human-pathogenic coronaviruses, which suggested that the LA–AA metabolism axis was significantly involved in the propagation of these viruses.

4. Discussion

In this study, a MS-based lipidomics approach was established to characterize the host cell lipid changes upon coronavirus infection. Univariate and multivariate statistical analyses were applied in data processing for the selection of significant lipid features. A total of 24 lipids including lysophospholipids and FAs were identified and were consistently up-regulated in HCoV-229E-infected cells. Seven representative lipids were confirmed by authentic standards, including lysoPC (16:0/0:0), PAF C-16, lysoPE (16:0/0:0), AA, LA, PA and OA. Subsequent pathway analysis indicated that the LA–AA metabolism axis, consisting of LA and AA as important precursor lipids, was substantially perturbed after HCoV-229E infection. Moreover, we demonstrated that exogenously supplied LA and AA were capable of inhibiting the replication of HCoV-229E and the highly pathogenic MERS-CoV, which suggested the LA–AA metabolism axis to be a conserved and essential pathway in the propagation of human coronaviruses.

A total of 24 lipids including lysoPCs, lysoPEs and unsaturated/saturated FAs were identified to be significantly upregulated after HCoV-229E infection. Twenty of these 24 (83.3%) lipids were lysoPC and lysoPE. LysoPC is the most abundant lysophospholipid in humans, with a high plasma concentration of several hundred micromoles. In addition, lysoPC was a potent inhibitor and could reversely arrest pore expansion during syncytium formation mediated by diverse viral fusogens [40]. Another lysophospholipid, lysoPE, is present at low concentrations in vivo but they induce various cellular responses such as activation of mitogen-activated protein kinase (MAPK) and neuronal differentiation when applied to cells in vitro [41]. Among the identified FAs, the LA and AA both belong to polyunsaturated omega-6 fatty acid and are essential fatty acids. In addition, LA is the metabolic precursor of AA, both of which are key components of the cell membrane. LA and AA also play fundamental roles in the biological function of many tissues by modulating enzymes, ion channels, receptors, as well as inflammation [42].

Coronavirus replication is associated with intracellular membrane rearrangement and depends on the formation of double membrane vesicles (DMVs) and other membranous structures as replicative organelles [16]. The cell membrane components consist mainly of glycerophospholipid components such as phosphatidylcholine (PC), phosphatidylethanolamine (PE), lysophosphatidylcholine (lysoPC), and lysophosphatidylethanolamine (lysoPE). A specific phospholipids composition is required by different viruses to form the optimal replicative organelles best suited for their replication [43]. Moreover, the lysoPC/PE was produced from PC/PE by cPLA2 activation, which simultaneously generated corresponding fatty acid moiety. In this regard, cPLA2 activation is commonly believed to be beneficial for virus replication [16,17].

In our study, we found that a number of lysophospholipids and FAs downstream of cPLA2 activation, were upregulated upon HCoV-229E infection. The upregulation of these lipid species including LA and AA were believed to promote efficient coronavirus replication. However, when we evaluate this hypothesis by exogenously supplementing additional LA and AA to HCoV-229E- or MERS-CoV-infected cells, we noticed a significant reduction in virus replication. Taken together, our data suggested that coronavirus infection did not randomly perturb the cellular lipid compositions. Instead, we speculate that coronaviruses precisely modulate and rearrange the host lipid profile to reach an intricate homeostasis optimized for its replication. Any exogenous manipulation that disrupts the equilibrium may interfere with the optimal replication of the viruses. Alternatively, supplementing LA and AA might disturb the LA–AA metabolism axis and result in feedback reversion of lysophospholipids into phospholipids through Land's cycle [44], thus limiting virus replication.

In addition, LA and AA are polyunsaturated fatty acids that are biological signaling precursors. They can be metabolized to important eicosanoids and metabolites, which play multiple roles in the host immune response and the pathogenesis of viral infections [45–47]. However, previous study had suggested that arachidonic acid (AA) downstream metabolites show no evidence of anti-coronaviral activity as observed through special inhibitors of cyclooxygenases (COX) 1/2 and 5-lipoxygenase (LOX), which are two key enzymes requiring AA as a precursor. The results indicated AA downstream products may not have a significant effect on coronaviruses replication, at least in vitro [16,48]. In this regard, the function of the downstream metabolites of LA and AA may play key roles in the pathogenesis of coronaviruses in vivo.

5. Conclusions

In the present study, we revealed that the cellular lipid profile was rearranged upon HCoV-229E infection. A total of 24 lipids including lysoPCs, lysoPEs and FAs were upregulated. Among them, LA and AA, which were mapped into the LA-AA metabolism axis, demonstrated strong modulatory effects on the replication of HCoV-229E and the highly pathogenic MERS-CoV. In this regard, our data suggested that optimal coronavirus replication required a specific composition of cellular lipids and any disruption could decrease the efficiency of coronavirus replication. Thus, the MS-based lipidomics strategy could be used to monitor virus-specific lipid requirement, to discover the perturbed pathways and identify novel lipids to interfere with virus replication. In further studies, combining lipidomics data with biological and immunological data may help to elucidate specific pathogenic mechanisms and identify novel treatment strategies for virus infections.

Supplementary Materials: The following are available online at http://www.mdpi.com/1999-4915/11/1/73/s1, Table S1: Gradient elution program applied for UPLC-MS analysis, Figure S1: OPLS-DA model validation and permutation test. (a) The cross-validated parameters (R2X = 0.83, R2Y = 0.98, Q2 = 0.97) are satisfactory for the OPLS-DA model; (b) The permutation test (100 times) indicates that the validated model is acceptable, Figure S2: The MS/MS mass spectra and predicted structures with expected fragmentation profiles of the 5 lipids in cell lysate: (A) Linoleic acid; (B) LysoPC(16:0/0:0); (C) LysoPE(16:0/0:0); (D) Arachidonic acid; and (E) PAF C-16.

Author Contributions: B.Y., H.C., and K.-Y.Y. conceived and designed research. B.Y., D.Y., P.-M.L., H.S., Y.W., S.Y., and K.-H.S. conducted experiments. B.Y., H.C. and D.Y. analyzed data. B.Y., H.C., D.Y., J.F.-W.C., and K.-Y.Y. wrote the manuscript. R.Y.-T.K. and J.F.-W.C. interpreted the results and gave advice. K.-H.S. and K.-Y.Y. supervised the study. All authors read and approved the manuscript.

Funding: This work was partly supported by the donations of Michael Seak-Kan Tong, Hui Ming, Hui Hoy and Chow Sin Lan Charity Fund Limited, Chan Yin Chuen Memorial Charitable Foundation, and Respiratory Viral Research Foundation Limited; and funding from the Health and Medical Research Fund (HKM-15-M05) of the Food and Health Bureau, Hong Kong Special Administrative Region Government; the Theme-based Research Scheme (T11-707/15-R) of the Research Grants Council, Hong Kong Special Administrative Region; the High Level Hospital-Summit Program in Guangdong, The University of Hong Kong-Shenzhen Hospital; and the Collaborative Innovation Center for Diagnosis and Treatment of Infectious Diseases, the Ministry of Education of China. The sponsors had no role in the design and conduct of the study, in the collection, analysis and interpretation of data, or in the preparation, review or approval of the manuscript.

Conflicts of Interest: J.F.-W.C. has received travel grants from Pfizer Corporation Hong Kong and Astellas Pharma Hong Kong Corporation Limited, and was an invited speaker for Gilead Sciences Hong Kong Limited and Luminex Corporation. The other authors declared no conflict of interest. The funding sources had no role in study design, data collection, analysis or interpretation or writing of the report. The corresponding author had full access to all the data in the study and had final responsibility for the decision to submit for publication.

References

1. Chan, J.F.; Li, K.S.; To, K.K.; Cheng, V.C.; Chen, H.; Yuen, K.Y. Is the discovery of the novel human betacoronavirus 2c EMC/2012 (HCoV-EMC) the beginning of another SARS-like pandemic? *J. Infect.* **2012**, *65*, 477–489. [CrossRef] [PubMed]
2. Chan, J.F.; To, K.K.; Tse, H.; Jin, D.Y.; Yuen, K.Y. Interspecies transmission and emergence of novel viruses: lessons from bats and birds. *Trends Microbiol.* **2013**, *21*, 544–555. [CrossRef]
3. Chan, J.F.; Lau, S.K.; Woo, P.C. The emerging novel Middle East respiratory syndrome coronavirus: The "knowns" and "unknowns". *J. Formos Med. Assoc.* **2013**, *112*, 372–381. [CrossRef]

4. Chan, J.F.; Lau, S.K.; To, K.K.; Cheng, V.C.; Woo, P.C.; Yuen, K.Y. Middle East respiratory syndrome coronavirus: Another zoonotic betacoronavirus causing SARS-like disease. *Clin. Microbiol. Rev.* **2015**, *28*, 465–522. [CrossRef] [PubMed]

5. Cheng, V.C.; Lau, S.K.; Woo, P.C.; Yuen, K.Y. Severe acute respiratory syndrome coronavirus as an agent of emerging and reemerging infection. *Clin. Microbiol. Rev.* **2007**, *20*, 660–694. [CrossRef] [PubMed]

6. Peiris, J.S.M.; Lai, S.T.; Poon, L.L.M.; Guan, Y.; Yam, L.Y.C.; Lim, W.; Nicholls, J.; Yee, W.K.S.; Yan, W.W.; Cheung, M.T.; et al. Coronavirus as a possible cause of severe acute respiratory syndrome. *Lancet* **2003**, *361*, 1319–1325. [CrossRef]

7. Zumla, A.; Chan, J.F.; Azhar, E.I.; Hui, D.S.; Yuen, K.Y. Coronaviruses—Drug discovery and therapeutic options. *Nat. Rev. Drug Discov.* **2016**, *15*, 327–347. [CrossRef] [PubMed]

8. Taube, S.; Jiang, M.; Wobus, C.E. Glycosphingolipids as receptors for non-enveloped viruses. *Viruses* **2010**, *2*, 1011–1049. [CrossRef] [PubMed]

9. Chazal, N.; Gerlier, D. Virus entry, assembly, budding, and membrane rafts. *Microbiol. Mol. Biol. Rev.* **2003**, *67*, 226–237, table of contents. [CrossRef] [PubMed]

10. Nagy, P.D.; Strating, J.R.P.M.; van Kuppeveld, F.J.M. Building Viral Replication Organelles: Close Encounters of the Membrane Types. *PLoS Pathog.* **2016**, *12*, e1005912. [CrossRef] [PubMed]

11. Hsu, N.Y.; Ilnytska, O.; Belov, G.; Santiana, M.; Chen, Y.H.; Takvorian, P.M.; Pau, C.; van der Schaar, H.; Kaushik-Basu, N.; Balla, T.; et al. Viral reorganization of the secretory pathway generates distinct organelles for RNA replication. *Cell* **2010**, *141*, 799–811. [CrossRef] [PubMed]

12. Diamond, D.L.; Syder, A.J.; Jacobs, J.M.; Sorensen, C.M.; Walters, K.A.; Proll, S.C.; McDermott, J.E.; Gritsenko, M.A.; Zhang, Q.; Zhao, R.; et al. Temporal proteome and lipidome profiles reveal hepatitis C virus-associated reprogramming of hepatocellular metabolism and bioenergetics. *PLoS Pathog.* **2010**, *6*, e1000719. [CrossRef] [PubMed]

13. Ono, A.; Ablan, S.D.; Lockett, S.J.; Nagashima, K.; Freed, E.O. Phosphatidylinositol (4,5) bisphosphate regulates HIV-1 Gag targeting to the plasma membrane. *Proc. Natl. Acad. Sci. USA* **2004**, *101*, 14889–14894. [CrossRef]

14. Zhang, J.; Pekosz, A.; Lamb, R.A. Influenza virus assembly and lipid raft microdomains: A role for the cytoplasmic tails of the spike glycoproteins. *J. Virol.* **2000**, *74*, 4634–4644. [CrossRef] [PubMed]

15. Knoops, K.; Kikkert, M.; Worm, S.H.; Zevenhoven-Dobbe, J.C.; van der Meer, Y.; Koster, A.J.; Mommaas, A.M.; Snijder, E.J. SARS-coronavirus replication is supported by a reticulovesicular network of modified endoplasmic reticulum. *PLoS Biol.* **2008**, *6*, e226. [CrossRef] [PubMed]

16. Muller, C.; Hardt, M.; Schwudke, D.; Neuman, B.W.; Pleschka, S.; Ziebuhr, J. Inhibition of Cytosolic Phospholipase A2alpha Impairs an Early Step of Coronavirus Replication in Cell Culture. *J. Virol.* **2018**, *92*. [CrossRef]

17. Vijay, R.; Hua, X.; Meyerholz, D.K.; Miki, Y.; Yamamoto, K.; Gelb, M.; Murakami, M.; Perlman, S. Critical role of phospholipase A2 group IID in age-related susceptibility to severe acute respiratory syndrome-CoV infection. *J. Exp. Med.* **2015**, *212*, 1851–1868. [CrossRef]

18. Yan, B.; Deng, Y.; Hou, J.; Bi, Q.; Yang, M.; Jiang, B.; Liu, X.; Wu, W.; Guo, D. UHPLC-LTQ-Orbitrap MS combined with spike-in method for plasma metabonomics analysis of acute myocardial ischemia rats and pretreatment effect of Danqi Tongmai tablet. *Mol. Biosyst.* **2015**, *11*, 486–496. [CrossRef] [PubMed]

19. Chan, J.F.; Chan, K.H.; Choi, G.K.; To, K.K.; Tse, H.; Cai, J.P.; Yeung, M.L.; Cheng, V.C.; Chen, H.; Che, X.Y.; et al. Differential cell line susceptibility to the emerging novel human betacoronavirus 2c EMC/2012: Implications for disease pathogenesis and clinical manifestation. *J. Infect. Dis* **2013**, *207*, 1743–1752. [CrossRef]

20. Chan, J.F.; Yao, Y.; Yeung, M.L.; Deng, W.; Bao, L.; Jia, L.; Li, F.; Xiao, C.; Gao, H.; Yu, P.; et al. Treatment With Lopinavir/Ritonavir or Interferon-beta1b Improves Outcome of MERS-CoV Infection in a Nonhuman Primate Model of Common Marmoset. *J. Infect. Dis.* **2015**, *212*, 1904–1913. [CrossRef]

21. Lau, S.K.; Lau, C.C.; Chan, K.H.; Li, C.P.; Chen, H.; Jin, D.Y.; Chan, J.F.; Woo, P.C.; Yuen, K.Y. Delayed induction of proinflammatory cytokines and suppression of innate antiviral response by the novel Middle East respiratory syndrome coronavirus: Implications for pathogenesis and treatment. *J. Gen. Virol.* **2013**, *94*, 2679–2690. [CrossRef] [PubMed]

22. Chu, H.; Chan, C.M.; Zhang, X.; Wang, Y.; Yuan, S.; Zhou, J.; Au-Yeung, R.K.; Sze, K.H.; Yang, D.; Shuai, H.; et al. Middle East respiratory syndrome coronavirus and bat coronavirus HKU9 both can utilize GRP78 for attachment onto host cells. *J. Biol. Chem.* **2018**, *293*, 11709–11726. [CrossRef]

23. Chan, C.M.; Chu, H.; Wang, Y.; Wong, B.H.; Zhao, X.; Zhou, J.; Yang, D.; Leung, S.P.; Chan, J.F.; Yeung, M.L.; et al. Carcinoembryonic Antigen-Related Cell Adhesion Molecule 5 Is an Important Surface Attachment Factor That Facilitates Entry of Middle East Respiratory Syndrome Coronavirus. *J. Virol.* **2016**, *90*, 9114–9127. [CrossRef] [PubMed]

24. Chan, J.F.; Choi, G.K.; Tsang, A.K.; Tee, K.M.; Lam, H.Y.; Yip, C.C.; To, K.K.; Cheng, V.C.; Yeung, M.L.; Lau, S.K.; et al. Development and Evaluation of Novel Real-Time Reverse Transcription-PCR Assays with Locked Nucleic Acid Probes Targeting Leader Sequences of Human-Pathogenic Coronaviruses. *J. Clin. Microbiol.* **2015**, *53*, 2722–2726. [CrossRef] [PubMed]

25. Chan, K.H.; Chan, J.F.; Tse, H.; Chen, H.; Lau, C.C.; Cai, J.P.; Tsang, A.K.; Xiao, X.; To, K.K.; Lau, S.K.; et al. Cross-reactive antibodies in convalescent SARS patients' sera against the emerging novel human coronavirus EMC (2012) by both immunofluorescent and neutralizing antibody tests. *J. Infect.* **2013**, *67*, 130–140. [CrossRef]

26. Chu, H.; Zhou, J.; Wong, B.H.; Li, C.; Chan, J.F.; Cheng, Z.S.; Yang, D.; Wang, D.; Lee, A.C.; Li, C.; et al. Middle East Respiratory Syndrome Coronavirus Efficiently Infects Human Primary T Lymphocytes and Activates the Extrinsic and Intrinsic Apoptosis Pathways. *J. Infect. Dis.* **2016**, *213*, 904–914. [CrossRef]

27. Zhou, J.; Chu, H.; Li, C.; Wong, B.H.; Cheng, Z.S.; Poon, V.K.; Sun, T.; Lau, C.C.; Wong, K.K.; Chan, J.Y.; et al. Active replication of Middle East respiratory syndrome coronavirus and aberrant induction of inflammatory cytokines and chemokines in human macrophages: Implications for pathogenesis. *J. Infect. Dis.* **2014**, *209*, 1331–1342. [CrossRef]

28. Chu, H.; Zhou, J.; Wong, B.H.; Li, C.; Cheng, Z.S.; Lin, X.; Poon, V.K.; Sun, T.; Lau, C.C.; Chan, J.F.; et al. Productive replication of Middle East respiratory syndrome coronavirus in monocyte-derived dendritic cells modulates innate immune response. *Virology* **2014**, *454–455*, 197–205. [CrossRef]

29. Chan, J.F.; Chan, K.H.; Kao, R.Y.; To, K.K.; Zheng, B.J.; Li, C.P.; Li, P.T.; Dai, J.; Mok, F.K.; Chen, H.; et al. Broad-spectrum antivirals for the emerging Middle East respiratory syndrome coronavirus. *J. Infect.* **2013**, *67*, 606–616. [CrossRef]

30. Tang, B.S.; Chan, K.H.; Cheng, V.C.; Woo, P.C.; Lau, S.K.; Lam, C.C.; Chan, T.L.; Wu, A.K.; Hung, I.F.; Leung, S.Y.; et al. Comparative host gene transcription by microarray analysis early after infection of the Huh7 cell line by severe acute respiratory syndrome coronavirus and human coronavirus 229E. *J. Virol.* **2005**, *79*, 6180–6193. [CrossRef]

31. Burnum-Johnson, K.E.; Kyle, J.E.; Eisfeld, A.J.; Casey, C.P.; Stratton, K.G.; Gonzalez, J.F.; Habyarimana, F.; Negretti, N.M.; Sims, A.C.; Chauhan, S.; et al. MPLEx: A method for simultaneous pathogen inactivation and extraction of samples for multi-omics profiling. *Analyst* **2017**, *142*, 442–448. [CrossRef] [PubMed]

32. Nakayasu, E.S.; Nicora, C.D.; Sims, A.C.; Burnum-Johnson, K.E.; Kim, Y.M.; Kyle, J.E.; Matzke, M.M.; Shukla, A.K.; Chu, R.K.; Schepmoes, A.A.; et al. MPLEx: A Robust and Universal Protocol for Single-Sample Integrative Proteomic, Metabolomic, and Lipidomic Analyses. *mSystems* **2016**, *1*. [CrossRef] [PubMed]

33. Sze, K.H.; Lam, W.H.; Zhang, H.; Ke, Y.H.; Tse, M.K.; Woo, P.C.Y.; Lau, S.K.P.; Lau, C.C.Y.; Cai, J.P.; Tung, E.T.K.; et al. Talaromyces marneffei Mp1p Is a Virulence Factor that Binds and Sequesters a Key Proinflammatory Lipid to Dampen Host Innate Immune Response. *Cell Chem. Biol.* **2017**, *24*, 182–194. [CrossRef] [PubMed]

34. Yuan, S.; Chu, H.; Chan, J.F.; Ye, Z.W.; Wen, L.; Yan, B.; Lai, P.M.; Tee, K.M.; Huang, J.; Chen, D.; et al. SREBP-dependent lipidomic reprogramming as a broad-spectrum antiviral target. *Nat. Commun.* **2019**, *10*, 120. [CrossRef] [PubMed]

35. Smith, C.A.; Want, E.J.; O'Maille, G.; Abagyan, R.; Siuzdak, G. XCMS: Processing mass spectrometry data for metabolite profiling using nonlinear peak alignment, matching, and identification. *Anal. Chem.* **2006**, *78*, 779–787. [CrossRef]

36. Kuhl, C.; Tautenhahn, R.; Bottcher, C.; Larson, T.R.; Neumann, S. CAMERA: An integrated strategy for compound spectra extraction and annotation of liquid chromatography/mass spectrometry data sets. *Anal. Chem.* **2012**, *84*, 283–289. [CrossRef] [PubMed]

37. Xia, J.; Sinelnikov, I.V.; Han, B.; Wishart, D.S. MetaboAnalyst 3.0—Making metabolomics more meaningful. *Nucleic Acids Res.* **2015**, *43*, W251–W257. [CrossRef]

38. Galindo-Prieto, B.; Eriksson, L.; Trygg, J. Variable influence on projection (VIP) for orthogonal projections to latent structures (OPLS). *J. Chemom.* **2014**, *28*, 623–632. [CrossRef]

Viruses **2019**, *11*, 73

39. Yang, Z.; Hou, J.J.; Qi, P.; Yang, M.; Yan, B.P.; Bi, Q.R.; Feng, R.H.; Yang, W.Z.; Wu, W.Y.; Guo, D.A. Colon-derived uremic biomarkers induced by the acute toxicity of Kansui radix: A metabolomics study of rat plasma and intestinal contents by UPLC-QTOF-MS(E). *J. Chromatogr. B Anal. Technol. Biomed. Life Sci.* **2016**, *1026*, 193–203. [CrossRef]

40. Ciechonska, M.; Duncan, R. Lysophosphatidylcholine reversibly arrests pore expansion during syncytium formation mediated by diverse viral fusogens. *J. Virol.* **2014**, *88*, 6528–6531. [CrossRef]

41. Makide, K.; Uwamizu, A.; Shinjo, Y.; Ishiguro, J.; Okutani, M.; Inoue, A.; Aoki, J. Novel lysophosphoplipid receptors: Their structure and function. *J. Lipid Res.* **2014**, *55*, 1986–1995. [CrossRef] [PubMed]

42. Tallima, H.; El Ridi, R. Arachidonic acid: Physiological roles and potential health benefits—A review. *J. Adv. Res.* **2018**, *11*, 33–41. [CrossRef] [PubMed]

43. Xu, K.; Nagy, P.D. RNA virus replication depends on enrichment of phosphatidylethanolamine at replication sites in subcellular membranes. *Proc. Natl. Acad. Sci. USA* **2015**, *112*, E1782–E1791. [CrossRef]

44. Wang, L.; Shen, W.; Kazachkov, M.; Chen, G.; Chen, Q.; Carlsson, A.S.; Stymne, S.; Weselake, R.J.; Zou, J. Metabolic interactions between the Lands cycle and the Kennedy pathway of glycerolipid synthesis in Arabidopsis developing seeds. *Plant. Cell* **2012**, *24*, 4652–4669. [CrossRef] [PubMed]

45. Demetz, E.; Schroll, A.; Auer, K.; Heim, C.; Patsch, J.R.; Eller, P.; Theurl, M.; Theurl, I.; Theurl, M.; Seifert, M.; et al. The arachidonic acid metabolome serves as a conserved regulator of cholesterol metabolism. *Cell Metab.* **2014**, *20*, 787–798. [CrossRef] [PubMed]

46. Persichini, T.; Mastrantonio, R.; Del Matto, S.; Palomba, L.; Cantoni, O.; Colasanti, M. The role of arachidonic acid in the regulation of nitric oxide synthase isoforms by HIV gp120 protein in astroglial cells. *Free Radic. Biol. Med.* **2014**, *74*, 14–20. [CrossRef] [PubMed]

47. Chandrasekharan, J.A.; Marginean, A.; Sharma-Walia, N. An insight into the role of arachidonic acid derived lipid mediators in virus associated pathogenesis and malignancies. *Prostaglandins Other Lipid Mediat.* **2016**, *126*, 46–54. [CrossRef] [PubMed]

48. Muller, C.; Karl, N.; Ziebuhr, J.; Pleschka, S. D, L-lysine acetylsalicylate + glycine Impairs Coronavirus Replication. *J. Antivir. Antiretrovir.* **2016**, *8*, 4. [CrossRef]

viruses

MDPI

Article

Lack of Middle East Respiratory Syndrome Coronavirus Transmission in Rabbits

W. Widagdo [1,†], Nisreen M. A. Okba [1,†], Mathilde Richard [1], Dennis de Meulder [1], Theo M. Bestebroer [1], Pascal Lexmond [1], Elmoubasher A. B. A. Farag [2], Mohammed Al-Hajri [2], Koert J. Stittelaar [3], Leon de Waal [3], Geert van Amerongen [3], Judith M. A. van den Brand [1,‡], Bart L. Haagmans [1,*] and Sander Herfst [1]

[1]   Department of Viroscience, Erasmus Medical Center, 3015GD Rotterdam, The Netherlands; w.widagdo@erasmusmc.nl (W.W.); n.okba@erasmusmc.nl (N.M.A.O.); m.richard@erasmusmc.nl (M.R.); d.demeulder@erasmusmc.nl (D.d.M.); t.bestebroer@erasmusmc.nl (T.M.B.); p.lexmond@erasmusmc.nl (P.L.); j.m.a.vandenbrand@uu.nl (J.M.A.v.d.B.); s.herfst@erasmusmc.nl (S.H.)
[2]   Ministry of Public Health, Doha, Qatar, PO Box. 42; eabdfarag@MOPH.GOV.QA (E.A.B.A.F.); malhajri1@MOPH.GOV.QA (M.A.-H.)
[3]   Viroclinics Biosciences BV, Rotterdam 3029 AK, The Netherlands; stittelaar@viroclinics.com (K.J.S.); dewaal@viroclinics.com (L.d.W.); amerongen@viroclinics.com (G.v.A.)
*   Correspondence: b.haagmans@erasmusmc.nl; Tel.: +31-10-704-4004; Fax: +31-10-704-4760
†   These authors contributed equally to this work.
‡   Current address: Department of Pathobiology, Utrecht University, 3584CS Utrecht, The Netherlands.

Received: 18 March 2019; Accepted: 22 April 2019; Published: 24 April 2019

Abstract: Middle East respiratory syndrome coronavirus (MERS-CoV) transmission from dromedaries to humans has resulted in major outbreaks in the Middle East. Although some other livestock animal species have been shown to be susceptible to MERS-CoV, it is not fully understood why the spread of the virus in these animal species has not been observed in the field. In this study, we used rabbits to further characterize the transmission potential of MERS-CoV. In line with the presence of MERS-CoV receptor in the rabbit nasal epithelium, high levels of viral RNA were shed from the nose following virus inoculation. However, unlike MERS-CoV-infected dromedaries, these rabbits did not develop clinical manifestations including nasal discharge and did shed only limited amounts of infectious virus from the nose. Consistently, no transmission by contact or airborne routes was observed in rabbits. Our data indicate that despite relatively high viral RNA levels produced, low levels of infectious virus are excreted in the upper respiratory tract of rabbits as compared to dromedary camels, thus resulting in a lack of viral transmission.

Keywords: MERS-coronavirus; transmission; rabbits

1. Introduction

Middle East respiratory syndrome coronavirus (MERS-CoV) is a novel pathogen that is known to infect dromedary camels and humans [1,2]. Seroepidemiological studies indicate that this virus has been circulating in dromedary camels in the Arabian Peninsula and Africa for decades [3–5]. MERS-CoV sequences obtained from these camels are largely similar to those obtained from human MERS cases in corresponding regions, thus providing evidence that dromedary camels act as the zoonotic source of this virus [6,7]. However, many primary human MERS cases do not have a history of direct contact with these animals [8]. This suggests the presence of unidentified routes of human-to-human transmission or the involvement of other animal species in spreading the virus to humans. Besides dromedary camels, other animal species, i.e. llamas, alpacas, and pigs have been shown to be susceptible and develop upper respiratory tract infection upon experimental intranasal

MERS-CoV inoculation [9–11]. This is in line with the expression of the MERS-CoV receptor, dipeptidyl peptidase-4 (DPP4), in their nasal epithelium [9]. MERS-CoV-seropositive llamas and alpacas have been reported in the field, and MERS-CoV-experimentally-inoculated alpacas have also been shown to transmit the virus via contact [12–14].

To further understand the zoonotic potential of MERS-CoV, it is crucial to delineate the factors involved in the spread of the virus among dromedaries as well as other animal species. In order to gain insight into these factors, we performed MERS-CoV transmission experiments in rabbits. We have previously shown that rabbits are susceptible to MERS-CoV and develop both upper and lower respiratory tract infection upon virus inoculation [15]. Naïve recipient rabbits were housed with MERS-CoV-inoculated donor rabbits either in the same or in adjacent cages to determine whether MERS-CoV can be transmitted via contact or airborne routes, respectively [16]. Donor rabbits were found to shed high levels of viral RNA but a limited amount of infectious virus thus potentially restricting MERS-CoV transmission in these animals.

2. Materials and Methods

2.1. Virus Stocks

In vivo experiments in this study were performed using Passage 7 human isolate MERS-CoV EMC strain (HCoV-EMC/2012) and passage 3 isolate MERS-CoV (Qatar15/2015; GenBank Acc. No. MK280984) that were propagated in Vero cells as described earlier [9]. Qatar15 was isolated from a 69 years old Qatari man that developed severe pneumonia and was PCR confirmed to have a MERS-CoV infection [17].

2.2. Animal Experiments

Animal experiments were approved and performed according to the guidelines from the Institutional Animal Welfare Committee (no. 201300121 approved on 17 July 2013, 122-17-01 approved on 28 September 2017, and AVD277002015283-WP01 approved on 2 November 2016). The studies were performed under biosafety level 3 (BSL3) conditions. To compare whether different routes of MERS-CoV inoculation generate similar clinical outcomes, twelve 6-month-old New Zealand rabbits (*Oryctolagus cuniculus*), specific pathogen free, and seronegative for MERS-CoV were divided into four groups. Animals were inoculated under ketamine-medetomidine anesthesia either (a) intranasally with 200 µL of 1×10^6 TCID$_{50}$/mL MERS-CoV; (b) intranasally with 1 mL 1×10^6 TCID$_{50}$/mL MERS-CoV; (c) intranasally with 200 µL of 1×10^6 TCID$_{50}$/mL MERS-CoV and intratracheally with 3 mL 4×10^6 TCID$_{50}$/mL MERS-CoV; or (d) intranasally with 1 mL PBS. The intranasal inoculums were divided equally over both nostrils. Nasal and throat swabs were obtained daily from day 1 up to day 4 post inoculation. These animals were then sacrificed on day 4 and respiratory tract tissues were collected. To compare the clinical outcomes of the MERS-CoV EMC strain and the Qatar15 strain, ten New Zealand rabbits were divided into two groups and inoculated with 1 mL of 1×10^6 TCID$_{50}$/mL of each MERS-CoV strain intranasally. Nasal and throat swabs were obtained from day 1 up to day 4 post inoculation and these animals were sacrificed on day 4.

To study MERS-CoV transmission, a modified version of the previously described influenza A virus ferret transmission set-up was used. This set-up consists of two clear polymethyl methacrylate cages of different sizes. Donor rabbits and direct contact recipients were housed in a cage of 35 cm × 30 cm × 65 cm (W × H × L), whereas airborne recipients were housed in a cage of 30 cm × 30 cm × 55 cm (W × H × L). These cages are separated by two stainless steel grids 10 cm apart to prevent direct contact but still allow airflow from the donor rabbit to the airborne recipient rabbit. These transmission cages allow the experiment to be conducted in negatively pressured isolators in the BSL3 facility, with HEPA-filtered airflow <0.1 m/s [18]. Since these cages were too small for New Zealand rabbits, we chose a smaller-sized breed, the Netherland dwarf rabbits (*Oryctolagus cuniculus*), for the MERS-CoV transmission experiment. We used both male and female rabbits with an age range of 6 months–3 years in these experiments. First, three MERS-CoV seronegative Netherland dwarf rabbits were inoculated intranasally with 1 mL of 1×10^6 TCID$_{50}$/mL MERS-CoV Qatar15 strain (500 µL per nostril) to show that they are equally susceptible to MERS-CoV as the New Zealand rabbits. Nasal and throat swabs were obtained daily up to 4 days post inoculation. These animals were then sacrificed on day 4 and their respiratory tract tissues were collected. For the virus transmission experiment, twelve Netherland dwarf rabbits were randomly distributed into four individually housed groups. One naïve rabbit from each group was inoculated intranasally with 1 mL of 1×10^6 TCID$_{50}$/mL MERS-CoV (500 µL per nostril), thus acting as donor rabbits. The other two rabbits were used as direct contact and airborne recipients, respectively. Donor and direct contact animals were of the same sex. All animals were sacrificed at 14 days post exposure and blood was collected to assess seroconversion.

2.3. Virological Analysis

Nasal swabs, throat swabs, and respiratory tract tissue samples were evaluated for the presence of infectious virus by virus titration, and for viral RNA by RT-qPCR against the UpE gene as previously described [9]. Samples with a cycle threshold less than forty were considered as positive for MERS-CoV RNA. Viral RNA was quantified as genome equivalents (GE) using MERS-CoV strain EMC (containing 10^6 TCID$_{50}$/mL) as a calibrator. Virus titration was performed in serial 10-fold dilutions on Vero cells. Cells were monitored under a light microscope at day 6 for cytopathic effect. The amount of infectious virus in swab samples was calculated by determining the TCID$_{50}$. Statistical analysis was performed using the GraphPad Prism program (La Jolla, CA, USA). Kruskal–Wallis test was applied due to the non-normal distribution of our data as priorly determined by Shapiro–Wilk test. The significant difference between groups was determined at a P-value < 0.05.

2.4. Histopathology and Histochemistry Analysis

Respiratory tract tissue samples were collected in formalin and embedded in paraffin for pathological analysis. Hematoxylin-eosin staining was performed for histopathological analysis. The presence of MERS-CoV nucleoprotein and MERS-CoV RNA was detected by immunohistochemistry and in-situ hybridization, respectively, using previously published protocols [15]. The localization of DPP4 in the respiratory tract of non-infected New Zealand rabbits was analyzed using an optimized immunohistochemical assay [15,19].

2.5. Serological Analysis

Collected serum samples were tested for MERS-CoV neutralizing antibodies using a virus neutralization assay and for MERS-CoV S1-specific antibodies using MERS-CoV S1 ELISA according to the previously published protocols [9]. Goat anti-rabbit IgG conjugated with HRP (1:2000, DAKO, Glostrup, Denmark) was used as a secondary antibody in the ELISA.

3. Results

3.1. Dipeptidyl peptidase-4 DPP4 is Expressed in the Upper and Lower Respiratory Tract of Rabbits

Rabbits are the smallest animal species that can be naturally infected by MERS-CoV. We previously reported that they develop both upper and lower respiratory tract infection upon MERS-CoV inoculation [15], suggesting the expression of the viral receptor at these locations. Using immunohistochemistry, we analyzed the DPP4 expression in rabbit respiratory tract tissues. In the upper respiratory tract, DPP4 is strongly expressed at the apical surface of both nasal respiratory and olfactory epithelium (Figure 1). In the lower respiratory tract, DPP4 is present in both bronchiolar and alveolar epithelial cells, although some variation in DPP4 expression was observed throughout the lungs. DPP4 is either absent, limitedly expressed on alveolar type II cells, or expressed on both alveolar type I and II cells (Figure 1). Thus, these data highlight a broad DPP4 expression in the respiratory tract tissues of rabbits. Our results show that rabbits express DPP4 in both the upper and lower respiratory tract epithelium, in line with MERS-CoV tropism in this species [9,19].

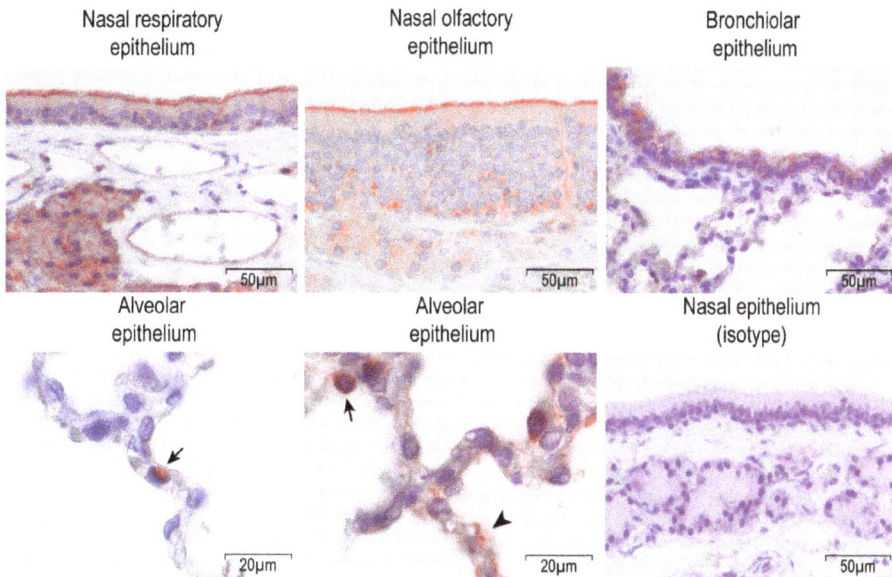

Figure 1. Dipeptidyl peptidase-1 (DPP4) expression in the respiratory tract tissues of rabbits. DPP4 is detected using immunohistochemistry and indicated in red in the figure. Type II cells are indicated with arrows, and type I cells with an arrowhead. Nasal epithelium and bronchiolar epithelium pictures were taken at a 400× magnification, and alveolar epithelium at 1000×. The isotype control showed no background signal in our assay.

3.2. Middle East Respiratory Syndrome Coronavirus (MERS-CoV) Infects Both Upper and Lower Respiratory Tract of Rabbits upon Intranasal Inoculation

In our previous study, we inoculated rabbits both intranasally and intratracheally [15]. Intratracheal inoculation is quite invasive, and thus requires a skillful operator to minimize procedure-related damage in the respiratory tract. In contrast, intranasal inoculation is less invasive and had been used in other studies to infect rabbits with MERS-CoV [20,21]. Here we investigated whether intranasal MERS-CoV inoculation is sufficient to induce both upper and lower respiratory tract infection in rabbits, in comparison to combined intranasal and intratracheal inoculation. Three New Zealand rabbits (*Oryctolagus cuniculus*) were inoculated with MERS-CoV EMC strain either intranasally with 200 μL of 1×10^6 TCID$_{50}$/mL (group a); intranasally with 1 mL of 1×10^6 TCID$_{50}$/mL (group b); intranasally with 200 μL of 1×10^6 TCID$_{50}$/mL and intratracheally with 3 mL of 4×10^6 TCID$_{50}$/mL (group c); or intranasally with 1ml of PBS as a negative control (group d). All three groups of MERS-CoV inoculated rabbits developed minimal clinical manifestations and histopathological lesions. The amount of viral RNA shed in the nasal and throat swabs did not vary among the inoculated groups (Figure 2A,B). However, in the lungs of the rabbits, the amount of viral RNA was significantly lower in group a than in groups b and c (Figure 2C). In line with these observations, fewer MERS-CoV-infected cells were observed in the lungs of group a animals compared to groups b and c (Figure 2D). Based on these results, we decided to use intranasal inoculation with 1 mL of 1×10^6 TCID$_{50}$/mL MERS-CoV for our subsequent experiments.

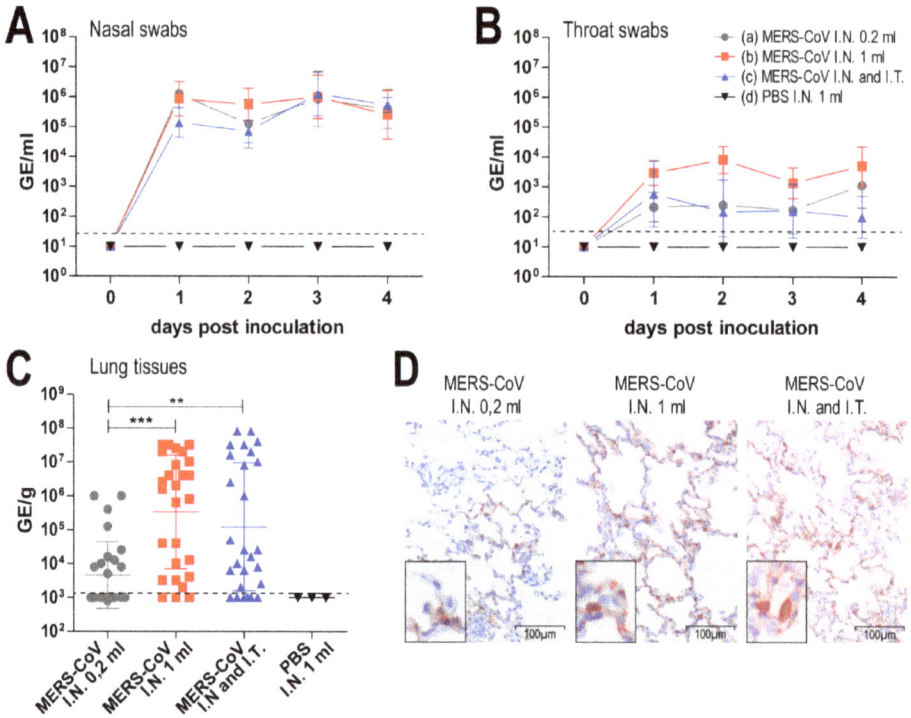

Figure 2. Middle East respiratory syndrome coronavirus (MERS-CoV) inoculation in rabbits with different routes and volumes. Three New Zealand rabbits were each infected either with (a) 200 µL of 1×10^6 $TCID_{50}$/mL MERS-CoV intranasal (I.N.); (b) 1 mL of 1×10^6 $TCID_{50}$/mL MERS-CoV I.N.; (c) 200 µL of 1×10^6 $TCID_{50}$/mL MERS-CoV I.N. combined with 3 mL of 4×10^6 $TCID_{50}$/mL MERS-CoV intratracheal (I.T.); or (d) 1mL of PBS I.N as negative control. These animals were sacrificed at day 4 post inoculation. Viral RNA shed by the MERS-CoV-inoculated rabbits in the nasal swabs (**A**) and throat swabs (**B**) are reported in genome equivalents per ml (GE/mL). Viral RNA detected in the lungs of these rabbits are reported in genome equivalents per gram tissues (GE/g). Dashed lines depict the detection limit of the assays. All error bars represent standard deviations. Statistical analysis is performed using Kruskal wallis test (**, *p* value < 0.01; ***, *p* value < 0.001). Representative figures of immunohistochemistry detecting MERS-CoV nucleoprotein (displayed in red) in the lungs of these rabbits were taken at a 200× magnification (**D**). All MERS-CoV-inoculated rabbits shed a relatively equal amount of viral RNA in the nasal and throat swabs (**A**,**B**). However, there was significantly less viral RNA in the lungs of rabbits in the group (a) in comparison to the other MERS-CoV-inoculated groups (**C**). A similar finding is observed in our immunohistochemistry analysis detecting MERS-CoV nucleoprotein (displayed in red) in the lungs of these rabbits (**D**). Pictures were taken in 200× magnification. The amount of viral RNA is displayed either in genome equivalent per mL (GE/mL) or per gram tissues (GE/g). Dashed lines depict the detection limit of the assays. All error bars represent standard deviations. Statistical analysis is performed using Kruskal–Wallis test (**, *p* value < 0.01; ***, *p* value < 0.001).

Different human MERS-CoV strains have been isolated since the EMC/2012 strain was first characterized [22,23]. However, studies that evaluate phenotypic differences between these strains in animals are currently lacking. We investigated whether a more recent MERS-CoV strain (Qatar15/2015) replicates differently in rabbits in comparison to the EMC strain. We found that rabbits inoculated with the MERS-CoV EMC strain and those with the Qatar15 strain developed an equally mild infection and shed similar levels of viral RNA in their nasal and throat swabs (Figure 3). Following this result, the MERS-CoV transmission experiment was performed using the Qatar15 strain, the more recent strain of these two.

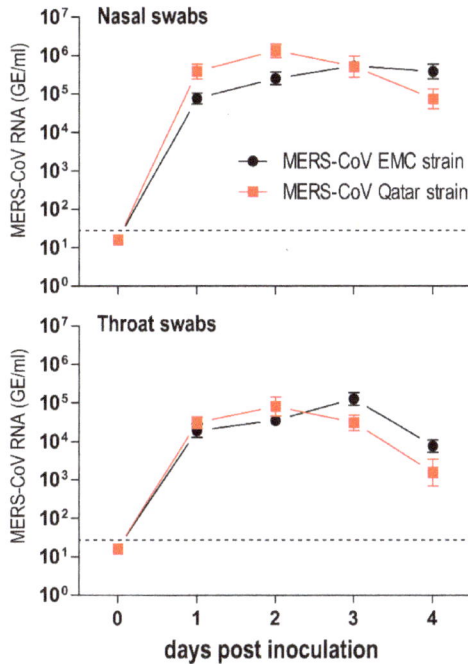

Figure 3. Middle East respiratory syndrome coronavirus (MERS-CoV) EMC strain and Qatar15 strain replicate equally in the upper respiratory tract of rabbits. Five New Zealand rabbits were each intranasally inoculated either with 1 mL of 1×10^6 TCID$_{50}$/mL MERS-CoV EMC strain or Qatar15 strain. Nasal and throat swabs were obtained from day 0 (before inoculation) until day 4 post inoculation. The amount of viral RNA is displayed in genome equivalents per mL (GE/mL). Dashed lines depict the detection limit of the assay. All error bars represent standard deviations.

3.3. Middle East Respiratory Syndrome Coronavirus (MERS-CoV) Transmission to Contact and Airborne-Exposed Rabbits

To study MERS-CoV transmission, an experimental set up previously used to investigate influenza A virus transmission between ferrets was used. This set up consists of two polymethyl methacrylate cages separated with two steel grids, 10 cm apart [16]. Because this set-up was too small to house New Zealand rabbits, we used a smaller-sized breed, the Netherland dwarf rabbits. Prior to the virus transmission experiment, Netherland dwarf rabbits were inoculated with MERS-CoV Qatar 15 strain to determine their susceptibility to the virus. Similar to the New Zealand rabbits [15], Netherland dwarf rabbits shed viral RNA in the nasal and throat swabs as well as in the respiratory tract tissues upon intranasal inoculation (Figure 4A,B). Identical to the New Zealand rabbits [15], the MERS-CoV-inoculated Netherland dwarf rabbits did not develop any clinical signs, including nasal discharge, and showed minimal histopathological lesions and immune cell infiltration in the respiratory tract.

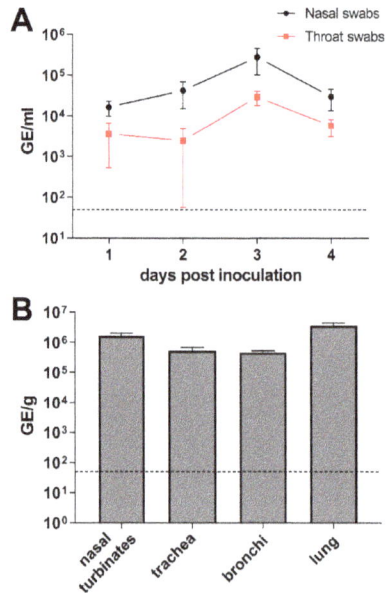

Figure 4. Middle East respiratory syndrome coronavirus (MERS-CoV) infects the upper and lower respiratory tract of Netherland dwarf rabbits. The amount of viral RNA in the nasal and throat swabs (**A**) as well as in the respiratory tract tissues (**B**) are displayed in genome equivalents per mL (GE/mL) and genome equivalents per gram tissues (GE/g), respectively. Dashed lines depict the detection limit of the assays. All error bars represent standard error of means.

To study MERS-CoV transmission, four Netherland dwarf rabbits were intranasally inoculated with MERS-CoV. Six-hours later, each of them was co-housed with one naïve rabbit in the same cage, and 24 h later another one was co-housed in an adjacent cage to determine whether MERS-CoV could be transmitted through contact and/or airborne routes. Nasal and throat swabs were collected every other day up to day 7 or 9 post inoculation/exposure for the donor and direct contact rabbits, respectively, and day 9 post exposure for the airborne recipient ones. Both viral RNA and infectious virus were quantified in these samples. We found that all donor rabbits shed high loads of viral RNA in both the nasal swabs ($\sim 10^5$–10^6 TCID$_{50}$ GE/mL) and the throat swabs ($\sim 10^3$–10^4 TCID$_{50}$ GE/mL). The amount of viral RNA shed by the inoculated rabbits remained high until day 7 post inoculation. On the other hand, recipient rabbits housed in the same cage (direct contact recipients), or adjacent cage (airborne recipients), shed limited amounts of viral RNA (~ 10 TCID$_{50}$ GE/mL) in both nasal and throat swabs. Among four animals in each group, only two direct contact recipient and two airborne recipient rabbits had detectable viral RNA up to day 5 post inoculation in the nasal swabs, while in the throat swabs, viral RNA was only detected in one direct contact recipient and one airborne recipient rabbit at day 1 post inoculation (Figure 5A,B). Infectious virus was detected at low level ($\sim 10^2$ TCID$_{50}$/mL) both in the nasal and throat swabs of the donor rabbits; in the nasal swabs of all donor rabbits at day 1 post inoculation, and in one of the donors up to day 7. In the throat swabs, infectious virus was only detected in two donors on day 1, up to day 5 in one of them. In contrast, none of the swabs from recipient rabbits was positive for virus titration (Figure 5C,D). Serological analysis of samples collected 14 days after exposure showed that only the donor rabbits seroconverted and developed neutralizing antibodies (Figure 6A,B). The antibody response of these directly inoculated rabbits was relatively low, confirming the results of previous studies [15,21]. This indicates that these rabbits developed MERS-CoV infection while the contact and airborne-exposed rabbits did not, supporting the results of the virus titration.

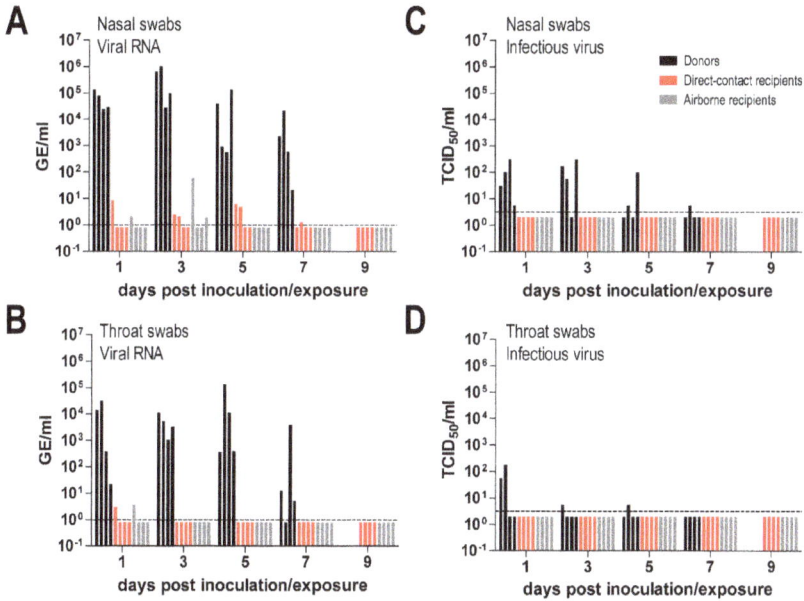

Figure 5. Shedding of Middle East respiratory syndrome coronavirus (MERS-CoV) RNA and infectious virus in directly inoculated, contact-exposed and airborne-exposed rabbits. MERS-CoV RNA and infectious virus were measured in the nasal (**A**,**B**) and throat swabs (**C**,**D**). The amount of viral RNA is displayed in genome equivalents per mL (GE/mL), while infectious virus is shown as 50% tissue culture infective dose per ml (TCID$_{50}$/mL). Dashed lines depict the detection limit of each assay.

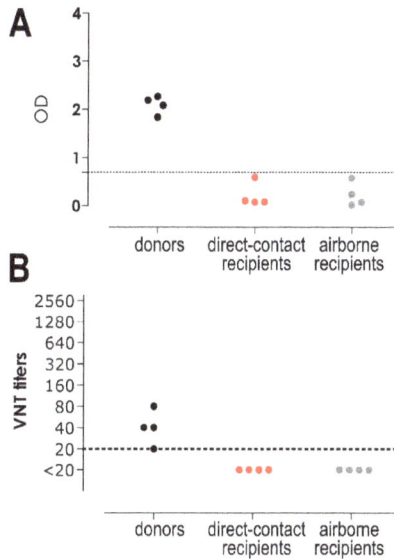

Figure 6. Middle East respiratory syndrome coronavirus (MERS-CoV)-specific antibody response in rabbits. S1-specific MERS-CoV antibodies were measured with ELISA and displayed as optical density (OD) value (**A**), while MERS-CoV neutralizing antibodies were measured with the virus neutralization test (VNT) and displayed in titers (**B**). Dashed lines depict the detection limit of the assay.

4. Discussion

Current data indicate that MERS-CoV is highly endemic in dromedary camels in the Arabian Peninsula and Africa and has been circulating in these animals for decades [3–5,24]. This suggests that this virus is easily transmitted between dromedary camels. From an epidemiological point of view, it is important to know whether other animal species in the region may also spread the virus to humans or other animal species. In vitro, MERS-CoV has been found to infect cells from a broad range of animal species including Old and New World camelids as well as primates, bats, cows, sheep, goats, pigs, horses, and rabbits [15,25,26]. The DPP4 viral receptor of these species, especially rabbits, has high similarity to that of humans and dromedary camels, especially in the region that interacts with the spike protein, and thus can facilitate MERS-CoV infection [25–27]. The New World camelids, i.e. llamas and alpacas, have been shown to seroconvert to MERS-CoV when present in regions where MERS-CoV is circulating and may transmit the virus [12–14]. It is currently unclear why, besides camelids, other livestock animals do not seem to transmit the virus to humans [24,28–30]. To further understand the transmission potential of MERS-CoV, we performed virus transmission experiments using rabbits as animal model.

To perform the virus transmission experiments, we housed MERS-CoV-inoculated rabbits together with naïve contact rabbits either in the same or adjacent cages. Rabbits developed both upper and lower respiratory tract infection upon MERS-CoV inoculation [15], either via intranasal or combined intranasal and intratracheal routes, in line with the localization of DPP4 in their respiratory tract epithelium. The amount of viral RNA being shed by the inoculated rabbits during the first three days post inoculation is almost as high as that of the MERS-CoV-inoculated dromedary camels [31,32]. However, none of the direct contact and airborne-exposed rabbits developed any clinical signs, shed significant levels of viral RNA, shed infectious virus, nor did they seroconvert. One possible reason for this lack of transmission is the limited amount of infectious virus being shed by the inoculated rabbits [21]. Previous studies have shown that in the nasal and lung tissues of experimentally infected rabbits, infectious virus was generally detected in a limited amount despite the abundant presence of viral RNA and virus nucleoprotein [15,21]. Alternatively, low levels of infectious virus transmitted to recipient animals may be unable to initiate a productive infection due to the presence of host proteins that restrict replication. Comparable to rabbits, MERS-CoV-infected pigs and goats develop minimal clinical signs, hardly shed infectious virus, and barely spread the virus to naïve animals [28,33]. In contrast, MERS-CoV-infected dromedary camels develop nasal discharge and shed a high amount infectious virus (10^4–10^5 TCID$_{50}$/mL), almost equal to the amount of viral RNA being shed (10^5–10^6 GE/mL), during the first 4 days post inoculation [31,32]. These interspecies differences may indicate presence of nasal discharge and infectious virus shedding as critical factors in MERS-CoV transmission. In humans, levels of infectious virus shed by MERS-CoV patients have rarely been reported. However, MERS-CoV patients that transmit the virus have been shown to shed a higher amount of viral RNA in their swabs compared to those that do not, supporting the quantity of virus shed as an important factor in the transmission of MERS-CoV between humans [34]. For influenza A viruses, infectious virus shedding has been documented as one of the main determinants of airborne virus transmission. Using ferrets as an animal model, it has been reported that a reduction in infectious virus shedding in the nasal swabs can subsequently limit virus transmission [35–37].

Our results show that despite the presence of DPP4 in the upper respiratory tract, accompanied by MERS-CoV replication at this site, a limited amount of infectious virus was shed. Similarly, titration of rabbit lung homogenates that show high levels of viral RNA and presence of nucleoprotein (Figure 2C,D) resulted in only low levels of infectious virus, in line with earlier observations [15,21]. At this stage, it is not clear which host mediated mechanisms limit the production of infectious virus while allowing viral RNA to still be shed at high levels. Since restriction of infectious virus shedding in the rabbits already occurred one day post inoculation, activation of host innate immune responses, including type I interferon induction, may be relevant. In vitro studies have shown that MERS-CoV is relatively sensitive to type I interferon-mediated antiviral activities [38,39]. In human plasmacytoid dendritic cells, MERS-CoV inoculation leads to secretion of large amount of type I and III interferons and production of viral RNA, but hardly any infectious virus is being produced [40]. It is also possible that most infectious virus shed by these rabbits is defective, lacking the capacity to efficiently infect target cells. Further studies are needed to elucidate the mechanisms that restrict MERS-CoV replication in rabbits compared to dromedary camels. Potentially, some of the MERS-CoV accessory proteins shown to antagonize immune responses including production of interferon, may not work efficiently in some MERS-CoV susceptible species, including rabbits. It is intriguing to investigate whether a similar phenomenon occurs in some human MERS-CoV infections and whether this is linked to the development of asymptomatic to mild clinical manifestations [41]. This might partly explain why MERS-CoV transmission in humans is rather inefficient in comparison to dromedary camels [42,43], and why camelids that secrete high levels of infectious virus are the only known zoonotic source of MERS-CoV [2,7,14,31]. Deciphering these mechanisms could potentially offer insight into understanding MERS-CoV transmission as well as developing novel treatments to tackle the ongoing outbreaks.

Author Contributions: Conceptualization, B.L.H., S.H.; supervision, B.L.H.; experiments, W.W., N.M.A.O., M.R., D.d.M., T.M.B., P.L., K.J.S., L.d.W., G.v.A., J.M.v.d.B. and S.H.; resources for Qatar15 strain virus, E.A.B.A.F., M.A.H.; writing—original draft preparation, W.W., N.M.A.O., B.L.H.; writing—review and editing, all authors.

Funding: This research was funded by Zoonotic Anticipation and Preparedness Initiative (Innovative Medicines Initiative grant 115760), with assistance and financial support from Innovative Medicines Initiative and the European Commission and contributions from European Federation of Pharmaceutical Industries and Associations partners. W.W. was supported by Nederlandse Organisatie voor Wetenschappelijk Onderzoek (grant 91213066). S.H is funded by an NWO VIDI grant (contract number 91715372). The research of M.R., T.M.B., and S.H. is supported by NIAID/NIH contract HHSN272201400008C.

Conflicts of Interest: K.S., L.d.W., and G.J.A. are full time employees at Viroclinics Biosciences BV. The other authors have no conflict of interest to disclose. The funders had no role in the design of the study; in the collection, analyses, or interpretation of data; in the writing of the manuscript, or in the decision to publish the results.

References

1. Zaki, A.M.; van Boheemen, S.; Bestebroer, T.M.; Osterhaus, A.D.; Fouchier, R.A. Isolation of a novel coronavirus from a man with pneumonia in Saudi Arabia. *N. Engl. J. Med.* **2012**, *367*, 1814–1820. [CrossRef] [PubMed]

2. Raj, V.S.; Farag, E.A.; Reusken, C.B.; Lamers, M.M.; Pas, S.D.; Voermans, J.; Smits, S.L.; Osterhaus, A.D.; Al-Mawlawi, N.; Al-Romaihi, H.E.; et al. Isolation of MERS coronavirus from a dromedary camel, Qatar, 2014. *Emerg. Infect. Dis.* **2014**, *20*, 1339–1342. [CrossRef]

3. Muller, M.A.; Corman, V.M.; Jores, J.; Meyer, B.; Younan, M.; Liljander, A.; Bosch, B.J.; Lattwein, E.; Hilali, M.; Musa, B.E.; et al. MERS coronavirus neutralizing antibodies in camels, Eastern Africa, 1983–1997. *Emerg. Infect. Dis.* **2014**, *20*, 2093–2095. [CrossRef] [PubMed]

4. Alagaili, A.N.; Briese, T.; Mishra, N.; Kapoor, V.; Sameroff, S.C.; Burbelo, P.D.; de Wit, E.; Munster, V.J.; Hensley, L.E.; Zalmout, I.S.; et al. Middle East respiratory syndrome coronavirus infection in dromedary camels in Saudi Arabia. *MBio* **2014**, *5*, e00884-14. [CrossRef]

5. Reusken, C.B.; Messadi, L.; Feyisa, A.; Ularamu, H.; Godeke, G.J.; Danmarwa, A.; Dawo, F.; Jemli, M.; Melaku, S.; Shamaki, D.; et al. Geographic distribution of MERS coronavirus among dromedary camels, Africa. *Emerg. Infect. Dis.* **2014**, *20*, 1370–1374. [CrossRef]

6. Paden, C.R.; Yusof, M.; Al Hammadi, Z.M.; Queen, K.; Tao, Y.; Eltahir, Y.M.; Elsayed, E.A.; Marzoug, B.A.; Bensalah, O.K.A.; Khalafalla, A.I.; et al. Zoonotic origin and transmission of Middle East respiratory syndrome coronavirus in the UAE. *Zoonoses Public Health* **2018**, *65*, 322–333. [CrossRef]

7. Haagmans, B.L.; Al Dhahiry, S.H.; Reusken, C.B.; Raj, V.S.; Galiano, M.; Myers, R.; Godeke, G.J.; Jonges, M.; Farag, E.; Diab, A.; et al. Middle East respiratory syndrome coronavirus in dromedary camels: An outbreak investigation. *Lancet Infect. Dis.* **2014**, *14*, 140–145. [CrossRef]

8. Conzade, R.; Grant, R.; Malik, M.R.; Elkholy, A.; Elhakim, M.; Samhouri, D.; Ben Embarek, P.K.; van Kerkhove, M.D. Reported Direct and Indirect Contact with Dromedary Camels among Laboratory-Confirmed MERS-CoV Cases. *Viruses* **2018**, *10*, 425. [CrossRef]

9. Vergara-Alert, J.; van den Brand, J.M.; Widagdo, W.; Munoz, M.T.; Raj, S.; Schipper, D.; Solanes, D.; Cordon, I.; Bensaid, A.; Haagmans, B.L.; et al. Livestock Susceptibility to Infection with Middle East Respiratory Syndrome Coronavirus. *Emerg. Infect. Dis.* **2017**, *23*, 232–240. [CrossRef]

10. Munster, V.J.; Adney, D.R.; van Doremalen, N.; Brown, V.R.; Miazgowicz, K.L.; Milne-Price, S.; Bushmaker, T.; Rosenke, R.; Scott, D.; Hawkinson, A.; et al. Replication and shedding of MERS-CoV in Jamaican fruit bats (Artibeus jamaicensis). *Sci. Rep.* **2016**, *6*, 21878. [CrossRef] [PubMed]

11. Crameri, G.; Durr, P.A.; Klein, R.; Foord, A.; Yu, M.; Riddell, S.; Haining, J.; Johnson, D.; Hemida, M.G.; Barr, J.; et al. Experimental Infection and Response to Rechallenge of Alpacas with Middle East Respiratory Syndrome Coronavirus. *Emerg. Infect. Dis.* **2016**, *22*, 1071–1074. [CrossRef]

12. Reusken, C.B.; Schilp, C.; Raj, V.S.; De Bruin, E.; Kohl, R.H.; Farag, E.A.; Haagmans, B.L.; Al-Romaihi, H.; Le Grange, F.; Bosch, B.J.; et al. MERS-CoV Infection of Alpaca in a Region Where MERS-CoV is Endemic. *Emerg. Infect. Dis.* **2016**, *22*, 1129–1131. [CrossRef]

13. David, D.; Rotenberg, D.; Khinich, E.; Erster, O.; Bardenstein, S.; van Straten, M.; Okba, N.M.A.; Raj, S.V.; Haagmans, B.L.; Miculitzki, M.; et al. Middle East respiratory syndrome coronavirus specific antibodies in naturally exposed Israeli llamas, alpacas and camels. *One Health* **2018**, *5*, 65–68. [CrossRef]

14. Adney, D.R.; Bielefeldt-Ohmann, H.; Hartwig, A.E.; Bowen, R.A. Infection, Replication, and Transmission of Middle East Respiratory Syndrome Coronavirus in Alpacas. *Emerg. Infect. Dis.* **2016**, *22*, 1031–1037. [CrossRef] [PubMed]

15. Haagmans, B.L.; van den Brand, J.M.; Provacia, L.B.; Raj, V.S.; Stittelaar, K.J.; Getu, S.; de Waal, L.; Bestebroer, T.M.; van Amerongen, G.; Verjans, G.M.; et al. Asymptomatic Middle East respiratory syndrome coronavirus infection in rabbits. *J. Virol.* **2015**, *89*, 6131–6135. [CrossRef] [PubMed]

16. Herfst, S.; Schrauwen, E.J.; Linster, M.; Chutinimitkul, S.; de Wit, E.; Munster, V.J.; Sorrell, E.M.; Bestebroer, T.M.; Burke, D.F.; Smith, D.J.; et al. Airborne transmission of influenza A/H5N1 virus between ferrets. *Science* **2012**, *336*, 1534–1541. [CrossRef]

17. ProMED-mail Qatar: New case-Qatari Supreme Council of Health. 8 March 2015. Available online: https://www.promedmail.org/index.php (accessed on 23 November 2018).

18. Munster, V.J.; de Wit, E.; van den Brand, J.M.; Herfst, S.; Schrauwen, E.J.; Bestebroer, T.M.; van de Vijver, D.; Boucher, C.A.; Koopmans, M.; Rimmelzwaan, G.F.; et al. Pathogenesis and transmission of swine-origin 2009 A(H1N1) influenza virus in ferrets. *Science* **2009**, *325*, 481–483. [CrossRef] [PubMed]

19. Widagdo, W.; Raj, V.S.; Schipper, D.; Kolijn, K.; van Leenders, G.J.; Bosch, B.J.; Bensaid, A.; Segales, J.; Baumgartner, W.; Osterhaus, A.D.; et al. Differential Expression of the Middle East Respiratory Syndrome Coronavirus Receptor in the Upper Respiratory Tracts of Humans and Dromedary Camels. *J. Virol.* **2016**, *90*, 4838–4842. [CrossRef]

20. Houser, K.V.; Gretebeck, L.; Ying, T.; Wang, Y.; Vogel, L.; Lamirande, E.W.; Bock, K.W.; Moore, I.N.; Dimitrov, D.S.; Subbarao, K. Prophylaxis With a Middle East Respiratory Syndrome Coronavirus (MERS-CoV)-Specific Human Monoclonal Antibody Protects Rabbits From MERS-CoV Infection. *J. Infect. Dis.* **2016**, *213*, 1557–1561. [CrossRef]

21. Houser, K.V.; Broadbent, A.J.; Gretebeck, L.; Vogel, L.; Lamirande, E.W.; Sutton, T.; Bock, K.W.; Minai, M.; Orandle, M.; Moore, I.N.; et al. Enhanced inflammation in New Zealand white rabbits when MERS-CoV reinfection occurs in the absence of neutralizing antibody. *PLoS Pathog.* **2017**, *13*, e1006565. [CrossRef]

22. Lamers, M.M.; Raj, V.S.; Shafei, M.; Ali, S.S.; Abdallh, S.M.; Gazo, M.; Nofal, S.; Lu, X.; Erdman, D.D.; Koopmans, M.P.; et al. Deletion Variants of Middle East Respiratory Syndrome Coronavirus from Humans, Jordan, 2015. *Emerg. Infect. Dis.* **2016**, *22*, 716–719. [CrossRef] [PubMed]

23. Assiri, A.M.; Midgley, C.M.; Abedi, G.R.; Bin Saeed, A.; Almasri, M.M.; Lu, X.; Al-Abdely, H.M.; Abdalla, O.; Mohammed, M.; Algarni, H.S.; et al. Epidemiology of a Novel Recombinant Middle East Respiratory Syndrome Coronavirus in Humans in Saudi Arabia. *J. Infect. Dis.* **2016**, *214*, 712–721. [CrossRef]

24. Reusken, C.B.; Haagmans, B.L.; Muller, M.A.; Gutierrez, C.; Godeke, G.J.; Meyer, B.; Muth, D.; Raj, V.S.; Smits-De Vries, L.; Corman, V.M.; et al. Middle East respiratory syndrome coronavirus neutralising serum antibodies in dromedary camels: A comparative serological study. *Lancet Infect. Dis.* **2013**, *13*, 859–866. [CrossRef]

25. Barlan, A.; Zhao, J.; Sarkar, M.K.; Li, K.; McCray, P.B., Jr.; Perlman, S.; Gallagher, T. Receptor variation and susceptibility to Middle East respiratory syndrome coronavirus infection. *J. Virol.* **2014**, *88*, 4953–4961. [CrossRef]

26. Van Doremalen, N.; Miazgowicz, K.L.; Milne-Price, S.; Bushmaker, T.; Robertson, S.; Scott, D.; Kinne, J.; McLellan, J.S.; Zhu, J.; Munster, V.J. Host species restriction of Middle East respiratory syndrome coronavirus through its receptor, dipeptidyl peptidase 4. *J. Virol.* **2014**, *88*, 9220–9232. [CrossRef] [PubMed]

27. Bosch, B.J.; Raj, V.S.; Haagmans, B.L. Spiking the MERS-coronavirus receptor. *Cell. Res.* **2013**, *23*, 1069–1070. [CrossRef] [PubMed]

28. Adney, D.R.; Brown, V.R.; Porter, S.M.; Bielefeldt-Ohmann, H.; Hartwig, A.E.; Bowen, R.A. Inoculation of Goats, Sheep, and Horses with MERS-CoV Does Not Result in Productive Viral Shedding. *Viruses* **2016**, *8*, 230. [CrossRef]

29. Hemida, M.G.; Perera, R.A.; Wang, P.; Alhammadi, M.A.; Siu, L.Y.; Li, M.; Poon, L.L.; Saif, L.; Alnaeem, A.; Peiris, M. Middle East Respiratory Syndrome (MERS) coronavirus seroprevalence in domestic livestock in Saudi Arabia, 2010 to 2013. *Euro Surveill.* **2013**, *18*, 20659. [CrossRef]

30. Ali, M.; El-Shesheny, R.; Kandeil, A.; Shehata, M.; Elsokary, B.; Gomaa, M.; Hassan, N.; El Sayed, A.; El-Taweel, A.; Sobhy, H.; et al. Cross-sectional surveillance of Middle East respiratory syndrome coronavirus (MERS-CoV) in dromedary camels and other mammals in Egypt, August 2015 to January 2016. *Euro Surveill* **2017**, *22*, 30487. [CrossRef]

31. Haagmans, B.L.; van den Brand, J.M.; Raj, V.S.; Volz, A.; Wohlsein, P.; Smits, S.L.; Schipper, D.; Bestebroer, T.M.; Okba, N.; Fux, R.; et al. An orthopoxvirus-based vaccine reduces virus excretion after MERS-CoV infection in dromedary camels. *Science* **2016**, *351*, 77–81. [CrossRef]

32. Adney, D.R.; van Doremalen, N.; Brown, V.R.; Bushmaker, T.; Scott, D.; de Wit, E.; Bowen, R.A.; Munster, V.J. Replication and shedding of MERS-CoV in upper respiratory tract of inoculated dromedary camels. *Emerg. Infect. Dis.* **2014**, *20*, 1999–2005. [CrossRef]

33. Vergara-Alert, J.; Raj, V.S.; Munoz, M.; Abad, F.X.; Cordon, I.; Haagmans, B.L.; Bensaid, A.; Segales, J. Middle East respiratory syndrome coronavirus experimental transmission using a pig model. *Transbound Emerg. Dis.* **2017**, *64*, 1342–1345. [CrossRef]

34. Kim, S.W.; Park, J.W.; Jung, H.D.; Yang, J.S.; Park, Y.S.; Lee, C.; Kim, K.M.; Lee, K.J.; Kwon, D.; Hur, Y.J.; et al. Risk factors for transmission of Middle East respiratory syndrome coronavirus infection during the 2015 outbreak in South Korea. *Clin. Infect. Dis.* **2017**, *64*, 551–557. [CrossRef] [PubMed]

35. Houser, K.V.; Pearce, M.B.; Katz, J.M.; Tumpey, T.M. Impact of prior seasonal H3N2 influenza vaccination or infection on protection and transmission of emerging variants of influenza A(H3N2)v virus in ferrets. *J. Virol.* **2013**, *87*, 13480–13489. [CrossRef]

36. Pearce, M.B.; Belser, J.A.; Houser, K.V.; Katz, J.M.; Tumpey, T.M. Efficacy of seasonal live attenuated influenza vaccine against virus replication and transmission of a pandemic 2009 H1N1 virus in ferrets. *Vaccine* **2011**, *29*, 2887–2894. [CrossRef]

37. Sorrell, E.M.; Schrauwen, E.J.; Linster, M.; de Graaf, M.; Herfst, S.; Fouchier, R.A. Predicting "airborne" influenza viruses: (trans-) mission impossible? *Curr. Opin. Virol.* **2011**, *1*, 635–642. [CrossRef] [PubMed]

38. De Wilde, A.H.; Raj, V.S.; Oudshoorn, D.; Bestebroer, T.M.; van Nieuwkoop, S.; Limpens, R.W.; Posthuma, C.C.; van der Meer, Y.; Barcena, M.; Haagmans, B.L.; et al. MERS-coronavirus replication induces severe in vitro cytopathology and is strongly inhibited by cyclosporin A or interferon-alpha treatment. *J. Gen. Virol.* **2013**, *94*, 1749–1760. [CrossRef] [PubMed]

39. Falzarano, D.; de Wit, E.; Martellaro, C.; Callison, J.; Munster, V.J.; Feldmann, H. Inhibition of novel beta coronavirus replication by a combination of interferon-alpha2b and ribavirin. *Sci. Rep.* **2013**, *3*, 1686. [CrossRef]

40. Scheuplein, V.A.; Seifried, J.; Malczyk, A.H.; Miller, L.; Hocker, L.; Vergara-Alert, J.; Dolnik, O.; Zielecki, F.; Becker, B.; Spreitzer, I.; et al. High secretion of interferons by human plasmacytoid dendritic cells upon recognition of Middle East respiratory syndrome coronavirus. *J. Virol.* **2015**, *89*, 3859–3869. [CrossRef] [PubMed]

41. World Health Organization MERS Situation Update. March 2018. Available online: http://www.emro.who.int/pandemic-epidemic-diseases/mers-cov/mers-situation-update-march-2018.html (accessed on 1 March 2019).

42. Drosten, C.; Meyer, B.; Muller, M.A.; Corman, V.M.; Al-Masri, M.; Hossain, R.; Madani, H.; Sieberg, A.; Bosch, B.J.; Lattwein, E.; et al. Transmission of MERS-coronavirus in household contacts. *N. Engl. J. Med.* **2014**, *371*, 828–835. [CrossRef]

43. Cho, S.Y.; Kang, J.M.; Ha, Y.E.; Park, G.E.; Lee, J.Y.; Ko, J.H.; Lee, J.Y.; Kim, J.M.; Kang, C.I.; Jo, I.J.; et al. MERS-CoV outbreak following a single patient exposure in an emergency room in South Korea: An epidemiological outbreak study. *Lancet* **2016**, *388*, 994–1001. [CrossRef]

viruses

MDPI

Article

A Human DPP4-Knockin Mouse's Susceptibility to Infection by Authentic and Pseudotyped MERS-CoV

Changfa Fan [1,†], Xi Wu [1,†], Qiang Liu [2], Qianqian Li [2], Susu Liu [1], Jianjun Lu [3], Yanwei Yang [3], Yuan Cao [1], Weijin Huang [2], Chunnan Liang [1], Tianlei Ying [4], Shibo Jiang [4] and Youchun Wang [2,*]

[1] Division of Animal Model Research, Institute for Laboratory Animal Resources, National Institutes for Food and Drug Control, Beijing 100050, China; fancf@nifdc.org.cn (C.F.); wuxi@nifdc.org.cn (X.W.); liususu@nifdc.org.cn (S.L.); caoyuan0512@163.com (Y.C.); chunnan_liang@nifdc.org.cn (C.L.)
[2] Division of HIV/AIDS and Sex-Transmitted Virus Vaccines, National Institutes for Food and Drug Control, Beijing 100050, China; liuqiang@nifdc.org.cn (Q.L.); liqianqian1199@163.com (Q.L.); huangweijin@nifdc.org.cn (W.H.)
[3] National Center for Safety Evaluation of Drugs, Institute for Food and Drug Safety Evaluation, National Institutes for Food and Drug Control, A8 Hongda Middle Street, Beijing Economic-Technological Development Area, Beijing 100176, China; lujianjun@nifdc.org.cn (J.L.); yangyanwei@nifdc.org.cn (Y.Y.)
[4] Key Laboratory of Medical Molecular Virology of the Ministries of Education and Health, Shanghai Medical College, Fudan University, Shanghai 200032, China; tlying@fudan.edu.cn (T.Y.); shibojiang@fudan.edu.cn (S.J.)
* Correspondence: wangyc@nifdc.org.cn; Tel.: +86-10-670955921; Fax: +86-10-6053754
† These authors contributed equally to this work.

Received: 24 June 2018; Accepted: 17 August 2018; Published: 23 August 2018

Abstract: Infection by the Middle East respiratory syndrome coronavirus (MERS-CoV) causes respiratory illness and has a high mortality rate (~35%). The requirement for the virus to be manipulated in a biosafety level three (BSL-3) facility has impeded development of urgently-needed antiviral agents. Here, we established a novel mouse model by inserting human dipeptidyl peptidase 4 (hDPP4) into the Rosa26 locus using CRISPR/Cas9, resulting in global expression of the transgene in a genetically stable mouse line. The mice were highly susceptible to infection by MERS-CoV clinical strain hCoV-EMC, which induced severe diffuse pulmonary disease in the animals, and could also be infected by an optimized pseudotyped MERS-CoV. Administration of the neutralizing monoclonal antibodies, H111-1 and m336, as well as a fusion inhibitor peptide, HR2P-M2, protected mice from challenge with authentic and pseudotyped MERS-CoV. These results confirmed that the hDPP4-knockin mouse is a novel model for studies of MERS-CoV pathogenesis and anti-MERS-CoV antiviral agents in BSL-3 and BSL-2 facilities, respectively.

Keywords: mouse model; hDPP4; MERS-CoV; pseudotyped virus; authentic virus

1. Introduction

According to the World Health Organization's (WHO) latest report, the Middle East respiratory syndrome coronavirus (MERS-CoV) has caused 2121 laboratory-confirmed infections and 740 deaths (~35% mortality rate) in 27 countries since the virus was first isolated in Saudi Arabia in September, 2012 (http://www.who.int/emergencies/mers-cov/en/). Although no cases of the closely-related severe acute respiratory syndrome coronavirus (SARS-CoV) have been reported since 2005 [1], the incidence of MERS-CoV infections and the number of affected countries both continue to increase. Epidemiological trends indicate that prevention of potential epidemic spread of MERS-CoV is likely to be a long and tough battle [2]. Unfortunately, no clinically-approved antiviral monoclonal antibodies (mAbs), small molecule drugs or vaccines against MERS-CoV are available.

Preclinical animal models that mimic the human pathogenesis of MERS-CoV infection are urgently needed for the development of vaccines and therapies. The classification of MERS-CoV as a biosafety level three (BSL-3) agent is an additional hurdle to progress in this area. Since the emergence of MERS-CoV, several animal species have been evaluated as potential models, including wild-type mice, mice deficient in innate immunity [3], Syrian hamsters [4], ferrets [5] and non-human primates (NHPs) [4,6]. Two NHPs (rhesus macaques and marmosets) were found to be susceptible to MERS-CoV infection, but other species including mice are naturally non-permissive to infection. Although human DPP4 (hDPP4) has been identified as the receptor for MERS-CoV [7] that mediates infection, mouse DPP4 (mDPP4) is not functional in this respect [8]. Mice adenovirally transduced with hDPP4 could be transiently infected by MERS-CoV but did not develop fatal MERS disease [9].Moreover, hDPP4-transgenic (Tg) mice expressing hDPP4 under the control of either the cytokeratin 18 promoter [10] or a ubiquitous promoter [11] were susceptible to MERS-CoV infection and developed fatal disease, but also developed unrelated lethal encephalitis. Two knock-in (KI) mice in which mDPP4 was replaced with hDPP4 using CRISPR/Cas9 [8,12] could be infected by high-titer MERS-CoV isolates but were more susceptible to mouse-adapted MERS-CoV strains. Since mDPP4 is central to normal glucose homeostasis [13] and plays an important role in immunity [14], altering its sequence may disrupt these functions.

Neutralizing antiviral mAbs are promising candidates for treatment and prevention of MERS-CoV infection [15]. Several highly potent mouse, human and humanized neutralizing mAbs targeting the receptor-binding domain (RBD) of the spike (S) protein have been reported [16–20], and antiviral compounds and prophylactic vaccines are also under development. These vaccines and therapeutic agents have typically been evaluated in in vitro systems using MERS-CoV pseudoviruses; however, their efficacy must eventually be confirmed in vivo using a suitable animal model. Several Tg or KI mice susceptible to MERS-CoV are available [9,11,12,21], and challenge experiments under BSL-3 conditions have been done for the evaluation of potential therapeutics and vaccines.

In this study, we established a novel KI mouse by inserting the full-length hDPP4 gene into the C57BL/6 mouse genome at the Rosa26 locus (used for constitutive, ubiquitous gene expression) using CRISPR/Cas9 gene editing technology. This mouse, termed R26-hDPP4, displayed severe lung disease related to acute respiratory symptoms (ARDS) as well as central nervous system (CNS) involvement after infection with authentic MERS-CoV clinical isolates at low dose. Moreover, high-titer MERS-CoV pseudovirus could also productively infect R26-hDPP4 mice, with effects comparable to those following authentic infection.

2. Materials and Methods

2.1. Ethics Statement

Wild-type C57BL/6 mice and genetically-modified mice were supplied by the Institute for Laboratory Animal Resources, National Institute for Food and Drug Control (Beijing, China). All studies were performed in compliance with animal protocols (#2017-B-004) approved by the Institutional Animal Care and Use Committee of the National Institute for Food and Drug Control, China Food and Drug Administration (CFDA, Beijing, China) and in compliance with the "Guide for the Care and Use of Laboratory Animals" (National Academies Press: Washington, DC, USA, 2011; 8th ed.). The license number of the Animal Use Certificate issued by the Science & Technology Department of China (Beijing, China) was SYXK 2016-004, approved on 18 February 2016

2.2. Construction of hDPP4-KIMice

To construct the targeting vector, cDNA encoding hDPP4 (also known as CD26 [14]) linked to the red fluorescent protein tdTomato by an internal ribosomal entry site (IRES) sequence was inserted between homologous recombination sequences at the Rosa26 locus. The targeting vector, sgRNA specific for the Rosa26 locus and Cas9 mRNA were microinjected into C57BL/6 zygotes, which were subsequently implanted into pseudo-pregnant mice. Offspring were genotyped using primers Dpp4f (5′-CTGCAGTACCCAAAGACTGTACGGG-3′) and Dpp4r (5′-GACACCTTTCCGGATTCAGCTCACA-3′). The expected amplicon size using these primers was 628bp. Genotype-positive founders were back-crossed with C57BL/6 mice to produce generation F1. Tail tips of four F1-positive mice were subjected to Southern blotting to confirm correct insertion of hDPP4.

2.3. Generation of Pseudoviruses

Replication-incompetent HIV virions pseudotyped with MERS-CoVS protein (human beta-coronavirus 2c EMC/2012, AFS88936.1) and expressing firefly luciferase (Fluc) were generated as previously described [22,23]. The MERS-CoV S protein sequence was codon optimized for *Homo sapiens* by GENEWIZ (Suzhou, China) and cloned into an expression vector to yield pcDNA3.1-opti-MERS-CoV-Spike. Briefly, human embryonic kidney (HEK) 293T or 293 cells (both from ATCC) were co-transfected with pcDNA3.1-opti-MERS-CoV-Spike and the lentiviral vector pSG3.Δenv.Fluc [22] using various transfection reagents including Lipofectamine 2000 (Invitrogen, 11668019, Carlsbad, CA, USA), Lipofectamine 3000 (Invitrogen, L3000015, USA), polyethylenimine (Alfa Aesar, 43896, Lancashire, UK), Neofect, VigoFect (Vigorous Biotechnology, T001, Beijing, China), and TurboFect (Thermo Scientific, R0531, Waltham, MA, USA). After incubation for 48 h, culture supernatants were centrifuged at 210× *g* for 5 min, filtered through 0.45-μM filters, and concentrated with 30-kDa ultrafiltration devices (Millipore, Boston, MA, USA). All experiments involving pseudotyped MERS-CoV were performed in a BSL-2facility.

2.4. In Vitro Neutralization Tests

To conduct high-throughput in vitro neutralization assays, an optimized dose of MERS-CoV pseudovirus was incubated with each mAb for 1 h at 37 °C, and then the mixture was added to human Huh7 hepatoma cells (Fengh Bio. Inc., Changsha, China) or other cell types in a 96-well plate and incubated for 48 h. The infectivity of MERS-CoV pseudoviruses was determined by measuring bioluminescence as described previously [24].

2.5. MERS-CoV-Neutralizing Monoclonal Antibodies

Recombinant extracellular domain of the MERS-CoV spike (S) protein (40069-V08B) from human beta-coronavirus 2c EMC/2012 and its recombinant RBD (40071-V08B1) were purchased from Sino Biological, Inc. (Beijing, China). Phage library construction and antibody selections were carried out as described below. Six-week-old female BALB/c mice were immunized intraperitoneally (I.P.) with 5 μg of recombinant S protein. RNA was extracted from spleen cells and used to construct a phage-displayed antibody library. Recombinant RBD protein was used to screen antibodies. After four rounds of panning, antibodies that specifically recognized the S protein were obtained. The variable regions of the light and heavy chains of these antibodies were fused to mouse kappa or IgG1 constant regions using PCR and cloned into the pSTEP2 HEK-293 transient expression vector. Suspension HEK-293 cells were transfected with light chain and heavy chain expression vectors and recombinant mouse mAbs were purified from culture supernatants using protein A affinity chromatography. The positive control mAbm336 was prepared and provided by Prof. Tianlei Ying [25]. The peptide inhibitor of MERS-CoV was provided by Prof. Shibo Jiang [26,27].

The variable regions of the light and heavy chains of mouse antibody MERS-H111 were aligned to human germline genes, and the most homologous human framework was chosen for humanization [28,29]. The murine sequence was preserved at key residues to maintain structural stability. Codon-optimized light-chain and heavy-chain variable region sequences were synthesized and cloned into expression vectors containing kappa and IgG1 constant regions. Suspension HEK-293 cells were transfected with expression vectors encoding the light and heavy chains of MERS-H111-1, and the expressed antibody was purified from the supernatant using protein A affinity chromatography after 7 days.

2.6. Murine Model of MERS-CoV Pseudovirus Infection

Mice were challenged by different routes with varying doses of pseudotyped MERS-CoV. The IVIS-Lumina II imaging system (Xenogen, Baltimore, MD, USA) was used to detect bioluminescence as described previously [30,31]. Prior to measuring luminescence, mice were anesthetized by I.P. injection of sodium pentobarbital (240 mg/kg). The exposure time was 60 s, and fluorescence intensity in regions of interest was analyzed using Living Image software (Caliper Life Sciences, Baltimore, MD, USA). Different wavelengths were used for detecting pseudovirus and tdTomato fluorescence. The substrate, D-luciferin (50 mg/kg, Xenogen-Caliper Corp., Alameda, CA, USA), was injected I.P. and imaging was conducted 10 min later. The relative intensities of emitted light were represented as colors ranging from red (intense) to blue (weak) and quantitatively presented as photon flux in photon/s/cm^2/sr.

2.7. Authentic Virus Infection of Mice and Plaque Assays

Four-week-old R26-hDPP4 and wild-type mice were challenged with authentic MERS-CoV strain hCoV-EMC (1.5×10^5 PFUs) by the intranasal (I.N.) route and body weight was measured daily. On the fifth day post-infection (p.i.), all mice were sacrificed for sample collection. The timing of sacrifice was based on the consideration of humane euthanasia for body weight loss exceeding 25% [8]. Viral titers in lung tissues were determined by plaque assays on Vero cells following a protocol described previously [18]. The authentic MERS-CoV strain hCoV-EMC was maintained and tested in the BSL-3 facility (facility No. ABSL-3059) of the Institute of Laboratory Animal Sciences, CAM & PUMC, Beijing China, with strict use of personal protective equipment as described previously [32].

2.8. RNA Extraction and Real-Time QuantitativePCR

Tissues were dissected, immediately immersed in RNAlater® stabilization reagent (Invitrogen, USA) and stored at −80 °C. Total RNA was extracted from individual tissues using TRIzol and quantified using a spectrophotometer at a wavelength of 260 nm. Random hexamers were used to prime reverse-transcription reactions using a reverse transcription (RT)-PCR kit containing SYBR green dye (Takara, Shiga, Japan). Real-time quantitative PCR (RT-qPCR) was performed using a Light Cycler 480 Real-Time PCR system (Roche, Indianapolis, IN, USA). Each reaction was performed in triplicate. Relative expression level of DPP4 was determined using primers DPP4-Q-F (5′-GGGTCACATGGTCACCAGTG-3′) and DPP4-Q-R (5′-TCTGTGTCGTTAAATTGGGCATA-3′) and normalized to expression of GAPDH (glyceraldehyde 3-phosphate dehydrogenase). Viral loads were quantified using primers gag-F1 (5′-AGCACAGCAAGCAGCAGC-3′), gag-R1 (5′-GTGGCTCCTTCTGATAATGCTGAA-3′) and a TaqMan probe (5′Fam-ACAGGAAACAGCAGCCAGGTCAGCCGA-3′Tamra). Data were presented as log viral RNA copies/GAPDH copies.

2.9. Western Blotting

Mouse tissues were homogenized in RIPA lysis buffer (50 mM Tris-HCl, pH 7.4, containing 150 mM NaCl, 100 mM EDTA, and 0.1% SDS) supplemented with 1× Proteinase Inhibitor (PI) (Roche, Basel, Switzerland). The denatured protein lysates were separated using 10% SDS-PAGE gels. After transfer, anti-hDPP4 mouse monoclonal antibody (1:3000 dilution, Origene, Rockville, MD, USA) and anti-GAPDH antibody (1:5000, Abcam, Cambridge, UK) were added, followed by horseradish peroxidase (HRP)-conjugated anti-mouse IgG (1:20,000, Santa Cruz Biotechnology, Santa Cruz, CA, USA). The Immobilon Western Chemiluminescent HRP Substrate kit (Millipore Corporation, Billerica, MA, USA) was used for development.

2.10. Immunohistochemistry

Mouse tissues were fixed in 10% neutral-buffered formalin, embedded in paraffin, and sectioned to about 3-μm thickness. Tissue sections were stained with hematoxylin and eosin for histopathological examination. For immunohistochemistry (IHC), tissue sections were rehydrated and incubated in Coplin jars filled with Citra buffer (pH 6.0) at 96 °C in a microwave oven for 10 min, cooled at room temperature for 60 min and then blocked with 10% normal goat serum at 37 °C for 60 min. The sections were incubated overnight at 4 °C with either 1:200 rabbit R723mAb (produced by phage display, with specificity against the receptor-binding domain of MERS-CoV; kindly provided by Beijing Wantai Biological Pharmacy Enterprise Co., Ltd., Beijing, China) or normal goat serum (control). Sections were washed with PBS and incubated with HRP-conjugated goat-anti-rabbit secondary antibody (Zhongshan Golden Bridge Biocompany, Beijing, China) for 40 min at room temperature, followed by development with 3,3'-diaminobenzidinesubstrate (Zhongshan Golden Bridge Biocompany, Beijing, China) and counterstaining with hematoxylin.

2.11. Data and Statistical Analysis

Sample sizes in each group were calculated based on anticipated effect sizes to yield statistically significant differences. Inhibition rates of mAbs were calculated using the following formula: (value of positive group−value of mAb treatment group)/value of positive group. The value represented either fluorescence derived from pseudovirus infection or authentic viral titers in lung tissues. Data were analyzed using SPSS (ver. 18.0; IBM, Armonk, NY, USA) or GraphPad Prism 5.0 (GraphPad Software, San Diego, CA, USA). Results from each experiment were presented as means plus standard deviations (SD) or standard error of mean (SEM). Student's *t*-tests were used to assess differences between groups. Dunnett's tests were performed for multiple comparisons. *p* values less than 0.05 were considered statistically significant.

3. Results

3.1. A HDPP4-Knockin Mouse (R26-hDPP4) Was Established Using CRISPR/Cas9

A KI mouse was generated by inserting the full-length hDPP4 gene [7] into the Rosa26 locus, which has been shown to be a safe harbor for developing genetically stable KI mouse lines [33]. Expression of hDDP4cDNA was driven by the splice acceptor, which allowed ubiquitous expression of hDPP4 under control of the Rosa26 promoter. The tdTomato gene was inserted downstream of hDPP4 with an internal ribosome entry site allowing co-expression of tdTomato and hDPP4. A poly(A) sequence and a woodchuck hepatitis virus posttranscriptional regulatory element (WPRE) were added to enhance mRNA stability and translation efficiency (Figure 1A). The targeting vector along with subgenomic RNA (sgRNA) and Cas9 mRNA were injected into zygotes of C57BL/6 mice. Successful insertion in nine of 41 offspring was positively confirmed by Southern blotting and the results for five mice are presented in Figure 1B. One male founder and one female founder died before weaning and two founders were too emaciated to mate. Five founders were backcrossed with C57BL/6 mice, and two matings produced F1-positive offspring. Tail tips of F1-positive mice were subjected to PCR genotyping (Figure 1C) and Southern blotting to confirm correct insertion of hDPP4. Homozygous mice were termed B6-Gt (Rosa) 26 Sortm1 (SA-hDPP4-tdTomato), abbreviated to R26-hDPP4. Significant hDPP4 expression could be detected by RT-PCR in all R26-hDPP4 mouse tissues tested (Figure 1D), but no expression was detected in wild-type mice. Western blotting of lung and brain tissues confirmed hDPP4 expression (Figure 1E). Unlike Tg mice, which showed the highest-level expression of hDPP4 in heart and brain [11], R26-hDPP4 mice had the highest hDPP4 expression in lung. Profiting from the insertion of tdTomato downstream of hDPP4, the global expression patterns of hDPP4 were examined visually (Figure 1F) via bioluminescent imaging (BLI). Since the liver, intestine and lung have bigger volumes, they have brighter images than other organs.

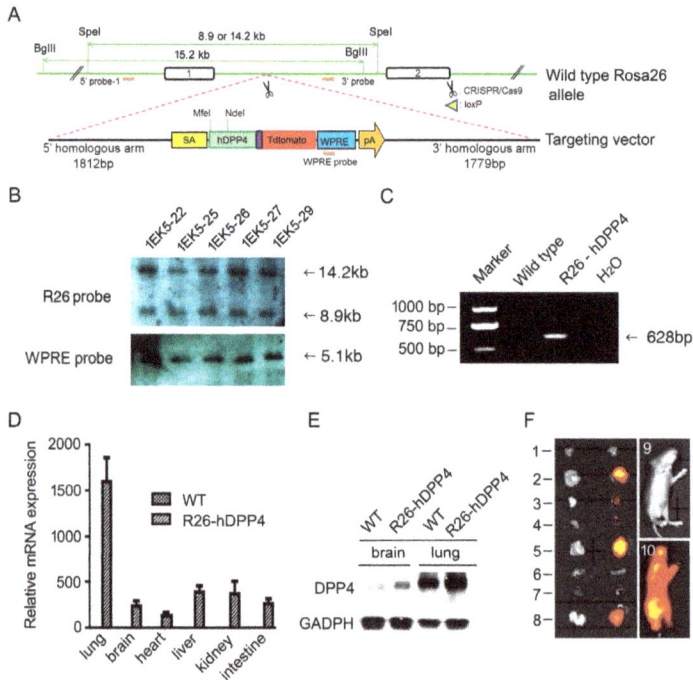

Figure 1. Establishment of a R26-hDPP4-knockin mouse model. (**A**) Schematic strategy for generation of R26-hDPP4-knockin mice via CRISPR/Cas9. (**B**) For the R26 probe, genomic DNA was digested with *Spe*I, and the expected sizes of wild-type and gene-targeted bands were 8.9 kb and 14.2 kb, respectively. For the WPRE probe, genomic DNA was digested with *Bgl*II, and the expected size of the gene-targeted band was 5.1 kb. Four representative results are shown. (**C**) PCR genotyping of R26-hDPP4 mice. The primer pair was designed to anneal in the coding region of hDPP4, and the expected PCR amplicon was 628 bp in length. (**D**) Quantitative reverse transcription PCR (RT-qPCR) of *hDPP4* mRNA in R26-hDPP4 mice and wild-type C57BL/6 mice. Values are presented as means ± SEMs of three independent experiments and were normalized to GAPDH levels. No expression of R26-hDPP4 in wild-type mice was detected. (**E**) Detection of DPP4 protein by western blotting in brain and lung. Weaker blotting signal was detected in wild-type mice, indicating the anti-hDPP4 antibody could recognize mDPP4. (**F**) Bioluminescence imaging (BLI) of newborn R26-hDPP4-knockin mice showing hDPP4 expression in their organs. The imaged organs were: (1) thymus, (2) liver, (3) stomach, (4) kidney, (5) intestine, (6) spleen, (7) heart, (8) lung, (9) and (10) the whole bodies of wild-type mice and R26-hDPP4 mice, respectively.

3.2. R26-hDPP4 Mice Were Susceptible to Authentic MERS-CoV Infection, with Infected Mice Exhibiting Disease Symptoms Similar to Those of MERS-CoV-Infected Human Patients

To test whether R26-hDPP4 mice could support efficient infection by and replication of MERS-CoV, 4- to 5-week-old R26-hDPP4 and wild-type mice were infected I.N. with MERS-CoV clinical strain hCoV-EMC [34] at a dose of 1.5×10^5 PFUs. The bodyweights of wild-type mice increased approximately 24% (Figure 2A) by day 5 p.i., while R26-hDPP4 mice experienced significant weight loss of up to 28% (23–33.5%). By day 4 p.i., all R26-hDPP4 mice had lost at least 20% of their bodyweights, meeting the typical cut-off used in humane euthanasia criteria [8]. Two mice (2/4) had lost more than 30% of their body weights by day 4 p.i.

Viral titers in the lungs of R26-hDPP4 mice were measured as approximately 10^3 PFU sat 5 days p.i. while no virus was detected in the lungs of wild-type mice (Figure 2B; mean difference 2.958 log

PFUs, 95% CI 1.266–4.699 log PFUs). Although weight loss and viral loads provide important measures of mouse susceptibility, these parameters do not directly reflect pathologic changes. Therefore, the lungs, brains, livers, and kidneys collected from infected mice were subjected to histopathological analysis and IHC (Figure 2C–O). By quantitative analysis of pathological scores, the symptoms of R26-hDPP4 mice were significantly different compared with wild-type mice (Figure 2C). Associated with these pathological changes, significant viral loads were detected in lung (Figure 2E). In the cerebellum, cerebral ganglia and cerebrum of R26-hDPP4 mice, perivascular gliosis was observed (Figure 2F; Supplementary Figure S1C) with accompanying virus detected in this organ (Figure 2G). No obvious lesions were observed in liver and kidney (Figure 2H) and no virus was detected in these organs (Figure 2I), despite the fact that a MERS-CoV-infected patient with acute nephritis has been reported [35]. By contrast, in the corresponding organs of wild-type mice, no or minimal pathological changes were observed and no virus could be detected (Figure 2J–O). These results indicated that R26-hDPP4 mice were permissive to infection by authentic MERS-CoV and that infection was accompanied by severe disease symptoms similar to those of MERS-CoV-infected human patients.

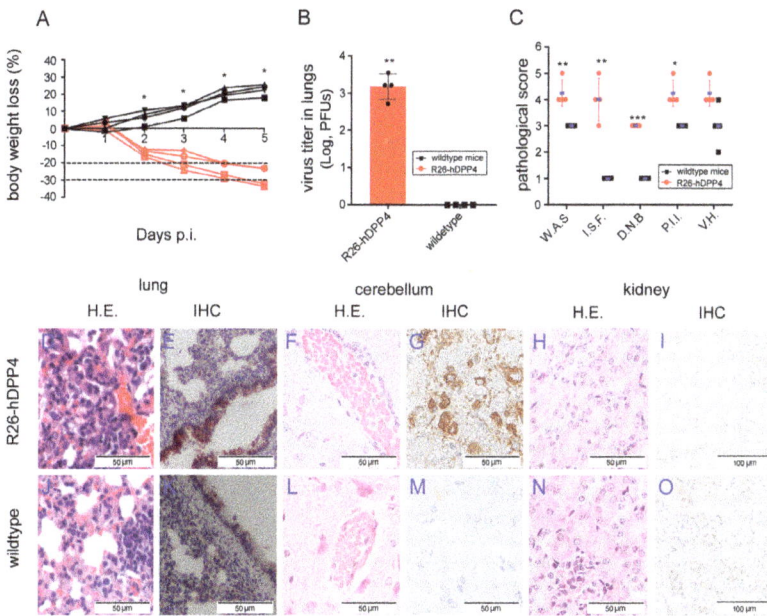

Figure 2. R26-hDPP4-knockin mice were susceptible to infection by authentic MERS-CoV at low dose. (**A**) Weight loss in R26-hDPP4 mice challenged with hCoV-EMC at a dose of 1.5×10^5 PFUs. (**B**) Viral titers in lungs of challenged R26-hDPP4 mice on day 5 p.i. LOD (limitation of detection): 0.85 PFU. (**C**) Quantitative analysis of pathological lesions in lungs. W. A. S. = widened alveolar septa; I. S. F. = inflammatory cells, serous and fibrinous exudation; D. N. B. = degeneration and necrosis of bronchial epithelial cells; P. I. I. = perivascular inflammatory cell infiltration; V. H. = vasodilator hyperemia. (**D–O**) Histopathological changes and viral loads in the lungs, brains, and kidneys of mice. R26-hDDP4 mice exhibited disease symptoms similar to those of MERS-CoV-infected human patients (**D**), while no or mild symptoms were observed in wild-type mice. (**J**). Perivascular gliosis in the cerebellum was observed in R26-hDPP4 mice (**F**) but not in wild-type mice (**L**). No pathological lesions were identified in the kidneys of either R26-hDPP4 or wild-type mice (**I,O**). IHC assays confirmed viral loads in lungs (**E**) and cerebella (**G**) of R26-hDPP4 mice; little or no virus was detected in the lungs (**K**) and cerebella (**M**) of wild-type mice. Four mice in each group were infected, and samples from all mice were subjected to tittering and histopathological analysis (* $p < 0.05$; ** $p < 0.01$; *** $p < 0.001$).

3.3. A MERS-CoV S-RBD-Specific Neutralizing Antibody Protected R26-hDPP4 Mice from Challenge with Authentic MERS-CoV

Next, we evaluated the effects of a neutralizing mAb on R26-hDPP4 mice infected with authentic virus. To this end, a humanized mAb, H111-1, was generated by phage display methods and administered intravenously (I.V.) at a single dose (either 1 mg/kg or 5 mg/kg) 6 h after infection with hCoV-EMC (1.5×10^5 PFUs). Administration of H111-1 significantly decreased viral titers in lungs (from 3 log PFUs to 0.9 log PFUs and 1.2 log PFUS for the 5 mg/kg and 1 mg/kg groups, respectively; Figure 3A), and reduced weight loss in a dose-dependent manner (Table 1). Viral titers in the higher mAb dose group were below the limit of detection. The average inhibition rates of authentic virus were 70% for 5 mg/kg mAb and 60% for 1 mg/kg mAb, respectively. Notably, mAb H111-1 completely ablated viral loads in the lungs in two of four mice in both low- and high- dose mAb groups (Figure 3A). IHC results confirmed lower viral loads in the lungs of mAb-treated mice (Figure 3(C❸,C❹)), while a reduction of viral load in cerebellum was less clear (Figure 3(C⑦,C⑧)). In addition, treatment with mAb H111-1 alleviated symptoms in both lung and brain (Figure 3(C❶,C❷,C❺,C❻)), with degeneration and necrosis of bronchial epithelial cells improving significantly in treated animals (Figure 3B, $p < 0.05$). These results indicated that humanized mAb H111 is a promising antiviral agent for preventing MERS-CoV infection. Importantly, these results also implied that the R26-hDPP4 mouse could be an effective model for evaluating antiviral agents against MERS-CoV in vivo.

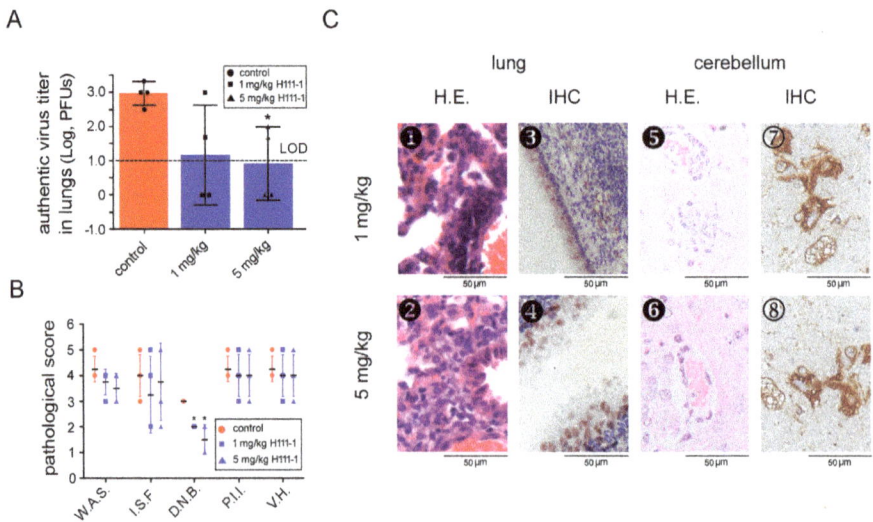

Figure 3. MERS-CoV S-RBD-specific humanized neutralizing antibody H111-1 protected R26-hDPP4 mice from challenge with authentic MERS-CoV. Four-week-old mice were administered either PBS (control), 1 mg/kg mAb H111-1 or 5 mg/kg mAb H111-1 via the I.P. route, and 6 h later they were challenged I.N. with hCoV-EMC (1.5×10^5 PFUs). On day 5 p.i., mice were sacrificed for virus titering and pathological analysis. (**A**) Treatment with mAb H111-1 significantly decreased viral titers in lungs. The dashed line indicates the LOD. (**B**) Efficacy of mAb H111-1 in abating pathological lesions caused by infection with authentic MERS-CoV. Explanation of pathological changes is given in Figure 2. (**C❶–C⑧**) Histopathological changes and viral loads in lungs and brains of R26-hDPP4 mice administered 1 mg/kg mAb H111-1 (**C❶,C❸,C❺,C⑦**) or 5 mg/kg mAb H111-1 (**C❷,C❹,C❻,C⑧**), * $p < 0.05$.

Table 1. Administration of neutralizing mAb H111-1 prevented weight loss in R26-hDPP4 mice infected with MERS-CoV *#.

Days p.i.	0	1	2	3	4	5
R26-hDPP4 + PBS	0	1.9 ± 2.2	-14.1 ± 1.9	-18.5 ± 5.1	-24.0 ± 4.5	-28.0 ± 5.6
R26-hDPP4 + 1 mg/kgmAbH111-1	0	-0.5 ± 1.4	-13.0 ± 3.5	-17.0 ± 3.0	-22.9 ± 2.7	-27.5 ± 1.9
R26-hDPP4 + 5 mg/kgmAbH111-1	0	-0.1 ± 2.3	-7.6 ± 4.8	-13.1 ± 3.0	-19.0 ± 3.6	-24.8 ± 3.6

Note: * All R26-hDPP4 mice were challenged with authentic MERS-CoV at a dose of 1.5×10^5 PFUs. # Four mice per group. Numbers represent percentage weight loss compared with weight prior to infection.

3.4. AMurine Model of Infection with Pseudotyped MERS-CoV was Established Using R26-hDPP4 Mice

3.4.1. Optimization of Pseudotyped MERS-CoV System and Establishment of a Model of Pseudotyped MERS-CoV Infection Using R26-hDPP4 Mice

Zhao et al [2] reported a MERS-CoV pseudovirus that allowed for single-round infection of several cell lines expressing hDPP4. However, despite using several proven methods [22], we at first failed to infect R26-hDPP4 mice, which were otherwise permissive to infection by authentic virus, using pseudotyped virus. At first, we attributed this failure to low pseudovirus titer and therefore we optimized the parameters of pseudovirus production (Supplementary Figure S2A–D). Finally, pseudovirus with titers up to $1.27 \times 10^{7.5}$ TCID$_{50}$/mL (Supplementary Figure S2E) was obtained. The pseudovirus was then titrated in 4-week-old R26-hDPP4 mice, and the 50% animal infectious dose (AID$_{50}$) was determined to be 10 TCID$_{50}$/mL viatheintrathoracic (I.T.) route.

To establish an in vitro neutralization assay, the cellular tropism of pHIV/MERSS/Fluc was tested in various cell lines. All 12 cell lines tested were permissive to pHIV/MERSS/Fluc infection, but Huh7 cells showed the highest infection efficiency (Supplementary Figure S2F), demonstrating the wide cellular tropism of the pseudotyped virions. As a result, the Huh7 cell line was chosen to establish a chemiluminescence-based, high-throughput antibody neutralization assay. The assay displayed good sensitivity and specificity.

Since the I.P. infection route was efficient for a pseudotyped Ebola virus model [36], groups of 4-week-old R26-hDPP4 mice were administered pseudotyped virus I.P. Mice were anaesthetized and observed with a BLI system at various days p.i. Bioluminescence of the transgene-encoded Fluc reporter was first observed at day 2 p.i. in the abdomen, close to the injection site (Figure 4A). The bioluminescence then spread to the thoracic cavity by day 6 p.i. Bioluminescence reached peak intensity and then gradually weakened up to day 20 p.i. (Figure 4B). To identify the sites of infection, mice were dissected and their organs were observed on the day of peak infection. Figure 4C showed that thymus, liver, spleen, kidney, lung and muscle, among other organs and tissues, were infected by pseudovirus. Higher pseudoviruscopy numbers were detected in these organs (Figure 4D), illustrating that R26-hDPP4 mice could be efficiently infected by pseudovirus. Since MERS-CoV infection mainly causes clinical ARDS [37,38], the primary site of infection should be the respiratory tract. Therefore, to maximize infection in the respiratory tract, we infected mice via the I.T. route. Live-animal imaging showed that infection was mainly localized to the chest, with maximal infection occurring on days 10 and 11 p.i. Using the I.T. route, the major infected organs were the thymus, heart and lung (Figure 3C). A uniform infection profile was observed in all mice, indicating that I.T. challenge had some advantages over I.P. challenge. I.N. and I.V. challenge were also attempted but did not result in productive infection. To test the influence of the animals' age on susceptibility to infection, 4- to 9-week-old mice were inoculated with pseudovirus. Both younger and older mice could be infected, but younger mice were more susceptible (Figure 4E,F). As reported previously [30], fluorescence intensity and virus copy number in tissues were strongly correlated, suggesting that fluorescence intensity could be used as an indicator of pseudovirus infection. Pseudovirus infection caused no histopathological changes, indicating that this tool was safe in mice (Figure 4G). Notably, through BLI visualization and confirmation by IHC, we found that pseudovirus tended to infect bronchi (Figure 4H), a tropism shared with authentic virus (Figure 2E).

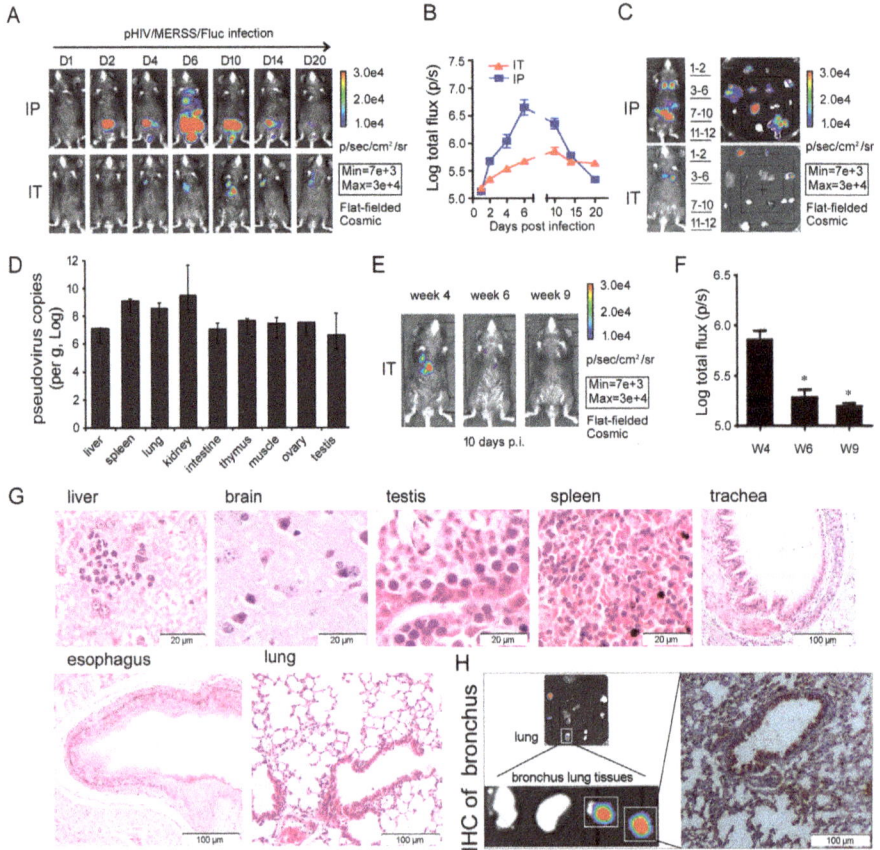

Figure 4. Establishment of the R26-hDDP4 knockin model of infection with MERS-CoV pseudovirus. (**A–C**) Four-week-old R26-hDPP4 mice were inoculated with $1.27 \times 10^{7.5}$ TCID$_{50}$ (I.P. route) or $3.8 \times 10^{6.5}$ TCID$_{50}$ (I.T. route) of pHIV/MERSS/Fluc per animal. (**B**) Bioluminescent images (BLI) are shown at different days p.i. Relative bioluminescence intensity was shown in pseudocolor, with red representing the strongest and blue representing the weakest photon fluxes. Data are shown as means ± deviation (**C**). Organs were examined for Fluc expression using BLI: 1 = thymus; 2 = heart; 3 = liver; 4 = spleen; 5 = kidney; 6 = lung; 7 = lymph node; 8 = muscle; 9 = skin; 10 = ovary or testis; 11 = brain; 12 = intestine. (**D**) The copy number of pHIV/MERSS/Flucmeasured by RT-qPCR (I.P. challenge route). (**E,F**) Susceptibility tests for mice at different ages. Four- to 9- week-old mice could be infected, but younger mice were more susceptible, * $p < 0.05$. (**G**) Histopathological examination of organs of pseudovirus-infected mice; no histopathological changes were observed. (**H**) Pseudotypedvirions mainly infected the bronchi as shown by IHC and BLI. Bright spots indicate bronchi separated from lung (n = 4–6/group).

3.4.2. Relevance of the R26-hDPP4 Mouse Model of Infection by Pseudotyped and Authentic MERS-CoV

We compared the infection profile between R26-hDDP4 mice infected with pseudotyped and authentic MERS-CoV (Table S1). Lungs were the major infected organs for both types of virus (Figure 5A,B), and pulmonary bronchial epithelial cells were the common sites of infection (Figure 5C,D). Lung tissue tropism was also supported by measurement of pseudovirus copy number (Figure 4D) and authentic virus titering (Figure 2B). The dose conversion of pseudovirus and authentic virus is shown in Figure 5E. Authentic virus was observed in the CNS (Figures 2G and 3(C⑦,C⑧)) of infected mice. Interestingly, we observed that pseudovirus could also enter the CNS, as it was detectable by qRT-PCR or BLI in the brain. How pseudovirus crosses the blood-brain barrier requires further investigation.

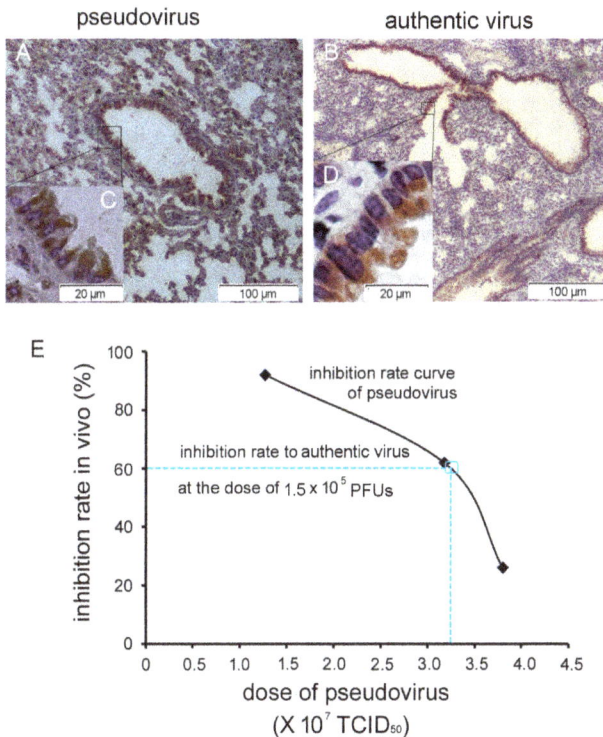

Figure 5. Relationship between pseudovirus and authentic virus models. (**A–D**) Pseudovirus and authentic virus infection in the lungs of R26-hDPP4 mice showed a similar pattern, as shown by IHC using mAb R723 against the RBD of the MERS-CoV S protein. (**C,D**) Both pseudovirus and authentic virus infected ciliated columnar epithelium of bronchi. (**E**) Dose conversion of pseudovirus and authentic virus. The full black line represented the inhibition rate (◆) of humanized mAb H111-1 (1 mg/kg) in vivo using different pseudovirus doses. The blue dashed line represents inhibition rate of H111-1 against authentic virus. When the dose of authentic virus was 1.5×10^5 PFUs, we calculated that the inhibition rate of H111-1 at a dose of 1 mg/kg would be 60% (see main text). From the inhibition rate curve of pseudovirus, the corresponding pseudovirus dose was 3.25×10^7 TCID$_{50}$. That is, 1 TCID$_{50}$ of pseudovirus corresponded to 0.0046 PFU of authentic virus ($1.5 \times 10^5 / 3.25 \times 10^7$). n = 4–6 mice per group.

3.5. MERS-CoV S-RBD-Specific Neutralizing Antibodies Protected R26-hDPP4 Mice against Challenge with Pseudotyped MERS-CoV

Monoclonal antibodies can provide robust strategy protection against infection caused by MERS-CoV [18,19,39]. In vitro pseudovirus neutralization assays are useful tools to test the efficacy of mAbs [2]. Therefore, the in vitro efficacy of the humanized neutralizing mAb H111-1 to protect R26-hDPP4 mice against infection by pseudotyped MERS-CoV was assessed. As a comparison, the in vitro efficacy of the fully human neutralizing mAb m336 [25] to protect R26-hDPP4 mice against infection was also determined. The 50% inhibitory concentrations (IC_{50}s) of mAbs H111-1 and m336 were determined to be 4.5 and 2.7 ng/mL, respectively, representing high in vitro inhibitory activity.

Both mAbs were administered I.T.to R26-hDPP4 mice at a dose of 1 mg/mouse prior to challenge with pseudovirus ($1.27 \times 10^{7.5}$ $TCID_{50}$). On day 8 and day 11 p.i., when peak signal could be detected, BLI images of the whole body and organs were obtained. Both mAbs demonstrated strong protective efficacy as shown in Figure 6A–F; H111-1 ablated pseudovirus infection when administered by either the I.P. or I.T. routes (Figure 6A,B,D). The pseudovirus reporter signal in mice administered m336 also decreased significantly in the whole body as well as in lung and thymus (Figure 6C,E,F, $p < 0.05$ or 0.01).

Next, a dose conversion between pseudovirus and authentic virus was calculated in order to establish a safety profile. Three groups of R26-hDPP4 mice were challenged with different doses of pseudovirus (1.25, 3.20, and 3.85×10^7 $TCID_{50}$) either with or without treatment with mAb H111-1 (1 mg/kg) 6 h prior to challenge. BLI of all mice was performed on day 6 p.i. and inhibition rates were calculated. As shown in Figure 4E, when mice were challenged with 1.5×10^5 PFUs of authentic virus, the inhibition rate of H111-1 (1 mg/kg) was 60%. From the inhibition rate curve for pseudovirus infection, we deduced that the pseudovirus dose corresponding to 60% inhibition was 3.25×10^7 $TCID_{50}$. That is, 1 $TCID_{50}$ of pseudovirus was equivalent to 0.0046 PFU of authentic virus. We assumed that this quantitative parameter would be useful for screening antiviral agents and evaluating vaccines against MERS-CoV. Since the R26-hDPP4 mouse provided a suitable model for infection by both pseudotyped and authentic MERS-CoV, the pseudovirus model is a convenient tool to evaluate the in vivo efficacy of anti-MERS-CoV agents or vaccines in most biological laboratories that lack BSL-3 facilities.

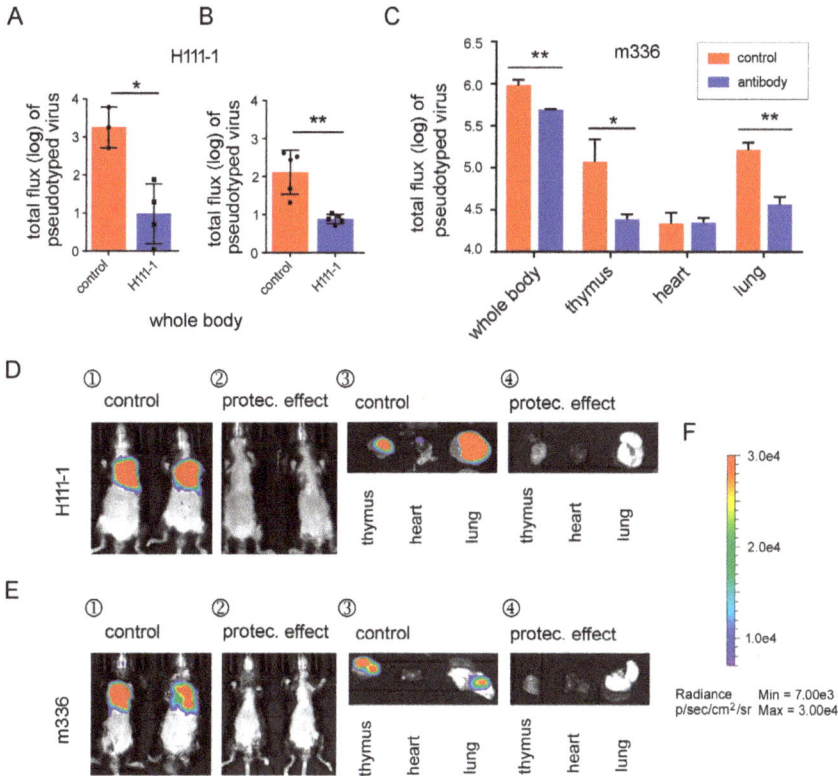

Figure 6. Inhibition of pseudotyped MERS-CoV infection in R26-hDPP4 mice by the novel mAb H111-1 and the well-characterized mAb m336. For evaluation of H111-1, mice were administered 1 mg/kg of mAb either I.P. (**A**) or I.T. (**B**) and 6 h later, challenged with pseudovirus I.T. at a dose of $3.8 \times 10^{6.5}$ TCID$_{50}$. On day 11 p.i., BLI of the whole body or specific organs was conducted (**D**). To evaluate the efficacy of m336 (**C**), mice were administered mAb and challenged using the same doses as for H111-1, and typical images (**E**) are shown. (**F**) Bar of photo flux; for details see Figure 2. $N = 4$ mice in each group, * means $p < 0.05$, ** means $p < 0.01$.

3.6. A MERS-CoV S-HR1-Specific Fusion Inhibitor Peptide Protected R26-DPP4 Mice from Infection by Pseudotyped MERS-CoV

Since peptides derived from the heptad repeat (HR)2 of the MERS-CoV S protein such as HR2P or its analogue HR2P-M2 [27] have been shown to protect RAG$^{-/-}$ mice from infection by MERS-CoV (EMC/2012 strain) [26], we investigated whether HR2P-M2 could also prevent pseudovirus infection in R26-hDPP4 mice. We tested the in vitro inhibitory effect of HR2P-M2 in the Huh7 cell line, and its IC$_{50}$ was determined to be 4504.5 ng/mL. In order to determine the in vivo protective effect of HR2P-M2, 5-week-old R26-DPP4 mice were administered 1000 µg of peptide per mouse I.T., and 30 min later challenged with $3.8 \times 10^{6.5}$ TCID$_{50}$ of pseudovirus using the same route. As shown in Figure 7, pseudovirus infection was clearly prevented by HR2P-M2: in one representative mouse, the pseudovirus signal decreased to levels similar to uninfected mice, representing full protection (Figure 7A,B). No or very weak pseudovirus signals were detected in lung, thymus and heart after dissection (Figure 7B), indicating that the R26-hDPP4 mouse model, like previous animal models, can be used for evaluation of antiviral peptides.

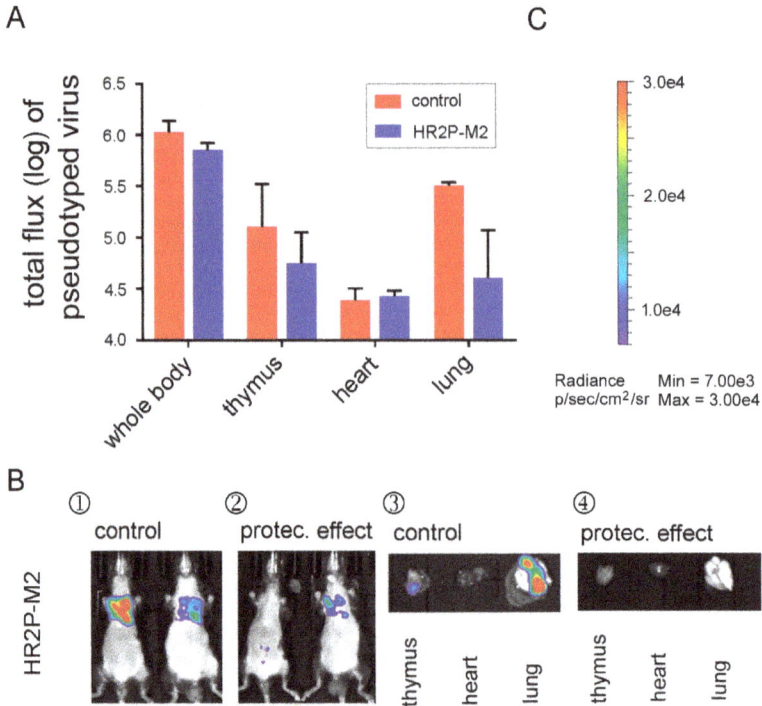

Figure 7. Inhibition of pseudotyped MERS-CoV infection in R26-hDPP4-knockin mice by peptide HR2P-M2. Mice were administered HR2P-M2 peptide or phosphate-buffered saline (PBS) I.T., respectively. Thirty min later, mice were infected I.T. with pseudotyped MERS-CoV ($3.8 \times 10^{6.5}$ TCID$_{50}$). BLI images were taken on day 11 p.i., and pseudovirus signals were recorded for the whole body or specific organs. (**A**) Flux value of pseudovirus for assessment of the protective efficacy of HR2P-M2. (**B**) BLI images of whole mice or their organs. (**C**) Bar of photo flux; for details see Figure 2. Four mice were used for each group, and representative images are shown.

4. Discussion

MERS-CoV infection causes acute respiratory distress [8,37], neurologic syndromes [40], and death. MAbs are promising tools for both therapeutic and prophylactic interventions against this pathogen [19,41–43]. However, the dearth of clinically-effective vaccines and therapeutics compelled us to develop a novel animal model that could faithfully mimic the characteristics of human disease and enable better evaluation of potential vaccines and therapies.

Respiratory disease (such as ARDS) and acute renal failure are the main symptoms of MERS-CoV infection, although patients with fatal disease in the CNS have been reported [40]. Therefore, an animal model of MERS-CoV disease that recapitulates the pathological characteristics in both respiratory tract and CNS, supports high-level virus replication in vivo, and exhibits lung pathology associated with ARDS is required. Moreover, the animal model must be economical, genetically stable, and show reproducible results [8]. Mouse-adapted viral strains have been widely used to establish infection models. Wherever possible, clinical isolates should be used in challenge experiments (http://www.gryphonscientific.com/gain-of-function/) [12]. Targeting transgenes to specific sites of the mouse genome has many advantages over the random insertion typically used in transgenic methods. Two hDPP4-KI mice created by CRISPR/Cas9 gene editing were recently reported [8,12]. In these mice, mDPP4 gene was either replaced by hDPP4 or mutated and its function was disrupted.

Since mDPP4 is endowed with different biological activities [44] in immunity and glycometabolism compared with hDPP4, we established KI mice by inserting full-length human DPP4 into the Rosa26 locus via CRISR/Cas9 to spare the function of mDPP4. Notably, the R26-hDPP4 mouse displayed an unusual expression profile, in which transgenic hDPP4 was stably and highly expressed in lung and also expressed to a lesser extent in brain and other organs (Figure 1D–F). We reasoned that this expression pattern might support robust infection and lesions in lung and brain, both using pseudotyped and authentic MERS-CoV. We further reasoned that insertion ofhDPP4 into the safe harbor of the Rosa26 locus and conserving mDPP4 function would be an appropriate strategy for establishing a genetically stable animal model. The insertion of hDPP4 mean, and whether it has implications for the expression of mDPP4 and its biological activities, is not clear.

We observed that R26-hDPP4 mice could be infected by the clinically authentic strain hCoV-EMC [34] at a dose of 1.5×10^5 PFUs, and that infection caused disease symptoms in both lung and brain similar to those observed in clinical patients. Infected R26-hDPP4 mice experienced significant loss of body weight, up to 33.6%, which was comparable to that observed in mice infected with a mouse-adapted strain [21]. Higher viral titers than in previous reports [8] were detected in lungs and were accompanied by diffuse pathogenesis including widened alveolar septa, vascular dilatation and hyperemia, perivascular infiltration of inflammatory cells, necrosis of bronchial epithelial cells, edema and hyaline membrane formation (Figure 2, Supplementary Figure S1). The pathophysiological findings were closely associated with clinical manifestations [45]. All infected R26-hDPP4mice had uniform disease symptoms with a significant difference observed between KI and wild-type mice, as revealed by quantitative pathological scores (Figure 2).

Very few reports have described the sites of MERS-CoV infection in patients, even though these data are crucial for understanding viral pathogenesis. For both cultural and religious reasons, no tissue samples have been made available [37]. However, studies using several animal models has been helpful in addressing this problem. In rhesus macaques, transient lower respiratory tract infection was accompanied by inflammatory cell infiltration [4]. By contrast, severe interstitial pneumonia within filtration and alveolar edema were observed in infected marmosets [6]. Research using small animal models, such as a mouse transduced with an Ad5 vector expressing hDPP4 [9], a Tg mouse model [11], and KI mice created with CRISPR/Cas9 technology [8,12],provide uncontested evidence that the respiratory system is the main site of MERS-CoV infection, in agreement with most clinical manifestations [37]. In this study, we also showed that the lungs, especially the bronchi, were the primary sites of infection both for pseudotyped and authentic virus (Figures 2–5; Table S1).

Whether MERS-CoV is able infect the nervous system remains controversial. Agrawal et al. [11] reported high viral loads in the brain, but no necrosis or inflammatory reactions were observed. In KI mice infected with a mouse-adapted viral strain, no viral replication was detected in brain, even at challenge doses of 5×10^6 PFUs. This animal model was heralded as having the advantage of mimicking the clinical manifestations of ARDS in the absence of CNS complications. However, even though respiratory infection and ARDS are common clinical symptoms of MERS-CoV infection, patients with severe neurological syndrome, including confusion, coma and ataxia as well as focal motor deficit, have been reported [40]. Here, when R26-hDDP4 mice were infected with the clinical strain hCoV-EMC at a dose of 1.5×10^5 PFUs, viral loads were detected using IHC assays (Figure 2) in the cerebellum, cerebral ganglia and cerebrum. In association with these viral loads, perivascular gliosis, an indicator of inflammatory reactions, was observed in these organs in all 12 R26-hDPP4mice irrespective of mAb administration or mAb ablation of inflammation (Figure 3). These phenomena were not observed in any wild-type mice challenged in the same manner. Since virus was mainly detected on the membranes of motor neurons, we reasoned that the dyskinesia observed in patients would also be manifested in mice. Additional studies to assess the possible nerve infection are being conducted, and viral loads in nervous system are considered to be titrated.

Pseudotyped virus, which has been widely used for many pathogens [2,30,46], is a useful, safe, and convenient tool for viral infection studies and therapeutic testing [23]. Pseudovirus cell models are popularly applied since they are relatively easy to develop; however, pseudovirus animal models are not as widely available, even though they have more extensive potential applications. Pseudovirus animal models may be difficult to achieve due to lack of susceptibility and the absence of suitably infectious pseudoviruses. Previously, an inhibition assay based on the pseudovirus cell model was established and used to detect neutralizing antibodies against MERS-CoV [2]. However, these pseudoviruses failed to infect the R26-hDPP4 mouse, which was shown to be particularly susceptible to authentic virus. We presumed that this effect might be caused by low pseudovirus titer. Therefore, the codons of the S protein, backbone plasmid [22], and production conditions were optimized systematically, resulting in an increase in titer of about 1000-fold (Supplementary Figure S2) and finally resulting in the successful infection of R26-hDPP4 mice.

To challenge mice, authentic virus is often administered by the I.N. route [18], which has the advantage of mimicking clinical respiratory tract infection. We tried to infect R26-hDPP4-KI mice with pseudovirus by the I.N. route but this was unsuccessful. This failure might have resulted from the use of less than 50 µL of pseudovirus inoculum for nasal dripping; this dosage may not have been sufficient for a single round of infection. We found that different infection routes might result in different biodistribution. Pseudovirus was detected by BLI in the thymus, liver, spleen, kidney, lung, muscle and intestine of mice infected via the I.P. route. However, for challenge via the I.T. route, pseudovirus was only observed in the thymus, heart and lung. Thus, the route of infection also had some effect on pseudovirus distribution. For example, viral loads might fail to be detected in the lungs of mice with minor I.P. infections. However, the I.T. route resulted in uniform infection patterns in the lungs, and especially in the bronchi (Figures 4 and 5). The pseudovirus distribution pattern (Figure 4) was similar to that of Tg mice infected with authentic virus [11]. Mice as old as 9 weeks could be infected, but younger mice (i.e., those 4 to 5 weeks old) had better susceptibility. The pseudovirus infection caused no pathological lesions in multiple organs, indicating that this model was safe (Figure 4).

To verify the potential of the R26-hDPP4 mouse model for evaluating different antiviral agents against MERS-CoV, we tested a newly generated humanized neutralizing mAb, H111-1, as well as a known MERS-CoV-neutralizing mAb, m336, in our pseudovirus model. As expected, m336 and H111-1 both showed outstanding inhibitory activity in the pseudovirus cell model (IC$_{50}$ = 4.5 ng/mL) and pseudovirus mouse model. Moreover, mAb H111-1 was also effective in protecting R26-hDPP4 mice from authentic virus infection. The protective effects of a well-characterized peptide inhibitor, HR2P-M2 [26], were assessed in the pseudovirus model with results similar to those described above. Overall, this pseudovirus mouse model showed outcomes that were consistent with those of the authentic virus mouse model, and the R26-hDPP4-KI mouse is, therefore, a unique animal model for evaluating the in vivo efficacy of anti-MERS-CoV therapeutics and vaccines in biological laboratories lacking BSL-3 facilities.

Some researchers have advanced skepticism regarding pseudovirus models, stating that infection by pseudovirus could not cause pathological changes. To allay these doubts, a combination analysis was deemed necessary in this study. As shown in Figure 5, the relevance of the pseudotyped and authentic virus models was demonstrated. We showed that the manifestations of infection in the lung, the main target of MERS-CoV infection, were very similar between the authentic and pseudovirus models and that in both cases infection was localized to the epithelial cells of the bronchi. A dose conversion between pseudotyped virus and authentic virus was calculated, indicating that this pseudovirus mouse model is reliable and can be used to evaluate the effects of antiviral agents. However, pseudovirus represents a single-cycle infection and without replication in vivo, it does not cause pathological changes. Authentic MERS-CoV infection model is a better choice when studying disease pathogenesis.

In summary, we have presented a novel KI mouse model (the R26-hDPP4 mouse) characterized by its genetic stability, its susceptibility to infection by both authentic and pseudotyed MERS-CoVs, its clear sites of infection, and its presentation of disease symptoms, both in the respiratory tract and

CNS, similar to those of clinical patients. These findings imply that this is a unique model for studying human pathogenesis. The established pseudovirus model has the advantages of strong signals, good repeatability and safety, making it as efficient as the authentic virus model for evaluating antiviral agents and vaccines in a typical biological laboratory.

Supplementary Materials: Supplementary materials can be found at http://www.mdpi.com/1999-4915/10/9/448/s1.

Author Contributions: Y.W., C.F., and S.J. designed the studies and wrote the paper, X.W. and C.F. established the KI mouse model, C.F. analyzed the data, X.W., Q.L. (Qiang Liu), and Q.L. (Qianqian Li) conducted the experiments and wrote parts of the paper, S.L., J.L., Y.Y., Y.C., W.H., C.F., and C.L. conducted the infection experiments and collected the data, S.J. provided the peptide HR2P-M2, and T.Y. produced and provided the neutralizing mAb m336, Y.W. decided the final version of the paper.

Funding: This research was supported by National Science and Technology Major Projects of Infectious Disease funds (grants 2017ZX103304402 and 2016YFC1201000).

Acknowledgments: We would like thank Beijing Biocytogen Co., Ltd. for help in establishing the knockin mouse model, Chunyun Sun for providing antibodies and Shuya Zhou, Chenfei Wang, Qin Zuo and Wenda Gu for supplying the mice. We thank the researchers, especially Linlin Bao and Jianglin Liu, at the Institute of Laboratory Sciences, CAM & PUMC, Beijing China, for performing the experiments of authentic virus infections. We thank LiwenBianji, Edanz Editing China (www.liwenbianji.cn/ac), for editing the English text of a draft of this manuscript.

Conflicts of Interest: The funders had no role in study design, data collection and analysis, decision to publish, or preparation of the manuscript. Therefore, they are no competing interests (financial or non-financial, professional, or personal) related to this work. Some mAbs were produced by Sino Biological Inc., which is an independent company with a long-term cooperative relationship with the National Institutes for Food and Drug Control. The company has been sufficiently informed of the work. They have no any financial or non-financial, professional, or personal competing interests, and they support the publication of this manuscript.

References

1. Du, L.; Zhao, G.; Chan, C.C.; Sun, S.; Chen, M.; Liu, Z.; Guo, H.; He, Y.; Zhou, Y.; Zheng, B.-J. Recombinant receptor-binding domain of SARS-CoV spike protein expressed in mammalian, insect and *E. coli* cells elicits potent neutralizing antibody and protective immunity. *Virology* **2009**, *393*, 144–150. [CrossRef] [PubMed]

2. Zhao, G.; Du, L.; Ma, C.; Li, Y.; Li, L.; Poon, V.K.; Wang, L.; Yu, F.; Zheng, B.-J.; Jiang, S. A safe and convenient pseudovirus-based inhibition assay to detect neutralizing antibodies and screen for viral entry inhibitors against the novel human coronavirus MERS-CoV. *Virol. J.* **2013**, *10*, 266. [CrossRef] [PubMed]

3. Coleman, C.M.; Matthews, K.L.; Goicochea, L.; Frieman, M.B. Wild-type and innate immune-deficient mice are not susceptible to the Middle East respiratory syndrome coronavirus. *J. Gen. Virol.* **2014**, *95*, 408–412. [CrossRef] [PubMed]

4. De Wit, E.; Rasmussen, A.L.; Falzarano, D.; Bushmaker, T.; Feldmann, F.; Brining, D.L.; Fischer, E.R.; Martellaro, C.; Okumura, A.; Chang, J.; et al. Middle East respiratory syndrome coronavirus (MERS-CoV) causes transient lower respiratory tract infection in rhesus macaques. *Proc. Natl. Acad. Sci. USA* **2013**, *110*, 16598–16603. [CrossRef] [PubMed]

5. Raj, V.S.; Smits, S.L.; Provacia, L.B.; van den Brand, J.M.; Wiersma, L.; Ouwendijk, W.J.; Bestebroer, T.M.; Spronken, M.I.; van Amerongen, G.; Rottier, P.J.; et al. Adenosine deaminase acts as a natural antagonist for dipeptidyl peptidase 4-mediated entry of the Middle East respiratory syndrome coronavirus. *J. Virol.* **2014**, *88*, 1834–1838. [CrossRef] [PubMed]

6. Falzarano, D.; de Wit, E.; Feldmann, F.; Rasmussen, A.L.; Okumura, A.; Peng, X.; Thomas, M.J.; van Doremalen, N.; Haddock, E.; Nagy, L.; et al. Infection with MERS-CoV causes lethal pneumonia in the common marmoset. *PLoS Pathog.* **2014**, *10*, e1004250. [CrossRef] [PubMed]

7. Raj, V.S.; Mou, H.; Smits, S.L.; Dekkers, D.H.; Müller, M.A.; Dijkman, R.; Muth, D.; Demmers, J.A.; Zaki, A.; Fouchier, R.A.; et al. Dipeptidyl peptidase 4 is a functional receptor for the emerging human coronavirus-EMC. *Nature* **2013**, *495*, 251–254. [CrossRef] [PubMed]

8. Cockrell, A.S.; Yount, B.L.; Scobey, T.; Jensen, K.; Douglas, M.; Beall, A.; Tang, X.-C.; Marasco, W.A.; Heise, M.T.; Baric, R.S. A mouse model for MERS coronavirus-induced acute respiratory distress syndrome. *Nat. Microbiol.* **2017**, *2*, 16226. [CrossRef] [PubMed]

9. Zhao, J.; Li, K.; Wohlford-Lenane, C.; Agnihothram, S.S.; Fett, C.; Zhao, J.; Gale, M.J.; Baric, R.S.; Enjuanes, L.; Gallagher, T. Rapid generation of a mouse model for Middle East respiratory syndrome. *Proc. Natl. Acad. Sci. USA* **2014**, *111*, 4970–4975. [CrossRef] [PubMed]

10. Li, K.; Wohlford-Lenane, C.; Perlman, S.; Zhao, J.; Jewell, A.K.; Reznikov, L.R.; Gibson-Corley, K.N.; Meyerholz, D.K.; McCray, P.B., Jr. Middle East respiratory syndrome coronavirus causes multiple organ damage and lethal disease in mice transgenic for human dipeptidyl peptidase 4. *J. Infect. Dis.* **2015**, *213*, 712–722. [CrossRef] [PubMed]

11. Agrawal, A.S.; Garron, T.; Tao, X.; Peng, B.-H.; Wakamiya, M.; Chan, T.-S.; Couch, R.B.; Tseng, C.-T.K. Generation of a transgenic mouse model of Middle East respiratory syndrome coronavirus infection and disease. *J. Virol.* **2015**, *89*, 3659–3670. [CrossRef] [PubMed]

12. Li, K.; Wohlford-Lenane, C.L.; Channappanavar, R.; Park, J.-E.; Earnest, J.T.; Bair, T.B.; Bates, A.M.; Brogden, K.A.; Flaherty, H.A.; Gallagher, T.; et al. Mouse-adapted MERS coronavirus causes lethal lung disease in human DPP4 knockin mice. *Proc. Natl. Acad. Sci. USA* **2017**, *114*, E3119–E3128. [CrossRef] [PubMed]

13. Lambeir, A.-M.; Durinx, C.; Scharpé, S.; De Meester, I. Dipeptidyl-peptidase IV from bench to bedside: An update on structural properties, functions, and clinical aspects of the enzyme DPP IV. *Crit. Rev. Clin. Lab. Sci.* **2003**, *40*, 209–294. [CrossRef] [PubMed]

14. Simeoni, L.; Rufini, A.; Moretti, T.; Forte, P.; Aiuti, A.; Fantoni, A. Human cd26 expression in transgenic mice affects murine t-cell populations and modifies their subset distribution. *Hum. Immunol.* **2002**, *63*, 719–730. [CrossRef]

15. Du, L.; Kou, Z.; Ma, C.; Tao, X.; Wang, L.; Zhao, G.; Chen, Y.; Yu, F.; Tseng, C.-T.K.; Zhou, Y. A truncated receptor-binding domain of MERS-CoV spike protein potently inhibits MERS-CoV infection and induces strong neutralizing antibody responses: Implication for developing therapeutics and vaccines. *PLoS ONE* **2013**, *8*, e81587. [CrossRef] [PubMed]

16. Ying, T.; Du, L.; Ju, T.W.; Prabakaran, P.; Lau, C.C.; Lu, L.; Liu, Q.; Wang, L.; Feng, Y.; Wang, Y.; et al. Exceptionally potent neutralization of Middle East respiratory syndrome coronavirus by human monoclonal antibodies. *J. Virol.* **2014**, *88*, 7796–7805. [CrossRef] [PubMed]

17. Du, L.; Zhao, G.; Yang, Y.; Qiu, H.; Wang, L.; Kou, Z.; Tao, X.; Yu, H.; Sun, S.; Tseng, C.-T.K.; et al. A conformation-dependent neutralizing monoclonal antibody specifically targeting receptor-binding domain in Middle East respiratory syndrome coronavirus spike protein. *J. Virol.* **2014**, *88*, 7045–7053. [CrossRef] [PubMed]

18. Tang, X.-C.; Agnihothram, S.S.; Jiao, Y.; Stanhope, J.; Graham, R.L.; Peterson, E.C.; Avnir, Y.; Tallarico, A.S.C.; Sheehan, J.; Zhu, Q.; et al. Identification of human neutralizing antibodies against MERS-CoV and their role in virus adaptive evolution. *Proc. Natl. Acad. Sci. USA* **2014**, *111*, E2018–E2026. [CrossRef] [PubMed]

19. Corti, D.; Zhao, J.; Pedotti, M.; Simonelli, L.; Agnihothram, S.; Fett, C.; Fernandez-Rodriguez, B.; Foglierini, M.; Agatic, G.; Vanzetta, F.; et al. Prophylactic and postexposure efficacy of a potent human monoclonal antibody against MERS coronavirus. *Proc. Natl. Acad. Sci. USA* **2015**, *112*, 10473–10478. [CrossRef] [PubMed]

20. Ying, T.; Prabakaran, P.; Du, L.; Shi, W.; Feng, Y.; Wang, Y.; Wang, L.; Li, W.; Jiang, S.; Dimitrov, D.S.; et al. Junctional and allele-specific residues are critical for MERS-CoV neutralization by an exceptionally potent germline-like antibody. *Nat. Commun.* **2015**, *6*, 8223. [CrossRef] [PubMed]

21. Tao, X.; Garron, T.; Agrawal, A.S.; Algaissi, A.; Peng, B.-H.; Wakamiya, M.; Chan, T.-S.; Lu, L.; Du, L.; Jiang, S.; et al. Characterization and demonstration of the value of a lethal mouse model of Middle East respiratory syndrome coronavirus infection and disease. *J. Virol.* **2016**, *90*, 57–67. [CrossRef] [PubMed]

22. Nie, J.; Wu, X.; Ma, J.; Cao, S.; Huang, W.; Liu, Q.; Li, X.; Li, Y.; Wang, Y. Development of In Vitro and In Vivo rabies virus neutralization assays based on a high-titer pseudovirus system. *Sci. Rep.* **2017**, *7*, 42769. [CrossRef] [PubMed]

23. Grehan, K.; Ferrara, F.; Temperton, N. An optimised method for the production of MERS-CoV spike expressing viral pseudotypes. *MethodsX* **2015**, *2*, 379–384. [CrossRef] [PubMed]

24. Liu, Q.; Huang, W.; Nie, J.; Zhu, R.; Gao, D.; Song, A.; Meng, S.; Xu, X.; Wang, Y. A novel high-throughput vaccinia virus neutralization assay and preexisting immunity in populations from different geographic regions in china. *PLoS ONE* **2012**, *7*, e33392. [CrossRef] [PubMed]

25. Agrawal, A.S.; Ying, T.; Tao, X.; Garron, T.; Algaissi, A.; Wang, Y.; Wang, L.; Peng, B.-H.; Jiang, S.; Dimitrov, D.S.; et al. Passive transfer of a germline-like neutralizing human monoclonal antibody protects transgenic mice against lethal Middle East respiratory syndrome coronavirus infection. *Sci. Rep.* **2016**, *6*, 31629. [CrossRef] [PubMed]

26. Channappanavar, R.; Lu, L.; Xia, S.; Du, L.; Meyerholz, D.K.; Perlman, S.; Jiang, S. Protective effect of intranasal regimens containing peptidic Middle East respiratory syndrome coronavirus fusion inhibitor against MERS-CoV infection. *J. Infect. Dis.* **2015**, *212*, 1894–1903. [CrossRef] [PubMed]

27. Lu, L.; Liu, Q.; Zhu, Y.; Chan, K.-H.; Qin, L.; Li, Y.; Wang, Q.; Chan, J.F.-W.; Du, L.; Yu, F.; et al. Structure-based discovery of Middle East respiratory syndrome coronavirus fusion inhibitor. *Nat. Commun.* **2014**, *5*, 3067. [CrossRef] [PubMed]

28. Steinberger, P.; Sutton, J.K.; Rader, C.; Elia, M.; Barbas, C.F. Generation and characterization of a recombinant human ccr5-specific antibody a phage display approach for rabbit antibody humanization. *J. Biol. Chem.* **2000**, *275*, 36073–36078. [CrossRef] [PubMed]

29. Yu, Y.; Lee, P.; Ke, Y.; Zhang, Y.; Yu, Q.; Lee, J.; Li, M.; Song, J.; Chen, J.; Dai, J.; et al. A humanized anti-VEGF rabbit monoclonal antibody inhibits angiogenesis and blocks tumor growth in xenograft models. *PLoS ONE* **2010**, *5*, e9072. [CrossRef] [PubMed]

30. Zhou, S.; Liu, Q.; Wu, X.; Chen, P.; Wu, X.; Guo, Y.; Liu, S.; Liang, Z.; Fan, C.; Wang, Y. A safe and sensitive enterovirus A71 infection model based on human SCARB2 knock-in mice. *Vaccine* **2016**, *34*, 2729–2736. [CrossRef] [PubMed]

31. Zaitseva, M.; Kapnick, S.M.; Scott, J.; King, L.R.; Manischewitz, J.; Sirota, L.; Kodihalli, S.; Golding, H. Application of bioluminescence imaging to the prediction of lethality in vaccinia virus-infected mice. *J. Virol.* **2009**, *83*, 10437–10447. [CrossRef] [PubMed]

32. Yao, Y.; Bao, L.; Deng, W.; Xu, L.; Li, F.; Lv, Q.; Yu, P.; Chen, T.; Xu, Y.; Zhu, H.; et al. An animal model of MERS produced by infection of rhesus macaques with MERS coronavirus. *J. Infect. Dis.* **2013**, *209*, 236–242. [CrossRef] [PubMed]

33. Zambrowicz, B.P.; Imamoto, A.; Fiering, S.; Herzenberg, L.A.; Kerr, W.G.; Soriano, P. Disruption of overlapping transcripts in the ROSA βgeo 26 gene trap strain leads to widespread expression of β-galactosidase in mouse embryos and hematopoietic cells. *Proc. Natl. Acad. Sci. USA* **1997**, *94*, 3789–3794. [CrossRef] [PubMed]

34. Zaki, A.M.; Van Boheemen, S.; Bestebroer, T.M.; Osterhaus, A.D.; Fouchier, R.A. Isolation of a novel coronavirus from a man with pneumonia in Saudi Arabia. *N. Engl. J. Med.* **2012**, *367*, 1814–1820. [CrossRef] [PubMed]

35. Cha, R.-H.; Yang, S.H.; Moon, K.C.; Joh, J.-S.; Lee, J.Y.; Shin, H.-S.; Kim, D.K.; Kim, Y.S. A case report of a Middle East respiratory syndrome survivor with kidney biopsy results. *J. Korean Med. Sci.* **2016**, *31*, 635–640. [CrossRef] [PubMed]

36. Liu, Q.; Zhou, S.; Fan, C.; Huang, W.; Li, Q.; Liu, S.; Wu, X.; Li, B.; Wang, Y. Biodistribution and residence time of adenovector serotype 5 in normal and immunodeficient mice and rats detected with bioluminescent imaging. *Sci. Rep.* **2017**, *7*, 3597. [CrossRef] [PubMed]

37. Zumla, A.; Hui, D.S.; Perlman, S. Middle East respiratory syndrome. *Lancet* **2015**, *386*, 995–1007. [CrossRef]

38. Munster, V.J.; de Wit, E.; Feldmann, H. Pneumonia from human coronavirus in a macaque model. *N. Engl. J. Med.* **2013**, *368*, 1560–1562. [CrossRef] [PubMed]

39. Pascal, K.E.; Coleman, C.M.; Mujica, A.O.; Kamat, V.; Badithe, A.; Fairhurst, J.; Hunt, C.; Strein, J.; Berrebi, A.; Sisk, J.M.; et al. Pre-and postexposure efficacy of fully human antibodies against spike protein in a novel humanized mouse model of MERS-CoV infection. *Proc. Natl. Acad. Sci. USA* **2015**, *112*, 8738–8743. [CrossRef] [PubMed]

40. Arabi, Y.; Harthi, A.; Hussein, J.; Bouchama, A.; Johani, S.; Hajeer, A.; Saeed, B.; Wahbi, A.; Saedy, A.; AlDabbagh, T.; et al. Severe neurologic syndrome associated with Middle East respiratory syndrome corona virus (MERS-CoV). *Infection* **2015**, *43*, 495–501. [CrossRef] [PubMed]

41. Luke, T.; Wu, H.; Zhao, J.; Channappanavar, R.; Coleman, C.M.; Jiao, J.-A.; Matsushita, H.; Liu, Y.; Postnikova, E.N.; Ork, B.L.; et al. Human polyclonal immunoglobulin G from transchromosomic bovines inhibits MERS-CoV In Vivo. *Sci. Transl. Med.* **2016**, *8*, 326ra21. [CrossRef] [PubMed]

42. Qiu, H.; Sun, S.; Xiao, H.; Feng, J.; Guo, Y.; Tai, W.; Wang, Y.; Du, L.; Zhao, G.; Zhou, Y. Single-dose treatment with a humanized neutralizing antibody affords full protection of a human transgenic mouse model from lethal Middle East respiratory syndrome (MERS)-coronavirus infection. *Antivir. Res.* **2016**, *132*, 141–148. [CrossRef] [PubMed]

43. Li, Y.; Wan, Y.; Liu, P.; Zhao, J.; Lu, G.; Qi, J.; Wang, Q.; Lu, X.; Wu, Y.; Liu, W.; et al. A humanized neutralizing antibody against MERS-CoV targeting the receptor-binding domain of the spike protein. *Cell Res.* **2015**, *25*, 1237–1249. [CrossRef] [PubMed]

44. Morimoto, C.; Schlossman, S.F. The structure and function of cd26 in the t-cell immune response. *Immunol. Rev.* **1998**, *161*, 55–70. [CrossRef] [PubMed]

45. Cauchemez, S.; Fraser, C.; Van Kerkhove, M.D.; Donnelly, C.A.; Riley, S.; Rambaut, A.; Enouf, V.; van der Werf, S.; Ferguson, N.M. Middle East respiratory syndrome coronavirus: Quantification of the extent of the epidemic, surveillance biases, and transmissibility. *Lancet Infect. Dis.* **2014**, *14*, 50–56. [CrossRef]

46. Lu, Y.; Jiang, T. Pseudovirus-based neuraminidase inhibition assays reveal potential H5N1 drug-resistant mutations. *Protein Cell* **2013**, *4*, 356–363. [CrossRef] [PubMed]

viruses

MDPI

Review

Advances in MERS-CoV Vaccines and Therapeutics Based on the Receptor-Binding Domain

Yusen Zhou [1,2], Yang Yang [3], Jingwei Huang [4], Shibo Jiang [4] and Lanying Du [4,*]

[1] State Key Laboratory of Pathogen and Biosecurity, Beijing Institute of Microbiology and Epidemiology, Beijing 100071, China; yszhou@bmi.ac.cn
[2] Institute of Medical and Pharmaceutical Sciences, Zhengzhou University, Zhengzhou 450052, China
[3] Department of Molecular Biophysics and Biochemistry, Yale University, New Haven, CT 06520, USA; y.yang@yale.edu
[4] Lindsley F. Kimball Research Institute, New York Blood Center, New York, NY 10065, USA; JHuang2@nybc.org (J.H.); SJiang@nybc.org (S.J.)
* Correspondence: ldu@nybc.org; Tel.: +1-212-570-3459

Received: 15 December 2018; Accepted: 10 January 2019; Published: 14 January 2019

Abstract: Middle East respiratory syndrome (MERS) coronavirus (MERS-CoV) is an infectious virus that was first reported in 2012. The MERS-CoV genome encodes four major structural proteins, among which the spike (S) protein has a key role in viral infection and pathogenesis. The receptor-binding domain (RBD) of the S protein contains a critical neutralizing domain and is an important target for development of MERS vaccines and therapeutics. In this review, we describe the relevant features of the MERS-CoV S-protein RBD, summarize recent advances in the development of MERS-CoV RBD-based vaccines and therapeutic antibodies, and illustrate potential challenges and strategies to further improve their efficacy.

Keywords: Coronavirus; MERS-CoV; spike protein; receptor-binding domain; vaccines; therapeutics

1. Introduction

Middle East respiratory syndrome (MERS) coronavirus (CoV) is an infectious virus that was first reported in June 2012 [1]. MERS-CoV may infect people of any age, but older age, underlying comorbidity (such as diabetes mellitus, renal disease, respiratory disease, heart disease, and hypertension), and delayed confirmation or late diagnosis are all factors that affect MERS disease outcomes and mortality [2–7]. Sex could be a factor in MERS epidemiology, as more males seem to be affected than females [8–10]. MERS-CoV infection of women during pregnancy has adverse outcomes, with fetal mortality of ~27%; however, only a limited number of pediatric MERS-CoV infections occur [11–14]. At the end of December 2018, 2,279 laboratory-confirmed MERS infections were reported globally (in 27 countries), leading to 806 deaths, and a mortality of 35.3%. Among these infections, 1,901 (83.4%) were reported in Saudi Arabia, with mortality in 732 individuals (38.5%) (http://www.emro.who.int/health-topics/mers-cov/mers-outbreaks.html). The largest MERS outbreak outside Saudi Arabia occurred in South Korea in 2015, with 186 cases and 38 deaths [9,15,16]. The most recent MERS cases were reported in 2018 in South Korea, the United Kingdom, and Malaysia, in addition to Saudi Arabia, the United Arab Emirates, and Oman (http://www.who.int/emergencies/mers-cov/en/).

MERS-CoV is thought to have originated in bats [17–20]. MERS-like viruses have been isolated from bats that use (at lower efficiency) the same receptor for cell entry as the MERS-CoV isolated from humans [21–23]. Dromedary camels are potential intermediates for long-term evolution of MERS-CoV and seasonal zoonotic transfer of virus to humans [24–27]. Antibodies specific to MERS-CoV, particularly neutralizing antibodies that neutralize MERS-CoV infection, have been detected in the sera of dromedary camels from a number of countries and regions, with high positive detection rates [28–34]. In addition to camel-to-camel transmission of MERS-CoV, camel-to-human transmission can occur via direct or indirect contact with sick animals [35]. There is a high prevalence of MERS-CoV infection in camel workers, particularly in Saudi Arabia [27]. Following infection of individuals, human-to-human transmission of MERS-CoV occurs frequently, and mainly contributes to community or healthcare-associated outbreaks [36–45]. Infection of humans by MERS-CoV, which mainly occurs through the lower respiratory tract, causes severe respiratory symptoms, leading to failure of the respiratory system and/or other organs [7,46,47]. On rare occasions, the human intestinal tract can be an alternative infection route [46,48].

Anti-MERS-CoV antibodies, including neutralizing antibodies, are found in infected patients, some of whom have persistently presented such antibodies for up to 34 months after the disease outbreak [49,50]. Antibody responses, and particularly neutralizing antibody responses, are crucial factors in the successful treatment of MERS-CoV infections in humans [50–52]. In this regard, the plasma of convalescent MERS-CoV-infected patients is useful for treatment of MERS-CoV infection clinically, but it requires a neutralizing antibody titer (e.g., 50% plaque-reduction neutralization titer: $PRNT_{50}$) ≥1:80 to obtain effective therapeutic results [51–53]. Currently, a human polyclonal IgG antibody (SAB-301) that is produced by transchromosomic cattle immunized with a MERS-CoV vaccine [54], as well as several vaccine candidates (https://clinicaltrials.gov/ct2/show/NCT03399578; https://clinicaltrials.gov/ct2/show/NCT03615911) [55], have been tested in human clinical trials. However, as yet no vaccines or therapeutic agents have been approved for the prevention or treatment of MERS, indicating the need for further development of novel and effective vaccines and therapeutics against MERS-CoV infection.

In the following review, we will briefly describe MERS-CoV spike (S) protein receptor-binding domain (RBD), and summarize recent advances in the development of RBD-based MERS-CoV vaccines and therapeutics, as well as the potential challenges and future expectations for their successful development.

2. MERS-CoV S Protein RBD

MERS-CoV belongs to the genus *Betacoronavirus*, as does the severe acute respiratory syndrome (SARS)-CoV [56,57]. MERS-CoV is a positive-sense, single-stranded RNA virus with a genome that encodes the structural proteins S, envelope (E), membrane (M), and nucleocapsid (N), as well as non-structural proteins (nsp1–16) translated from open reading frames (ORFs) 1a and 1b, and several accessory proteins, such as 3, 4a, 4b, 5, and 8b (Figure 1A,B) [58–61]. The functions of these proteins have been elucidated. As an example, nsp1 is known to have an endonucleolytic RNA-cleavage function in the promotion of infectious virus particle production in human cells [62], and it also participates in efficient viral replication through interaction with viral RNA, probably via a cis-acting element at the 5'-terminal coding region [63]. Evidence indicates that nsp16 is necessary for interferon resistance and viral pathogenesis [64], whereas nsp15 is a nidoviral uridylate-specific endoribonuclease, the structure of which has recently been resolved, and its activity may be mediated by catalytic residues, oligomeric assembly, or RNA-binding efficiency [65]. In addition, 4a protein recognizes and binds double-stranded (ds)RNA, in which four residues of the protein might be crucial for 4a-dsRNA stability [66], and it also inhibits protein kinase R-mediated antiviral stress responses [67]. 4a mediated-inhibition of stress-granule formation may facilitate viral translation, resulting in efficient MERS-CoV replication [68]. The MERS-CoV 4b protein interferes with the NF-κB-dependent innate immune response during infection [69].

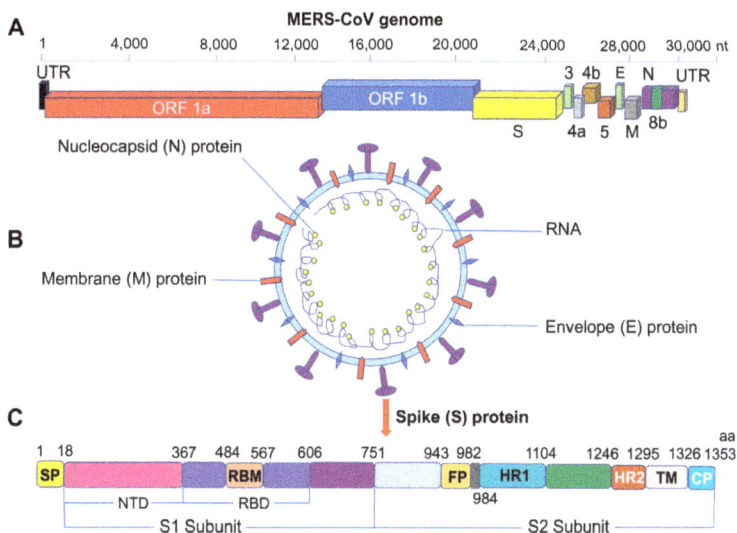

Figure 1. Schematic structures of MERS-CoV S protein. (**A**) MERS-CoV genomic structure, with the untranslated region (UTR), open reading frame regions ORF1a and ORF1b, spike (S), envelope (E), membrane (M), and nucleocapsid (N) genes. (**B**) Schematic structure of the MERS-CoV virion and its major structural proteins. (**C**) Schematic structure of the MERS-CoV S protein and its functional domains, including the N-terminal domain (NTD), receptor-binding domain (RBD), receptor-binding motif (RBM), fusion peptide (FP), heptad repeat region 1 and 2 (HR1 and HR2), transmembrane region (TM), and cytoplasmic tail (CP). aa, amino acid; MERS-CoV, Middle East respiratory syndrome coronavirus; nt, nucleotide.

MERS-CoV S protein has an important role in viral pathogenesis, determining host tropism and entry into host cells [58,70,71]. The S protein contains an S1 subunit at the N terminus and an S2 subunit at the C terminus. The S1 subunit is composed of the N-terminal domain (NTD) and RBD [58,72,73]. The RBD has a key role in the mediation of binding of MERS-CoV to cells expressing dipeptidyl peptidase 4 (DPP4) receptor, enabling the virus to enter into target cells by fusing with cell membranes through the formation of a fusion core (Figure 1C) [74–77]. The S protein requires host cellular proteases for its activity in viral entry, but although evidence initially indicated that cellular furin activates S protein, subsequent results have demonstrated no evidence for the involvement of furin during viral entry [71,78]. The DPP4 receptor varies among different host species, and MERS-CoV is thought to use multiple pathways to enable rapid adaptation to species-specific variations [79–81].

In addition to DPP4, MERS-CoV can bind to sialic acid via the S1 subunit of S protein, or utilize the membrane-associated 78 kDa glucose-regulated protein (GRP78) to attach to target cells, suggesting that these proteins may also have roles in virion attachment [82,83].

The structures of MERS-CoV RBD alone and complexed with DPP4 have been determined (Figure 2) [77,84,85]. The RBD has a fold-rich tertiary structure, which consists of a core and a receptor-binding motif (RBM), with stabilization provided by four disulfide bonds and two glycans [77]. A number of RBD residues are located at the DPP4-binding interface, and they have a critical role in RBD–DPP4 binding [77,84,85]. Structural analysis of MERS-CoV trimeric S protein has identified specific features of the RBD and its complex with DPP4. Notably, in the prefusion conformation of the S trimer, individual RBDs are either buried (lying state) or exposed (standing state), and this flexibility presumably facilitates recognition by DPP4 [86]. Other structural studies have revealed four S-trimer conformational states, in which each RBD is either tightly packed at the membrane-distal apex or rotated into a receptor-accessible conformation, suggesting fusion initiation through sequential RBD events [87]. In configurations with one, two, or three RBDs rotated out, RBD determinants are exposed at the apex of the RBD–DPP4 complex, and they are accessible for interaction with DPP4 (Figure 3) [87].

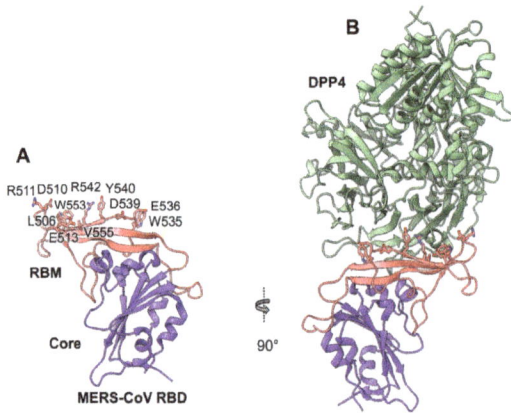

Figure 2. Structural basis of MERS-CoV S-protein RBD–DPP4 interaction. Structural data for the complex of MERS-CoV S-protein RBD bound to DPP4 are from the protein data bank (PDB) (ID: 4KR0). The MERS-CoV RBD core is colored in blue, the RBM is colored in red, and DPP4 is colored in green. The RBM residues directly involved in DPP4 binding are shown as sticks. DPP4, dipeptidyl peptidase 4; RBD, receptor-binding domain; RBM, receptor-binding motif; S, spike protein.

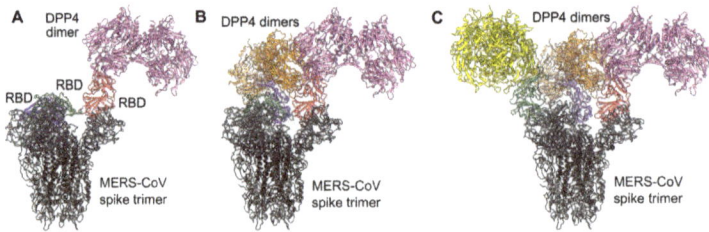

Figure 3. Models of the MERS-CoV S-protein trimer bound to DPP4. The models were generated by superimposing the MERS-CoV RBD in the structure of the MERS-CoV S-protein RBD–DPP4 complex (PDB ID: 4KR0) onto the RBDs in the structures of MERS-CoV S-protein trimers with (**A**) one RBD (PDB ID: 5X5F), (**B**) two RBDs (PDB ID: 5X5C), or (**C**) three RBDs (PDB ID: 5X59) in the "standing" conformation. The MERS-CoV S-protein trimers are colored in gray, with three RBDs colored in red, blue, and green. Three DPP4 dimers are colored in plum, orange, and yellow. DPP4, dipeptidyl peptidase 4; PDB, protein data bank; RBD, receptor-binding domain; S, spike protein.

The function and structure of the S-protein RBD demonstrate that it is an important target for development of vaccines and therapeutic agents against MERS-CoV.

3. Recent Advances in the Development of Vaccines Based on the MERS-CoV S-Protein RBD

A number of MERS vaccines have been developed based on viral RBD, including nanoparticles, virus-like particles (VLPs), and recombinant proteins, and their protective efficacy has been evaluated in animal models, including mice with adenovirus 5 (Ad5)-directed expression of human DPP4 (Ad5/hDPP4), hDPP4-transgenic (hDPP4-Tg) mice, and non-human primates (NHPs) [88–94]. Features of these RBD-based vaccines, in terms of functionality, antigenicity, immunogenicity, and protective ability, are shown in Table 1.

Table 1. Vaccines based on MERS-CoV RBD. Live MERS-CoV strains used for neutralization and challenge experiments, as well as vaccine-induced neutralizing-antibody titers, are described in parentheses.

Name	Functionality and Antigenicity	Immunogenicity in Induction of Antibody Response	Immunogenicity in Induction of Cellular Immune Response	Protective Immunity	Ref.
RBD-[SSG]-FR and RBD-FR nanoparticles	Bind to DPP4 receptor; antisera block RBD-hDPP4 binding	Induce MERS-CoV RBD-specific antibodies (IgG, IgG1, IgG2a, IgG2b, IgA) in mice	Elicit MERS-CoV RBD-specific T-cell responses (IFN-γ, TNF-α) in mouse splenocytes	N/A	[89]
sVLP (spherical virus-like particle)	N/A	Induces MERS-CoV RBD-specific antibodies (IgG) in mice, neutralizing pseudotyped MERS-CoV (1:320)	Elicits MERS-CoV RBD-specific cellular immune response (IFN-γ, IL-2, IL-4) in mouse splenocytes	N/A	[90]
rRBD (recombinant RBD)	N/A	Induces MERS-CoV RBD-specific antibodies (IgG, IgG1, IgG2a) in mice or NHPs, neutralizing pseudotyped (1:800 to 1:1,600) and live (EMC2012: 1:269 to 1:363) MERS-CoV	Induces MERS-CoV RBD-specific cellular immune response (TNF-α, IFN-γ, IL-2, IL-4, IL-6) in mouse splenocytes or monkey PBMCs	Partially protects NHPs from MERS-CoV (EMC2012: 6.5×10^7 TCID$_{50}$) infection with alleviated pneumonia and decreased viral load	[88,91]
RBD (S377-588)-Fc	Binds strongly to soluble and cell-associated hDPP4 or cDPP4 receptors and MERS-CoV RBD-specific neutralizing mAbs (Mersmab1, m336, m337, m338)	Induces MERS-CoV S1-specific antibodies (IgG, IgG1, IgG2a) in mice and rabbits, cross-neutralizing 17 pseudotyped (>1:10^4), 2 live (EMC2012, London1-2012: ≥1:10^3) MERS-CoV, and 5 mAb escape mutants (>1:10^4)	Elicits MERS-CoV S1-specific cellular immune responses (IFN-γ, IL-2) in mouse splenocytes	Protects Ad5/hDPP4-transduced BALB/c mice and hDPP4-Tg mice (67% survival rate) from challenge by MERS-CoV (EMC2012: 10^5 PFU for BALB/c; 10^3–10^4 TCID$_{50}$ for Tg), without toxicity or immune enhancement	[92,95–99]
2012-RBD 2013-RBD 2014-RBD 2015-RBD Camel-RBD	Bind strongly to hDPP4 and cDPP4 receptors and MERS-CoV RBD-specific mAbs (Mersmab1, m336, m337, m338) with high affinity	Induce MERS-CoV S1-specific antibodies (IgG, IgG1, IgG2a) in mice, potently cross-neutralizing 17 pseudotyped (≥1:10^4), 2 live (EMC2012, London1-2012: ≥1:10^2) MERS-CoV, and 5 mAb escape mutants (≥1:10^4)	N/A	N/A	[100]
RBD-Fd	Binds strongly to soluble and cell-associated hDPP4 receptors and MERS-CoV RBD-specific mAbs (Mersmab1, m336, m337, m338)	Induces robust and long-term MERS-CoV S1-specific antibodies (IgG, IgG1, IgG2a) in mice, neutralizing at least 9 pseudotyped (>1:10^4) and live (EMC2012: >1:10^3) MERS-CoV	N/A	Protects hDPP4-Tg mice (83% survival rate) from lethal MERS-CoV (EMC2012: 10^4 TCID$_{50}$) challenge	[94]
RBD (T579N)	Binds strongly to soluble and cell-associated hDPP4 receptors and MERS-CoV RBD-specific mAbs (hHS-1, m336, m337, m338)	Induces highly potent neutralizing antibodies in mice against live MERS-CoV (EMC2012: >1:3 × 10^3)	N/A	Significantly enhances efficacy in fully protecting hDPP4-Tg mice (100% survival rate) from lethal MERS-CoV (EMC2012: 10^4 TCID$_{50}$) challenge	[93]

Note: Ad5, adenovirus 5; DPP4, dipeptidyl peptidase 4; cDPP4, camel DPP4; hDPP4, human DPP4; hDPP4-Tg, hDPP4-transgenic; Fd, foldon; FR, ferritin; mAb, monoclonal antibody; N/A, not available or not applicable; NHP, non-human primate; PBMCs, peripheral blood mononuclear cells; PFU, plaque-forming unit; RBD, receptor-binding domain; S, spike; TCID$_{50}$, median tissue-culture infection dose.

A soluble nanoparticle vaccine formed in *Escherichia coli* by the RNA-mediated folding of a RBD-ferritin (FR) hybrid elicits robust RBD-specific antibody and cellular immune responses in mice, producing antisera that effectively block the binding of RBD to hDPP4 in vitro [89]. The adjuvants alum and the squalene-based MF59 significantly augment the antibody titers and T-cell responses induced by RBD–FR nanoparticle vaccines engineered with or without a SSG linker [89]. Similarly, a chimeric, spherical VLP (sVLP) vaccine expressing MERS-CoV RBD induces specific antibody and cellular immune responses in mice, preventing pseudotyped MERS-CoV entry into susceptible cells [90]. The protective efficacy of these two types of MERS vaccine does not yet seem to have been investigated in a viral-challenge animal model.

Recombinant vaccines involving RBD subunits have been extensively studied for protection against MERS-CoV infection in MERS-CoV-susceptible animal models [93,95–97,100,101]. A recombinant RBD (rRBD) fragment (residues 367–606) expressed in insect cells elicits an antibody response and the production of neutralizing antibodies in mice and NHPs [88,91]. It gives incomplete protection in MERS-CoV-challenged NHPs, with the alleviation of pneumonia and clinical manifestations, as well as the reduction of viral load in lung, trachea, and oropharyngeal swabs [91].

A MERS-CoV S-protein RBD fragment containing residues 377–588 has been identified as a critical neutralizing domain [95]. A treatment regimen involving two doses of a fusion of this fragment and the Fc region of human IgG (S377-588-Fc) four weeks apart is able to induce strong, long-term antibody responses (including production of neutralizing antibodies) in mice [98]. These responses are significantly greater than those with a single dose or two doses at intervals of one, two, or three weeks [98]. rRBDs with single or multiple mutations corresponding to S-protein sequences of MERS-CoV strains isolated from humans or camels from 2012 to 2015 have also been studied [100]. All these rRBDs bind RBD-specific neutralizing monoclonal antibodies (mAbs) and DPP4, and are highly immunogenic, eliciting the production of S1-specific antibodies in mice, which cross-neutralizes multiple MERS pseudoviruses and live MERS-CoV [100]. A trimeric RBD-Fd protein formed by fusing a MERS-CoV RBD fragment (residues 377–588) to the foldon trimerization motif, binds strongly to DPP4, and elicits robust and long-term responses with the production of MERS-CoV S1-specific antibodies and neutralizing antibodies in mice, and protects hDPP4-Tg mice against MERS-CoV infection [94].

The protection provided by existing subunit vaccines based on wild-type MERS-CoV RBD is not complete, with survival rates in hDPP4-Tg mice after a MERS-CoV challenge of ~67% for S377-588-Fc and 83% for RBD-Fd [94,98]. However, a variant RBD (T579N) vaccine produced by masking a non-neutralizing epitope at residue 579 with a glycan probe has both functionality in binding DPP4, and antigenicity in binding four potent MERS-CoV RBD-specific neutralizing mAbs (hHS-1, m336, m337, and m338) [93]. The T579N vaccine has significantly greater efficacy than the wild-type RBD vaccine, and it fully protects against a lethal MERS-CoV challenge in immunized hDPP4-Tg mice [93], demonstrating the possibility of developing RBD-based MERS-CoV vaccines with high efficacy.

4. Recent Advances in the Development of Therapeutics Based on the MERS-CoV S-Protein RBD

MERS-CoV RBD-targeting antibodies have been developed as effective tools to prevent and treat MERS-CoV infections [102–109]. These antibodies generally have greater neutralizing activity against MERS-CoV infection than non-RBD S1-based or S2-based antibodies [58,103,110,111]. The prophylactic and therapeutic efficacies of RBD-targeting antibodies have been tested in Ad5/hDPP4 mice, hDPP4-Tg mice, and NHPs [102,104,112–114].

In an earlier review, we described the antiviral mechanisms, in vivo protection, and crystal structures of previously reported MERS-CoV RBD-specific mAbs, including mouse mAbs Mersmab1, 2E6, 4C2, F11, and D12, and human mAbs LCA60, MERS-4, MERS-27, REGN3048, REGN3051, 1E9, 1F8, 3A1, 3B11, 3B12, 3B11-N, 3C12, M14D3, m336, m337, m338, hMS-1, and 4C2h [58]. In this review, we focus on newly reported antibodies targeting MERS-CoV S-protein RBD, or on newly identified features of existing mAbs that were not described previously (Table 2) [102,112–115].

Table 2. Therapeutic antibodies targeting MERS-CoV RBD. Live MERS-CoV strains used for neutralization and challenge experiments are indicated in parentheses.

Name	Source	Binding MERS-CoV RBD	Structure Available	In vitro Anti-MERS-CoV Activity	In vivo Protection	Ref.
MERS-GD27 MERS-GD33 mAbs	Human	K_d: 0.775 nM (for MERS-GD27) and 0.575 nM (for MERS-GD33). Recognize RBD residues L506, D510, E513, W535, E536, D539, E540, W553 (for MERS-GD27), R511, and A556 (for MERS-GD33)	Yes, crystal structure for the Fab–RBD complex	IC_{50}: 0.001 μg/mL against pseudotyped MERS-CoV; 0.001 μg/mL against live MERS-CoV; both mAbs have synergistic effect against pseudotyped MERS-CoV with reduced IC_{50} by 0.499-fold (for MERS-GD27) or 6.05-fold (for MERS-GD33) vs individual mAbs	MERS-GD27 prophylactically and therapeutically protects hDPP4-Tg mice against MERS-CoV (EMC2012: 3 LD_{50}) with 60% and 40% survival rates, respectively	[112,113]
MCA1 mAb	Human	Recognizes RBD residues D510, W535, E536, D539, Y540, R542, and Q544	Yes, crystal structure for the Fab–RBD complex	IC_{50}: 0.39 μg/mL against live MERS-CoV (EMC2012)	Prophylactically and therapeutically (5–20 mg/kg) inhibits MERS-CoV (EMC2012: 5×10^6 $TCID_{50}$) replication in common marmosets, improving clinical outcomes and reducing lung disease and viral replication	[114]
JC57-14 mAb	Macaque	Recognizes RBD residues W535, E536, D539, Y540, and R542	Yes, crystal structure for the Fab–RBD complex	IC_{50}: 0.0084 μg/mL against pseudotyped MERS-CoV and 0.07 μg/mL against live MERS-CoV (EMC2012), cross-neutralizing 8 pseudotyped MERS-CoVs	N/A	[102]
CDC2-C2 mAb	Human	Recognizes RBD residues F506, D509, W535, E536, D539, Y540, and R542	Yes, crystal structure for the Fab–RBD complex	IC_{50}: 0.0057 μg/mL against pseudotyped MERS-CoV and 0.058 μg/mL against live MERS-CoV (EMC2012), cross-neutralizing 10 pseudotyped MERS-CoVs	Prophylactically (20 mg/kg) protects hDPP4-Tg mice against MERS-CoV (EMC2012: 10^6 $TCID_{50}$) in lungs with 100% survival rate	[102]
MERS-4 mAb	Human	Recognizes RBD residues L507, L545, S546, P547, C549; binds RBD from outside of the RBD DPP4-binding interface	Yes, crystal structure for the Fab–RBD complex	Has synergistic neutralization effect with MERS-27, m336, and 5F9 mAbs against pseudotyped MERS-CoV, with the reduction of IC_{50} by 2.6-fold (for MERS-4 + m336) and 15.21-fold (for MERS-4 + 5F9)	N/A	[115]
VHH-83 HCAb-83 VHHs	Dromedary	K_d: 0.1 nM (for VHH-83) and 2.5 pM (for HCAb-83). Recognizes RBD residue D539	N/A	$PRNT_{50}$: 0.0012–0.0014 μg/mL against live MERS-CoV (EMC2012)	HCAb-83 (200 μg) prophylactically protects hDPP4-Tg mice (K18) against MERS-CoV (EMC2012: 10^5 PFU) in lungs with 100% survival rate	[116]
NbMS10 NbMS10-Fc VHHs	Llama	K_d: 0.87 nM (for NbMS10) and 0.35 nM (for NbMS10-Fc). Recognize RBD residue D539	N/A	IC_{50}: 0.003–0.979 μg/ml (for NbMS10) and 0.003–0.067 μg/ml (for NbMS10-Fc) in cross-neutralizing ≥ 11 pseudotyped MERS-CoVs	NbMS10-Fc (10 mg/kg) prophylactically and therapeutically protects hDPP4-Tg mice against MERS-CoV (EMC2012, $10^{5.3}$ $TCID_{50}$) with 100% survival rate	[117]

Note: DPP4, dipeptidyl peptidase 4; Fab, antigen-binding fragment; hDPP4-Tg, human DPP4-transgenic; IC_{50}, half-maximal inhibitory concentration; K_d, antibody binding affinity; LD_{50}, 50% lethal dose; N/A, not available or not applicable; $PRNT_{50}$, 50% plaque-reduction neutralization titer; RBD, receptor-binding domain; $TCID_{50}$, median tissue-culture infectious dose; VHH, single-domain antibody fragment.

4.1. MERS-CoV S-Protein RBD-Targeting mAbs

RBD-targeting human mAbs have been extensively reported. Most of these mAbs can neutralize pseudotyped or live MERS-CoV in vitro, and some have shown protection against MERS-CoV infection in animal models in vivo [102,112–115]. The structures of several of these mAbs with their antigen-binding fragments (Fabs) or single-chain variable fragments (scFvs) complexed with RBD are known (Figure 4) [102,112–115]. Binding of these mAbs to RBD involves two major recognition modes, with binding to RBD residues contacted by or overlapping with DPP4 (as is the case for GD-27, MCA1, and CDC2-C2), or with binding to the RBD residues outside of the DPP4-binding interface (as seen with MERS-4) (Table 2).

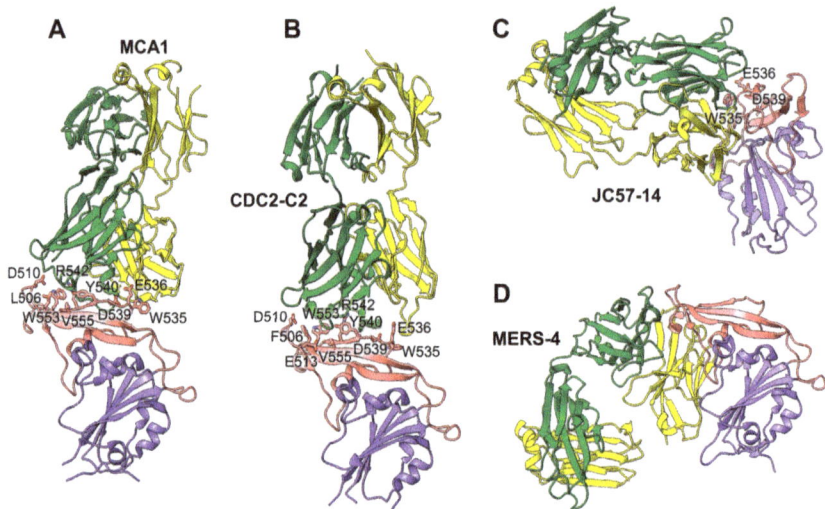

Figure 4. Structural basis for MERS-CoV RBD recognition by neutralizing mAbs. (**A**) Crystal structure of MERS-CoV RBD complexed with MCA1 mAb (PDB ID: 5GMQ). (**B**) Crystal structure of the MERS-CoV RBD complexed with CDC2-C2 mAb (PDB ID: 6C6Z). (**C**) Crystal structure of the MERS-CoV RBD complexed with JC57-14 mAb (PDB ID: 6C6Y). (**D**) Crystal structure of MERS-CoV RBD complexed with MERS-4 mAb (PDB ID: 5ZXV). The MERS-CoV RBD core is colored in blue, the RBM is colored in red, and the heavy chains and light chains of each mAb are colored in green and yellow, respectively. The DPP4-binding residues that are blocked by each mAb are shown as sticks. mAb, monoclonal antibody; MERS-CoV, Middle East respiratory syndrome coronavirus; PDB, protein data bank; RBD, receptor-binding domain; RBM, receptor-binding motif.

The human mAbs MERS-GD27 and MERS-GD33 each recognize distinct regions of the RBD [113]. These mAbs have a synergistic effect in the neutralization of pseudotyped MERS-CoV in vitro, with a much lower half-maximal inhibitory concentration (IC_{50}) for their use in combination than separately [113]. An analysis of crystal structures has indicated that MERS-GD27 binds RBD at the DPP4-binding site, and that the neutralization and recognition epitopes almost completely overlap this site, as seen previously for MERS-CoV RBD-targeting neutralizing mAbs, such as m336 [106,113]. The MERS-GD27 mAb protects hDPP4-Tg mice from MERS-CoV challenge, both preventively and therapeutically, with significantly lower lung virus titers and RNA copy numbers at day 5 post-challenge, and higher survival rates (60% for pre-challenge vaccination and 40% for post-challenge vaccination) relative to control mice treated with an irrelevant mAb [112].

The human mAb MCA1 was isolated from a MERS survivor via the construction of a phage-display antibody library from peripheral B cells [114]. Crystal structure analysis indicates that MCA1 binds MERS-CoV S-protein RBD at residues involved in receptor binding, thus interfering with RBD binding to hDPP4 (Figure 4A) [114]. This mAb prophylactically and therapeutically inhibits MERS-CoV replication in common marmosets, resulting in significantly improved outcomes and reduced lung disease, compared with unvaccinated controls, and undetectable virus titers 3 days post-challenge [114].

A probe-based single-B-cell cloning strategy has been used for the isolation of CDC2-C2 and CDC2-C5 mAbs from a patient convalescing from MERS, as well as for the isolation of JC57-11 and JC57-14 mAbs from NHPs immunized with MERS-CoV full-length S DNA and protein [102]. All these antibodies have neutralizing activities against both pseudotyped and live MERS-CoV. Among them, CDC2-C2 is the most potent against 10 pseudotyped MERS-CoV strains, with neutralization IC_{50} values ranging from 0.002 µg/mL to 0.011 µg/mL [102]. Crystal-structure analysis of the CDC2-C2 and JC57-14 Fab–RBD complexes indicates that both mAbs bind RBD in the "out" (exposed) position, with the CDC2-C2 RBD binding overlapping with the DPP4-contacting residues (Figure 4B,C) [102]. In addition, CDC2-C2 prophylactically protects hDPP4-Tg mice from MERS-CoV infection, resulting in no detectable viral replication in the lungs three days post-challenge, and no fatalities over 28 days of observation [102].

The human mAb MERS-4 also neutralizes pseudotyped MERS-CoV and, notably, displays synergistic neutralization in combination with the MERS-CoV S-protein RBD-targeting MERS-27 and m336 mAbs [106,118], as well as the S-protein NTD-targeting 5F9 mAb, in each case with dramatic reduction of the IC_{50} compared with individual mAbs [115]. Structural analysis of a MERS-4-Fab–RBD complex revealed that MERS-4 binds the RBD from outside the DPP4-binding interface, rather than competing with DPP4 (Figure 4D). Unlike MERS-27, which binds RBD regardless of its conformational state within the S trimer, MERS-4 binds RBD in the "standing" position where its epitopes are readily exposed and accessible [115]. Thus, MERS-4 displays unique epitope specificity, and an unusual mechanism of action involving indirect interference with DPP4 binding through conformational changes, which may explain the observation of synergistic neutralization in combination with other mAbs [115].

4.2. Nanobodies Targeting the MERS-CoV S-protein RBD

Single-domain antibody fragments (VHHs), or nanobodies, are the antigen-recognition regions of camelid heavy-chain-only antibodies (HcAbs), which do not contain light chains. VHHs are easily expressed with high yield, and they have intrinsic stability, strong binding affinity, and specificity to target antigens, and they have therefore been developed as important therapeutic tools against viral infection, including that of MERS-CoV [116,117,119–123].

Four VHHs (VHH-1, VHH-4, VHH-83, and VHH-101) have been identified from bone marrow cells of dromedary camels immunized with modified vaccinia virus (MVA) expressing MERS-CoV S protein, and challenged with MERS-CoV [116]. These VHHs bind MERS-CoV S protein with low K_d values (0.1–1 nM), recognize an epitope at residue D539 of RBD, and neutralize MERS-CoV (PRNT$_{50}$, 0.0014–0.012 µg/mL) [116]. These four monomeric VHHs have each been fused with a C-terminal human IgG2 tag to generate four HCAbs (HCAb-1, HCAb-4, HCAb-83, and HCAb-101), with a higher binding affinity and a longer half-life than the free VHHs [116]. Studies of protective efficacy show that hDPP4-Tg mice (K18) injected with monomeric VHH-83 (20 or 200 µg per mouse) lose weight, and die within seven days post-infection, possibly because of the short half-life of the VHH. However, when the mice are injected with HCAb-83 (200 µg per mouse), which has an extended half-life (~4.5 days), protection against MERS-CoV is complete, with no viral titers or pathological changes in the lungs of virally challenged mice [116].

By immunizing llamas with a recombinant RBD fragment (residues 377–588) fused to a C-terminal human IgG Fc tag (S377-588-Fc), we constructed a VHH library, and we used it to generate a monomeric VHH, NbMS10, and a human Fc-fused VHH, NbMS10-Fc [117]. Both VHHs can be expressed in a yeast expression system to high purity, and bind RBD with high affinity, recognizing a conformational epitope (residue 539) at the RBD–DPP4 interface, and blocking the binding of RBD to DPP4. These VHHs, particularly NbMS10-Fc, potently cross-neutralize pseudotyped MERS-CoV strains isolated from different countries, hosts, and time periods [117]. Importantly, the Fc-fused NbMS10-Fc significantly improves the serum half-life of NbMS10, and a single-dose treatment of hDPP4-Tg mice with this agent completely protects them against lethal MERS-CoV challenge [117]. These single-domain VHHs demonstrate the feasibility of developing cost-effective, potent, and broad-spectrum therapeutic antibodies against MERS-CoV infection.

5. Potential Challenges and Future Perspectives

Compared with vaccines based on MERS-CoV full-length S protein, which have the potential to attenuate neutralizing activity or enhance immune pathology, vaccines developed from MERS-CoV S-protein RBD are safer, and they do not cause immunological toxicity or eosinophilic immune enhancement [55,99,110,124]. Moreover, RBD-based therapeutic antibodies are generally more potent than non-RBD S1-based or S2-based antibodies [58,104,111]. Hence, RBD-based vaccines and therapeutic antibodies have the potential for further development as effective tools to prevent and treat MERS-CoV infection.

Despite their acknowledged advantages, there are some issues associated with RBD-based interventions that need to be addressed. For example, RBD is under a high level of pressure of positive selection, and mutations occur in the RBD-DPP4 binding interface that might reduce the efficacy of these treatments [100,125–127]. One possible way to avoid this effect, and to delay the emergence of escape mutants is to combine RBD-targeting therapeutics with those targeting other regions of the S protein, or to combine antibodies recognizing distinct epitopes within the RBD [102,128]. Such combinatorial strategies could also dramatically reduce antibody neutralization doses, providing feasible means to combat the continual threat of MERS-CoV.

Some recent advances have been made in the structure-guided design of anti-MERS-CoV interventions. Structurally designed inhibitors of the 3CL protease have demonstrated potency against MERS-CoV [129]. Also, a structurally designed S-protein trimer in the optimal prefusion conformation is shown to elicit production of high titers of anti-MERS-CoV neutralizing antibodies [87]. Indeed, based on the previous studies on the structural design of MERS-CoV RBD, non-neutralizing epitopes in the RBD can be masked, to refocus the immunogenicity of the RBD on the neutralizing epitopes, and thus to enhance its ability to confer immune protection [93]. Results from these structure-based studies will help to inform the design of innovative RBD-based anti-MERS-CoV vaccines and therapeutics with improved efficacy.

Author Contributions: Y.Z. and L.D. drafted the manuscript. Y.Y. and J.H. prepared the figures. Y.Z., S.J., and L.D. reviewed and revised the manuscript.

Acknowledgments: This study was supported by the NSFC grant 81571983, and the NIH grants R21AI128311, R01AI137472, and R01AI139092. The funders had no role in study design, data collection and interpretation, or the decision to submit the work for publication.

Conflicts of Interest: The authors declare no competing interests.

References

1. Zaki, A.M.; van Boheemen, S.; Bestebroer, T.M.; Osterhaus, A.D.; Fouchier, R.A. Isolation of a novel coronavirus from a man with pneumonia in Saudi Arabia. *N. Engl. J. Med.* **2012**, *367*, 1814–1820. [CrossRef] [PubMed]

2. Sha, J.; Li, Y.; Chen, X.; Hu, Y.; Ren, Y.; Geng, X.; Zhang, Z.; Liu, S. Fatality risks for nosocomial outbreaks of Middle East respiratory syndrome coronavirus in the Middle East and South Korea. *Arch. Virol.* **2017**, *162*, 33–44. [CrossRef] [PubMed]

3. Rivers, C.M.; Majumder, M.S.; Lofgren, E.T. Risks of death and severe disease in patients with Middle East respiratory syndrome coronavirus, 2012-2015. *Am. J. Epidemiol.* **2016**, *184*, 460–464. [CrossRef] [PubMed]

4. Yang, Y.M.; Hsu, C.Y.; Lai, C.C.; Yen, M.F.; Wikramaratna, P.S.; Chen, H.H.; Wang, T.H. Impact of comorbidity on fatality rate of patients with Middle East respiratory syndrome. *Sci. Rep.* **2017**, *7*, 11307. [CrossRef] [PubMed]

5. Matsuyama, R.; Nishiura, H.; Kutsuna, S.; Hayakawa, K.; Ohmagari, N. Clinical determinants of the severity of Middle East respiratory syndrome (MERS): A systematic review and meta-analysis. *BMC Public Health* **2016**, *16*, 1203. [CrossRef] [PubMed]

6. Badawi, A.; Ryoo, S.G. Prevalence of comorbidities in the Middle East respiratory syndrome coronavirus (MERS-CoV): A systematic review and meta-analysis. *Int. J. Infect. Dis.* **2016**, *49*, 129–133. [CrossRef] [PubMed]

7. Assiri, A.; Al-Tawfiq, J.A.; Al-Rabeeah, A.A.; Al-Rabiah, F.A.; Al-Hajjar, S.; Al-Barrak, A.; Flemban, H.; Al-Nassir, W.N.; Balkhy, H.H.; Al-Hakeem, R.F.; et al. Epidemiological, demographic, and clinical characteristics of 47 cases of Middle East respiratory syndrome coronavirus disease from Saudi Arabia: A descriptive study. *Lancet Infect. Dis.* **2013**, *13*, 752–761. [CrossRef]

8. Jansen, A.; Chiew, M.; Konings, F.; Lee, C.K.; Ailan, L. Sex matters - a preliminary analysis of Middle East respiratory syndrome in the Republic of Korea, 2015. *West. Pac. Surveill Response J.* **2015**, *6*, 68–71. [CrossRef]

9. Kim, K.H.; Tandi, T.E.; Choi, J.W.; Moon, J.M.; Kim, M.S. Middle East respiratory syndrome coronavirus (MERS-CoV) outbreak in South Korea, 2015: Epidemiology, characteristics and public health implications. *J. Hosp. Infect.* **2017**, *95*, 207–213. [CrossRef]

10. Assiri, A.; McGeer, A.; Perl, T.M.; Price, C.S.; Al Rabeeah, A.A.; Cummings, D.A.; Alabdullatif, Z.N.; Assad, M.; Almulhim, A.; Makhdoom, H.; et al. Hospital outbreak of Middle East respiratory syndrome coronavirus. *N. Engl. J. Med.* **2013**, *369*, 407–416. [CrossRef]

11. Alfaraj, S.H.; Al-Tawfiq, J.A.; Memish, Z.A. Middle East respiratory syndrome coronavirus (MERS-CoV) infection during pregnancy: Report of two cases & review of the literature. *J. Microbiol. Immunol. Infect.* **2018**. [CrossRef]

12. Jeong, S.Y.; Sung, S.I.; Sung, J.H.; Ahn, S.Y.; Kang, E.S.; Chang, Y.S.; Park, W.S.; Kim, J.H. MERS-CoV infection in a pregnant woman in Korea. *J. Korean Med. Sci.* **2017**, *32*, 1717–1720. [CrossRef] [PubMed]

13. Alfaraj, S.H.; Al-Tawfiq, J.A.; Altuwaijri, T.A.; Memish, Z.A. Middle East respiratory syndrome coronavirus in pediatrics: A report of seven cases from Saudi Arabia. *Front. Med.* **2018**. [CrossRef] [PubMed]

14. Assiri, A.; Abedi, G.R.; Al, M.M.; Bin, S.A.; Gerber, S.I.; Watson, J.T. Middle East respiratory syndrome coronavirus infection during pregnancy: A Report of 5 cases from Saudi Arabia. *Clin. Infect. Dis.* **2016**, *63*, 951–953. [CrossRef] [PubMed]

15. Choi, J.Y. An outbreak of Middle East respiratory syndrome coronavirus infection in South Korea, 2015. *Yonsei Med. J.* **2015**, *56*, 1174–1176. [CrossRef] [PubMed]

16. Lee, S.S.; Wong, N.S. Probable transmission chains of Middle East respiratory syndrome coronavirus and the multiple generations of secondary infection in South Korea. *Int. J. Infect. Dis.* **2015**, *38*, 65–67. [CrossRef] [PubMed]

17. Anthony, S.J.; Gilardi, K.; Menachery, V.D.; Goldstein, T.; Ssebide, B.; Mbabazi, R.; Navarrete-Macias, I.; Liang, E.; Wells, H.; Hicks, A.; et al. Further evidence for bats as the evolutionary source of Middle East respiratory syndrome coronavirus. *MBio* **2017**, *8*, e00373-17. [CrossRef]

18. Wang, Q.; Qi, J.; Yuan, Y.; Xuan, Y.; Han, P.; Wan, Y.; Ji, W.; Li, Y.; Wu, Y.; Wang, J.; et al. Bat origins of MERS-CoV supported by bat coronavirus HKU4 usage of human receptor CD26. *Cell Host Microbe* **2014**, *16*, 328–337. [CrossRef] [PubMed]

19. Yang, Y.; Du, L.; Liu, C.; Wang, L.; Ma, C.; Tang, J.; Baric, R.S.; Jiang, S.; Li, F. Receptor usage and cell entry of bat coronavirus HKU4 provide insight into bat-to-human transmission of MERS coronavirus. *Proc. Natl. Acad. Sci. USA* **2014**, *111*, 12516–12521. [CrossRef]

20. Munster, V.J.; Adney, D.R.; Van Doremalen, N.; Brown, V.R.; Miazgowicz, K.L.; Milne-Price, S.; Bushmaker, T.; Rosenke, R.; Scott, D.; Hawkinson, A.; et al. Replication and shedding of MERS-CoV in Jamaican fruit bats (Artibeus jamaicensis). *Sci. Rep.* **2016**, *6*, 21878. [CrossRef]

21. Luo, C.M.; Wang, N.; Yang, X.L.; Liu, H.Z.; Zhang, W.; Li, B.; Hu, B.; Peng, C.; Geng, Q.B.; Zhu, G.J.; et al. Discovery of novel bat coronaviruses in south China that use the same receptor as Middle East respiratory syndrome coronavirus. *J. Virol.* **2018**, *92*, e00116-18. [CrossRef]

22. Woo, P.C.Y.; Lau, S.K.P.; Chen, Y.; Wong, E.Y.M.; Chan, K.H.; Chen, H.; Zhang, L.; Xia, N.; Yuen, K.Y. Rapid detection of MERS coronavirus-like viruses in bats: Pote1ntial for tracking MERS coronavirus transmission and animal origin. *Emerg. Microbes Infect.* **2018**, *7*, 18. [CrossRef] [PubMed]

23. Lau, S.K.P.; Zhang, L.; Luk, H.K.H.; Xiong, L.; Peng, X.; Li, K.S.M.; He, X.; Zhao, P.S.; Fan, R.Y.Y.; Wong, A.C.P.; et al. Receptor usage of a novel bat lineage C Betacoronavirus reveals evolution of Middle East respiratory syndrome-related coronavirus spike proteins for human dipeptidyl peptidase 4 binding. *J. Infect. Dis.* **2018**, *218*, 197–207. [CrossRef] [PubMed]

24. Dudas, G.; Carvalho, L.M.; Rambaut, A.; Bedford, T. MERS-CoV spillover at the camel-human interface. *Elife* **2018**, *7*, e31257. [CrossRef]

25. Yusof, M.F.; Eltahir, Y.M.; Serhan, W.S.; Hashem, F.M.; Elsayed, E.A.; Marzoug, B.A.; Abdelazim, A.S.; Bensalah, O.K.; Al Muhairi, S.S. Prevalence of Middle East respiratory syndrome coronavirus (MERS-CoV) in dromedary camels in Abu Dhabi Emirate, United Arab Emirates. *Virus Genes* **2015**, *50*, 509–513. [CrossRef] [PubMed]

26. Haagmans, B.L.; Al Dhahiry, S.H.; Reusken, C.B.; Raj, V.S.; Galiano, M.; Myers, R.; Godeke, G.J.; Jonges, M.; Farag, E.; Diab, A.; et al. Middle East respiratory syndrome coronavirus in dromedary camels: An outbreak investigation. *Lancet Infect. Dis.* **2014**, *14*, 140–145. [CrossRef]

27. Alshukairi, A.N.; Zheng, J.; Zhao, J.; Nehdi, A.; Baharoon, S.A.; Layqah, L.; Bokhari, A.; Al Johani, S.M.; Samman, N.; Boudjelal, M.; et al. High prevalence of MERS-CoV infection in camel workers in Saudi Arabia. *MBio* **2018**, *9*, e01985-18. [CrossRef]

28. Harcourt, J.L.; Rudoler, N.; Tamin, A.; Leshem, E.; Rasis, M.; Giladi, M.; Haynes, L.M. The prevalence of Middle East respiratory syndrome coronavirus (MERS-CoV) antibodies in dromedary camels in Israel. *Zoonoses Public Health* **2018**. [CrossRef] [PubMed]

29. Saqib, M.; Sieberg, A.; Hussain, M.H.; Mansoor, M.K.; Zohaib, A.; Lattwein, E.; Muller, M.A.; Drosten, C.; Corman, V.M. Serologic evidence for MERS-CoV infection in dromedary camels, Punjab, Pakistan, 2012-2015. *Emerg. Infect. Dis.* **2017**, *23*, 550–551. [CrossRef] [PubMed]

30. Harrath, R.; Abu Duhier, F.M. Sero-prevalence of Middle East respiratory syndrome coronavirus (MERS-CoV) specific antibodies in dromedary camels in Tabuk, Saudi Arabia. *J. Med. Virol.* **2018**, *90*, 1285–1289. [CrossRef] [PubMed]

31. Falzarano, D.; Kamissoko, B.; de Wit, E.; Maiga, O.; Cronin, J.; Samake, K.; Traore, A.; Milne-Price, S.; Munster, V.J.; Sogoba, N.; et al. Dromedary camels in northern Mali have high seropositivity to MERS-CoV. *One Health* **2017**, *3*, 41–43. [CrossRef] [PubMed]

32. Ali, M.; El-Shesheny, R.; Kandeil, A.; Shehata, M.; Elsokary, B.; Gomaa, M.; Hassan, N.; El, S.A.; El-Taweel, A.; Sobhy, H.; et al. Cross-sectional surveillance of Middle East respiratory syndrome coronavirus (MERS-CoV) in dromedary camels and other mammals in Egypt, August 2015 to January 2016. *Euro. Surveill* **2017**, *22*, 30487. [CrossRef]

33. Ali, M.A.; Shehata, M.M.; Gomaa, M.R.; Kandeil, A.; El-Shesheny, R.; Kayed, A.S.; El-Taweel, A.N.; Atea, M.; Hassan, N.; Bagato, O.; et al. Systematic, active surveillance for Middle East respiratory syndrome coronavirus in camels in Egypt. *Emerg. Microbes Infect.* **2017**, *6*. [CrossRef] [PubMed]

34. Deem, S.L.; Fevre, E.M.; Kinnaird, M.; Browne, A.S.; Muloi, D.; Godeke, G.J.; Koopmans, M.; Reusken, C.B. Serological evidence of MERS-CoV antibodies in dromedary camels (Camelus dromedaries) in Laikipia county, Kenya. *PLoS ONE* **2015**, *10*, e0140125. [CrossRef] [PubMed]

35. Conzade, R.; Grant, R.; Malik, M.R.; Elkholy, A.; Elhakim, M.; Samhouri, D.; Ben Embarek, P.K.; Van Kerkhove, M.D. Reported direct and indirect contact with dromedary camels among laboratory-confirmed MERS-CoV cases. *Viruses* **2018**, *10*, 425. [CrossRef]

36. Hui, D.S.; Azhar, E.I.; Kim, Y.J.; Memish, Z.A.; Oh, M.D.; Zumla, A. Middle East respiratory syndrome coronavirus: Risk factors and determinants of primary, household, and nosocomial transmission. *Lancet Infect. Dis.* **2018**, *18*, e217–e227. [CrossRef]

37. Amer, H.; Alqahtani, A.S.; Alzoman, H.; Aljerian, N.; Memish, Z.A. Unusual presentation of Middle East respiratory syndrome coronavirus leading to a large outbreak in Riyadh during 2017. *Am. J. Infect. Control.* **2018**, *46*, 1022–1025. [CrossRef] [PubMed]

38. Park, S.H.; Kim, Y.S.; Jung, Y.; Choi, S.Y.; Cho, N.H.; Jeong, H.W.; Heo, J.Y.; Yoon, J.H.; Lee, J.; Cheon, S.; et al. Outbreaks of Middle East respiratory syndrome in two hospitals initiated by a single patient in Daejeon, South Korea. *Infect. Chemother.* **2016**, *48*, 99–107. [CrossRef]

39. Cho, S.Y.; Kang, J.M.; Ha, Y.E.; Park, G.E.; Lee, J.Y.; Ko, J.H.; Lee, J.Y.; Kim, J.M.; Kang, C.I.; Jo, I.J.; et al. MERS-CoV outbreak following a single patient exposure in an emergency room in South Korea: An epidemiological outbreak study. *Lancet* **2016**, *388*, 994–1001. [CrossRef]
40. Hastings, D.L.; Tokars, J.I.; Abdel Aziz, I.Z.; Alkhaldi, K.Z.; Bensadek, A.T.; Alraddadi, B.M.; Jokhdar, H.; Jernigan, J.A.; Garout, M.A.; Tomczyk, S.M.; et al. Outbreak of Middle East respiratory syndrome at tertiary care hospital, Jeddah, Saudi Arabia, 2014. *Emerg. Infect. Dis.* **2016**, *22*, 794–801. [CrossRef]
41. Hunter, J.C.; Nguyen, D.; Aden, B.; Al, B.Z.; Al, D.W.; Abu, E.K.; Khudair, A.; Al, M.M.; El, S.F.; Imambaccus, H.; et al. Transmission of Middle East respiratory syndrome coronavirus infections in healthcare settings, Abu Dhabi. *Emerg. Infect. Dis.* **2016**, *22*, 647–656. [CrossRef]
42. Harriman, K.; Brosseau, L.; Trivedi, K. Hospital-associated Middle East respiratory syndrome coronavirus infections. *N. Engl. J. Med.* **2013**, *369*, 1761. [PubMed]
43. Memish, Z.A.; Al-Tawfiq, J.A.; Assiri, A. Hospital-associated Middle East respiratory syndrome coronavirus infections. *N. Engl. J. Med.* **2013**, *369*, 1761–1762. [CrossRef] [PubMed]
44. Memish, Z.A.; Zumla, A.I.; Al-Hakeem, R.F.; Al-Rabeeah, A.A.; Stephens, G.M. Family cluster of Middle East respiratory syndrome coronavirus infections. *N. Engl. J. Med.* **2013**, *368*, 2487–2494. [CrossRef] [PubMed]
45. Al-Tawfiq, J.A.; Auwaerter, P.G. Healthcare-associated infections: The hallmark of Middle East respiratory syndrome coronavirus with review of the literature. *J. Hosp. Infect.* **2019**, *101*, 20–29. [CrossRef] [PubMed]
46. Guery, B.; Poissy, J.; El, M.L.; Sejourne, C.; Ettahar, N.; Lemaire, X.; Vuotto, F.; Goffard, A.; Behillil, S.; Enouf, V.; et al. Clinical features and viral diagnosis of two cases of infection with Middle East respiratory syndrome coronavirus: A report of nosocomial transmission. *Lancet* **2013**, *381*, 2265–2272. [CrossRef]
47. Arabi, Y.M.; Arifi, A.A.; Balkhy, H.H.; Najm, H.; Aldawood, A.S.; Ghabashi, A.; Hawa, H.; Alothman, A.; Khaldi, A.; Al, R.B. Clinical course and outcomes of critically ill patients with Middle East respiratory syndrome coronavirus infection. *Ann. Intern. Med.* **2014**, *160*, 389–397. [CrossRef]
48. Zhou, J.; Li, C.; Zhao, G.; Chu, H.; Wang, D.; Yan, H.H.; Poon, V.K.; Wen, L.; Wong, B.H.; Zhao, X.; et al. Human intestinal tract serves as an alternative infection route for Middle East respiratory syndrome coronavirus. *Sci. Adv.* **2017**, *3*, eaao4966. [CrossRef]
49. Payne, D.C.; Iblan, I.; Rha, B.; Alqasrawi, S.; Haddadin, A.; Al, N.M.; Alsanouri, T.; Ali, S.S.; Harcourt, J.; Miao, C.; et al. Persistence of antibodies against Middle East respiratory syndrome coronavirus. *Emerg. Infect. Dis.* **2016**, *22*, 1824–1826. [CrossRef]
50. Muller, M.A.; Meyer, B.; Corman, V.M.; Al-Masri, M.; Turkestani, A.; Ritz, D.; Sieberg, A.; Aldabbagh, S.; Bosch, B.J.; Lattwein, E.; et al. Presence of Middle East respiratory syndrome coronavirus antibodies in Saudi Arabia: A nationwide, cross-sectional, serological study. *Lancet Infect. Dis.* **2015**, *15*, 559–564. [CrossRef]
51. Arabi, Y.M.; Hajeer, A.H.; Luke, T.; Raviprakash, K.; Balkhy, H.; Johani, S.; Al-Dawood, A.; Al-Qahtani, S.; Al-Omari, A.; Al-Hameed, F.; et al. Feasibility of using convalescent plasma immunotherapy for MERS-CoV infection, Saudi Arabia. *Emerg. Infect. Dis.* **2016**, *22*, 1554–1561. [CrossRef] [PubMed]
52. Arabi, Y.; Balkhy, H.; Hajeer, A.H.; Bouchama, A.; Hayden, F.G.; Al-Omari, A.; Al-Hameed, F.M.; Taha, Y.; Shindo, N.; Whitehead, J.; et al. Feasibility, safety, clinical, and laboratory effects of convalescent plasma therapy for patients with Middle East respiratory syndrome coronavirus infection: A study protocol. *Springerplus* **2015**, *4*, 709. [CrossRef] [PubMed]
53. Ko, J.H.; Seok, H.; Cho, S.Y.; Ha, Y.E.; Baek, J.Y.; Kim, S.H.; Kim, Y.J.; Park, J.K.; Chung, C.R.; Kang, E.S.; et al. Challenges of convalescent plasma infusion therapy in Middle East respiratory coronavirus infection: A single centre experience. *Antivir. Ther.* **2018**, *23*, 617–622. [CrossRef] [PubMed]
54. Beigel, J.H.; Voell, J.; Kumar, P.; Raviprakash, K.; Wu, H.; Jiao, J.A.; Sullivan, E.; Luke, T.; Davey, R.T., Jr. Safety and tolerability of a novel, polyclonal human anti-MERS coronavirus antibody produced from transchromosomic cattle: A phase 1 randomised, double-blind, single-dose-escalation study. *Lancet Infect. Dis.* **2018**, *18*, 410–418. [CrossRef]
55. Zhou, Y.; Jiang, S.; Du, L. Prospects for a MERS-CoV spike vaccine. *Expert. Rev. Vaccines* **2018**, *17*, 677–686. [CrossRef] [PubMed]
56. Zhang, N.; Jiang, S.; Du, L. Current advancements and potential strategies in the development of MERS-CoV vaccines. *Expert. Rev. Vaccines* **2014**, *13*, 761–774. [CrossRef] [PubMed]
57. Chan, J.F.; Li, K.S.; To, K.K.; Cheng, V.C.; Chen, H.; Yuen, K.Y. Is the discovery of the novel human betacoronavirus 2c EMC/2012 (HCoV-EMC) the beginning of another SARS-like pandemic? *J. Infect.* **2012**, *65*, 477–489. [CrossRef] [PubMed]

58. Du, L.; Yang, Y.; Zhou, Y.; Lu, L.; Li, F.; Jiang, S. MERS-CoV spike protein: A key target for antivirals. *Expert. Opin. Ther. Targets* **2017**, *21*, 131–143. [CrossRef]

59. Van Boheemen, S.; de Graaf, M.; Lauber, C.; Bestebroer, T.M.; Raj, V.S.; Zaki, A.M.; Osterhaus, A.D.; Haagmans, B.L.; Gorbalenya, A.E.; Snijder, E.J.; et al. Genomic characterization of a newly discovered coronavirus associated with acute respiratory distress syndrome in humans. *MBio* **2012**, *3*, e00473-12. [CrossRef]

60. Almazan, F.; DeDiego, M.L.; Sola, I.; Zuniga, S.; Nieto-Torres, J.L.; Marquez-Jurado, S.; Andres, G.; Enjuanes, L. Engineering a replication-competent, propagation-defective Middle East respiratory syndrome coronavirus as a vaccine candidate. *MBio* **2013**, *4*, e00650-13. [CrossRef]

61. Scobey, T.; Yount, B.L.; Sims, A.C.; Donaldson, E.F.; Agnihothram, S.S.; Menachery, V.D.; Graham, R.L.; Swanstrom, J.; Bove, P.F.; Kim, J.D.; et al. Reverse genetics with a full-length infectious cDNA of the Middle East respiratory syndrome coronavirus. *Proc. Natl. Acad. Sci. USA* **2013**, *110*, 16157–16162. [CrossRef]

62. Nakagawa, K.; Narayanan, K.; Wada, M.; Popov, V.L.; Cajimat, M.; Baric, R.S.; Makino, S. The endonucleolytic RNA cleavage function of nsp1 of Middle East respiratory syndrome coronavirus promotes the production of infectious virus particles in specific human cell lines. *J. Virol.* **2018**, *92*, e01157-18. [CrossRef]

63. Terada, Y.; Kawachi, K.; Matsuura, Y.; Kamitani, W. MERS coronavirus nsp1 participates in an efficient propagation through a specific interaction with viral RNA. *Virology* **2017**, *511*, 95–105. [CrossRef] [PubMed]

64. Menachery, V.D.; Gralinski, L.E.; Mitchell, H.D.; Dinnon, K.H., 3rd; Leist, S.R.; Yount, B.L., Jr.; Graham, R.L.; McAnarney, E.T.; Stratton, K.G.; Cockrell, A.S.; et al. Middle East respiratory syndrome coronavirus nonstructural protein 16 Is necessary for interferon resistance and viral pathogenesis. *mSphere* **2017**, *2*, e00346-17. [CrossRef]

65. Zhang, L.; Li, L.; Yan, L.; Ming, Z.; Jia, Z.; Lou, Z.; Rao, Z. Structural and biochemical characterization of endoribonuclease Nsp15 encoded by Middle East respiratory syndrome coronavirus. *J. Virol.* **2018**, *92*, e00893-18. [CrossRef]

66. Batool, M.; Shah, M.; Patra, M.C.; Yesudhas, D.; Choi, S. Structural insights into the Middle East respiratory syndrome coronavirus 4a protein and its dsRNA binding mechanism. *Sci. Rep.* **2017**, *7*, 11362. [CrossRef] [PubMed]

67. Rabouw, H.H.; Langereis, M.A.; Knaap, R.C.; Dalebout, T.J.; Canton, J.; Sola, I.; Enjuanes, L.; Bredenbeek, P.J.; Kikkert, M.; de Groot, R.J.; et al. Middle East respiratory coronavirus accessory protein 4a inhibits PKR-mediated antiviral stress responses. *PLoS Pathog.* **2016**, *12*, e1005982. [CrossRef] [PubMed]

68. Nakagawa, K.; Narayanan, K.; Wada, M.; Makino, S. Inhibition of stress granule formation by Middle East respiratory syndrome coronavirus 4a accessory protein facilitates viral translation, leading to efficient virus replication. *J. Virol.* **2018**, *92*, e00902-18. [CrossRef]

69. Canton, J.; Fehr, A.R.; Fernandez-Delgado, R.; Gutierrez-Alvarez, F.J.; Sanchez-Aparicio, M.T.; Garcia-Sastre, A.; Perlman, S.; Enjuanes, L.; Sola, I. MERS-CoV 4b protein interferes with the NF-kappaB-dependent innate immune response during infection. *PLoS Pathog.* **2018**, *14*, e1006838. [CrossRef]

70. Park, J.E.; Li, K.; Barlan, A.; Fehr, A.R.; Perlman, S.; McCray, P.B., Jr.; Gallagher, T. Proteolytic processing of Middle East respiratory syndrome coronavirus spikes expands virus tropism. *Proc. Natl. Acad. Sci. USA* **2016**, *113*, 12262–12267. [CrossRef]

71. Millet, J.K.; Whittaker, G.R. Host cell entry of Middle East respiratory syndrome coronavirus after two-step, furin-mediated activation of the spike protein. *Proc. Natl. Acad. Sci. USA* **2014**, *111*, 15214–15219. [CrossRef] [PubMed]

72. Li, F. Structure, function, and evolution of coronavirus spike proteins. *Annu. Rev. Virol.* **2016**, *3*, 237–261. [CrossRef] [PubMed]

73. Wang, Q.; Wong, G.; Lu, G.; Yan, J.; Gao, G.F. MERS-CoV spike protein: Targets for vaccines and therapeutics. *Antiviral Res.* **2016**, *133*, 165–177. [CrossRef] [PubMed]

74. Raj, V.S.; Mou, H.; Smits, S.L.; Dekkers, D.H.; Muller, M.A.; Dijkman, R.; Muth, D.; Demmers, J.A.; Zaki, A.; Fouchier, R.A.; et al. Dipeptidyl peptidase 4 is a functional receptor for the emerging human coronavirus-EMC. *Nature* **2013**, *495*, 251–254. [CrossRef] [PubMed]

75. Gao, J.; Lu, G.; Qi, J.; Li, Y.; Wu, Y.; Deng, Y.; Geng, H.; Li, H.; Wang, Q.; Xiao, H.; et al. Structure of the fusion core and inhibition of fusion by a heptad repeat peptide derived from the S protein of Middle East respiratory syndrome coronavirus. *J. Virol.* **2013**, *87*, 13134–13140. [CrossRef] [PubMed]

76. Lu, L.; Liu, Q.; Zhu, Y.; Chan, K.H.; Qin, L.; Li, Y.; Wang, Q.; Chan, J.F.; Du, L.; Yu, F.; et al. Structure-based discovery of Middle East respiratory syndrome coronavirus fusion inhibitor. *Nat. Commun.* **2014**, *5*, 3067. [CrossRef] [PubMed]

77. Chen, Y.; Rajashankar, K.R.; Yang, Y.; Agnihothram, S.S.; Liu, C.; Lin, Y.L.; Baric, R.S.; Li, F. Crystal structure of the receptor-binding domain from newly emerged Middle East respiratory syndrome coronavirus. *J. Virol.* **2013**, *87*, 10777–10783. [CrossRef] [PubMed]

78. Matsuyama, S.; Shirato, K.; Kawase, M.; Terada, Y.; Kawachi, K.; Fukushi, S.; Kamitani, W. Middle East respiratory syndrome coronavirus spike protein is not activated directly by cellular furin during viral entry into target cells. *J. Virol.* **2018**, *92*, e00683-18. [CrossRef]

79. Barlan, A.; Zhao, J.; Sarkar, M.K.; Li, K.; McCray, P.B., Jr.; Perlman, S.; Gallagher, T. Receptor variation and susceptibility to Middle East respiratory syndrome coronavirus infection. *J. Virol.* **2014**, *88*, 4953–4961. [CrossRef]

80. Van Doremalen, N.; Miazgowicz, K.L.; Milne-Price, S.; Bushmaker, T.; Robertson, S.; Scott, D.; Kinne, J.; McLellan, J.S.; Zhu, J.; Munster, V.J. Host species restriction of Middle East respiratory syndrome coronavirus through its receptor, dipeptidyl peptidase 4. *J. Virol.* **2014**, *88*, 9220–9232.

81. Letko, M.; Miazgowicz, K.; McMinn, R.; Seifert, S.N.; Sola, I.; Enjuanes, L.; Carmody, A.; van Doremalen, N.; Munster, V. Adaptive evolution of MERS-CoV to species variation in DPP4. *Cell Rep.* **2018**, *24*, 1730–1737. [CrossRef]

82. Chu, H.; Chan, C.M.; Zhang, X.; Wang, Y.; Yuan, S.; Zhou, J.; Au-Yeung, R.K.; Sze, K.H.; Yang, D.; Shuai, H.; et al. Middle East respiratory syndrome coronavirus and bat coronavirus HKU9 both can utilize GRP78 for attachment onto host cells. *J. Biol. Chem.* **2018**, *293*, 11709–11726. [CrossRef] [PubMed]

83. Li, W.; Hulswit, R.J.G.; Widjaja, I.; Raj, V.S.; McBride, R.; Peng, W.; Widagdo, W.; Tortorici, M.A.; van Dieren, B.; Lang, Y.; et al. Identification of sialic acid-binding function for the Middle East respiratory syndrome coronavirus spike glycoprotein. *Proc. Natl. Acad. Sci. USA* **2017**, *114*, E8508–E8517. [CrossRef] [PubMed]

84. Wang, N.; Shi, X.; Jiang, L.; Zhang, S.; Wang, D.; Tong, P.; Guo, D.; Fu, L.; Cui, Y.; Liu, X.; et al. Structure of MERS-CoV spike receptor-binding domain complexed with human receptor DPP4. *Cell Res.* **2013**, *23*, 986–993. [CrossRef] [PubMed]

85. Lu, G.; Hu, Y.; Wang, Q.; Qi, J.; Gao, F.; Li, Y.; Zhang, Y.; Zhang, W.; Yuan, Y.; Bao, J.; et al. Molecular basis of binding between novel human coronavirus MERS-CoV and its receptor CD26. *Nature* **2013**, *500*, 227–231. [CrossRef] [PubMed]

86. Yuan, Y.; Cao, D.; Zhang, Y.; Ma, J.; Qi, J.; Wang, Q.; Lu, G.; Wu, Y.; Yan, J.; Shi, Y.; et al. Cryo-EM structures of MERS-CoV and SARS-CoV spike glycoproteins reveal the dynamic receptor binding domains. *Nat. Commun.* **2017**, *8*, 15092. [CrossRef]

87. Pallesen, J.; Wang, N.; Corbett, K.S.; Wrapp, D.; Kirchdoerfer, R.N.; Turner, H.L.; Cottrell, C.A.; Becker, M.M.; Wang, L.; Shi, W.; et al. Immunogenicity and structures of a rationally designed prefusion MERS-CoV spike antigen. *Proc. Natl Acad. Sci. USA* **2017**, *114*, E7348–E7357. [CrossRef]

88. Lan, J.; Deng, Y.; Chen, H.; Lu, G.; Wang, W.; Guo, X.; Lu, Z.; Gao, G.F.; Tan, W. Tailoring subunit vaccine immunity with adjuvant combinations and delivery routes using the Middle East respiratory coronavirus (MERS-CoV) receptor-binding domain as an antigen. *PLoS ONE* **2014**, *9*, e112602. [CrossRef] [PubMed]

89. Kim, Y.S.; Son, A.; Kim, J.; Kwon, S.B.; Kim, M.H.; Kim, P.; Kim, J.; Byun, Y.H.; Sung, J.; Lee, J.; et al. Chaperna-mediated assembly of ferritin-based Middle East respiratory syndrome-coronavirus nanoparticles. *Front. Immunol.* **2018**, *9*, 1093. [CrossRef]

90. Wang, C.; Zheng, X.; Gai, W.; Wong, G.; Wang, H.; Jin, H.; Feng, N.; Zhao, Y.; Zhang, W.; Li, N.; et al. Novel chimeric virus-like particles vaccine displaying MERS-CoV receptor-binding domain induce specific humoral and cellular immune response in mice. *Antiviral Res.* **2017**, *140*, 55–61. [CrossRef]

91. Lan, J.; Yao, Y.; Deng, Y.; Chen, H.; Lu, G.; Wang, W.; Bao, L.; Deng, W.; Wei, Q.; Gao, G.F.; et al. Recombinant receptor binding domain protein induces partial protective immunity in Rhesus Macaques against Middle East respiratory syndrome coronavirus challenge. *EBioMedicine* **2015**, *2*, 1438–1446. [CrossRef] [PubMed]

92. Zhang, N.; Channappanavar, R.; Ma, C.; Wang, L.; Tang, J.; Garron, T.; Tao, X.; Tasneem, S.; Lu, L.; Tseng, C.T.; et al. Identification of an ideal adjuvant for receptor-binding domain-based subunit vaccines against Middle East respiratory syndrome coronavirus. *Cell Mol. Immunol.* **2016**, *13*, 180–190. [CrossRef] [PubMed]

93. Du, L.; Tai, W.; Yang, Y.; Zhao, G.; Zhu, Q.; Sun, S.; Liu, C.; Tao, X.; Tseng, C.K.; Perlman, S.; et al. Introduction of neutralizing immunogenicity index to the rational design of MERS coronavirus subunit vaccines. *Nat. Commun.* **2016**, *7*, 13473. [CrossRef] [PubMed]

94. Tai, W.; Zhao, G.; Sun, S.; Guo, Y.; Wang, Y.; Tao, X.; Tseng, C.K.; Li, F.; Jiang, S.; Du, L.; et al. A recombinant receptor-binding domain of MERS-CoV in trimeric form protects human dipeptidyl peptidase 4 (hDPP4) transgenic mice from MERS-CoV infection. *Virology* **2016**, *499*, 375–382. [CrossRef] [PubMed]

95. Ma, C.; Wang, L.; Tao, X.; Zhang, N.; Yang, Y.; Tseng, C.T.; Li, F.; Zhou, Y.; Jiang, S.; Du, L. Searching for an ideal vaccine candidate among different MERS coronavirus receptor-binding fragments–the importance of immunofocusing in subunit vaccine design. *Vaccine* **2014**, *32*, 6170–6176. [CrossRef]

96. Zhang, N.; Tang, J.; Lu, L.; Jiang, S.; Du, L. Receptor-binding domain-based subunit vaccines against MERS-CoV. *Virus Res.* **2015**, *202*, 151–159. [CrossRef]

97. Tang, J.; Zhang, N.; Tao, X.; Zhao, G.; Guo, Y.; Tseng, C.T.; Jiang, S.; Du, L.; Zhou, Y. Optimization of antigen dose for a receptor-binding domain-based subunit vaccine against MERS coronavirus. *Hum. Vaccin Immunother.* **2015**, *11*, 1244–1250. [CrossRef]

98. Wang, Y.; Tai, W.; Yang, J.; Zhao, G.; Sun, S.; Tseng, C.K.; Jiang, S.; Zhou, Y.; Du, L.; Gao, J. Receptor-binding domain of MERS-CoV with optimal immunogen dosage and immunization interval protects human transgenic mice from MERS-CoV infection. *Hum. Vaccin Immunother.* **2017**, *13*, 1615–1624. [CrossRef]

99. Nyon, M.P.; Du, L.; Tseng, C.K.; Seid, C.A.; Pollet, J.; Naceanceno, K.S.; Agrawal, A.; Algaissi, A.; Peng, B.H.; Tai, W.; et al. Engineering a stable CHO cell line for the expression of a MERS-coronavirus vaccine antigen. *Vaccine* **2018**, *36*, 1853–1862. [CrossRef]

100. Tai, W.; Wang, Y.; Fett, C.A.; Zhao, G.; Li, F.; Perlman, S.; Jiang, S.; Zhou, Y.; Du, L. Recombinant receptor-binding domains of multiple Middle East respiratory syndrome coronaviruses (MERS-CoVs) induce cross-neutralizing antibodies against divergent human and camel MERS-CoVs and antibody escape mutants. *J. Virol.* **2017**, *91*, e01651-16. [CrossRef]

101. Ma, C.; Li, Y.; Wang, L.; Zhao, G.; Tao, X.; Tseng, C.T.; Zhou, Y.; Du, L.; Jiang, S. Intranasal vaccination with recombinant receptor-binding domain of MERS-CoV spike protein induces much stronger local mucosal immune responses than subcutaneous immunization: Implication for designing novel mucosal MERS vaccines. *Vaccine* **2014**, *32*, 2100–2108. [CrossRef] [PubMed]

102. Wang, L.; Shi, W.; Chappell, J.D.; Joyce, M.G.; Zhang, Y.; Kanekiyo, M.; Becker, M.M.; van Doremalen, N.; Fischer, R.; Wang, N.; et al. Importance of neutralizing monoclonal antibodies targeting multiple antigenic sites on MERS-CoV Spike to avoid neutralization escape. *J. Virol* **2018**, JVI.02002-17.

103. Li, Y.; Wan, Y.; Liu, P.; Zhao, J.; Lu, G.; Qi, J.; Wang, Q.; Lu, X.; Wu, Y.; Liu, W.; et al. A humanized neutralizing antibody against MERS-CoV targeting the receptor-binding domain of the spike protein. *Cell. Res.* **2015**, *25*, 1237–1249. [CrossRef] [PubMed]

104. Corti, D.; Zhao, J.; Pedotti, M.; Simonelli, L.; Agnihothram, S.; Fett, C.; Fernandez-Rodriguez, B.; Foglierini, M.; Agatic, G.; Vanzetta, F.; et al. Prophylactic and postexposure efficacy of a potent human monoclonal antibody against MERS coronavirus. *Proc. Natl. Acad. Sci. USA* **2015**, *112*, 10473–10478. [CrossRef]

105. Pascal, K.E.; Coleman, C.M.; Mujica, A.O.; Kamat, V.; Badithe, A.; Fairhurst, J.; Hunt, C.; Strein, J.; Berrebi, A.; Sisk, J.M.; et al. Pre- and postexposure efficacy of fully human antibodies against Spike protein in a novel humanized mouse model of MERS-CoV infection. *Proc. Natl. Acad. Sci. USA* **2015**, *112*, 8738–8743. [CrossRef] [PubMed]

106. Ying, T.; Prabakaran, P.; Du, L.; Shi, W.; Feng, Y.; Wang, Y.; Wang, L.; Li, W.; Jiang, S.; Dimitrov, D.S.; et al. Junctional and allele-specific residues are critical for MERS-CoV neutralization by an exceptionally potent germline-like antibody. *Nat. Commun.* **2015**, *6*, 8223. [CrossRef] [PubMed]

107. Qiu, H.; Sun, S.; Xiao, H.; Feng, J.; Guo, Y.; Tai, W.; Wang, Y.; Du, L.; Zhao, G.; Zhou, Y. Single-dose treatment with a humanized neutralizing antibody affords full protection of a human transgenic mouse model from lethal Middle East respiratory syndrome (MERS)-coronavirus infection. *Antiviral Res.* **2016**, *132*, 141–148. [CrossRef] [PubMed]

108. Du, L.; Zhao, G.; Yang, Y.; Qiu, H.; Wang, L.; Kou, Z.; Tao, X.; Yu, H.; Sun, S.; Tseng, C.T.; et al. A conformation-dependent neutralizing monoclonal antibody specifically targeting receptor-binding domain in Middle East respiratory syndrome coronavirus spike protein. *J. Virol.* **2014**, *88*, 7045–7053. [CrossRef] [PubMed]

109. Ying, T.; Du, L.; Ju, T.W.; Prabakaran, P.; Lau, C.C.; Lu, L.; Liu, Q.; Wang, L.; Feng, Y.; Wang, Y.; et al. Exceptionally potent neutralization of middle East respiratory syndrome coronavirus by human monoclonal antibodies. *J. Virol.* **2014**, *88*, 7796–7805. [CrossRef] [PubMed]

110. Du, L.; Jiang, S. Middle East respiratory syndrome: Current status and future prospects for vaccine development. *Expert. Opin. Biol. Ther.* **2015**, *15*, 1647–1651. [CrossRef]

111. Wang, L.; Shi, L.; Joyce, M.G.; Modjarrad, K.; Zhang, Y.; Leung, K.; Lees, C.R.; Zhou, T.; Yassine, H.M.; Kanekiyo, M.; et al. Evaluation of candidate vaccine approaches for MERS-CoV. *Nat. Commun.* **2015**, *6*, 7712. [CrossRef] [PubMed]

112. Niu, P.; Zhao, G.; Deng, Y.; Sun, S.; Wang, W.; Zhou, Y.; Tan, W. A novel human mAb (MERS-GD27) provides prophylactic and postexposure efficacy in MERS-CoV susceptible mice. *Sci. China Life Sci.* **2018**, *61*, 1280–1282. [CrossRef] [PubMed]

113. Niu, P.; Zhang, S.; Zhou, P.; Huang, B.; Deng, Y.; Qin, K.; Wang, P.; Wang, W.; Wang, X.; Zhou, J.; et al. Ultrapotent human neutralizing antibody repertoires against Middle East respiratory syndrome coronavirus from a recovered patient. *J. Infect Dis.* **2018**, *218*, 1249–1260. [CrossRef] [PubMed]

114. Chen, Z.; Bao, L.; Chen, C.; Zou, T.; Xue, Y.; Li, F.; Lv, Q.; Gu, S.; Gao, X.; Cui, S.; et al. Human neutralizing monoclonal antibody inhibition of Middle East respiratory syndrome coronavirus replication in the common marmoset. *J. Infect Dis.* **2017**, *215*, 1807–1815. [CrossRef] [PubMed]

115. Zhang, S.; Zhou, P.; Wang, P.; Li, Y.; Jiang, L.; Jia, W.; Wang, H.; Fan, A.; Wang, D.; Shi, X.; et al. Structural definition of a unique neutralization epitope on the receptor-binding domain of MERS-CoV spike glycoprotein. *Cell Rep.* **2018**, *24*, 441–452. [CrossRef] [PubMed]

116. Raj, V.S.; Okba, N.M.A.; Gutierrez-Alvarez, J.; Drabek, D.; van Dieren, B.; Widagdo, W.; Lamers, M.M.; Widjaja, I.; Fernandez-Delgado, R.; Sola, I.; et al. Chimeric camel/human heavy-chain antibodies protect against MERS-CoV infection. *Sci. Adv.* **2018**, *4*, eaas9667. [CrossRef] [PubMed]

117. Zhao, G.; He, L.; Sun, S.; Qiu, H.; Tai, W.; Chen, J.; Li, J.; Chen, Y.; Guo, Y.; Wang, Y.; et al. A novel nanobody targeting Middle East respiratory syndrome coronavirus (MERS-CoV) receptor-binding domain has potent cross-neutralizing activity and protective efficacy against MERS-CoV. *J. Virol* **2018**, *92*, e00837-18. [CrossRef]

118. Yu, X.; Zhang, S.; Jiang, L.; Cui, Y.; Li, D.; Wang, D.; Wang, N.; Fu, L.; Shi, X.; Li, Z.; et al. Structural basis for the neutralization of MERS-CoV by a human monoclonal antibody MERS-27. *Sci Rep.* **2015**, *5*, 13133. [CrossRef]

119. Wilken, L.; McPherson, A. Application of camelid heavy-chain variable domains (VHHs) in prevention and treatment of bacterial and viral infections. *Int. Rev. Immunol.* **2018**, *37*, 69–76. [CrossRef] [PubMed]

120. Van Heeke, G.; Allosery, K.; De Brabandere, V.; De Smedt, T.; Detalle, L.; de Fougerolles, A. Nanobodies(R) as inhaled biotherapeutics for lung diseases. *Pharmacol. Ther.* **2017**, *169*, 47–56.

121. Detalle, L.; Stohr, T.; Palomo, C.; Piedra, P.A.; Gilbert, B.E.; Mas, V.; Millar, A.; Power, U.F.; Stortelers, C.; Allosery, K.; et al. Generation and characterization of ALX-0171, a potent novel therapeutic nanobody for the treatment of respiratory syncytial virus infection. *Antimicrob. Agents Chemother.* **2015**, *60*, 6–13. [CrossRef] [PubMed]

122. Steeland, S.; Vandenbroucke, R.E.; Libert, C. Nanobodies as therapeutics: Big opportunities for small antibodies. *Drug Discov. Today* **2016**, *21*, 1076–1113. [CrossRef] [PubMed]

123. Muyldermans, S. Nanobodies: Natural single-domain antibodies. *Annu. Rev. Biochem.* **2013**, *82*, 775–797. [CrossRef] [PubMed]

124. Du, L.; Tai, W.; Zhou, Y.; Jiang, S. Vaccines for the prevention against the threat of MERS-CoV. *Expert Rev. Vaccines* **2016**, *15*, 1123–1134. [CrossRef] [PubMed]

125. Zhang, Z.; Shen, L.; Gu, X. Evolutionary dynamics of MERS-CoV: Potential recombination, positive selection and transmission. *Sci. Rep.* **2016**, *6*, 25049. [CrossRef] [PubMed]

126. Kim, Y.; Cheon, S.; Min, C.K.; Sohn, K.M.; Kang, Y.J.; Cha, Y.J.; Kang, J.I.; Han, S.K.; Ha, N.Y.; Kim, G.; et al. Spread of mutant Middle East respiratory syndrome coronavirus with reduced affinity to human CD26 during the South Korean outbreak. *MBio* **2016**, *7*, e00019. [CrossRef] [PubMed]

127. Kleine-Weber, H.; Elzayat, M.T.; Wang, L.; Graham, B.S.; Muller, M.A.; Drosten, C.; Pohlmann, S.; Hoffmann, M. Mutations in the spike protein of Middle East respiratory syndrome coronavirus transmitted in Korea increase resistance to antibody-mediated neutralization. *J. Virol.* **2019**, *93*, e01381-18. [CrossRef]

128. Wang, C.; Hua, C.; Xia, S.; Li, W.; Lu, L.; Jiang, S. Combining a fusion inhibitory peptide targeting the MERS-CoV S2 protein HR1 domain and a neutralizing antibody specific for the S1 protein receptor-binding domain (RBD) showed potent synergism against pseudotyped MERS-CoV with or without mutations in RBD. *Viruses* **2019**, *11*, 31. [CrossRef]
129. Galasiti Kankanamalage, A.C.; Kim, Y.; Damalanka, V.C.; Rathnayake, A.D.; Fehr, A.R.; Mehzabeen, N.; Battaile, K.P.; Lovell, S.; Lushington, G.H.; Perlman, S.; et al. Structure-guided design of potent and permeable inhibitors of MERS coronavirus 3CL protease that utilize a piperidine moiety as a novel design element. *Eur. J. Med. Chem* **2018**, *150*, 334–346. [CrossRef]

Review

Middle East Respiratory Syndrome Vaccine Candidates: Cautious Optimism

Craig Schindewolf and Vineet D. Menachery *

Department of Microbiology and Immunology, University of Texas Medical Branch, Galveston, 77555 TX, USA; crschind@UTMB.EDU
* Correspondence: vimenach@utmb.edu; Tel.: +1-409-266-6934

Received: 18 December 2018; Accepted: 12 January 2019; Published: 17 January 2019

Abstract: Efforts towards developing a vaccine for Middle East respiratory syndrome coronavirus (MERS-CoV) have yielded promising results. Utilizing a variety of platforms, several vaccine approaches have shown efficacy in animal models and begun to enter clinical trials. In this review, we summarize the current progress towards a MERS-CoV vaccine and highlight potential roadblocks identified from previous attempts to generate coronavirus vaccines.

Keywords: Middle East respiratory syndrome coronavirus; severe acute respiratory syndrome coronavirus; coronavirus spike glycoprotein; vaccine platforms; correlates of immunity; animal models

---

1. Introduction

In recent years, viral zoonotic diseases have caused outbreaks marked by rapid spread and high mortality, including the 2002 emergence of severe acute respiratory syndrome coronavirus (SARS-CoV) [1], the 2009 H1N1 swine flu pandemic [2], and the 2013 Ebola outbreak in West Africa [3]. Such outbreaks are difficult to predict as new strains emerge or reemerge from zoonotic reservoirs [4]. Coronaviruses (CoVs), large positive-stranded RNA viruses of the order *Nidovirales* [5], were considered minor human pathogens, causing cold-like symptoms and occasionally associated with pneumonia and more severe disease [6]. However, the emergence of SARS-CoV in 2002 and Middle East respiratory syndrome coronavirus (MERS-CoV) in 2012, members of βCoV lineages B and C, respectively, marked a shift in our understanding of the pathogenic potential of coronaviruses [7]. As these more virulent viruses are genetically similar to those currently circulating in bats [8,9], CoVs may pose a threat for future zoonoses [10].

Since the emergence of MERS-CoV in Saudi Arabia in 2012, over 2200 confirmed cases have been reported in at least 27 countries, with an overall mortality rate of 35% (https://www.who.int/emergencies/mers-cov). Additionally, more severe disease has been noted in the aged, immunocompromised, and those with chronic health conditions [11]. Camels, which show seropositivity to MERS-CoV in archived sera dating back to 1983 [12], serve as intermediate hosts and are able to spread the virus to humans [13], who may then spread the infection person-to-person [14]. While a range of therapeutics have been explored for CoV disease [15–17], a MERS-CoV vaccine remains the most scalable, cost-effective prophylactic measure. Currently, a vaccine for MERS-CoV is not available, although several candidates have been developed using a variety of approaches. Vaccine studies were initially hampered by a lack of small animal models of MERS-CoV disease [18]. While rodents possess homologues for dipeptidyl peptidase 4 (DPP4), the human receptor for MERS-CoV [19], rodent DPP4 homologues are incompatible with MERS-CoV infection [20–22]. However, several in vivo approaches have been developed to overcome these barriers and facilitate MERS-CoV vaccine testing in small animal models [23–27].

Preclinical vaccine development for both SARS-CoV and MERS-CoV has largely aimed to stimulate a robust immune response against the viral envelope-protruding spike (S) glycoprotein [28,29], a class I fusion protein, and/or the nucleocapsid (N) protein [30,31]. MERS-CoV S is proteolytically cleaved by host furin [32] during maturation into an S1 domain responsible for binding to DPP4 as well as an S2 domain containing two heptad-repeat regions that facilitate membrane fusion (Figure 1). The S1 domain can be further divided into the N-terminal domain (NTD), or S1^A, associated with binding sialic acid [33], and the receptor binding domain (RBD), comprising the majority of the C-terminal domain of S1. Cryo-electron microscopy studies have shown that the RBD is flexible and opens upward or away from the viral envelope in order to establish contact with DPP4, which may expose S2' [34], a second protease cleavage site within S2. Cleavage at S2' is necessary for membrane fusion upon viral entry [32]. The centrality of S to viral entry helps explain why antibodies that target it are potently neutralizing [35]. On the other hand, while CoV N proteins are abundantly expressed during infection [36], immunization with SARS-CoV N did not induce strongly neutralizing antibodies [37], likely because N is not displayed on the viral surface. However, N is more conserved than S within CoV lineage [38], and vaccination with SARS-CoV N was shown to induce cytotoxic T cell responses in mice [39]. Therefore, N may help induce cell-mediated immunity to CoV infection [40], as may S [41]. CD4+ and CD8+ T cell responses from recovered MERS-CoV patients were particularly strong towards N peptides [42]. However, vaccination with N-based immunogens may carry risks associated with T$_h$2-related eosinophilic immune enhancement [43], as may S-based vaccines [44]. Notwithstanding, because of the protection afforded by the robust immune response it generates, S has been the target of most vaccine candidates for MERS-CoV. In this review, we summarize the current state of MERS-CoV vaccine candidates and also describe potential barriers to MERS-CoV vaccine efficacy that first surfaced during research on developing a SARS-CoV vaccine.

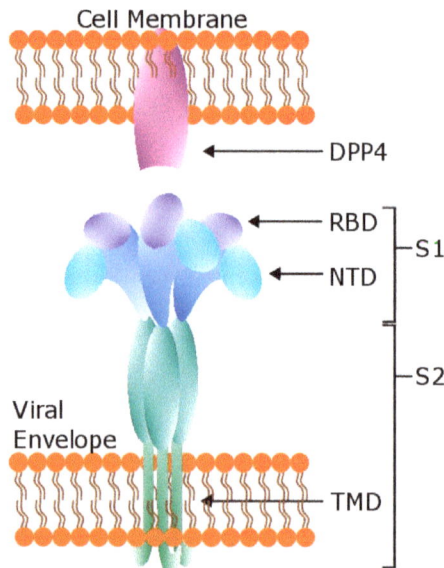

Figure 1. The MERS-CoV spike (S) glycoprotein, a Class I fusion protein and the target of the majority of vaccine candidates, exists naturally in trimer form as shown in this simplified diagram. DPP4: dipeptidyl peptidase 4, the receptor for S. S1: S1 domain of S. S2: S2 domain of S. RBD: receptor binding domain. NTD: N-terminal domain of S1. TMD: transmembrane domain. Structural configurations adapted from [34,45,46].

2. Subunit Vaccines: Immunogenically Focused

Subunit vaccines comprise one or more immunogenic components derived from a pathogen [47]. They have gained popularity in recent decades due to the relative ease of their production and their reduced risks in vivo compared to vaccine types that involve live virus, namely live attenuated vaccines, viral vector vaccines, and even improperly prepared inactivated vaccines.

2.1. Receptor Binding Domain

It is known from studies of recovered SARS-CoV patients that antibodies generated against the receptor binding domain (RBD) are both long-lasting (>3 years) and neutralizing [48]. The RBD in the MERS-CoV S glycoprotein was initially mapped to a region spanning residues 358 to 662 [49,50]; antisera from RBD protein-immunized mice or rabbits protected against in vitro infection with MERS-CoV. In further studies exploring immune correlates of protection, intranasal administration of RBD protein induced S1-specific immunoglobulin (Ig) G1, IgG2a, IgG3, lung IgA, and neutralizing antibodies (NAb, that is, untyped antibodies shown to functionally inhibit free virus from infecting cells) [51], as well as cell-mediated responses as measured by IL-2 and IFN-γ production in antigen-stimulated CD4+ and CD8+ splenocytes [52]. In this and subsequent studies, the RBD protein was fused to a fragment crystallizable (Fc) region of human IgG1 to increase the in vivo half-life of the immunogen [53]. Profiling of the immunogenic region of the RBD indicated residues 377-588 bound with the highest affinity to soluble DPP4 and induced the highest NAb titers, when administered to both mice and rabbits [54]. This range-refined RBD protein vaccine has been stably expressed in a modified high-yield CHO cell line [55]. Purified and adjuvanted with AddaVax™ (MF59-like), the RBD protein vaccine was protective when administered intramuscularly to transgenic mice expressing human DPP4 (hDPP4), with no evidence of immunological toxicity or eosinophilic immune enhancement.

Since the S glycoprotein exists in trimeric form on the virion, a quality lost in shortened forms of the protein, a trimer of this RBD protein vaccine has been generated containing a foldon trimerization motif [56]. This RBD protein trimer has been shown to elicit long-lasting NAb and be protective in challenged hDPP4-transgenic mice [57]. Independently, a monomeric RBD protein vaccine has been developed and tested in rhesus macaques, where it reduced MERS-associated lung pathology and reduced viral loads when adjuvanted with alum and administered intramuscularly in a three-dose regimen prior to challenge [58]. Finally, RBD proteins encoding sequences from different strains of MERS-CoV have been shown to induce cross-neutralizing antibodies against divergent human and camel MERS-CoV strains as well as monoclonal antibody (mAb) escape mutants, confirming the promise of the RBD as a valid vaccine target [59].

2.2. Full-Length S

Targeting the entire S glycoprotein has the advantage of including non-RBD neutralizing epitopes, including those in the more conserved domains. S protein "nanoparticles," protein aggregates containing full-length S, have been proposed as a subunit vaccine, as nanoparticle vaccination adjuvanted with Matrix-M1™ elicited NAb in mice [60] protected adenovirally hDPP4-transduced mice [24] from MERS-CoV challenge [61]. Measurements of viral titer and viral RNA were near the limit of detection in these vaccinated mice. However, despite these promising results, antibody-dependent enhancement (ADE) of infection was previously noted in the context of vaccination with full-length SARS-CoV S protein vaccine [62]. While ADE has not been demonstrated with full length S from MERS-CoV, further studies must consider this as a potential issue.

A variation of the full-length S glycoprotein vaccine is a trimer of S ectodomain (all but the transmembrane domain) conformationally locked in the prefusion state by the substitution of two proline residues in the S2 domain [34]. Work on related fusion proteins such as the F glycoprotein of respiratory syncytial virus has shown that stabilizing the glycoprotein in its prefusion state helps elicit a stronger neutralizing antibody response [63]. This "prefusion" S administered to mice elicited sera with greater neutralization activity against a panel of pseudoviruses bearing strain-specific variants of MERS-CoV S as compared to wild-type S. Since the S2 domain of CoVs is more conserved than the S1 domain [64], targeting epitopes in S2 may provide broader protection against different MERS-CoV strains and other lineage C βCoVs.

2.3. N-Terminal Domain

RBD- and full-length S-based protein vaccines build upon prior vaccine efforts for SARS-CoV. In contrast, vaccines targeting the N-terminal domain of S1 (NTD) offer a novel target. The NTD of S1 does not contain the RBD for MERS-CoV; however, the NTD binds sialic acid and is key to infecting certain cell types [33]. Immunization with NTD protein protected against MERS-CoV challenge in adenovirally hDPP4-transduced mice, inducing cell-mediated responses in splenocytes (CD8+ IFN-γ production, CD4+ IL-2 production, and IL-17A production) as well as humoral responses (IgG and NAb), although NAb titer was lower compared to that of an RBD protein vaccine [65]. Overall, these results suggest that targeting S1 domains outside the RBD may be a viable strategy for MERS-CoV vaccines.

3. DNA Vaccines: Efficient Protection

DNA vaccines offer a rapid platform to design and deliver immunogenic proteins, typically encoded on plasmid vectors and injected into tissue with accompanying electroporation [66]. DNA vaccine administration via electroporation has been tested in clinical trials with immunogenicity comparable to other vaccine types and predominantly low-grade adverse events reported [67]. The in vivo expression of plasmid-encoded proteins recapitulates native post-translational modifications while maintaining the capacity to stimulate both humoral and cell-mediated immunity [68]. While concerns about the safety of DNA vaccines and their potential to integrate into host cell chromosomes were voiced early in their development [69], integration using various plasmids and inserts appears to be extremely rare [70]. Several DNA vaccines for MERS-CoV have been reported to date.

3.1. Full-Length S

pVax1^{TM} is a proprietary, optimized plasmid vaccine vector that has been developed as a MERS-CoV vaccine by encoding a consensus MERS-CoV S glycoprotein containing codon and other proprietary optimizations, as well as an IgE leader sequence to promote expression and mRNA export [71]. Intramuscular administration of this construct with electroporation induced antibodies with cross-MERS-CoV-strain neutralization and antigen-specific, polyfunctional T cell responses in rhesus macaques. These humoral and cell-mediated immune responses correlated with minimal lung pathology and reduced lung viral loads upon MERS-CoV challenge. The same study reported NAb induction in dromedary camels, which indicates that the vaccine could be used in zoonotic reservoirs. Building upon these preclinical results, the pVax1^{TM} vaccine (GLS-5300) has completed a Phase I clinical trial (clinicaltrials.gov/ct2/show/NCT02670187).

A second vaccine platform utilizes pVRC8400 [72,73], a plasmid vector engineered for high transgene expression and enhanced cell-mediated responses. A vaccine regimen consisting of intramuscular administration of MERS-CoV strain England1 full-length S encoded on pVRC8400, with electroporation, and an AlPO$_4$-adjuvanted S1 protein booster, induced NAb in rhesus macaques up to 10 weeks following booster [74]. This vaccine resulted in lower lung pathology upon challenge with the MERS-CoV strain JordanN3. In the same study, the full-length S DNA/S1 protein vaccine induced higher NAb titer in mice than other prime/boost combinations involving constructs encoding either S1 or S with the transmembrane domain (TMD) deleted. Consistent with this finding, the study reported that mAbs induced against domains outside the RBD were able to neutralize MERS-CoV pseudovirus. Together, these results reiterated the immunogenic potential of non-RBD epitopes including those derived from S quaternary (trimer) structure, which could help generate immune responses able to minimize escape variants derived from immunization targeting either the RBD or the S1 domain alone.

3.2. S1 Domain

The S1 domain from MERS-CoV strain Al-Hasa_15_2013 encoded on pcDNA^{TM}3.1(+), a proprietary plasmid from which pVax1^{TM} is derived, has also been tested as a vaccine platform. This vaccine induced NAb in mice when given intramuscularly and antigen-specific cytokine production including CD4+ and CD8+ production of both IL-4 and IFN-γ in murine splenocytes [75]. In addition, the vaccine protected adenovirally hDPP4-transduced mice against challenge with MERS-CoV strain EMC/2012. Moreover, this S1 vaccine elicited more NAb than did full-length S in the same vector. This result was attributed to increased secretion of the S1 protein, which lacked a TMD, and greater uptake by antigen-presenting cells [76]. A separate group's study comparing full-length S- and S1-encoding pcDNA^{TM}3.1 vaccines found that the S1 vaccine elicited a more balanced IgG2a/IgG1 ratio in mice compared to that elicited by full-length S [77] suggesting a balanced T$_h$1/T$_h$2 response [78]. Together, these results show that multiple plasmid vaccine vectors encoding either full-length S or the S1 domain induce adaptive immunity and protect against MERS-CoV challenge.

4. Viral Vector Vaccines: Optimized Delivery

Viral vector vaccines contain one or more immunogenic proteins of the pathogen of interest in the context of an attenuated virus backbone. This approach takes advantage of cellular entry by the virus as well as adjuvantation from viral components, and induces both humoral and strong cell-mediated responses [79]. Early studies into viral vector vaccines for MERS-CoV built upon established platforms and have subsequently transitioned to newer viral vector approaches.

Venezuelan equine encephalitis (VEE) virus replicon particles (VRPs), an alphavirus-based platform that replaces the VEE structural genes with a foreign transgene, has been shown to induce strong humoral and cellular immune responses [80,81]. A VRP encoding MERS-CoV S elicited NAb in both young and aged mice [38]. Additional studies with this vector have shown that immunization with an N protein-expressing VRP protected adenovirally hDPP4-transduced mice from MERS-CoV challenge in a CD4+ T cell- and IFN-γ-dependent manner [82]. Moreover, a specific N protein epitope was stimulatory in mice transgenic for human leukocyte antigen DR2 and DR3, highlighting the relevance of this epitope to human antigen recognition and to promoting cell-mediated immunity in humans.

Modified vaccinia virus Ankara (MVA) [83,84], a well-established vaccine platform, has been developed to encode full-length MERS-CoV S. This vaccine induced NAb and CD8+ T cell responses in mice [85] and also protected against MERS-CoV-induced histopathology in adenovirally hDPP4-transduced mice before challenge [86]. Moreover, minimal inflammation and lymph node hyperplasia was observed at the site of injection [87]. This same MVA-MERS-CoV S vaccine injected intramuscularly into dromedary camels was shown to induce NAb and to limit excretion of infectious virus upon intranasal challenge with MERS-CoV [88]. A Phase I clinical trial is underway (clinicaltrials.gov/ct2/show/NCT03615911).

Adenoviruses compose a third platform of viral vectors for MERS-CoV vaccines. Adenovirus-vectored vaccines have been tested in clinical trials for a wide variety of diseases, notably HIV [89]. However, their efficacy may be hampered by pre-existing immunity to prevalent adenovirus serotypes [90,91]. For example, pre-existing immunity to human adenovirus serotype 5 (Ad5) was shown to result in reduced CD8+ T cell responses against an Ad5-vectored transgene [92]. To take advantage of the shared respiratory route of infection of both MERS-CoV and adenovirus, Ad5-vectored full-length S and S1 vaccines have been developed [93]. These elicited antigen-specific IgG and NAb when administered intramuscularly to mice with subsequent intranasal boosting. Importantly, this study did not detect immunity against the Ad5 vector in dromedary camels, the intended vaccination population. Moreover, camel peripheral blood mononuclear cells and a camel-derived fibroblast cell line were able to be infected with Ad5. Another Ad5-MERS-CoV S vaccine has been separately developed, as has a human adenovirus type 41 (Ad41)-MERS-CoV S vaccine [94]. Adenovirus type 41 (Ad41) is an enteric pathogen with potential use as an orally administered vaccine [95]. Both of these vaccines, Ad5-MERS-CoV S and Ad41-MERS-CoV S, were reported to induce humoral responses when administered intragastrically in mice. In addition to humoral responses, they also induced long-lasting cell-mediated responses in the lung and spleen when administered intramuscularly. One final Ad5-based MERS-CoV immunization regimen has been reported [96]. Immunizing with Ad5-vectored S followed by boosting with S nanoparticles induced S-specific IgG, NAb, and both T_h1 and T_h2 cell-mediated responses in mice, and also protected adenovirally hDPP4-transduced mice from MERS-CoV challenge.

To circumvent the seroprevalence of circulating human adenoviruses, chimpanzee adenoviruses have also been developed as viral vaccine vectors [97] and have entered clinical trials [98]. A MERS-CoV S-encoding vaccine based on a chimpanzee adenoviral vector (ChAdOx1) was shown to induce high levels of NAb and cell-mediated responses (CD8+ IFN-γ, TNFα, and IL-17 production) in mice 4 weeks post-immunization [99]. This vaccine was constructed with a codon-optimized S glycoprotein sequence and the tissue plasminogen activator (tPA) gene leader sequence to promote secretion [100,101]. The ChAdOx1-MERS-CoV S vaccine protected against lethal challenge in a transgenic hDPP4 mouse model [102]. Based on previous work with the ChAdOx1 vector demonstrating its safety in humans, the ChAdOx1-MERS-CoV S vaccine is undergoing a Phase I clinical trial (clinicaltrials.gov/ct2/show/NCT03399578).

Several additional viral vectors have been employed as MERS-CoV vaccines. Measles virus vector platforms have been developed over the past two decades [103]. A full-length or soluble form of S encoded in measles vaccine strain MV_{vac2} induced NAb, proliferation of T cells, S-specific IFN-γ production, and cytotoxic activity [104]. The vaccine also protected against MERS-CoV challenge in adenovirally hDPP4-transduced mice that were transgenic for a measles virus receptor. Further characterization of the T cell responses induced by this vaccine has been performed [105]. Of note, 5-fold higher numbers of reactive T cells were induced by vaccination with MV_{vac2}-S than those induced by N protein using the same vector. Additionally, antigen-specific IFN-γ production by T cells could be induced in older mice (7 months old) at levels near those induced in younger mice (6–12 weeks old).

Newcastle disease virus (NDV) has been explored as a vaccine vector as it infects the respiratory tract and can induce systemic and mucosal immunity in non-human primates [106]. An NVD vector expressing MERS-CoV S was shown to induce long-lasting (up 14 weeks post-immunization) NAb titers in camels [107]. The research group behind this study also examined vesicular stomatitis virus (VSV) [108] as a viral vector. S expressed from a VSV reverse genetics system was shown to incorporate onto the surface of virions rescued in cell culture. Purified vaccine was able to infect cells in an hDPP4-dependent manner, induced S-specific IgG and NAb in mice, and stimulated humoral and cell-mediated (IFN-γ-production) responses in rhesus macaques [109].

Similar to the VSV platform, a rabies virus (RABV) vector has been explored. Inspired by studies combining rabies and Ebola vaccine platforms [110], a β-propiolactone-inactivated dual rabies/MERS vaccine has been proposed which incorporates the MERS-CoV S1 domain fused to rabies virus G protein on the RABV virion [111]. This vaccine elicited S-specific IgG and NAb and fully protected adenovirally hDPP4-transduced mice from MERS-CoV challenge. The VSV and RABV approaches described here are unique in that they encode S (or S1) in the vector genome and also display it on the virion surface.

Finally, virus-like particles (VLPs), which comprise self-assembling immunogenic proteins, but no genome [112], have also been used as viral vectors. A baculovirus VLP containing S as well as MERS-CoV envelope and matrix proteins elicited RBD-specific IgG and IFN-γ responses in rhesus macaques [113]. A subsequent baculovirus VLP vaccine was developed that focused only on a fusion of the RBD from S and the immunogenic VP2 protein of canine parvovirus. This vaccine induced

RBD-specific IgG, NAb, and cell-mediated responses including IFN-γ, IL-2, and IL-4 production in mice, and also activated dendritic cells in inguinal lymph nodes [114]. In summary, a variety of viral vector vaccines for MERS-CoV induce promising immune responses in animal models and often demonstrate protection from challenge.

5. Live Attenuated and Inactivated Vaccines: Situationally Useful

A final approach to developing MERS-CoV vaccines delivers the whole virus, either inactivated or live but attenuated. Both of these vaccine types resemble the original virus, preserving structural features and a full or nearly-full repertoire of immunogenic components. Inactivated viruses may contain structural deformations introduced by inactivation, but, unlike attenuated viruses, they pose no risks, if properly inactivated, either of reversion to a virulent state or persistent infection in immunocompromised patients. Fewer examples of whole virus vaccines, compared to the other vaccine types, have been developed for MERS-CoV.

5.1. Inactivated

Development of inactivated vaccines for MERS-CoV has been stymied by prior concerns with SARS-CoV inactivated vaccines. Eosinophil-related lung pathology was observed for a SARS-CoV vaccine doubly inactivated with both formalin and UV irradiation [115]. This response was particularly notable in aged mice versus young mice, and following heterologous versus homologous challenge. Similarly, immunization with a gamma-irradiated MERS-CoV vaccine adjuvanted with either alum or MF59 elicited NAb and reduced viral titer upon challenge in hDPP4-transgenic mice, but induced eosinophil-related lung pathology in vaccinated mice after challenge [116].

A different inactivation method was tried for a second MERS-CoV inactivated vaccine. Formalin-inactivated MERS-CoV adjuvanted with alum and oligodeoxynucleotides containing unmethylated CpG motifs was shown to elicit levels of NAb on par with those elicited by an S glycoprotein-only vaccine [117]. Moreover, the vaccine offered better protection than S alone based on reduction of lung viral titer in adenovirally hDPP4-transduced mice after MERS-CoV challenge. Remarkably, eosinophil-mediated vaccine-related pathology was not observed in this animal model. Interestingly, it has also been shown that including Toll-like receptor agonists in a UV-inactivated SARS-CoV vaccine reduced T_h2-associated pathology in lungs after challenge [118]. These results suggest that inactivated CoV vaccines may remain viable options for further development with the right inactivation method and adjuvants.

5.2. Live Attenuated

Live attenuated vaccines for MERS-CoV show efficacy in animal models, but so far have not been pursued in subsequent studies. While riskier than other vaccine types, live attenuated vaccines have historically offered protection against a variety of threatening illnesses [119] and may be reserved for outbreak scenarios where they offer an immunogenically robust solution. The CoV envelope (E) protein is important in virion assembly and egress, and has also been shown to inhibit the host cell stress response [120]. An E-deletion mutant of SARS-CoV was previously found to be protective in vivo against SARS-CoV challenge [121]. An initial study into a MERS-CoV reverse genetics system reported on a replication-competent, but propagation-defective mutant lacking the E protein that could be rescued in cell culture with E expressed in *trans* [122]. However, the E-deletion mutant was rescued at 100-fold lower titer compared to wild-type MERS-CoV, perhaps explaining why this mutant has not been further developed as a live attenuated vaccine candidate.

Other CoV components have been targeted in live attenuated vaccine development. Nonstructural protein 14 (nsp14) contains an exoribonuclease (ExoN) essential to replication fidelity that is found in all known nidoviruses with genome sizes greater than 20 kb [5]. A stable deficiency in nsp14 attenuated SARS-CoV in young, aged, and immunocompromised mice, and was able to induce protection following vaccination [123]. However, ExoN mutants for MERS-CoV have not been reported.

CoV gene nonstructural protein 16 (nsp16) is a 2′-O-methyl-transferase involved in viral mRNA capping [124] and was previously inactivated in a SARS-CoV live attenuated vaccine [125,126]. A live nsp16-deficient MERS-CoV vaccine was similarly attenuated in a type I interferon- and IFIT1-dependent manner. Immunization with the MERS-CoV nsp16 mutant induced NAb and protected CRISPR-engineered hDPP4-transgenic mice [23] from challenge with a mouse-adapted MERS-CoV strain [127].

Finally, live attenuated vaccines lacking CoV accessory proteins have also been considered. CoV accessory proteins are dispensable for viral replication but have been shown to modulate interferon signaling and pro-inflammatory cytokine production [128]. A MERS-CoV strain lacking accessory open reading frames (ORFs) 3, 4, and 5 was attenuated in vivo, induced NAb, and like the nsp16 mutant, protected CRISPR-engineered hDPP4-transgenic mice from challenge with a mouse-adapted MERS-CoV strain [129]. Overall, whole vaccine approaches to MERS-CoV vaccination appear both protective and safe in animal models.

6. Conclusions and Future Directions

Aided by knowledge gained from vaccine development against SARS-CoV and other contemporary viral diseases, MERS-CoV vaccine development efforts have multiplied since its emergence, yielding promising vaccine candidates spanning multiple platforms (Table 1). Nevertheless, key barriers to vaccine efficacy first noted for SARS-CoV may also hold true for MERS-CoV. As with SARS-CoV [130], mortality from MERS-CoV has disproportionately affected the aged. Additionally, immunocompromised individuals and those with chronic conditions are at greater risk of mortality from MERS-CoV infection [11]. A universal MERS-CoV vaccine must offer protection to these vulnerable classes of people.

More studies on the effectiveness of the proposed MERS-CoV vaccines in models of immunosenescence, immunocompromise, and chronic conditions are needed. In this regard, vaccination studies with SARS-CoV have indicated that vaccines may be capable of inducing protection in young animals while failing to protect aged animals [115]. In light of the threat of related coronavirus strains emerging, MERS-CoV vaccine studies must also consider heterologous challenge models to ensure safety from vaccine-induced immunopathology, especially in older individuals [115]. In short, vaccine-induced immunopathology, especially T_h2-related eosinophilic immune enhancement [131], from both homologous and heterologous challenge should be specifically monitored in vulnerable populations. Interestingly, it was shown that pathology resulting from MERS-CoV challenge in the lungs of immunosuppressed rhesus macaques was lower compared to that of non-immunosuppressed macaques, underscoring the immunopathogenic component of respiratory disease caused by CoVs [132].

The different vaccine platforms described herein have unique advantages and disadvantages. Since severe CoV disease maintains an immunopathogenic component, a successful vaccine must strike a balance between protection and excessive immune activation. As seen with full-length S [44] as well as inactivated virus [116], vaccination may produce immunopathology under certain conditions. Alternatively, protection must be thorough enough to prevent NAb escape, a phenomenon inversely correlated with the number of immunogenic epitopes. While antibodies induced against S of either SARS-CoV or MERS-CoV poorly cross-neutralize across their respective lineages [38], a vaccine that contains multiple immunogenic epitopes would perhaps also offer greater cross-protection against heterologous strains within lineage as they emerge, especially if conserved epitopes are included in vaccine design. A greater understanding of MERS-CoV pathology will also help guide future vaccine development efforts by illuminating possibly critical differences in vaccine responses between MERS-CoV and SARS-CoV, the latter of which has largely influenced vaccine development against the former.

Overall, MERS-CoV vaccines have shown encouraging results in preclinical studies and we hope these vaccines stand up to safety considerations in order to proceed through clinical trials. While development of therapeutic treatment is critical, vaccination carries the promise of mitigating future outbreaks and alleviating disease burden from the most vulnerable populations including the aged, the immunosuppressed, healthcare workers, family members of infected patients, and those in endemic areas.

Table 1. MERS-CoV vaccine candidates grouped according to category.

Vaccine Type	Humoral Response in:	Cell-Mediated Response in:	Protective in:	Clinical Trial	Source (s)
Subunit					
RBD	M, P	M, P	M, P		[50,55,58]
S nanoparticles	M		M		[61]
Prefusion-locked S	M				[34]
NTD	M	M	M		[65]
DNA					
pVax1-S	M, P, C	M, P	P	Phase I	[71]
pVRC8400-S [1]	M, P		P		[74]
pcDNA3.1(+)-S1 or S	M	M	M		[75,77]
Viral Vector					
VEEV-S	M				[38]
VEEV-N		M	M		[82]
MVA-S	M, C	M	M, C	Phase I	[88]
Ad5-S or S1	M	M			[93,94]
Ad5-S [2]	M	M	M		[96]
Ad41-S	M	M			[94]
ChAdOx1-S	M	M	M	Phase I	[102]
MVvac2-S	M	M	M		[104]
Newcastle-S	M, C				[107]
VSV-S	M, P	P			[109]
Rabies-S1	M		M		[111]
Bac-S,E,M	P	P			[113]
Bac-RBD+VP2	M	M			[114]
Whole					
Formalin inactivated	M		M		[117]
MERS-ΔE					[122]
MERS-dNSP16	M		M		[127]
MERS-dORF3-5	M		M		[129]

Humoral response denotes any antibody response generated, in most cases a NAb response. Cell-mediated responses denote T cell activation markers including IFN-γ. S: MERS-CoV spike protein. N: MERS-CoV nucleocapsid. RBD: receptor binding domain. NTD: N-terminal domain. S1: spike subdomain S1. Bac: baculovirus VLP. E: MERS-CoV envelope protein. M (under *Vaccine Type*): MERS-CoV membrane protein. VP2: canine parvovirus VP2 protein. M (under *Protective in:*): mouse. P: non-human primate. C: camel. [1] With S1 protein booster; [2] with S nanoparticles booster.

Funding: Research in this manuscript was supported by grants from the National Institute of Allergy & Infectious Disease and the National Institute of Aging of the NIH under awards U19AI100625 (VDM) and R00AG049092 (VDM). This work was also supported by the McLaughlin Endowment through the Institute of Human Infection and Immunity at the University of Texas Medical Branch. The content is solely the responsibility of the authors and does not necessarily represent the official views of the NIH.

Conflicts of Interest: The authors declare no conflict of interest.

References

1. Lau, S.K.; Woo, P.C.; Li, K.S.; Huang, Y.; Tsoi, H.W.; Wong, B.H.; Wong, S.S.; Leung, S.Y.; Chan, K.H.; Yuen, K.Y. Severe acute respiratory syndrome coronavirus-like virus in Chinese horseshoe bats. *Proc. Natl. Acad. Sci. USA* **2005**, *102*, 14040–14045. [CrossRef] [PubMed]

2. Trifonov, V.; Khiabanian, H.; Rabadan, R. Geographic dependence, surveillance, and origins of the 2009 influenza A (H1N1) virus. *N. Engl. J. Med.* **2009**, *361*, 115–119. [CrossRef] [PubMed]

3. Gire, S.K.; Goba, A.; Andersen, K.G.; Sealfon, R.S.; Park, D.J.; Kanneh, L.; Jalloh, S.; Momoh, M.; Fullah, M.; Dudas, G.; et al. Genomic surveillance elucidates Ebola virus origin and transmission during the 2014 outbreak. *Science* **2014**, *345*, 1369–1372. [CrossRef] [PubMed]

4. Baize, S.; Pannetier, D.; Oestereich, L.; Rieger, T.; Koivogui, L.; Magassouba, N.; Soropogui, B.; Sow, M.S.; Keita, S.; De Clerck, H.; et al. Emergence of Zaire Ebola virus disease in Guinea. *N. Engl. J. Med.* **2014**, *371*, 1418–1425. [CrossRef] [PubMed]

5. Snijder, E.J.; Decroly, E.; Ziebuhr, J. The Nonstructural Proteins Directing Coronavirus RNA Synthesis and Processing. *Adv. Virus Res.* **2016**, *96*, 59–126. [PubMed]

6. Riski, H.; Hovi, T. Coronavirus infections of man associated with diseases other than the common cold. *J. Med. Virol.* **1980**, *6*, 259–265. [CrossRef] [PubMed]

7. Chan, J.F.; Lau, S.K.; To, K.K.; Cheng, V.C.; Woo, P.C.; Yuen, K.Y. Middle East respiratory syndrome coronavirus: Another zoonotic betacoronavirus causing SARS-like disease. *Clin. Microbiol. Rev.* **2015**, *28*, 465–522. [CrossRef]

8. Li, W.; Shi, Z.; Yu, M.; Ren, W.; Smith, C.; Epstein, J.H.; Wang, H.; Crameri, G.; Hu, Z.; Zhang, H.; et al. Bats are natural reservoirs of SARS-like coronaviruses. *Science* **2005**, *310*, 676–679. [CrossRef]

9. Anthony, S.J.; Gilardi, K.; Menachery, V.D.; Goldstein, T.; Ssebide, B.; Mbabazi, R.; Navarrete-Macias, I.; Liang, E.; Wells, H.; Hicks, A.; et al. Further Evidence for Bats as the Evolutionary Source of Middle East Respiratory Syndrome Coronavirus. *mBio* **2017**, *8*. [CrossRef]

10. Menachery, V.D.; Yount, B.L., Jr.; Debbink, K.; Agnihothram, S.; Gralinski, L.E.; Plante, J.A.; Graham, R.L.; Scobey, T.; Ge, X.Y.; Donaldson, E.F.; et al. A SARS-like cluster of circulating bat coronaviruses shows potential for human emergence. *Nat. Med.* **2015**, *21*, 1508–1513. [CrossRef]

11. Zumla, A.; Hui, D.S.; Perlman, S. Middle East respiratory syndrome. *Lancet* **2015**, *386*, 995–1007. [CrossRef]

12. Muller, M.A.; Corman, V.M.; Jores, J.; Meyer, B.; Younan, M.; Liljander, A.; Bosch, B.J.; Lattwein, E.; Hilali, M.; Musa, B.E.; et al. MERS coronavirus neutralizing antibodies in camels, Eastern Africa, 1983–1997. *Emerg. Infect. Dis.* **2014**, *20*, 2093–2095. [CrossRef] [PubMed]

13. Azhar, E.I.; El-Kafrawy, S.A.; Farraj, S.A.; Hassan, A.M.; Al-Saeed, M.S.; Hashem, A.M.; Madani, T.A. Evidence for camel-to-human transmission of MERS coronavirus. *N. Engl. J. Med.* **2014**, *370*, 2499–2505. [CrossRef] [PubMed]

14. The Health Protection Agency (HPA) UK Novel Coronavirus Investigation Team. Evidence of person-to-person transmission within a family cluster of novel coronavirus infections, United Kingdom, February 2013. *Euro Surveill.* **2013**, *18*, 20427.

15. Zumla, A.; Chan, J.F.; Azhar, E.I.; Hui, D.S.; Yuen, K.Y. Coronaviruses-drug discovery and therapeutic options. *Nat. Rev. Drug Discov.* **2016**, *15*, 327–347. [CrossRef] [PubMed]

16. Dyall, J.; Coleman, C.M.; Hart, B.J.; Venkataraman, T.; Holbrook, M.R.; Kindrachuk, J.; Johnson, R.F.; Olinger, G.G., Jr.; Jahrling, P.B.; Laidlaw, M.; et al. Repurposing of clinically developed drugs for treatment of Middle East respiratory syndrome coronavirus infection. *Antimicrob. Agents Chemother.* **2014**, *58*, 4885–4893. [CrossRef] [PubMed]

17. Public Health England/ISARIC. *Treatment of MERS-CoV: Information for Clinicians, Clinical Decision-Making Support for Treatment of MERS-CoV Patients*; Public Health England: London, UK, 2015.

18. Vergara-Alert, J.; Vidal, E.; Bensaid, A.; Segales, J. Searching for animal models and potential target species for emerging pathogens: Experience gained from Middle East respiratory syndrome (MERS) coronavirus. *One Health* **2017**, *3*, 34–40. [CrossRef]

19. Raj, V.S.; Mou, H.; Smits, S.L.; Dekkers, D.H.; Muller, M.A.; Dijkman, R.; Muth, D.; Demmers, J.A.; Zaki, A.; Fouchier, R.A.; et al. Dipeptidyl peptidase 4 is a functional receptor for the emerging human coronavirus-EMC. *Nature* **2013**, *495*, 251–254. [CrossRef]

20. Peck, K.M.; Cockrell, A.S.; Yount, B.L.; Scobey, T.; Baric, R.S.; Heise, M.T. Glycosylation of mouse DPP4 plays a role in inhibiting Middle East respiratory syndrome coronavirus infection. *J. Virol.* **2015**, *89*, 4696–4699. [CrossRef]

21. Cockrell, A.S.; Peck, K.M.; Yount, B.L.; Agnihothram, S.S.; Scobey, T.; Curnes, N.R.; Baric, R.S.; Heise, M.T. Mouse dipeptidyl peptidase 4 is not a functional receptor for Middle East respiratory syndrome coronavirus infection. *J. Virol.* **2014**, *88*, 5195–5199. [CrossRef]

22. Coleman, C.M.; Matthews, K.L.; Goicochea, L.; Frieman, M.B. Wild-type and innate immune-deficient mice are not susceptible to the Middle East respiratory syndrome coronavirus. *J. Gen. Virol.* **2014**, *95 Pt 2*, 408–412. [CrossRef]

23. Cockrell, A.S.; Yount, B.L.; Scobey, T.; Jensen, K.; Douglas, M.; Beall, A.; Tang, X.C.; Marasco, W.A.; Heise, M.T.; Baric, R.S. A mouse model for MERS coronavirus-induced acute respiratory distress syndrome. *Nat. Microbiol.* **2016**, *2*, 16226. [CrossRef] [PubMed]

24. Zhao, J.; Li, K.; Wohlford-Lenane, C.; Agnihothram, S.S.; Fett, C.; Gale, M.J., Jr.; Baric, R.S.; Enjuanes, L.; Gallagher, T.; McCray, P.B., Jr.; et al. Rapid generation of a mouse model for Middle East respiratory syndrome. *Proc. Natl. Acad. Sci. USA* **2014**, *111*, 4970–4975. [CrossRef] [PubMed]

25. Pascal, K.E.; Coleman, C.M.; Mujica, A.O.; Kamat, V.; Badithe, A.; Fairhurst, J.; Hunt, C.; Strein, J.; Berrebi, A.; Sisk, J.M.; et al. Pre- and postexposure efficacy of fully human antibodies against Spike protein in a novel humanized mouse model of MERS-CoV infection. *Proc. Natl. Acad. Sci. USA* **2015**, *112*, 8738–8743. [CrossRef] [PubMed]

26. Tao, X.; Garron, T.; Agrawal, A.S.; Algaissi, A.; Peng, B.H.; Wakamiya, M.; Chan, T.S.; Lu, L.; Du, L.; Jiang, S.; et al. Characterization and Demonstration of the Value of a Lethal Mouse Model of Middle East Respiratory Syndrome Coronavirus Infection and Disease. *J. Virol.* **2016**, *90*, 57–67. [CrossRef] [PubMed]

27. Li, K.; Wohlford-Lenane, C.L.; Channappanavar, R.; Park, J.E.; Earnest, J.T.; Bair, T.B.; Bates, A.M.; Brogden, K.A.; Flaherty, H.A.; Gallagher, T.; et al. Mouse-adapted MERS coronavirus causes lethal lung disease in human DPP4 knockin mice. *Proc. Natl. Acad. Sci. USA* **2017**, *114*, E3119–E3128. [CrossRef] [PubMed]

28. Du, L.; He, Y.; Zhou, Y.; Liu, S.; Zheng, B.J.; Jiang, S. The spike protein of SARS-CoV—A target for vaccine and therapeutic development. *Nat. Rev. Microbiol.* **2009**, *7*, 226–236. [CrossRef] [PubMed]

29. Zhang, N.; Jiang, S.; Du, L. Current advancements and potential strategies in the development of MERS-CoV vaccines. *Expert Rev. Vaccines* **2014**, *13*, 761–774. [CrossRef]

30. Surjit, M.; Lal, S.K. The SARS-CoV nucleocapsid protein: A protein with multifarious activities. *Infect. Genet. Evolut.* **2008**, *8*, 397–405. [CrossRef]

31. Chang, C.K.; Lo, S.C.; Wang, Y.S.; Hou, M.H. Recent insights into the development of therapeutics against coronavirus diseases by targeting N protein. *Drug Discov. Today* **2016**, *21*, 562–572. [CrossRef]

32. Millet, J.K.; Whittaker, G.R. Host cell entry of Middle East respiratory syndrome coronavirus after two-step, furin-mediated activation of the spike protein. *Proc. Natl. Acad. Sci. USA* **2014**, *111*, 15214–15219. [CrossRef] [PubMed]

33. Li, W.; Hulswit, R.J.G.; Widjaja, I.; Raj, V.S.; McBride, R.; Peng, W.; Widagdo, W.; Tortorici, M.A.; van Dieren, B.; Lang, Y.; et al. Identification of sialic acid-binding function for the Middle East respiratory syndrome coronavirus spike glycoprotein. *Proc. Natl. Acad. Sci. USA* **2017**, *114*, E8508–E8517. [CrossRef] [PubMed]

34. Pallesen, J.; Wang, N.; Corbett, K.S.; Wrapp, D.; Kirchdoerfer, R.N.; Turner, H.L.; Cottrell, C.A.; Becker, M.M.; Wang, L.; Shi, W.; et al. Immunogenicity and structures of a rationally designed prefusion MERS-CoV spike antigen. *Proc. Natl. Acad. Sci. USA* **2017**, *114*, E7348–E7357. [CrossRef] [PubMed]

35. Jiang, L.; Wang, N.; Zuo, T.; Shi, X.; Poon, K.M.; Wu, Y.; Gao, F.; Li, D.; Wang, R.; Guo, J.; et al. Potent neutralization of MERS-CoV by human neutralizing monoclonal antibodies to the viral spike glycoprotein. *Sci. Transl. Med.* **2014**, *6*, 234ra59. [CrossRef] [PubMed]

36. Irigoyen, N.; Firth, A.E.; Jones, J.D.; Chung, B.Y.; Siddell, S.G.; Brierley, I. High-Resolution Analysis of Coronavirus Gene Expression by RNA Sequencing and Ribosome Profiling. *PLoS Pathog.* **2016**, *12*, e1005473. [CrossRef] [PubMed]

37. Buchholz, U.J.; Bukreyev, A.; Yang, L.; Lamirande, E.W.; Murphy, B.R.; Subbarao, K.; Collins, P.L. Contributions of the structural proteins of severe acute respiratory syndrome coronavirus to protective immunity. *Proc. Natl. Acad. Sci. USA* **2004**, *101*, 9804–9809. [CrossRef] [PubMed]

38. Agnihothram, S.; Gopal, R.; Yount, B.L., Jr.; Donaldson, E.F.; Menachery, V.D.; Graham, R.L.; Scobey, T.D.; Gralinski, L.E.; Denison, M.R.; Zambon, M.; et al. Evaluation of serologic and antigenic relationships between middle eastern respiratory syndrome coronavirus and other coronaviruses to develop vaccine platforms for the rapid response to emerging coronaviruses. *J. Infect. Dis.* **2014**, *209*, 995–1006. [CrossRef] [PubMed]

39. Zhu, M.S.; Pan, Y.; Chen, H.Q.; Shen, Y.; Wang, X.C.; Sun, Y.J.; Tao, K.H. Induction of SARS-nucleoprotein-specific immune response by use of DNA vaccine. *Immunol. Lett.* **2004**, *92*, 237–243. [CrossRef] [PubMed]

40. Channappanavar, R.; Zhao, J.; Perlman, S. T cell-mediated immune response to respiratory coronaviruses. *Immunol. Res.* **2014**, *59*, 118–128. [CrossRef]

41. Wang, Y.D.; Sin, W.Y.; Xu, G.B.; Yang, H.H.; Wong, T.Y.; Pang, X.W.; He, X.Y.; Zhang, H.G.; Ng, J.N.; Cheng, C.S.; et al. T-cell epitopes in severe acute respiratory syndrome (SARS) coronavirus spike protein elicit a specific T-cell immune response in patients who recover from SARS. *J. Virol.* **2004**, *78*, 5612–5618. [CrossRef]

42. Zhao, J.; Alshukairi, A.N.; Baharoon, S.A.; Ahmed, W.A.; Bokhari, A.A.; Nehdi, A.M.; Layqah, L.A.; Alghamdi, M.G.; Al Gethamy, M.M.; Dada, A.M.; et al. Recovery from the Middle East respiratory syndrome is associated with antibody and T-cell responses. *Sci. Immunol.* **2017**, *2*. [CrossRef] [PubMed]

43. Deming, D.; Sheahan, T.; Heise, M.; Yount, B.; Davis, N.; Sims, A.; Suthar, M.; Harkema, J.; Whitmore, A.; Pickles, R.; et al. Vaccine efficacy in senescent mice challenged with recombinant SARS-CoV bearing epidemic and zoonotic spike variants. *PLoS Med.* **2006**, *3*, e525. [CrossRef]

44. Tseng, C.T.; Sbrana, E.; Iwata-Yoshikawa, N.; Newman, P.C.; Garron, T.; Atmar, R.L.; Peters, C.J.; Couch, R.B. Immunization with SARS coronavirus vaccines leads to pulmonary immunopathology on challenge with the SARS virus. *PLoS ONE* **2012**, *7*, e35421. [CrossRef]

45. Weihofen, W.A.; Liu, J.; Reutter, W.; Saenger, W.; Fan, H. Crystal structure of CD26/dipeptidyl-peptidase IV in complex with adenosine deaminase reveals a highly amphiphilic interface. *J. Biol. Chem.* **2004**, *279*, 43330–43335. [CrossRef] [PubMed]

46. Wang, Q.; Wong, G.; Lu, G.; Yan, J.; Gao, G.F. MERS-CoV spike protein: Targets for vaccines and therapeutics. *Antivir. Res.* **2016**, *133*, 165–177. [CrossRef] [PubMed]

47. Hansson, M.; Nygren, P.A.; Stahl, S. Design and production of recombinant subunit vaccines. *Biotechnol. Appl. Biochem.* **2000**, *32 (Pt 2)*, 95–107. [CrossRef] [PubMed]

48. Cao, Z.; Liu, L.; Du, L.; Zhang, C.; Jiang, S.; Li, T.; He, Y. Potent and persistent antibody responses against the receptor-binding domain of SARS-CoV spike protein in recovered patients. *Virol. J.* **2010**, *7*, 299. [CrossRef]

49. Du, L.; Zhao, G.; Kou, Z.; Ma, C.; Sun, S.; Poon, V.K.; Lu, L.; Wang, L.; Debnath, A.K.; Zheng, B.J.; et al. Identification of a receptor-binding domain in the S protein of the novel human coronavirus Middle East respiratory syndrome coronavirus as an essential target for vaccine development. *J. Virol.* **2013**, *87*, 9939–9942. [CrossRef]

50. Mou, H.; Raj, V.S.; van Kuppeveld, F.J.; Rottier, P.J.; Haagmans, B.L.; Bosch, B.J. The receptor binding domain of the new Middle East respiratory syndrome coronavirus maps to a 231-residue region in the spike protein that efficiently elicits neutralizing antibodies. *J. Virol.* **2013**, *87*, 9379–9383. [CrossRef]

51. Burton, D.R.; Williamson, R.A.; Parren, P.W. Antibody and virus: Binding and neutralization. *Virology* **2000**, *270*, 1–3. [CrossRef]

52. Ma, C.; Li, Y.; Wang, L.; Zhao, G.; Tao, X.; Tseng, C.T.; Zhou, Y.; Du, L.; Jiang, S. Intranasal vaccination with recombinant receptor-binding domain of MERS-CoV spike protein induces much stronger local mucosal immune responses than subcutaneous immunization: Implication for designing novel mucosal MERS vaccines. *Vaccine* **2014**, *32*, 2100–2108. [CrossRef] [PubMed]

53. Zhang, M.Y.; Wang, Y.; Mankowski, M.K.; Ptak, R.G.; Dimitrov, D.S. Cross-reactive HIV-1-neutralizing activity of serum IgG from a rabbit immunized with gp41 fused to IgG1 Fc: Possible role of the prolonged half-life of the immunogen. *Vaccine* **2009**, *27*, 857–863. [CrossRef] [PubMed]

54. Ma, C.; Wang, L.; Tao, X.; Zhang, N.; Yang, Y.; Tseng, C.K.; Li, F.; Zhou, Y.; Jiang, S.; Du, L. Searching for an ideal vaccine candidate among different MERS coronavirus receptor-binding fragments—The importance of immunofocusing in subunit vaccine design. *Vaccine* **2014**, *32*, 6170–6176. [CrossRef] [PubMed]

55. Nyon, M.P.; Du, L.; Tseng, C.K.; Seid, C.A.; Pollet, J.; Naceanceno, K.S.; Agrawal, A.; Algaissi, A.; Peng, B.H.; Tai, W.; et al. Engineering a stable CHO cell line for the expression of a MERS-coronavirus vaccine antigen. *Vaccine* **2018**, *36*, 1853–1862. [CrossRef] [PubMed]

56. Letarov, A.V.; Londer, Y.Y.; Boudko, S.P.; Mesyanzhinov, V.V. The carboxy-terminal domain initiates trimerization of bacteriophage T4 fibritin. *Biochem. Biokhimiia* **1999**, *64*, 817–823.

57. Tai, W.; Zhao, G.; Sun, S.; Guo, Y.; Wang, Y.; Tao, X.; Tseng, C.K.; Li, F.; Jiang, S.; Du, L.; et al. A recombinant receptor-binding domain of MERS-CoV in trimeric form protects human dipeptidyl peptidase 4 (hDPP4) transgenic mice from MERS-CoV infection. *Virology* **2016**, *499*, 375–382. [CrossRef] [PubMed]

58. Lan, J.; Yao, Y.; Deng, Y.; Chen, H.; Lu, G.; Wang, W.; Bao, L.; Deng, W.; Wei, Q.; Gao, G.F.; et al. Recombinant Receptor Binding Domain Protein Induces Partial Protective Immunity in Rhesus Macaques Against Middle East Respiratory Syndrome Coronavirus Challenge. *EBioMedicine* **2015**, *2*, 1438–1446. [CrossRef]

59. Tai, W.; Wang, Y.; Fett, C.A.; Zhao, G.; Li, F.; Perlman, S.; Jiang, S.; Zhou, Y.; Du, L. Recombinant Receptor-Binding Domains of Multiple Middle East Respiratory Syndrome Coronaviruses (MERS-CoVs) Induce Cross-Neutralizing Antibodies against Divergent Human and Camel MERS-CoVs and Antibody Escape Mutants. *J. Virol.* **2017**, *91*, e01651-16. [CrossRef]

60. Coleman, C.M.; Liu, Y.V.; Mu, H.; Taylor, J.K.; Massare, M.; Flyer, D.C.; Smith, G.E.; Frieman, M.B. Purified coronavirus spike protein nanoparticles induce coronavirus neutralizing antibodies in mice. *Vaccine* **2014**, *32*, 3169–3174. [CrossRef]

61. Coleman, C.M.; Venkataraman, T.; Liu, Y.V.; Glenn, G.M.; Smith, G.E.; Flyer, D.C.; Frieman, M.B. MERS-CoV spike nanoparticles protect mice from MERS-CoV infection. *Vaccine* **2017**, *35*, 1586–1589. [CrossRef]

62. Kam, Y.W.; Kien, F.; Roberts, A.; Cheung, Y.C.; Lamirande, E.W.; Vogel, L.; Chu, S.L.; Tse, J.; Guarner, J.; Zaki, S.R.; et al. Antibodies against trimeric S glycoprotein protect hamsters against SARS-CoV challenge despite their capacity to mediate FcgammaRII-dependent entry into B cells in vitro. *Vaccine* **2007**, *25*, 729–740. [CrossRef] [PubMed]

63. McLellan, J.S.; Chen, M.; Joyce, M.G.; Sastry, M.; Stewart-Jones, G.B.; Yang, Y.; Zhang, B.; Chen, L.; Srivatsan, S.; Zheng, A.; et al. Structure-based design of a fusion glycoprotein vaccine for respiratory syncytial virus. *Science* **2013**, *342*, 592–598. [CrossRef] [PubMed]

64. Rota, P.A.; Oberste, M.S.; Monroe, S.S.; Nix, W.A.; Campagnoli, R.; Icenogle, J.P.; Penaranda, S.; Bankamp, B.; Maher, K.; Chen, M.H.; et al. Characterization of a novel coronavirus associated with severe acute respiratory syndrome. *Science* **2003**, *300*, 1394–1399. [CrossRef] [PubMed]

65. Jiaming, L.; Yanfeng, Y.; Yao, D.; Yawei, H.; Linlin, B.; Baoying, H.; Jinghua, Y.; Gao, G.F.; Chuan, Q.; Wenjie, T. The recombinant N-terminal domain of spike proteins is a potential vaccine against Middle East respiratory syndrome coronavirus (MERS-CoV) infection. *Vaccine* **2017**, *35*, 10–18. [CrossRef] [PubMed]

66. Aihara, H.; Miyazaki, J. Gene transfer into muscle by electroporation in vivo. *Nat. Biotechnol.* **1998**, *16*, 867–870. [CrossRef] [PubMed]

67. Sardesai, N.Y.; Weiner, D.B. Electroporation delivery of DNA vaccines: Prospects for success. *Curr. Opin. Immunol.* **2011**, *23*, 421–429. [CrossRef]

68. Liu, M.A. DNA vaccines: An historical perspective and view to the future. *Immunol. Rev.* **2011**, *239*, 62–84. [CrossRef]

69. Nichols, W.W.; Ledwith, B.J.; Manam, S.V.; Troilo, P.J. Potential DNA vaccine integration into host cell genome. *Ann. N. Y. Acad. Sci.* **1995**, *772*, 30–39. [CrossRef]

70. Sheets, R.L.; Stein, J.; Manetz, T.S.; Duffy, C.; Nason, M.; Andrews, C.; Kong, W.P.; Nabel, G.J.; Gomez, P.L. Biodistribution of DNA plasmid vaccines against HIV-1, Ebola, Severe Acute Respiratory Syndrome, or West Nile virus is similar, without integration, despite differing plasmid backbones or gene inserts. *Toxicol. Sci.* **2006**, *91*, 610–619. [CrossRef]

71. Muthumani, K.; Falzarano, D.; Reuschel, E.L.; Tingey, C.; Flingai, S.; Villarreal, D.O.; Wise, M.; Patel, A.; Izmirly, A.; Aljuaid, A.; et al. A synthetic consensus anti-spike protein DNA vaccine induces protective immunity against Middle East respiratory syndrome coronavirus in nonhuman primates. *Sci. Transl. Med.* **2015**, *7*, 301ra132. [CrossRef]

72. Barouch, D.H.; Yang, Z.Y.; Kong, W.P.; Korioth-Schmitz, B.; Sumida, S.M.; Truitt, D.M.; Kishko, M.G.; Arthur, J.C.; Miura, A.; Mascola, J.R.; et al. A human T-cell leukemia virus type 1 regulatory element enhances the immunogenicity of human immunodeficiency virus type 1 DNA vaccines in mice and nonhuman primates. *J. Virol.* **2005**, *79*, 8828–8834. [CrossRef] [PubMed]

73. Cayabyab, M.J.; Kashino, S.S.; Campos-Neto, A. Robust immune response elicited by a novel and unique Mycobacterium tuberculosis protein using an optimized DNA/protein heterologous prime/boost protocol. *Immunology* **2012**, *135*, 216–225. [CrossRef] [PubMed]

74. Wang, L.; Shi, W.; Joyce, M.G.; Modjarrad, K.; Zhang, Y.; Leung, K.; Lees, C.R.; Zhou, T.; Yassine, H.M.; Kanekiyo, M.; et al. Evaluation of candidate vaccine approaches for MERS-CoV. *Nat. Commun.* **2015**, *6*, 7712. [CrossRef] [PubMed]

75. Chi, H.; Zheng, X.; Wang, X.; Wang, C.; Wang, H.; Gai, W.; Perlman, S.; Yang, S.; Zhao, J.; Xia, X. DNA vaccine encoding Middle East respiratory syndrome coronavirus S1 protein induces protective immune responses in mice. *Vaccine* **2017**, *35*, 2069–2075. [CrossRef] [PubMed]

76. Lanzavecchia, A. Mechanisms of antigen uptake for presentation. *Curr. Opin. Immunol.* **1996**, *8*, 348–354. [CrossRef]

77. Al-Amri, S.S.; Abbas, A.T.; Siddiq, L.A.; Alghamdi, A.; Sanki, M.A.; Al-Muhanna, M.K.; Alhabbab, R.Y.; Azhar, E.I.; Li, X.; Hashem, A.M. Immunogenicity of Candidate MERS-CoV DNA Vaccines Based on the Spike Protein. *Sci. Rep.* **2017**, *7*, 44875. [CrossRef]

78. Stevens, T.L.; Bossie, A.; Sanders, V.M.; Fernandez-Botran, R.; Coffman, R.L.; Mosmann, T.R.; Vitetta, E.S. Regulation of antibody isotype secretion by subsets of antigen-specific helper T cells. *Nature* **1988**, *334*, 255–258. [CrossRef]

79. Rollier, C.S.; Reyes-Sandoval, A.; Cottingham, M.G.; Ewer, K.; Hill, A.V. Viral vectors as vaccine platforms: Deployment in sight. *Curr. Opin. Immunol.* **2011**, *23*, 377–382. [CrossRef]

80. Pushko, P.; Parker, M.; Ludwig, G.V.; Davis, N.L.; Johnston, R.E.; Smith, J.F. Replicon-helper systems from attenuated Venezuelan equine encephalitis virus: Expression of heterologous genes in vitro and immunization against heterologous pathogens in vivo. *Virology* **1997**, *239*, 389–401. [CrossRef]

81. Agnihothram, S.; Menachery, V.D.; Yount, B.L., Jr.; Lindesmith, L.C.; Scobey, T.; Whitmore, A.; Schafer, A.; Heise, M.T.; Baric, R.S. Development of a Broadly Accessible Venezuelan Equine Encephalitis Virus Replicon Particle Vaccine Platform. *J. Virol.* **2018**, *92*, e00027-18. [CrossRef]

82. Zhao, J.; Mangalam, A.K.; Channappanavar, R.; Fett, C.; Meyerholz, D.K.; Agnihothram, S.; Baric, R.S.; David, C.S.; Perlman, S. Airway Memory CD4(+) T Cells Mediate Protective Immunity against Emerging Respiratory Coronaviruses. *Immunity* **2016**, *44*, 1379–1391. [CrossRef] [PubMed]

83. Sutter, G.; Moss, B. Nonreplicating vaccinia vector efficiently expresses recombinant genes. *Proc. Natl. Acad. Sci. USA* **1992**, *89*, 10847–10851. [CrossRef] [PubMed]

84. Stittelaar, K.J.; Kuiken, T.; de Swart, R.L.; van Amerongen, G.; Vos, H.W.; Niesters, H.G.; van Schalkwijk, P.; van der Kwast, T.; Wyatt, L.S.; Moss, B.; et al. Safety of modified vaccinia virus Ankara (MVA) in immune-suppressed macaques. *Vaccine* **2001**, *19*, 3700–3709. [CrossRef]

85. Song, F.; Fux, R.; Provacia, L.B.; Volz, A.; Eickmann, M.; Becker, S.; Osterhaus, A.D.; Haagmans, B.L.; Sutter, G. Middle East respiratory syndrome coronavirus spike protein delivered by modified vaccinia virus Ankara efficiently induces virus-neutralizing antibodies. *J. Virol.* **2013**, *87*, 11950–11954. [CrossRef] [PubMed]

86. Volz, A.; Kupke, A.; Song, F.; Jany, S.; Fux, R.; Shams-Eldin, H.; Schmidt, J.; Becker, C.; Eickmann, M.; Becker, S.; et al. Protective Efficacy of Recombinant Modified Vaccinia Virus Ankara Delivering Middle East Respiratory Syndrome Coronavirus Spike Glycoprotein. *J. Virol.* **2015**, *89*, 8651–8656. [CrossRef] [PubMed]

87. Langenmayer, M.C.; Lulf-Averhoff, A.T.; Adam-Neumair, S.; Fux, R.; Sutter, G.; Volz, A. Distribution and absence of generalized lesions in mice following single dose intramuscular inoculation of the vaccine candidate MVA-MERS-S. *Biologicals* **2018**, *54*, 58–62. [CrossRef] [PubMed]

88. Haagmans, B.L.; van den Brand, J.M.; Raj, V.S.; Volz, A.; Wohlsein, P.; Smits, S.L.; Schipper, D.; Bestebroer, T.M.; Okba, N.; Fux, R.; et al. An orthopoxvirus-based vaccine reduces virus excretion after MERS-CoV infection in dromedary camels. *Science* **2016**, *351*, 77–81. [CrossRef]

89. Hammer, S.M.; Sobieszczyk, M.E.; Janes, H.; Karuna, S.T.; Mulligan, M.J.; Grove, D.; Koblin, B.A.; Buchbinder, S.P.; Keefer, M.C.; Tomaras, G.D.; et al. Efficacy trial of a DNA/rAd5 HIV-1 preventive vaccine. *N. Engl. J. Med.* **2013**, *369*, 2083–2092. [CrossRef]

90. Chirmule, N.; Propert, K.; Magosin, S.; Qian, Y.; Qian, R.; Wilson, J. Immune responses to adenovirus and adeno-associated virus in humans. *Gene Ther.* **1999**, *6*, 1574–1583. [CrossRef]

91. Mast, T.C.; Kierstead, L.; Gupta, S.B.; Nikas, A.A.; Kallas, E.G.; Novitsky, V.; Mbewe, B.; Pitisuttithum, P.; Schechter, M.; Vardas, E.; et al. International epidemiology of human pre-existing adenovirus (Ad) type-5, type-6, type-26 and type-36 neutralizing antibodies: Correlates of high Ad5 titers and implications for potential HIV vaccine trials. *Vaccine* **2010**, *28*, 950–957. [CrossRef]

92. Mercier, S.; Rouard, H.; Delfau-Larue, M.H.; Eloit, M. Specific antibodies modulate the interactions of adenovirus type 5 with dendritic cells. *Virology* **2004**, *322*, 308–317. [CrossRef] [PubMed]

93. Kim, E.; Okada, K.; Kenniston, T.; Raj, V.S.; AlHajri, M.M.; Farag, E.A.; AlHajri, F.; Osterhaus, A.D.; Haagmans, B.L.; Gambotto, A. Immunogenicity of an adenoviral-based Middle East Respiratory Syndrome coronavirus vaccine in BALB/c mice. *Vaccine* **2014**, *32*, 5975–5982. [CrossRef]

94. Guo, X.; Deng, Y.; Chen, H.; Lan, J.; Wang, W.; Zou, X.; Hung, T.; Lu, Z.; Tan, W. Systemic and mucosal immunity in mice elicited by a single immunization with human adenovirus type 5 or 41 vector-based vaccines carrying the spike protein of Middle East respiratory syndrome coronavirus. *Immunology* **2015**, *145*, 476–484. [CrossRef] [PubMed]

95. Lemiale, F.; Haddada, H.; Nabel, G.J.; Brough, D.E.; King, C.R.; Gall, J.G. Novel adenovirus vaccine vectors based on the enteric-tropic serotype 41. *Vaccine* **2007**, *25*, 2074–2084. [CrossRef] [PubMed]

96. Jung, S.Y.; Kang, K.W.; Lee, E.Y.; Seo, D.W.; Kim, H.L.; Kim, H.; Kwon, T.; Park, H.L.; Lee, S.M.; Nam, J.H. Heterologous prime-boost vaccination with adenoviral vector and protein nanoparticles induces both Th1 and Th2 responses against Middle East respiratory syndrome coronavirus. *Vaccine* **2018**, *36*, 3468–3476. [CrossRef] [PubMed]

97. Farina, S.F.; Gao, G.P.; Xiang, Z.Q.; Rux, J.J.; Burnett, R.M.; Alvira, M.R.; Marsh, J.; Ertl, H.C.; Wilson, J.M. Replication-defective vector based on a chimpanzee adenovirus. *J. Virol.* **2001**, *75*, 11603–11613. [CrossRef]

98. Ledgerwood, J.E.; DeZure, A.D.; Stanley, D.A.; Coates, E.E.; Novik, L.; Enama, M.E.; Berkowitz, N.M.; Hu, Z.; Joshi, G.; Ploquin, A.; et al. Chimpanzee Adenovirus Vector Ebola Vaccine. *N. Engl. J. Med.* **2017**, *376*, 928–938. [CrossRef] [PubMed]

99. Alharbi, N.K.; Padron-Regalado, E.; Thompson, C.P.; Kupke, A.; Wells, D.; Sloan, M.A.; Grehan, K.; Temperton, N.; Lambe, T.; Warimwe, G.; et al. ChAdOx1 and MVA based vaccine candidates against MERS-CoV elicit neutralising antibodies and cellular immune responses in mice. *Vaccine* **2017**, *35*, 3780–3788. [CrossRef]

100. Li, Z.; Howard, A.; Kelley, C.; Delogu, G.; Collins, F.; Morris, S. Immunogenicity of DNA vaccines expressing tuberculosis proteins fused to tissue plasminogen activator signal sequences. *Infect. Immunity* **1999**, *67*, 4780–4786.

101. Zhang, Y.; Feng, L.; Li, L.; Wang, D.; Li, C.; Sun, C.; Li, P.; Zheng, X.; Liu, Y.; Yang, W.; et al. Effects of the fusion design and immunization route on the immunogenicity of Ag85A-Mtb32 in adenoviral vectored tuberculosis vaccine. *Hum. Vaccines Immunother.* **2015**, *11*, 1803–1813. [CrossRef]

102. Munster, V.J.; Wells, D.; Lambe, T.; Wright, D.; Fischer, R.J.; Bushmaker, T.; Saturday, G.; van Doremalen, N.; Gilbert, S.C.; de Wit, E.; et al. Protective efficacy of a novel simian adenovirus vaccine against lethal MERS-CoV challenge in a transgenic human DPP4 mouse model. *NPJ Vaccines* **2017**, *2*, 28. [CrossRef] [PubMed]

103. Zuniga, A.; Wang, Z.; Liniger, M.; Hangartner, L.; Caballero, M.; Pavlovic, J.; Wild, P.; Viret, J.F.; Glueck, R.; Billeter, M.A.; et al. Attenuated measles virus as a vaccine vector. *Vaccine* **2007**, *25*, 2974–2983. [CrossRef] [PubMed]

104. Malczyk, A.H.; Kupke, A.; Prufer, S.; Scheuplein, V.A.; Hutzler, S.; Kreuz, D.; Beissert, T.; Bauer, S.; Hubich-Rau, S.; Tondera, C.; et al. A Highly Immunogenic and Protective Middle East Respiratory Syndrome Coronavirus Vaccine Based on a Recombinant Measles Virus Vaccine Platform. *J. Virol.* **2015**, *89*, 11654–11667. [CrossRef]

105. Bodmer, B.S.; Fiedler, A.H.; Hanauer, J.R.H.; Prufer, S.; Muhlebach, M.D. Live-attenuated bivalent measles virus-derived vaccines targeting Middle East respiratory syndrome coronavirus induce robust and multifunctional T cell responses against both viruses in an appropriate mouse model. *Virology* **2018**, *521*, 99–107. [CrossRef] [PubMed]

106. Kim, S.H.; Samal, S.K. Newcastle Disease Virus as a Vaccine Vector for Development of Human and Veterinary Vaccines. *Viruses* **2016**, *8*, 183. [CrossRef] [PubMed]

107. Liu, R.-Q.; Ge, J.-Y.; Wang, J.-L.; Yu, S.; Zhang, H.-L.; Wang, J.-L.; Wen, Z.-Y.; Bu, Z.-G. Newcastle disease virus-based MERS-CoV candidate vaccine elicits high-level and lasting neutralizing antibodies in Bactrian camels. *J. Integr. Agric.* **2017**, *16*, 2264–2273. [CrossRef]

108. Lichty, B.D.; Power, A.T.; Stojdl, D.F.; Bell, J.C. Vesicular stomatitis virus: Re-inventing the bullet. *Trends Mol. Med.* **2004**, *10*, 210–216. [CrossRef]

109. Liu, R.; Wang, J.; Shao, Y.; Wang, X.; Zhang, H.; Shuai, L.; Ge, J.; Wen, Z.; Bu, Z. A recombinant VSV-vectored MERS-CoV vaccine induces neutralizing antibody and T cell responses in rhesus monkeys after single dose immunization. *Antivir. Res.* **2018**, *150*, 30–38. [CrossRef]

110. Willet, M.; Kurup, D.; Papaneri, A.; Wirblich, C.; Hooper, J.W.; Kwilas, S.A.; Keshwara, R.; Hudacek, A.; Beilfuss, S.; Rudolph, G.; et al. Preclinical Development of Inactivated Rabies Virus-Based Polyvalent Vaccine Against Rabies and Filoviruses. *J. Infect. Dis.* **2015**, *212* (Suppl. 2), S414–S424. [CrossRef]

111. Wirblich, C.; Coleman, C.M.; Kurup, D.; Abraham, T.S.; Bernbaum, J.G.; Jahrling, P.B.; Hensley, L.E.; Johnson, R.F.; Frieman, M.B.; Schnell, M.J. One-Health: A Safe, Efficient, Dual-Use Vaccine for Humans and Animals against Middle East Respiratory Syndrome Coronavirus and Rabies Virus. *J. Virol.* **2017**, *91*, e02040-16. [CrossRef]

112. Zeltins, A. Construction and characterization of virus-like particles: A review. *Mol. Biotechnol.* **2013**, *53*, 92–107. [CrossRef]

113. Wang, C.; Zheng, X.; Gai, W.; Zhao, Y.; Wang, H.; Feng, N.; Chi, H.; Qiu, B.; Li, N.; Wang, T.; et al. MERS-CoV virus-like particles produced in insect cells induce specific humoural and cellular imminity in rhesus macaques. *Oncotarget* **2017**, *8*, 12686–12694. [CrossRef] [PubMed]

114. Wang, C.; Zheng, X.; Gai, W.; Wong, G.; Wang, H.; Jin, H.; Feng, N.; Zhao, Y.; Zhang, W.; Li, N.; et al. Novel chimeric virus-like particles vaccine displaying MERS-CoV receptor-binding domain induce specific humoral and cellular immune response in mice. *Antivir. Res.* **2017**, *140*, 55–61. [CrossRef] [PubMed]

115. Bolles, M.; Deming, D.; Long, K.; Agnihothram, S.; Whitmore, A.; Ferris, M.; Funkhouser, W.; Gralinski, L.; Totura, A.; Heise, M.; et al. A double-inactivated severe acute respiratory syndrome coronavirus vaccine provides incomplete protection in mice and induces increased eosinophilic proinflammatory pulmonary response upon challenge. *J. Virol.* **2011**, *85*, 12201–12215. [CrossRef] [PubMed]

116. Agrawal, A.S.; Tao, X.; Algaissi, A.; Garron, T.; Narayanan, K.; Peng, B.H.; Couch, R.B.; Tseng, C.T. Immunization with inactivated Middle East Respiratory Syndrome coronavirus vaccine leads to lung immunopathology on challenge with live virus. *Hum. Vaccines Immunother.* **2016**, *12*, 2351–2356. [CrossRef] [PubMed]

117. Deng, Y.; Lan, J.; Bao, L.; Huang, B.; Ye, F.; Chen, Y.; Yao, Y.; Wang, W.; Qin, C.; Tan, W. Enhanced protection in mice induced by immunization with inactivated whole viruses compare to spike protein of middle east respiratory syndrome coronavirus. *Emerg. Microbes Infect.* **2018**, *7*, 60. [CrossRef]

118. Iwata-Yoshikawa, N.; Uda, A.; Suzuki, T.; Tsunetsugu-Yokota, Y.; Sato, Y.; Morikawa, S.; Tashiro, M.; Sata, T.; Hasegawa, H.; Nagata, N. Effects of Toll-like receptor stimulation on eosinophilic infiltration in lungs of BALB/c mice immunized with UV-inactivated severe acute respiratory syndrome-related coronavirus vaccine. *J. Virol.* **2014**, *88*, 8597–8614. [CrossRef]

119. Plotkin, S.A. Vaccines: Past, present and future. *Nat. Med.* **2005**, *11* (Suppl. 4), S5–S11. [CrossRef]

120. Ruch, T.R.; Machamer, C.E. The coronavirus E protein: Assembly and beyond. *Viruses* **2012**, *4*, 363–382. [CrossRef]

121. Lamirande, E.W.; DeDiego, M.L.; Roberts, A.; Jackson, J.P.; Alvarez, E.; Sheahan, T.; Shieh, W.J.; Zaki, S.R.; Baric, R.; Enjuanes, L.; et al. A live attenuated severe acute respiratory syndrome coronavirus is immunogenic and efficacious in golden Syrian hamsters. *J. Virol.* **2008**, *82*, 7721–7724. [CrossRef]

122. Almazan, F.; DeDiego, M.L.; Sola, I.; Zuniga, S.; Nieto-Torres, J.L.; Marquez-Jurado, S.; Andres, G.; Enjuanes, L. Engineering a replication-competent, propagation-defective Middle East respiratory syndrome coronavirus as a vaccine candidate. *mBio* **2013**, *4*, e00650-13. [CrossRef] [PubMed]

123. Graham, R.L.; Becker, M.M.; Eckerle, L.D.; Bolles, M.; Denison, M.R.; Baric, R.S. A live, impaired-fidelity coronavirus vaccine protects in an aged, immunocompromised mouse model of lethal disease. *Nat. Med.* **2012**, *18*, 1820–1826. [CrossRef] [PubMed]

124. Zust, R.; Cervantes-Barragan, L.; Habjan, M.; Maier, R.; Neuman, B.W.; Ziebuhr, J.; Szretter, K.J.; Baker, S.C.; Barchet, W.; Diamond, M.S.; et al. Ribose 2′-O-methylation provides a molecular signature for the distinction of self and non-self mRNA dependent on the RNA sensor Mda5. *Nat. Immunol.* **2011**, *12*, 137–143. [CrossRef] [PubMed]

125. Menachery, V.D.; Yount, B.L., Jr.; Josset, L.; Gralinski, L.E.; Scobey, T.; Agnihothram, S.; Katze, M.G.; Baric, R.S. Attenuation and restoration of severe acute respiratory syndrome coronavirus mutant lacking 2′-o-methyltransferase activity. *J. Virol.* **2014**, *88*, 4251–4264. [CrossRef] [PubMed]

126. Menachery, V.D.; Gralinski, L.E.; Mitchell, H.D.; Dinnon, K.H., III; Leist, S.R.; Yount, B.L., Jr.; McAnarney, E.T.; Graham, R.L.; Waters, K.M.; Baric, R.S. Combination Attenuation Offers Strategy for Live Attenuated Coronavirus Vaccines. *J. Virol.* **2018**, *92*, e00710-18. [CrossRef] [PubMed]

127. Menachery, V.D.; Gralinski, L.E.; Mitchell, H.D.; Dinnon, K.H., III; Leist, S.R.; Yount, B.L., Jr.; Graham, R.L.; McAnarney, E.T.; Stratton, K.G.; Cockrell, A.S.; et al. Middle East Respiratory Syndrome Coronavirus Nonstructural Protein 16 Is Necessary for Interferon Resistance and Viral Pathogenesis. *mSphere* **2017**, *2*, e00346-17. [CrossRef] [PubMed]

128. Liu, D.X.; Fung, T.S.; Chong, K.K.; Shukla, A.; Hilgenfeld, R. Accessory proteins of SARS-CoV and other coronaviruses. *Antivir. Res.* **2014**, *109*, 97–109. [CrossRef] [PubMed]

129. Menachery, V.D.; Mitchell, H.D.; Cockrell, A.S.; Gralinski, L.E.; Yount, B.L., Jr.; Graham, R.L.; McAnarney, E.T.; Douglas, M.G.; Scobey, T.; Beall, A.; et al. MERS-CoV Accessory ORFs Play Key Role for Infection and Pathogenesis. *mBio* **2017**, *8*. [CrossRef] [PubMed]

130. World Health Organization. *Consensus Document on the Epidemiology of Severe Acute Respiratory Syndrome (SARS)*; World Health Organization: Geneva, Switzerland, 2003.

131. Hotez, P.J.; Bottazzi, M.E.; Tseng, C.T.; Zhan, B.; Lustigman, S.; Du, L.; Jiang, S. Calling for rapid development of a safe and effective MERS vaccine. *Microbes Infect.* **2014**, *16*, 529–531. [CrossRef]

132. Prescott, J.; Falzarano, D.; de Wit, E.; Hardcastle, K.; Feldmann, F.; Haddock, E.; Scott, D.; Feldmann, H.; Munster, V.J. Pathogenicity and Viral Shedding of MERS-CoV in Immunocompromised Rhesus Macaques. *Front. Immunol.* **2018**, *9*, 205. [CrossRef]

viruses

MDPI

Article

Efficacy of an Adjuvanted Middle East Respiratory Syndrome Coronavirus Spike Protein Vaccine in Dromedary Camels and Alpacas

Danielle R. Adney [1], Lingshu Wang [2], Neeltje van Doremalen [3], Wei Shi [2], Yi Zhang [2], Wing-Pui Kong [2], Megan R. Miller [3], Trenton Bushmaker [3], Dana Scott [3], Emmie de Wit [3], Kayvon Modjarrad [4], Nikolai Petrovsky [5], Barney S. Graham [2], Richard A. Bowen [1,*,†] and Vincent J. Munster [1,3,*,†]

1   Department of Microbiology, Immunology, and Pathology, Colorado State University, Fort Collins, CO 80521, USA; Danielle.Adney@colostate.edu
2   Vaccine Research Center, National Institute of Allergy and Infectious Diseases, National Institute of Health, Bethesda, MD 20892, USA; wangling@niaid.nih.gov (L.W.); shiw@mail.nih.gov (W.S.); yi.zhang3@nih.gov (Y.Z.); wkong@mail.nih.gov (W.-P.K.); bgraham@mail.nih.gov (B.S.G.)
3   Laboratory of Virology, Division of Intramural Research, National Institute of Allergy and Infectious Diseases, National Institutes of Health, Hamilton, MT 59840, USA; neeltje.vandoremalen@nih.gov (N.v.D.); megan.r.miller@colostate.edu (M.R.M.); bushmakertj@niaid.nih.gov (T.B.); dana.scott@nih.gov (D.S.); Emmie.deWit@nih.gov (E.d.W.)
4   Military HIV Research, Program, Walter Reed Army Institute of Research, Silver Spring, MD 20910, USA; kmodjarrad@hivresearch.org
5   Flinders University and Vaxine Pty Ltd, Flinders Medical Centre, Bedford Park, SA 5042, Australia; nikolai.petrovsky@flinders.edu.au
*   Correspondence: rbowen@rams.colostate.edu (R.A.B.); Vincent.munster@nih.gov (V.J.M.)
†   These authors contributed equally to this article.

Received: 27 December 2018; Accepted: 28 February 2019; Published: 2 March 2019

Abstract: MERS-CoV is present in dromedary camels throughout the Middle East and Africa. Dromedary camels are the primary zoonotic reservoir for human infections. Interruption of the zoonotic transmission chain from camels to humans, therefore, may be an effective strategy to control the ongoing MERS-CoV outbreak. Here we show that vaccination with an adjuvanted MERS-CoV Spike protein subunit vaccine confers complete protection from MERS-CoV disease in alpaca and results in reduced and delayed viral shedding in the upper airways of dromedary camels. Protection in alpaca correlates with high serum neutralizing antibody titers. Lower titers of serum neutralizing antibodies correlate with delayed and significantly reduced shedding in the nasal turbinates of dromedary camels. Together, these data indicate that induction of robust neutralizing humoral immune responses by vaccination of naïve animals reduces shedding that potentially could diminish the risk of zoonotic transmission.

Keywords: MERS-CoV; camels; vaccines; One Health

1. Introduction

Since the emergence of the Middle East respiratory syndrome coronavirus (MERS-CoV) in 2012, there have been more than 2279 confirmed human cases and 806 deaths, yielding a case fatality ratio of 35% [1,2]. Nearly all cases either have occurred in or been exported from the Middle East. Dromedary camels have been identified as the primary reservoir of MERS-CoV through multiple studies that have employed combinations of diagnostic platforms, to include serology, molecular detection of viral RNA, and virus isolation [3–15]. Serologic studies suggest that MERS-CoV has been present in dromedary camels for at least three decades in regions as wide ranging as the Middle East, North Africa, and East Africa [16]. The outbreak among human populations has persisted—albeit in low numbers—despite a lack of sustained human-to-human transmission [17,18]. Data from molecular epidemiologic studies suggest that the outbreak has endured through multiple zoonotic spillover events from dromedary camels to humans. Additionally, several studies in Saudi Arabia and Qatar have demonstrated a clear epidemiologic link between camel and human populations [4,19–26].

Although MERS-CoV typically causes severe pneumonia in humans, the virus primarily infects the upper respiratory tract of dromedary camels, from where viable virus is shed in high quantities for approximately 7 days, and where viral RNA can be detected up to 35 days [27]. Occupational exposure that involves close proximity to infected dromedary camels is a major risk factor for human acquisition. This risk profile has been confirmed by studies in Saudi Arabia where camel herders and slaughterhouse workers were much more likely than the general population to either have serologic evidence of past exposure to MERS-CoV or develop MERS [20,26].

The high case fatality rate, continued zoonotic transmission, and sustained animal reservoir of MERS-CoV underscore the need for intervention strategies focused on preventing zoonotic transmission. Here, we evaluated the efficacy of an intramuscular, adjuvanted S1 subunit vaccine to prevent infection or reduce shedding in dromedary camels (*Camelus dromedarius*) and alpaca (*Vicugna pacos*).

2. Materials and Methods

2.1. Vaccine Development

DNA expression vectors were produced encoding the S1 portion of the Spike glycoprotein of MERS-CoV England1 strain (strain England1, GenBank ID: AFY13307) (Figure 1). Sequences were reverse-translated and codon-optimized for human cell expression as described [28]. Proteins were produced by transfecting Expi293 cells with the mammalian expression vector VRC8400 expressing the codon-optimized S1 gene. Protein was purified from transfected cell culture supernatants with HisTrap HP Hiload 16/60 Superdex columns (GE Healthcare, Piscataway, NJ, USA), and stored at −80 °C in PBS until use.

2.2. Ethics Statement

All experiments were approved by the Colorado State University Institutional Animal Care and Use Committee. Work with infectious MERS-CoV strains under BSL3 conditions was approved by the Institutional Biosafety Committee (IBC). Inactivation and removal of samples from high containment was performed according to IBC-approved standards.

2.3. Study Design

To evaluate the effect of vaccination on MERS-CoV shedding, we used both the actual animal reservoir (camels) and a surrogate infection model (alpaca) [27,29]. Five MERS-CoV seronegative dromedary camels and four alpaca were purchased by private sale for use in this study. All animals were born in the United States and were all adult intact males for the Alpaca; one intact adult male, one castrated adult male, and one adult female for the vaccinated dromedary camels; and three intact male dromedary camels for the unvaccinated control animals. Animals were fed ad libitum, housed in outdoor pens during immunization and moved into an Animal Biosafety Level 3 facility one week prior to challenge to allow for acclimation to the facility.

Vaccinated animals were given 400 µg S1 protein combined with 40 mg Advax HCXL adjuvant (a delta inulin formulation, Vaxine Pty Ltd., Bedford Park, SA, Australia), in two 1 mL intramuscular injections given on each shoulder on day 0 and day 28, and then given 400 µg S1 protein in the Sigma Adjuvant System (an oil-in-water emulsion, Sigma-Aldrich, St. Louis, MO, USA) on day 105.

The Advax adjuvant was chosen for stimulating broad immunogenicity in dromedary camels [30], the Sigma Adjuvant System for known induction of MERS-CoV neutralizing antibodies [28].

Animals were challenged with MERS-CoV strain HCoV/EMC/2012 in three groups along a staggered schedule, due to facility limitations. Group 1 was inoculated 130 days after first vaccination and consisted of vaccinated camel 3 (CA3), group 2 was inoculated 131 days after first vaccination and consisted of unvaccinated camels 4 and 5 (CA4 and CA5) and vaccinated alpaca 1 and 2 (A1 and A2), and group 3 was inoculated 132 days after first vaccination and consisted of vaccinated camels 1 and 2 (CA1 and CA2) and unvaccinated alpaca 3 and 4 (A3 and A4). Animals were sedated with xylazine and then inoculated intranasally with a total dose of 10^7 TCID$_{50}$ of a human isolate of MERS-CoV (strain HCoV-EMC/2012) as described previously [27]. Camels were inoculated with 5 mL per nare, and alpaca were inoculated with 3 mL per nare. All animals were evaluated at least once daily for temperature, nasal discharge, activity level, and food consumption. Nasal swabs were collected daily and placed immediately into viral transport medium or virus lysis buffer and then frozen until processing. Animals were sedated on day 5 post-inoculation and humanely euthanized with intravenous pentobarbital. The following tissues were collected for viral isolation and formalin-fixed for immunohistochemistry and histopathology: nasal turbinates, larynx, trachea, lung (right apical, left apical, right caudal, left caudal lobes), kidney, spleen, mediastinal and mesenteric lymph nodes and muscle. Historical samples from a previously infected camel euthanized on day 5 post inoculation [27] were included as an additional control (CA6). CA6 in the present study corresponds to camel 1 in the publication by Adney et al. [27].

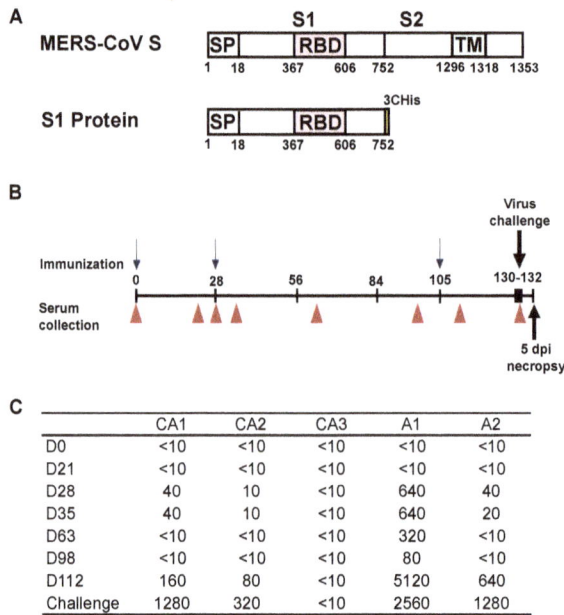

	CA1	CA2	CA3	A1	A2
D0	<10	<10	<10	<10	<10
D21	<10	<10	<10	<10	<10
D28	40	10	<10	640	40
D35	40	10	<10	640	20
D63	<10	<10	<10	320	<10
D98	<10	<10	<10	80	<10
D112	160	80	<10	5120	640
Challenge	1280	320	<10	2560	1280

Figure 1. Vaccine immunogens and neutralizing antibody responses. (**A**) Schematic representation of the full MERS-CoV spike protein (top) and the S1 subunit used as antigen for vaccination (bottom). (**B**) Schematic overview of the immunization timeline with the immunization (blue arrows), serum sampling (red triangle), and challenge and euthanasia (black arrows) time points indicated. (**C**) Neutralizing antibody titers in dromedary camels and alpacas vaccinated with adjuvanted MERS-CoV S1 as determined by PRNT (90% neutralization).

2.4. Immunization and Serology

Blood was collected on day 0, 21, 28, 35, 63, 98, 112 after first vaccination and at time of challenge (day 130–132) into serum-separating tubes. Serum was analyzed for MERS-CoV neutralization activity by plaque reduction neutralization test (PRNT) as described previously using a 90% neutralization cutoff [27].

2.5. Virus Titration

Nasal swabs and tissue samples were titrated by plaque assay in Vero E6 cells. In short, homogenized tissue was titrated in BA-1 medium and plaques were counted on both days 1 and 3 following the second overlay as described previously [27].

2.6. RNA Extraction and qRT-PCR

RNA was extracted from swabs, urine and fecal samples using the QiaAmp Viral RNA kit (Qiagen). For detection of viral RNA, 5 μL of RNA was used in a one-step real-time RT-PCR upE assay using the Rotor-GeneTM probe kit (Qiagen) as described previously. Standard dilutions of a titered virus stock were run in parallel, to calculate $TCID_{50}$ equivalents in the samples.

2.7. MERS-CoV Spike Glycoprotein Sequencing

Total RNA from nasal turbinate tissue samples from dromedary camels and alpaca were extracted using the RNeasy Mini Kit (Qiagen) and cDNAs were synthesized using random hexamers and the High Capacity RNA to cDNA Kit (Thermo Fisher, Waltham, MA, USA). cDNA was subsequently used to PCR-amplify the MERS-CoV S using iProof High-Fidelity DNA Polymerase (Biorad, Hercules, CA, USA) according to the manufacturer's protocol; primer sequences are available upon request. Sequences were assembled on SeqMan Pro (DNASTAR, Madison, WI, USA) and analyzed on MegAlign (DNASTAR, Madison, WI, USA) by comparison to the MERS-CoV (strain HCoV-EMC/2012) input sequence.

2.8. Histopathology and Immunohistochemistry

Dromedary camel and alpaca tissues were evaluated for pathology and presence of viral antigens. Tissues were fixed for >7 days in 10% neutral-buffered formalin and placed in cassettes and processed with a Sakura VIP-5 Tissue Tek, on a 12-hour automated schedule, using a graded series of ethanol, xylene, and ParaPlast Extra. Embedded tissues are sectioned at 5 μm and dried overnight at 42 degrees C prior to staining. All tissues were processed for immunohistochemistry using the Discovery XT automated processor (Ventana Medical Systems) with a DABMap (Ventana Medical Systems) kit. To detect MERS-CoV antigen, immunohistochemistry was performed using a rabbit polyclonal antiserum against HCoV-EMC/2012 (1:1000) as a primary antibody as described previously [27]. Grading of histopathology and immunohistochemistry was done blinded by a board-certified veterinary pathologist.

2.9. Statistical Analysis

p values were calculated in a 2-way ANOVA with Sidak's multiple comparisons test using the Prism software (Version 6.04, GraphPad, La Jolla, CA, USA). Statistically significant differences met a threshold (α) of 0.05.

3. Results

3.1. Humoral Responses in Dromedary Camels and Alpaca Vaccinated against MERS-CoV

Three dromedary camels (CA1, CA2, CA3) and two alpaca (A1, A2) were vaccinated with an adjuvanted S1-protein subunit vaccine. Three camels (CA4, CA5, CA6) served as unvaccinated controls;

CA6 is a historical control [27]. Two alpaca (A3, A4) served as unvaccinated controls for the alpaca group. Animals were vaccinated on days 0 and 28 with 400 µg of S1 protein (Figure 1A) co-formulated with 40 mg Advax[TM] HCXL adjuvant (Vaxine Pty Ltd, Adelaide, Australia) [30]. The admixed product was delivered at each time point as two 1 mL intramuscular injections in each shoulder. All animals were boosted on day 105 with 400 µg of S1 protein emulsified in Sigma Adjuvant System (Sigma Aldrich Co. LLC. St. Louis, MO, USA) to complete a 0, 4-, and 15-week immunization schedule (Figure 1B). Serum from vaccinated animals was collected and evaluated by plaque reduction neutralization test (PRNT) with MERS-CoV strain HCoV-EMC/2012 (Figure 1C). On day 28 after priming vaccination, low levels of MERS-CoV neutralizing antibodies were detected in two of three the camels (virus neutralizing titers of 1:40, 1:10, and <10 respectively; Figure 1C). The day 28 boost did not result in an increase in neutralizing titers in the PRNT assay and neutralizing titers decreased between the second and third boost. The third immunization resulted in a quick increase in neutralizing titer in the two camels that responded to the vaccine; neutralizing titers were high in these two camels by the time of challenge. Neutralizing antibodies were not detected in CA3 (<1:10) at any point during the experiment (Figure 1C). A stronger neutralizing response was observed in the alpaca after vaccination, with virus neutralizing titers of 1:640 and 1:40 four weeks after the initial vaccination to end titers of 1:2560 and 1:640 at the time of challenge (Figure 1C).

3.2. Vaccine Efficacy in Preventing MERS-CoV Disease and Virus Shedding

Animals were challenged intranasally with 10^7 50% tissue culture infectious dose ($TCID_{50}$) of MERS-CoV (strain HCoV-EMC/2012) and euthanized and necropsied at 5 days post-inoculation (dpi). Only the unvaccinated control camels developed mild clinical disease. Nasal discharge was observed in concurrent control camels at 2 dpi, but quickly resolved. Observable nasal discharge was not detected in the vaccinated camels nor in any of the alpaca. Minor temperature fluctuations were detected in several of the camels and alpaca; however, there was no appreciable fever associated with infection (Figure S1).

Virus shedding was first detected in nasal swabs from unvaccinated control camels at 1 dpi and every day thereafter until euthanasia at 5 dpi (Figure 2A). Despite the presence of neutralizing antibodies in two of the three vaccinated camels, all animals shed virus after challenge. The two vaccinated camels with a detectable humoral immune response (CA1 and CA2) exhibited reduced viral shedding through 4 dpi. At 5 dpi, there were no differences in MERS-CoV titers measured in nasal swabs between vaccinated and control camels. The vaccinated camel without detectable neutralizing antibody titers (CA3) did not shed virus on 1 dpi but otherwise exhibited virus shedding kinetics that were similar to those of unvaccinated controls. Overall, infectious MERS-CoV titers in nasal swabs differed statistically significantly between vaccinated and unvaccinated camels on 1 and 3 dpi ($p < 0.05$). In contrast to the camels, vaccination of alpaca resulted in complete protection from MERS-CoV shedding. No infectious virus was detected in the nasal swabs collected from the vaccinated alpaca (A1, A2), while infectious virus was detected in unvaccinated alpaca every day until necropsy at 5 dpi (Figure 2B).

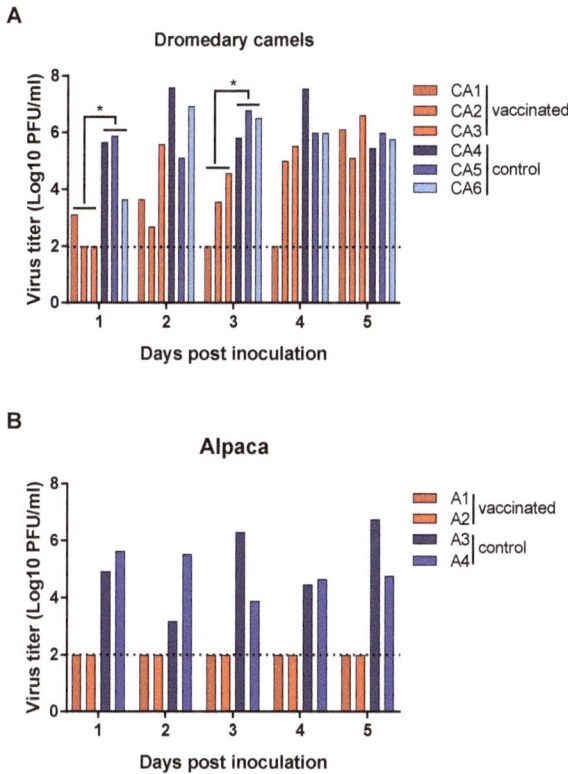

Figure 2. Virus shedding in nasal swabs of vaccinated and unvaccinated dromedary camels and alpaca after challenge with MERS-CoV. Nasal swabs were obtained from dromedary camels (**A**) and alpaca (**B**) daily after MERS-CoV challenge and virus titers in swabs were determined in a plaque assay. Red bars indicate vaccinated animals and blue bars indicate unvaccinated control animals. A dashed line indicates the detection limit of the assay. Asterisks indicate statistically significant difference between geometric mean titers ($p < 0.05$).

Infectious virus was detected in the nasal turbinates, larynx, and trachea of all unvaccinated camels (Figure 3A), confirming that the virus stock prepared for challenge resulted in productive infection. Additionally, infectious virus was detected in 1 of 4 lung lobes of CA6. All three vaccinated camels had infectious virus in their nasal turbinates at titers similar to those of the unvaccinated camels. However, virus titers in the larynx and trachea were significantly lower in the vaccinated camels than in unvaccinated controls (Figure 3A). Viral RNA was also detected outside the respiratory tract of camels, predominantly in the lymph nodes (Figure S2A).

Both unvaccinated alpaca had high titers of infectious virus present in their nasal turbinates and lower titers in the larynx and trachea. Neither infectious virus nor viral RNA could be detected in the respiratory tract tissues of the vaccinated alpaca (Figure 3B and Figure S2B). In the unvaccinated control alpaca, low amounts of viral genomic RNA were detected outside the respiratory tract (Figure S2B).

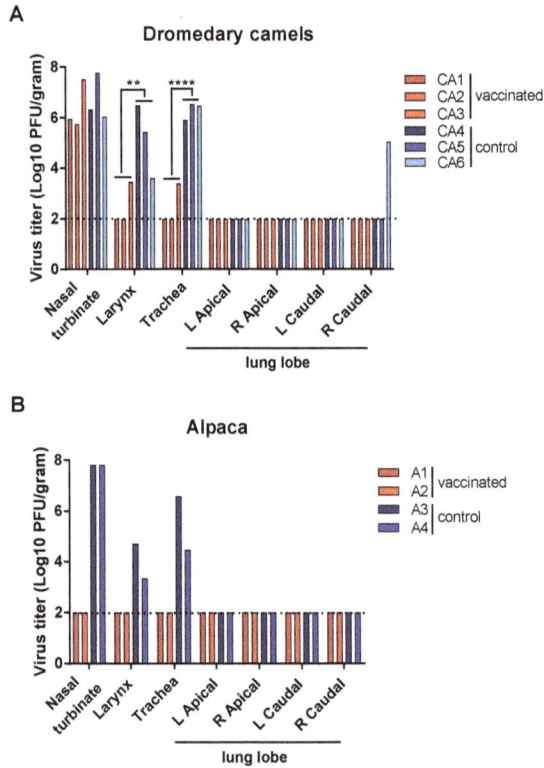

Figure 3. Infectious virus in respiratory tissues of vaccinated and unvaccinated dromedary camels and alpaca after challenge with MERS-CoV. Dromedary camels (**A**) and alpaca (**B**) were euthanized on 5 dpi, tissues were collected, and virus titers were determined in a plaque assay. Red bars indicate vaccinated animals and blue bars indicate unvaccinated control animals. A dashed line indicates the detection limit of the assay. Asterisks indicate statistically significant difference between geometric mean titers (** $p < 0.01$; **** $p < 0.0001$). L: left; R: right.

3.3. Histopathological Changes in Vaccinated Versus Control Animals

All camels displayed, on histologic section, a multifocal, minimal-to-mild, acute rhinitis (Figure 4A,G). The nasal turbinates exhibited a loss of cilia and cellular polarity, occasional necrotic epithelial cells admixed with small numbers of neutrophils, and rare erosions of respiratory epithelium. The underlying submucosa was infiltrated by small-to-moderate numbers of lymphocytes and plasma cells with fewer macrophages and neutrophils. The non-responsive vaccinated camel, CA3, and control CA4 displayed a minimal-to-mild multifocal acute tracheitis, as evidenced by a loss of individual epithelial cells and cell polarity. Necrosis and infiltration of the epithelium by small numbers of neutrophils were observed. None of the camels showed pathological changes in the lower respiratory tract (Figure 4B,C,H,I). Immunohistochemistry (IHC) demonstrated the presence of MERS-CoV antigen in the epithelial cells of the nasal turbinates of CA1-5 (Figure 4D,J). Viral antigen was also detected in the tracheal epithelial cells of CA3 and CA4, but not in CA1 or CA2 (Figure 4E).

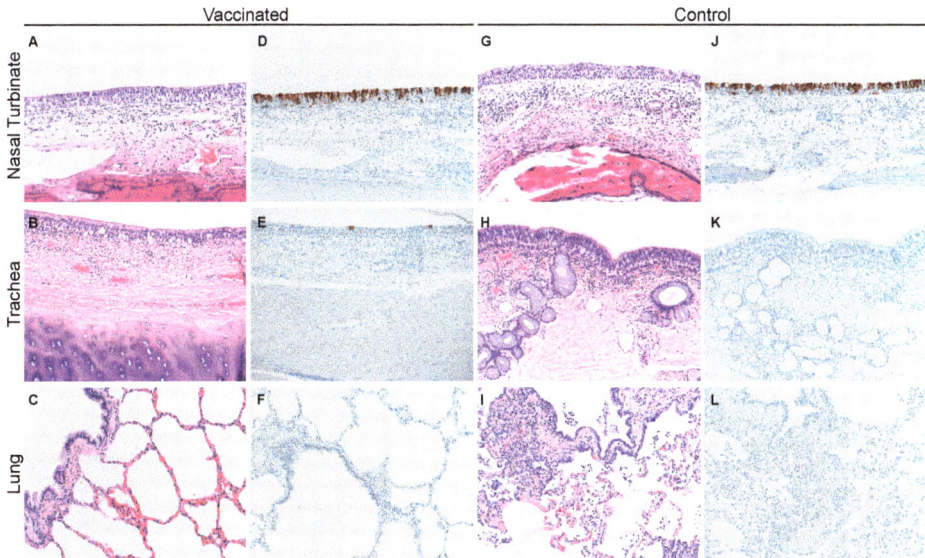

Figure 4. Histopathological changes in the respiratory tract of vaccinated and unvaccinated dromedary camels after challenge with MERS-CoV. Nasal turbinate, trachea, and lung were collected from vaccinated (**A–F**; CA3 shown) and unvaccinated control animals (**G–L**; CA5 shown) on 5 dpi and stained with hematoxylin and eosin (**A–C** and **G–I**), or a polyclonal anti-MERS-CoV antibody panels (**D–F** and **J–L**). MERS-CoV antigen is visible as a red brown staining in the immunohistochemistry panel. Vaccinated and unvaccinated camel groups displayed multifocal, minimal-to-mild, acute rhinitis. MERS-CoV antigen was primarily detected in the nasal turbinates. Magnification: 400×.

Similar to the camels, the unvaccinated alpaca displayed a multifocal, minimal-to-mild, acute rhinitis (Figure 5G) with loss of cilia and cellular polarity, occasional necrotic epithelial cells admixed with small numbers of neutrophils, and rare erosions of respiratory epithelium. The subjacent submucosa was infiltrated by small-to-moderate numbers of lymphocytes and plasma cells with fewer macrophages and neutrophils. IHC confirmed the presence of MERS-CoV antigen within the epithelial cells of the nasal turbinates of these control alpaca (Figure 5J). No histopathological changes or MERS-CoV antigen were detected in any of the vaccinated alpaca (Figure 5G–L).

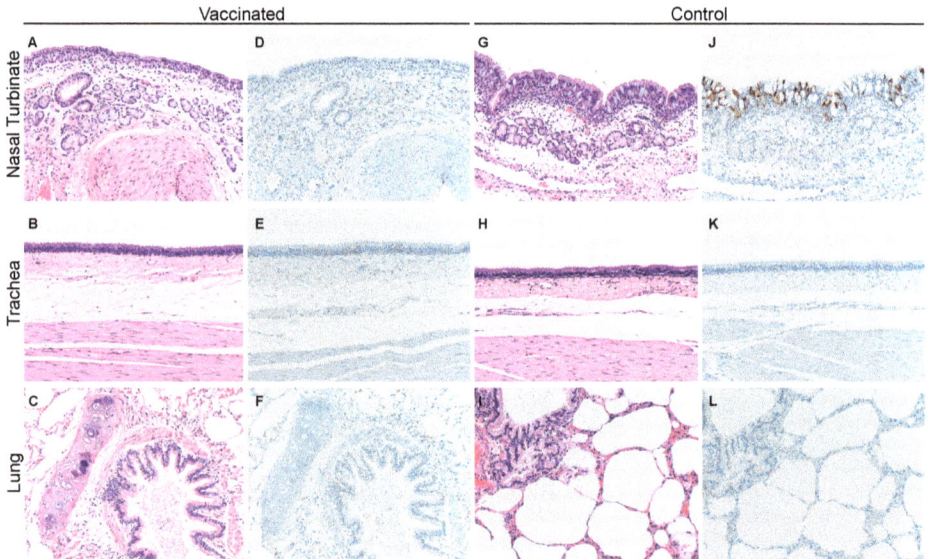

Figure 5. Histopathological changes in the respiratory tract of vaccinated and unvaccinated alpaca after challenge with MERS-CoV. Nasal turbinate, trachea, and lung were collected from vaccinated (**A–F**; A1 shown) and unvaccinated control animals (**G–L**; A3 shown) on 5 dpi and stained with hematoxylin and eosin (**A–C** and **G–I**), or a polyclonal anti-MERS-CoV antibody panels (**D–F** and **J–L**). MERS-CoV antigen is visible as a red brown staining in the immunohistochemistry panel. No histopathological changes were observed in the vaccinated alpaca whereas the unvaccinated control alpaca displayed multifocal, minimal-to-mild, acute rhinitis. MERS-CoV antigen was only detected in the nasal turbinates. Magnification: 400×.

4. Discussion

Given the importance of the zoonotic reservoir of MERS-CoV in sustaining the current human epidemic, strategies to interrupt the transmission from dromedary camels to humans may serve as an effective countermeasure [31]. This strategy has been implemented in the setting of Hendra virus infection in Australia, where vaccination of horses is aimed at interrupting the transmission chain of Hendra virus from fruit bats to horses to humans [32]. Likewise, dromedary camels are the drivers of zoonotic transmission of MERS-CoV and vaccination of dromedary camels to reduce shedding of MERS-CoV would likely limit its potential for zoonotic transmission.

Large fractions of dromedary camels in the Middle East and North and East Africa have antibodies specific to MERS-CoV [3,6,8,9,12,14,33,34]. The degree of protection from reinfection provided by these antibodies is currently unclear. Evidence points to the ability of MERS-CoV to re-infect dromedary camels despite the presence of antibodies in serum, as MERS-CoV seropositive camels that still shed detectable MERS-CoV from the respiratory tract have been found in the Middle East [8]. However, a correlation between strong humoral responses in seropositive adult dromedary camels and low prevalence of MERS-CoV was observed in Jordan, whereas seropositive young camels <3 years old with weaker humoral responses displayed high MERS-CoV prevalence [34], suggesting that antibodies against MERS-CoV afford some protection from reinfection or virus shedding. This observation also suggests that younger animals are the main driving force behind the circulation of MERS-CoV and should be the primary target for vaccination strategies.

This study highlights several important issues to consider when developing a dromedary camel vaccine against MERS-CoV. First of all, virus shedding was reduced, but not completely absent in the presence of antibodies in dromedary camels after vaccination. However, virus replication in vaccinated animals was largely confined to the nasal turbinates, whereas the MERS-CoV replication in unvaccinated animals extended deeper down the respiratory tract into the trachea. In a study using intranasal administration of a modified vaccinia virus Ankara virus vector vaccine expressing full length Spike protein, delayed shedding was also observed in one of the vaccinated dromedary camels [35]. After vaccination, low neutralizing antibody titers were detected in the upper respiratory tract. The apparent low neutralizing capacity of the upper respiratory tract combined with continuous exposure to MERS-CoV (from co-housed animals) could potentially deplete the neutralizing capacity by scavenging the available neutralizing antibodies until the neutralizing capacity is sufficiently reduced to allow for virus replication. Therefore, vaccine platforms aimed specifically at inducing high levels of mucosal immunity or those that induce humoral as well as cellular immunity may be required for complete protection. In-depth studies to determine the correlates of protection in dromedary camels would aid the development of better-targeted vaccines, especially the correlation of levels of neutralizing antibodies in the serum vs. protection against MERS-CoV in the upper respiratory tract. Second, the response to vaccination in dromedary camels varies widely, as one out of three camels in the present study did not respond to vaccination at all. This camel was a ten-year-old male, housed in a pen adjacent to a mature female. Most of the vaccination schedule coincided with the mating period of this camel. Thus, it is possible, though unclear, that reproductive status impacted the camel's immune response. Variable neutralizing responses after MERS-CoV vaccination have been reported in a previous dromedary camel vaccine study. One animal in a group of three failed to mount a detectable neutralizing response after vaccination with a DNA vaccine expressing full length Spike protein [36]. Vaccine development efforts should thus focus on vaccine platforms that uniformly induce immunity in dromedary camels. Finally, intranasal challenge with MERS-CoV yielded distinctly different outcomes in the alpaca and dromedary camels. Upon viral challenge, both immunized alpaca were completely protected from MERS-CoV infection. Thus, although alpaca are an attractive alternative experimental model to dromedary camels since MERS-CoV replication dynamics in alpaca appear to be similar to those in dromedary camels [29,37] but alpaca are easier to obtain and house, vaccine efficacy ultimately has to be confirmed in the dromedary camel model.

Supplementary Materials: The following are available online at http://www.mdpi.com/1999-4915/11/3/212/s1.

Author Contributions: Conceptualization, K.M., B.S.G., R.A.B. and V.J.M.; Methodology, D.R.A., L.W., W.S., Y.Z., W.-P.K., N.v.D., M.R.M., T.B., D.S. and E.d.W.; Software, D.R.A. and E.d.W.; Validation, K.M., B.S.G., R.A.B. and V.J.M.; Formal Analysis, D.R.A.; Resources, N.P.; Data Curation, D.R.A., N.v.D., L.W., E.d.W. and D.S.; Writing—Original Draft Preparation, D.R.A., E.d.W., R.A.B. and V.J.M.; Writing—D.R.A., E.d.W., K.M., B.S.G., R.A.B. and V.J.M.; Visualization, D.R.A., E.d.W. and D.S.; Supervision, R.A.B. and V.J.M.

Funding: This work was supported by the Intramural Research Program of the National Institute of Allergy and Infectious Diseases (NIAID), National Institutes of Health (NIH), and the Animal Models Core at Colorado State University. D.R.A. was supported by the Infectious Disease Translational Research Training Program at Colorado State University.

Acknowledgments: The authors would like to thank Rick Brandes, Greg Harding, and Amy Richardson for their assistance with animal care (Colorado State University); Tina Thomas, Dan Long, and Rebecca Rosenke for histopathology, and Anita Mora for figure preparation (NIAID).

Conflicts of Interest: The authors declare no conflict of interest.

References

1. WHO. Middle East Respiratory Syndrome Coronavirus (MERS-CoV). Available online: http://www.who.int/emergencies/mers-cov/en/ (accessed on 1 December 2018).

2. Zaki, A.M.; van Boheemen, S.; Bestebroer, T.M.; Osterhaus, A.D.; Fouchier, R.A. Isolation of a novel coronavirus from a man with pneumonia in Saudi Arabia. *N. Engl. J. Med.* **2012**, *367*, 1814–1820. [CrossRef] [PubMed]

3. Alagaili, A.N.; Briese, T.; Mishra, N.; Kapoor, V.; Sameroff, S.C.; de Wit, E.; Munster, V.J.; Hensley, L.E.; Zalmout, I.S.; Kapoor, A.; et al. Middle East Respiratory Syndrome Coronavirus Infection in Dromedary Camels in Saudi Arabia. *mBio* **2014**, *5*, e00884-14. [CrossRef] [PubMed]

4. Azhar, E.I.; Hashem, A.M.; El-Kafrawy, S.A.; Sohrab, S.S.; Aburizaiza, A.S.; Farraj, S.A.; Hassan, A.M.; Al-Saeed, M.S.; Jamjoom, G.A.; Madani, T.A. Detection of the Middle East Respiratory Syndrome Coronavirus Genome in an Air Sample Originating from a Camel Barn Owned by an Infected Patient. *mBio* **2014**, *5*, e01450-14. [CrossRef] [PubMed]

5. Chu, D.K.; Poon, L.L.; Gomaa, M.M.; Shehata, M.M.; Perera, R.A.; Abu Zeid, D.; El Rifay, A.S.; Siu, L.Y.; Guan, Y.; Webby, R.J.; et al. MERS coronaviruses in dromedary camels, Egypt. *Emerg. Infect. Dis.* **2014**, *20*, 1049–1053. [CrossRef] [PubMed]

6. Corman, V.M.; Jores, J.; Meyer, B.; Younan, M.; Liljander, A.; Said, M.Y.; Gluecks, I.; Lattwein, E.; Bosch, B.J.; Drexler, J.F.; et al. Antibodies against MERS coronavirus in dromedary camels, Kenya, 1992–2013. *Emerg. Infect. Dis.* **2014**, *20*, 1319–1322. [CrossRef] [PubMed]

7. Haagmans, B.L.; Al Dhahiry, S.H.; Reusken, C.B.; Raj, V.S.; Galiano, M.; Myers, R.; Godeke, G.J.; Jonges, M.; Farag, E.; Diab, A.; et al. Middle East respiratory syndrome coronavirus in dromedary camels: An outbreak investigation. *Lancet Infect. Dis.* **2014**, *14*, 140–145. [CrossRef]

8. Hemida, M.G.; Chu, D.K.; Poon, L.L.; Perera, R.A.; Alhammadi, M.A.; Ng, H.Y.; Siu, L.Y.; Guan, Y.; Alnaeem, A.; Peiris, M. MERS Coronavirus in Dromedary Camel Herd, Saudi Arabia. *Emerg. Infect. Dis.* **2014**, *20*, 1231–1234. [CrossRef] [PubMed]

9. Hemida, M.G.; Perera, R.A.; Al Jassim, R.A.; Kayali, G.; Siu, L.Y.; Wang, P.; Chu, K.W.; Perlman, S.; Ali, M.A.; Alnaeem, A.; et al. Seroepidemiology of Middle East respiratory syndrome (MERS) coronavirus in Saudi Arabia (1993) and Australia (2014) and characterisation of assay specificity. *Euro Surveill.* **2014**, *19*, 20828. [CrossRef] [PubMed]

10. Meyer, B.; Muller, M.A.; Corman, V.M.; Reusken, C.B.; Ritz, D.; Godeke, G.J.; Lattwein, E.; Kallies, S.; Siemens, A.; van Beek, J.; et al. Antibodies against MERS coronavirus in dromedary camels, United Arab Emirates, 2003 and 2013. *Emerg. Infect. Dis.* **2014**, *20*, 552–559. [CrossRef] [PubMed]

11. Muller, M.A.; Corman, V.M.; Jores, J.; Meyer, B.; Younan, M.; Liljander, A.; Bosch, B.J.; Lattwein, E.; Hilali, M.; Musa, B.E.; et al. MERS coronavirus neutralizing antibodies in camels, Eastern Africa, 1983–1997. *Emerg. Infect. Dis.* **2014**, *20*, 2093–2095. [CrossRef] [PubMed]

12. Perera, R.; Wang, P.; Gomaa, M.; El-Shesheny, R.; Kandeil, A.; Bagato, O.; Siu, L.; Shehata, M.; Kayed, A.; Moatasim, Y.; et al. Seroepidemiology for MERS coronavirus using microneutralisation and pseudoparticle virus neutralisation assays reveal a high prevalence of antibody in dromedary camels in Egypt, June 2013. *Euro Surveill.* **2013**, *18*, 20574. [CrossRef] [PubMed]

13. Raj, V.S.; Farag, E.A.; Reusken, C.B.; Lamers, M.M.; Pas, S.D.; Voermans, J.; Smits, S.L.; Osterhaus, A.D.; Al-Mawlawi, N.; Al-Romaihi, H.E.; et al. Isolation of MERS coronavirus from a dromedary camel, Qatar, 2014. *Emerg. Infect. Dis.* **2014**, *20*, 1339–1342. [CrossRef] [PubMed]

14. Reusken, C.B.; Messadi, L.; Feyisa, A.; Ularamu, H.; Godeke, G.J.; Danmarwa, A.; Dawo, F.; Jemli, M.; Melaku, S.; Shamaki, D.; et al. Geographic distribution of MERS coronavirus among dromedary camels, Africa. *Emerg. Infect. Dis.* **2014**, *20*, 1370–1374. [CrossRef] [PubMed]

15. Sabir, J.S.; Lam, T.T.; Ahmed, M.M.; Li, L.; Shen, Y.; Abo-Aba, S.E.; Qureshi, M.I.; Abu-Zeid, M.; Zhang, Y.; Khiyami, M.A.; et al. Co-circulation of three camel coronavirus species and recombination of MERS-CoVs in Saudi Arabia. *Science* **2016**, *351*, 81–84. [CrossRef] [PubMed]

16. de Wit, E.; van Doremalen, N.; Falzarano, D.; Munster, V.J. SARS and MERS: Recent insights into emerging coronaviruses. *Nat. Rev. Microbiol.* **2016**, *14*, 523–534. [CrossRef] [PubMed]

17. Cauchemez, S.; Nouvellet, P.; Cori, A.; Jombart, T.; Garske, T.; Clapham, H.; Moore, S.; Mills, H.L.; Salje, H.; Collins, C.; et al. Unraveling the drivers of MERS-CoV transmission. *Proc. Natl. Acad. Sci USA* **2016**, *113*, 9081–9086. [CrossRef] [PubMed]

18. Nishiura, H.; Miyamatsu, Y.; Chowell, G.; Saitoh, M. Assessing the risk of observing multiple generations of Middle East respiratory syndrome (MERS) cases given an imported case. *Euro Surveill* **2015**, *20*, 21181. [CrossRef] [PubMed]

19. Al Hammadi, Z.M.; Chu, D.K.; Eltahir, Y.M.; Al Hosani, F.; Al Mulla, M.; Tarnini, W.; Hall, A.J.; Perera, R.A.; Abdelkhalek, M.M.; Peiris, J.S.; et al. Asymptomatic MERS-CoV Infection in Humans Possibly Linked to Infected Dromedaries Imported from Oman to United Arab Emirates, May 2015. *Emerg. Infect. Dis.* **2015**, *21*, 2197–2200. [CrossRef] [PubMed]

20. Reusken, C.B.; Farag, E.A.; Haagmans, B.L.; Mohran, K.A.; Godeke, G.J.t.; Raj, S.; Alhajri, F.; Al-Marri, S.A.; Al-Romaihi, H.E.; Al-Thani, M.; et al. Occupational Exposure to Dromedaries and Risk for MERS-CoV Infection, Qatar, 2013–2014. *Emerg. Infect. Dis.* **2015**, *21*, 1422–1425. [CrossRef] [PubMed]

21. Azhar, E.I.; El-Kafrawy, S.A.; Farraj, S.A.; Hassan, A.M.; Al-Saeed, M.S.; Hashem, A.M.; Madani, T.A. Evidence for camel-to-human transmission of MERS coronavirus. *N. Engl. J. Med.* **2014**, *370*, 2499–2505. [CrossRef] [PubMed]

22. Memish, Z.A.; Cotten, M.; Meyer, B.; Watson, S.J.; Alsahafi, A.J.; Al Rabeeah, A.A.; Corman, V.M.; Sieberg, A.; Makhdoom, H.Q.; Assiri, A.; et al. Human infection with MERS coronavirus after exposure to infected camels, Saudi Arabia, 2013. *Emerg. Infect. Dis.* **2014**, *20*, 1012–1015. [CrossRef] [PubMed]

23. Drosten, C.; Kellam, P.; Memish, Z.A. Evidence for camel-to-human transmission of MERS coronavirus. *N. Engl. J. Med.* **2014**, *371*, 1359–1360. [CrossRef] [PubMed]

24. Farag, E.A.; Reusken, C.B.; Haagmans, B.L.; Mohran, K.A.; Stalin Raj, V.; Pas, S.D.; Voermans, J.; Smits, S.L.; Godeke, G.J.; Al-Hajri, M.M.; et al. High proportion of MERS-CoV shedding dromedaries at slaughterhouse with a potential epidemiological link to human cases, Qatar 2014. *Infect. Ecol. Epidemiol.* **2015**, *5*, 28305. [CrossRef] [PubMed]

25. Madani, T.A.; Azhar, E.I.; Hashem, A.M. Evidence for camel-to-human transmission of MERS coronavirus. *N. Engl. J. Med.* **2014**, *371*, 1359–1360. [CrossRef]

26. Muller, M.A.; Meyer, B.; Corman, V.M.; Al-Masri, M.; Turkestani, A.; Ritz, D.; Sieberg, A.; Aldabbagh, S.; Bosch, B.J.; Lattwein, E.; et al. Presence of Middle East respiratory syndrome coronavirus antibodies in Saudi Arabia: A nationwide, cross-sectional, serological study. *Lancet Infect. Dis.* **2015**, *15*, 559–564. [CrossRef]

27. Adney, D.R.; van Doremalen, N.; Brown, V.R.; Bushmaker, T.; Scott, D.; de Wit, E.; Bowen, R.A.; Munster, V.J. Replication and shedding of MERS-CoV in upper respiratory tract of inoculated dromedary camels. *Emerg. Infect. Dis.* **2014**, *20*, 1999–2005. [CrossRef] [PubMed]

28. Wang, L.; Shi, W.; Joyce, M.G.; Modjarrad, K.; Zhang, Y.; Leung, K.; Lees, C.R.; Zhou, T.; Yassine, H.M.; Kanekiyo, M.; et al. Evaluation of candidate vaccine approaches for MERS-CoV. *Nat. Commun.* **2015**, *6*, 7712. [CrossRef] [PubMed]

29. Adney, D.R.; Bielefeldt-Ohmann, H.; Hartwig, A.E.; Bowen, R.A. Infection, Replication, and Transmission of Middle East Respiratory Syndrome Coronavirus in Alpacas. *Emerg. Infect. Dis.* **2016**, *22*, 1031–1037. [CrossRef] [PubMed]

30. Eckersley, A.M.; Petrovsky, N.; Kinne, J.; Wernery, R.; Wernery, U. Improving the Dromedary Antibody Response: The Hunt for the Ideal Camel Adjuvant. *J. Camel Pract. Res.* **2011**, *18*, 35–46.

31. Modjarrad, K.; Moorthy, V.S.; Ben Embarek, P.; van Kerkhove, M.; Kim, J.; Kieny, M.P. A roadmap for MERS-CoV research and product development: Report from a World Health Organization consultation. *Nat. Med.* **2016**, *22*, 701–705. [CrossRef] [PubMed]

32. Middleton, D.; Pallister, J.; Klein, R.; Feng, Y.R.; Haining, J.; Arkinstall, R.; Frazer, L.; Huang, J.A.; Edwards, N.; Wareing, M.; et al. Hendra virus vaccine, a one health approach to protecting horse, human, and environmental health. *Emerg. Infect. Dis.* **2014**, *20*, 372–379. [CrossRef] [PubMed]

33. Hemida, M.G.; Perera, R.A.; Wang, P.; Alhammadi, M.A.; Siu, L.Y.; Li, M.; Poon, L.L.; Saif, L.; Alnaeem, A.; Peiris, M. Middle East Respiratory Syndrome (MERS) coronavirus seroprevalence in domestic livestock in Saudi Arabia, 2010 to 2013. *Euro Surveill.* **2013**, *18*, 20659. [CrossRef] [PubMed]

34. Van Doremalen, N.; Hijazeen, Z.S.; Holloway, P.; Al Omari, B.; McDowell, C.; Adney, D.; Talafha, H.A.; Guitian, J.; Steel, J.; Amarin, N.; et al. High Prevalence of Middle East Respiratory Coronavirus in Young Dromedary Camels in Jordan. *Vector-Borne Zoonotic Dis.* **2017**, *17*, 155–159. [CrossRef] [PubMed]

35. Haagmans, B.L.; van den Brand, J.M.; Raj, V.S.; Volz, A.; Wohlsein, P.; Smits, S.L.; Schipper, D.; Bestebroer, T.M.; Okba, N.; Fux, R.; et al. An orthopoxvirus-based vaccine reduces virus excretion after MERS-CoV infection in dromedary camels. *Science* **2016**, *351*, 77–81. [CrossRef] [PubMed]

36. Muthumani, K.; Falzarano, D.; Reuschel, E.L.; Tingey, C.; Flingai, S.; Villarreal, D.O.; Wise, M.; Patel, A.; Izmirly, A.; Aljuaid, A.; et al. A synthetic consensus anti-spike protein DNA vaccine induces protective immunity against Middle East respiratory syndrome coronavirus in nonhuman primates. *Sci. Transl. Med.* **2015**, *7*, 301ra132. [CrossRef] [PubMed]

37. Crameri, G.; Durr, P.A.; Klein, R.; Foord, A.; Yu, M.; Riddell, S.; Haining, J.; Johnson, D.; Hemida, M.G.; Barr, J.; et al. Experimental Infection and Response to Rechallenge of Alpacas with Middle East Respiratory Syndrome Coronavirus. *Emerg. Infect. Dis.* **2016**, *22*, 1071–1074. [CrossRef] [PubMed]

viruses

MDPI

Article

CD8+ T Cells Responding to the Middle East Respiratory Syndrome Coronavirus Nucleocapsid Protein Delivered by Vaccinia Virus MVA in Mice

Svenja Veit [1], Sylvia Jany [1], Robert Fux [1], Gerd Sutter [1,2,*] and Asisa Volz [1,2]

[1]   Institute for Infectious Diseases and Zoonoses, LMU Munich, 80539 Munich, Germany;
      svenja.veit@micro.vetmed.uni-muenchen.de (S.V.); sylvia.jany@micro.vetmed.uni-muenchen.de (S.J.);
      robert.fux@micro.vetmed.uni-muenchen.de (R.F.); asisa.volz@micro.vetmed.uni-muenchen.de (A.V.)
[2]   German Center for Infection Research (DZIF), partner site Munich, 80539 Munich, Germany
*   Correspondence: gerd.sutter@lmu.de; Tel.: +49-89-2180-2514

Received: 22 November 2018; Accepted: 14 December 2018; Published: 16 December 2018

Abstract: Middle East respiratory syndrome coronavirus (MERS-CoV), a novel infectious agent causing severe respiratory disease and death in humans, was first described in 2012. Antibodies directed against the MERS-CoV spike (S) protein are thought to play a major role in controlling MERS-CoV infection and in mediating vaccine-induced protective immunity. In contrast, relatively little is known about the role of T cell responses and the antigenic targets of MERS-CoV that are recognized by CD8+ T cells. In this study, the highly conserved MERS-CoV nucleocapsid (N) protein served as a target immunogen to elicit MERS-CoV-specific cellular immune responses. Modified Vaccinia virus Ankara (MVA), a safety-tested strain of vaccinia virus for preclinical and clinical vaccine research, was used for generating MVA-MERS-N expressing recombinant N protein. Overlapping peptides spanning the whole MERS-CoV N polypeptide were used to identify major histocompatibility complex class I/II-restricted T cell responses in BALB/c mice immunized with MVA-MERS-N. We have identified a H2-d restricted decamer peptide epitope in the MERS-N protein with CD8+ T cell antigenicity. The identification of this epitope, and the availability of the MVA-MERS-N candidate vaccine, will help to evaluate MERS-N-specific immune responses and the potential immune correlates of vaccine-mediated protection in the appropriate murine models of MERS-CoV infection.

Keywords: MERS-CoV; MERS-CoV nucleocapsid protein; murine CD8+ T cell epitope; MVA vaccine

1. Introduction

The Middle East respiratory syndrome coronavirus (MERS-CoV), a hitherto unknown β-coronavirus, emerged as a causative agent of a severe respiratory disease in humans in 2012. This new coronavirus was first isolated from the sputum of a patient suffering from severe pneumonia and renal failure [1]. To date, the MERS-CoV still causes disease and death in humans with a total of 2260 confirmed cases including 803 fatalities [2,3]. Epidemiological data suggest that the MERS-CoV is endemic in Saudi Arabia, which accounts for the majority of primary community-acquired cases. Many of those primary cases are due to virus exposure through direct contact with dromedary camels, the primary animal reservoir of MERS-CoV. Alternatively, camel workers undergoing subclinical infections are suggested to mediate virus transmission to other susceptible individuals [4,5]. Other outbreaks of MERS have been caused by nosocomial transmissions in health care settings [6–9]. Most of the MERS-CoV infections occur within the Arabian Peninsula, i.e., Saudi Arabia, Qatar, and United Arab Emirates, however MERS cases have been reported in various other countries around the world [9,10]. At present, there is still relatively little known about the pathogenesis of MERS-CoV. The highest incidence of severe disease is observed in elderly and immunocompromised individuals [11]. Those at general risk for infections are health care workers and people in close contact with dromedary camels [12,13]. These groups are therefore considered relevant target populations for prophylactic vaccination against MERS-CoV infection and prevention of MERS. The World Health Organization (WHO) and the Coalition for Epidemic Preparedness Innovations (CEPI) have listed MERS as priority target for vaccine development [14]. However, so far, no candidate vaccines have proceeded beyond phase I/IIa clinical testing [15]. One of these candidate vaccines is based on Modified Vaccinia virus Ankara (MVA), a safety-tested and replication-deficient vaccinia virus serving as an advanced viral vector platform for the development of new vaccines against infectious diseases and cancer (for review see [16]). In that context, we still know relatively little about the correlates of vaccine induced protection against MERS-CoV. It is well-known that virus-neutralizing antibodies directed against the spike glycoprotein (S protein) correlate with protective immunity against coronavirus infections in general [17–20]. Since the S protein is present on the cell surface, S protein is considered as the major antigen to induce virus neutralizing antibodies and as a key immunogen for the development of MERS-CoV candidate vaccines [21–25]. However, based on current knowledge from the biology of β-coronaviruses, we hypothesize that also other viral proteins warrant consideration as immunogens and targets of virus-specific antibodies and T cells. Among those, the nucleocapsid protein (N protein) is produced at high levels in infected cells and has been proposed as useful candidate protein for clinical diagnosis [26–30]. The coronavirus N proteins have been associated with multiple functions in the virus life cycle including the regulation of viral RNA synthesis, the packaging of the viral RNA in helical nucleocapsids, and in virion assembly through interaction with the viral M protein [31–34]. Furthermore, several reports suggest that the severe acute respiratory syndrome coronavirus (SARS-CoV) N protein functions as an immune evasion protein and an antagonist of the host interferon response [35–37]. Recently, the overexpression of MERS-CoV N in human A549 cells was found to be linked to an up-regulation of antiviral host gene expression including the synthesis of the inflammatory chemokine CXCL10 [38]. Despite this possible immune modulatory activity, SARS-CoV N-specific immune responses are reported to be long-lived and more broadly reactive when compared to SARS-CoV S-specific immunity [39]. Likewise, we were curious as to the suitability of the MERS-CoV N protein to serve as a vaccine antigen. The N protein is not present on the surface of MERS-CoV particles nor is it predicted to be expressed on the surface membrane of MVA infected cells. From this we hypothesized that the most relevant part of MERS-CoV N-specific immunity is based on CD8+ T cell responses relying on the processing and presentation of intracellular antigens. Currently, there is little information about MERS-CoV N-specific immune responses including the in vivo induction of N-specific cellular immunity.

In this study, we investigated the synthesis and delivery of the MERS-CoV N protein as a privileged antigen by a MVA vector virus. The recombinant MVA expressing a synthetic gene sequence of full-length MERS-CoV N (MVA-MERS-N) proved genetically stable and fully replication-competent in chicken embryo fibroblasts, an established cell substrate for MVA vaccine manufacturing. Upon in vitro infection, MVA-MERS-N produced high amounts of the heterologous protein that were detectable with MERS-CoV N-specific antibodies. Furthermore, MVA-MERS-N was tested as an experimental vaccine in BALB/c mice and elicited MERS-CoV N-specific interferon γ (IFN-γ)-producing CD8+ T cells. Using peptide library covering the whole MERS-CoV N polypeptide, we identified new H2-d restricted peptide epitopes of MERS-CoV N in BALB/c mice. This data will be highly relevant for further assessment of N antigen-specific immune responses in the well-established MERS-CoV-BALB/c mouse immunization/challenge model [40–44].

2. Materials and Methods

2.1. Mice

Female BALB/c mice (6 to 10-week-old) were purchased from Charles River Laboratories (Sulzfeld, Germany). For experimental work, mice were housed in an isolated (ISO) cage unit (Tecniplast, Hohenpeißenberg, Germany) and had free access to food and water. All animal experiments were handled in compliance with the German regulations for animal experimentation (Animal Welfare Act, approved by the Government of Upper Bavaria, Munich, Germany).

2.2. Cells

Primary chicken embryo fibroblasts (CEF) were prepared from 10-day-old chicken embryos (SPF eggs, VALO, Cuxhaven, Germany) and maintained in Minimum Essential Medium Eagle (MEM) (SIGMA-ALDRICH, Taufkirchen, Germany) containing 10% heat-inactivated fetal bovine serum (FBS) (SIGMA-ALDRICH, Taufkirchen, Germany), 1% Penicillin-Streptomycin (SIGMA-ALDRICH, Taufkirchen, Germany), and 1% MEM non-essential amino acid solution (SIGMA-ALDRICH, Taufkirchen, Germany). Human HeLa (ATCC CCL-2) cells were maintained in MEM containing 10% FBS and 1% Penicillin-Streptomycin. Human HaCat (CLS Cell Lines Service GmbH, Eppelheim, Germany) cells were cultured in Dulbecco's Modified Eagle's Medium (SIGMA-ALDRICH, Taufkirchen, Germany) supplemented with 10% heat-inactivated FBS, 2% HEPES-solution (SIGMA-ALDRICH, Taufkirchen, Germany) and antibiotics as described above. All cells were maintained at 37 °C and 5% CO_2 atmosphere.

2.3. Plasmid Constructions

The cDNA encoding the entire amino acid (aa) sequence (413 aa) of the MERS-CoV N protein was in silico modified by introducing silent codon alterations to remove three termination signals (TTTTTNT) for vaccinia virus early transcription and two G/C nucleotide runs from the original MERS-CoV gene sequence (Human betacoronavirus 2c EMC/2012, GenBank accession no. JX869059). A cDNA fragment was generated by DNA synthesis (Invitrogen Life Technology, Regensburg, Germany) and cloned into the MVA transfer plasmid pIIIH5red [45] to place the MERS-CoV N gene sequence under the transcriptional control of the vaccinia virus early/late promoter PmH5 [46] resulting in the MVA vector plasmid pIIIH5red-MERS-N.

2.4. Generation of Recombinant Virus

Recombinant MVA was generated using standard methodology as described previously [45]. Briefly, monolayers of nearly confluent CEF grown in six-well tissue culture plates (Sarstedt, Nürnbrecht, Germany) were infected with non-recombinant MVA (clonal isolate MVA F6) at 0.05 multiplicity of infection (MOI) and, 45 min after infection, CEF cells were transfected with plasmid pIIIH5red-MERS-N DNA using X-tremeGENE DNA Transfection Reagent Lipofectamine (Roche Diagnostics, Penzberg, Germany) as recommended by the manufacturer. At 48 h after infection, the cell cultures were harvested and recombinant MVA expressing the MERS-CoV N protein was clonally isolated by consecutive rounds of plaque purification in CEF by screening for transient co-expression of the red fluorescent marker protein mCherry.

Quality control experiments were performed using standard methodology [45]. Genetic identity and genetic stability of the recombinant virus were assessed via polymerase chain reaction (PCR) analysis of the genomic viral DNA. Replicative capacities of the MVA vector virus was tested in multi-step growth experiments in CEF and human HaCat and HeLa cells.

To generate high titer vaccine preparations for preclinical studies, recombinant MVA was amplified in CEF, purified by ultracentrifugation through 36% sucrose cushions, resuspended in 10 mM Tris-HCl buffer, pH 9.0, and stored at −80 °C. The sucrose purified MVA-MERS-N vaccine preparations corresponded in total protein/total DNA content to the purity profile of a MVA candidate vaccine for human use.

2.5. Western Blot Analysis of Recombinant Proteins

Confluent cell monolayers of CEF or HaCat cells were infected at a MOI of 5 with recombinant MVA expressing the MERS-CoV N or S protein [23]. Non-infected (mock) or wild-type MVA-infected cells served as controls. Cell lysates were prepared at different time points after infection and stored at −80 °C. Total cell proteins were resolved by electrophoresis in a sodium dodecyl sulfate (SDS)-10% polyacrylamide gel (SDS-PAGE) and subsequently transferred onto a nitrocellulose membrane via electroblotting. After 1 h blocking in a phosphate buffered saline (PBS) buffer containing 1% (w/v) non-fat dried milk and 0.1% (v/v) NP-40 detergent, the blots were incubated with monoclonal mouse anti-MERS-CoV Nucleocapsid antibody (Sino Biological, Beijing, China, 1:1000), monoclonal rabbit anti-MERS-CoV Spike Protein S1 Antibody (Sino Biological, 1:500), or polyclonal sera from MERS-CoV infected rabbits or cynomolgus macaques (kindly provided by Dr. Bart Haagmans, Erasmus Medical Center, Rotterdam, 1:1000) [23] as primary antibodies. After washing with 0.1% NP-40 in PBS, the blots were incubated with anti-mouse IgG (1:5000), or anti-rabbit IgG antibody (1:5000), or protein A (1:1000) conjugated to horseradish peroxidase (Cell Signaling Technology, Frankfurt am Main, Germany). After further washing, blots were developed using SuperSignal®West Dura Extended Duration substrate (Thermo Fisher Scientific, Planegg, Gemany).

2.6. Immunization Experiments in Mice

Groups of female BALB/c mice (n = 2 to 5) were immunized twice within a 21-day interval with 10^8 plaque-forming-units (PFU) of recombinant MVA-MERS-N or non-recombinant MVA (MVA) or PBS as mock vaccine. Vaccinations were given via the intramuscular (i.m.) or intraperitoneal (i.p.) route using 25 µL (i.m.) or 200 µL (i.p.) volumes per inoculation. All mice were monitored daily for welfare and potential adverse events of immunization. At day 8 post prime-boost immunization, animals were sacrificed by cervical dislocation and spleens were taken for T cell analysis.

2.7. Synthetic Peptides and Design of Peptide Pools

For T cell immune monitoring, we identified 101 individual synthetic peptides (assigned as 1 to 101) in silico spanning the entire MERS-CoV N protein sequence (Human betacoronavirus 2c EMC/2012, GenBank accession no. JX869059). This peptide library was designed to contain 15-mer peptides overlapping by 11 aa. Eighty-four peptides could be synthesized (Thermo Fisher Scientific) and were organized into two-dimensional matrix peptide pools (V1 to V9 and H1 to H9) containing 9 or 10 peptides as described previously [47,48]. For further T cell epitope mapping, the 11 aa sequence shared between peptide #89 and #90 was trimmed into 8-10-mer peptides, which were also obtained from Thermo Fisher Scientific. All peptides were dissolved in PBS to a concentration of 2 mg/mL and stored at −20 °C until use.

2.8. T cell Analysis by Enzyme-Linked Immunospot (ELISPOT)

Spleens were harvested on day 8 post prime-boost vaccination. Splenocytes were prepared by passing through a 70 µm strainer (Falcon®, A Corning Brand, Corning, USA) and incubating with Red Blood Cell Lysis Buffer (SIGMA-ALDRICH, Taufkirchen, Germany). Cells were washed and resuspended in RPMI 1640 medium (SIGMA-ALDRICH) containing 10% heat inactivated FBS and 1% Penicillin-Streptomycin. Splenocytes were further processed by using the QuadroMACS Kit (Miltenyi Biotec, Bergisch Gladbach, Germany) to separate CD8+ and CD4+ splenocytes with MACS Micro Beads (Miltenyi Biotec, Bergisch Gladbach, Germany).

IFN-γ-producing T cells were measured using ELISPOT assays (Elispot kit for mouse IFN-γ, MABTECH, Germany) following the manufacturer's instructions. Briefly, 1×10^6 splenocytes were seeded in 96-well plates (Sarstedt, Nürnbrecht, Germany) and stimulated with peptide pools or individual peptides (2 µg peptide/mL RPMI 1640 medium) at 37 °C for 48 h. Non-stimulated cells and cells treated with phorbol myristate acetate (PMA) (SIGMA-ALDRICH) and ionomycin (SIGMA-ALDRICH, Taufkirchen, Germany) or MVA F2L$_{26-34}$ peptide (F2L, SPGAAGYDL, Thermo Fisher Scientific, Planegg, Germany) [49] served as negative and positive controls, respectively. Automated ELISPOT plate reader software (A.EL.VIS Eli.Scan, A.EL.VIS ELISPOT Analysis Software, Hannover, Germany) was used to count and analyze spots.

2.9. T Cell Analysis by Intracellular Cytokine Staining (ICS) and Flow Cytometry

Splenocytes were prepared as described above. Splenocytes were added to 96-well plates (1×10^6 cells/well) and stimulated for 6 h with MERS-CoV N-specific peptide (at 8 µg peptide/mL RPMI 1640 medium) in presence of the protein transport inhibitor Brefeldin A (Biolegend, San Diego, CA, USA; 5 µg/mL). Non-stimulated cells served as a background control and cells stimulated with 5 ng/mL PMA and 500 ng/mL ionomycin or with F2L peptide (8 µg/mL RPMI 1640 medium) were used as positive controls. After stimulation, cell surface antigens were stained using PE-conjugated anti-mouse CD3 (clone: 17A2, Biolegend, San Diego, CA, USA), PE/Cy7-conjugated anti-mouse CD4 (clone: GK1.5, Biolegend, San Diego, CA, USA), or FITC-conjugated anti-mouse CD8a (clone: 5H10-1, Biolegend, San Diego, CA, USA) antibody and incubated for 30 min on ice. The surface-stained cells were washed with staining buffer (MACS QuantTM Running Buffer, Miltenyi Biotec), then fixed and permeabilized with Fixation- and Perm/Wash-Buffer (Biolegend, San Diego, CA, USA), and finally stained for intracellular IFN-γ expression using APC-conjugated anti-mouse-IFN-γ antibody (clone: XMG1.2, Biolegend, San Diego, CA, USA) for 30 min on ice. Following final washes, cells were resuspended in staining buffer and analyzed using the MACS Quant®VYB flow cytometer (Miltenyi Biotec, Bergisch Gladbach, Germany).

2.10. Statistical Analysis

Statistical analysis was performed by t-test using GraphPad Prism version 5 software (GraphPad software, San Diego CA, USA); P-values less than 0.05 were considered to be statistically significant.

3. Results

3.1. Construction and Characterization of Recombinant MVA Expressing MERS-CoV N Gene Encoding Sequences

Recombinant virus MVA-MERS-N was formed in CEF that were infected with MVA and transfected with the MVA vector plasmid pIIIH5red-MERS-N (Figure 1a). The MVA DNA sequences in pIIIH5red-MERS-N (flank-1, flank-2) targeted the insertion of the N gene sequences into the site of deletion III within the MVA genome. The clonal isolation was facilitated by co-production of the red fluorescent reporter protein mCherry allowing for the convenient detection of MVA-MERS-N infected cells during plaque purification. The repetitive DNA sequence of flank-1 (FR) served to remove the marker gene mCherry from the genome of the final recombinant virus through initiating an intragenomic homologous recombination (marker gene deletion). After PCR analysis confirmed the presence of more than 95% MVA-MERS-N recombinant viruses in the cultures, we selected the final marker-free recombinant viruses by plaque purification and screening for plaques without mCherry fluorescence. To confirm genetic integrity and proper insertion of the heterologous N gene sequences within the MVA-MERS-N genome, we analyzed viral genomic DNA by PCR using specific oligonucleotide primers specific for MVA sequences adjacent to the deletion III insertion site (Figure 1b). Additional PCRs specific for MVA sequences within the C7L gene locus or adjacent to the major deletion sites I, II, IV, V, and VI served to control for the genetic identity and genomic stability of MVA-MERS-N (Figure 1c, and data not shown). Next, we evaluated the recombinant virus MVA-MERS-N by multi-step growth analysis in different cell lines (Figure 1d). In CEF, the cell culture routinely used to propagate recombinant MVA vaccines, MVA-MERS-N efficiently replicated to titers similar to those obtained with non-recombinant MVA. In contrast, the human cell lines HaCat and HeLa proved non-permissive for productive virus growth confirming the well-preserved replication deficiency of the recombinant MVA-MERS-N in cells of mammalian origin.

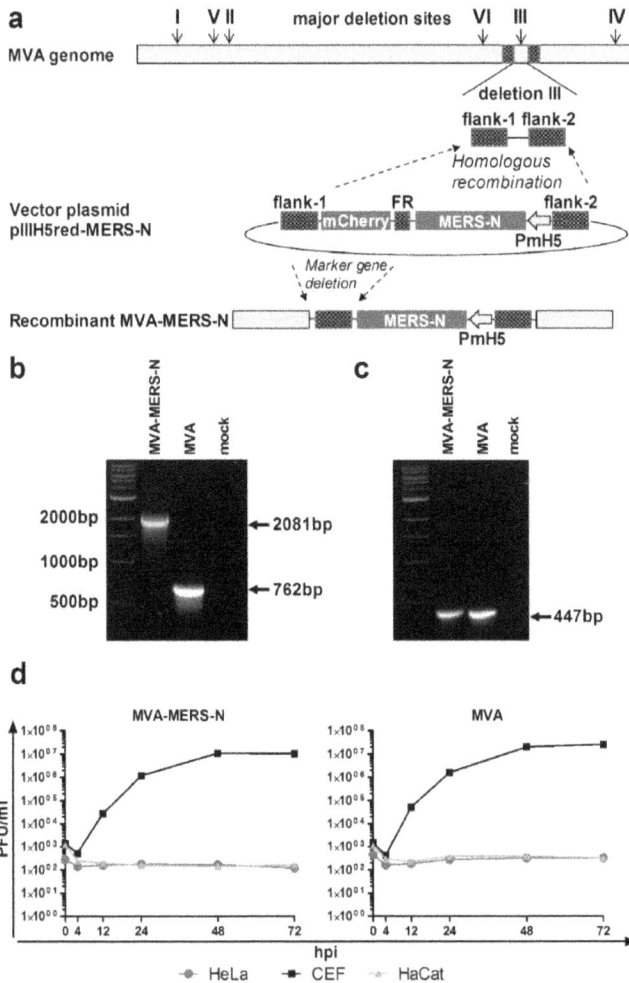

Figure 1. Generation and characterization of recombinant Modified Vaccinia virus Ankara expressing the Middle East respiratory syndrome coronavirus N protein (MVA-MERS-N); (**a**) Schematic diagram of the MVA genome indicating the major deletion sites I–VI on the top. Flank-1 and flank-2 refer to MVA DNA sequences adjacent to corresponding insertion site. Deletion III was used to insert MERS-N encoding gene sequences under the transcriptional control of the vaccinia virus promoter PmH5. Repetitive sequences (FR) were designed to remove the mCherry marker by intragenomic homologous recombination (marker gene deletion); (**b**,**c**) PCR analyses of genomic viral DNA using oligonucleotide primers to confirm the correct insertion of recombinant MERS-N gene into deletion III (**b**), and the genetic integrity of the MVA genome for the C7L gene locus (**c**); (**d**) Multi-step growth analysis of recombinant MVA-MERS-N and non-recombinant MVA (MVA); Chicken embryo fibroblasts (CEF) and human HaCat or HeLa cells were infected at a multiplicity of infection (MOI) of 0.05 with MVA-MERS-N or MVA. Infected cells were collected at different time points after infection and titrated on CEF cells.

To confirm the synthesis of MERS-CoV N protein upon MVA-MERS-N infection, total cell proteins from infected CEF and HaCat cells were separated by SDS-PAGE and analyzed by immunoblotting (Figure 2). Consistent with the expected molecular mass of the MERS-CoV N protein we readily detected a ~45 kDa polypeptide using the N-specific mouse monoclonal antibody. At 24 hours post infection (hpi) a prominent band of N protein was visible in the lysates from both cell lines suggesting efficient synthesis of the recombinant protein under permissive and non-permissive growth conditions for MVA-MERS-N (Figure 2a).

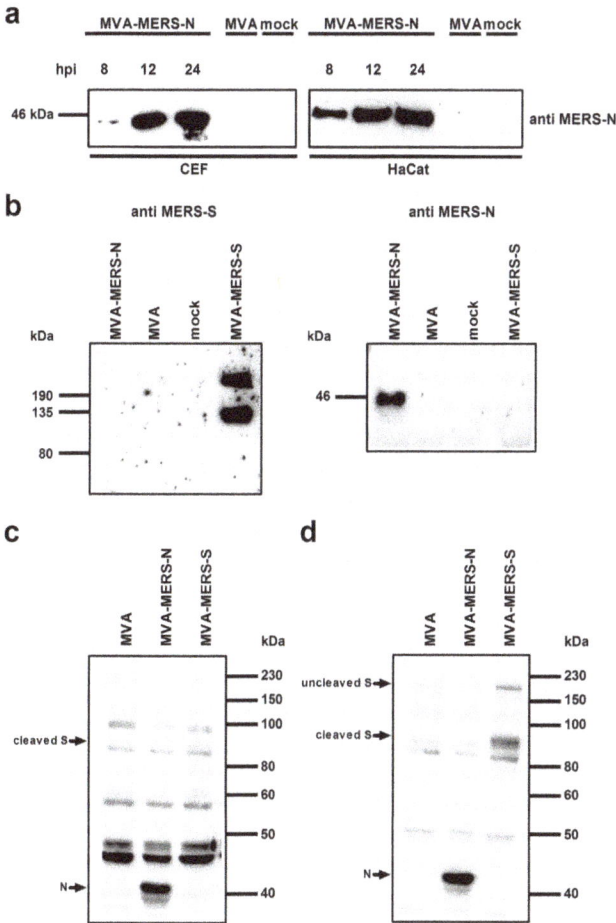

Figure 2. Analysis of recombinant MVA-MERS proteins; (**a**) Western Blot analysis of MERS-CoV N protein produced in CEF or HaCat cells. Lysates from cells infected with recombinant MVA (MVA-MERS-N, MVA-MERS-S) or non-recombinant MVA (MVA) at a MOI of five, or from non-infected cells (mock) were prepared at eight, 12, or 24 hpi. Proteins were analyzed by immunoblotting with a monoclonal anti-MERS-N antibody; (**b**–**d**) Western Blot analysis of MERS-CoV N and S proteins produced in CEF. Total cell extracts from CEF infected with recombinant MVA (MVA-MERS-N, MVA-MERS-S) or non-recombinant MVA (MVA) at a MOI of five, or from non-infected cells (mock) were prepared at 24 hpi. Cell lysates and proteins were tested by immunoblotting using monoclonal anti MERS-N and anti MERS-S antibody (**b**) or polyclonal sera from MERS-CoV infected rabbits (**c**) or cynomolgus macaques (**d**). Arrows indicate the N- or S-specific protein bands.

In addition, the comparative Western blot analysis of cell lysates from MVA-MERS-N or MVA-MERS-S infected CEF with antigen-specific mouse monoclonal antibodies suggested the production of comparable amounts of both MERS-CoV candidate antigens (Figure 2b). This observation is in line with the fact that the MVA-MERS-S candidate vector vaccine expresses the MERS-CoV S gene sequences using the identical PmH5 promoter system [23]. As shown in previous studies, we detected two MERS-CoV S-specific protein bands upon infection with MVA-MERS-S indicating the authentic proteolytic cleavage of the full-length S glycoprotein (~210 kDa) into an N-terminal (~120 kDa S1 domain) and a C-terminal (~85 kDa S2 domain; not detected) subunit [23,50,51]. Following this, we used the total protein lysates from MVA-MERS-N or MVA-MERS-S infected CEF to assess the recognition of the MERS-CoV N and S antigens by sera from experimentally MERS-CoV infected animals. The Western blot analysis of sera from an infected rabbit (Figure 2c) or a cynomolgus monkey (Figure 2d) revealed the presence of antibodies specific for the MERS-CoV N protein. The recognition of the MERS-CoV N protein was at least as prominent as the MERS-CoV S antigen, which was suggestive of the induction of substantial N-specific antibody responses after experimental MERS-CoV infections.

3.2. Characterization of MERS-CoV N-specific T Cell Responses

3.2.1. Initial Screen of MERS-CoV N Epitopes Using Overlapping Peptide Pools

T cell responses against coronaviruses are known to be long lived and mostly target the more conserved CoV internal structural N protein. However, information on MERS-CoV N antigen-specific T cell specificities is still limited. Thus, we aimed first to identify N polypeptide-specific T cell epitopes in BALB/c mice immunized twice with recombinant virus MVA-MERS-N or non-recombinant MVA as a control via the intraperitoneal and intramuscular routes. Eight days after the final immunization, splenocytes were prepared and the purified CD4+ and CD8+ T cells were restimulated in vitro with overlapping peptides corresponding to the N protein. Overlapping peptides were pooled using a two-dimensional, pooled-peptide matrix system (Table S1) and screened by IFN-γ ELISPOT. The stimulation of splenoctyes from MVA-MERS-N immunized mice with the peptides from 16 out of the total 18 peptide pools did not result in the detection of IFN-γ producing T cells above background numbers obtained with splenocytes from mock or MVA-control vaccinated animals. Stimulation with the peptides from pools H8 (n = 10) and V8 (n = 10) as well as the use of the vaccinia virus positive peptide F2L [49] (data not shown) showed elevated numbers of IFN-γ spot forming cells (SFC) in CD8+ T cell cultures (Figure 3a,b). MVA-MERS-N immunizations given by i.p. and i.m. routes resulted in comparable T cell stimulatory capacities of overlapping N-specific peptides from pools V8 and H8. In contrast, peptides from other pools showed no or only minor stimulatory activities, as exemplified for peptides in pools V4 and V6.

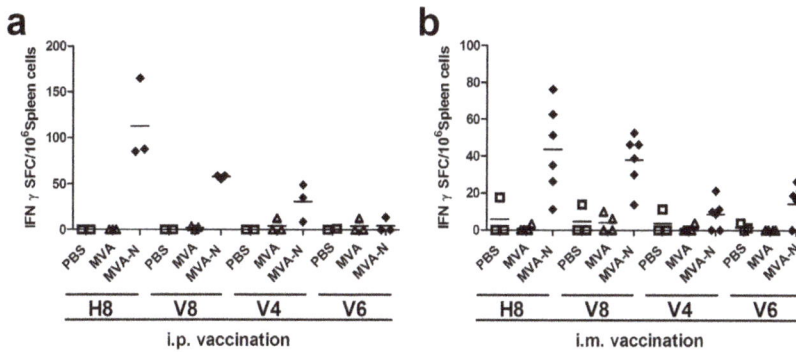

Figure 3. Screening for H2-d restricted T cell epitopes in MERS-CoV N protein using matrix peptide pools; (**a–b**) groups of BALB/c mice (*n* = 2 to 6) were vaccinated twice (21-day interval) by i.p. (**a**) or i.m. (**b**) application with 10^8 plaque-forming-units (PFU) of recombinant MVA-MERS-N (MVA-N). Mice inoculated with non-recombinant MVA (MVA) or phosphate-buffered saline (PBS) were used as controls. Splenocytes were restimulated in vitro with pools of overlapping peptides corresponding to MERS-CoV N protein. IFN-γ spot-forming CD8+ T cells (IFN-γ SFC) were measured by ELISPOT. The lines represent means.

3.2.2. Reassessment of Positive Reacting Peptides Pools for MERS-CoV N T Cell Epitopes

Following this, the peptides within the V8 and H8 peptide pools were used to elucidate in more detail the T cell epitope specificities. We subdivided the peptides from H8 and V8 in four new pools each containing five peptides (H8.1, H8.2, V8.1, V8.2). In addition, we separately tested the two 15-mer peptides #89 and #90, which were shared between pools H8 and V8. We again vaccinated BALB/c mice with MVA-MERS-N using i.p. or i.m. inoculations in a prime-boost regime. Splenocytes were prepared at day eight after the last vaccination and purified CD8+ T cells were restimulated with subpools V8.1., V8.2., H8.1, and H8.2 (Figure 4a). Stimulation with peptides from pools V8.1 or H8.1 activated only minor levels of IFN-γ producing cells with mean levels about 23 SFC/10^6 splenocytes. Yet, the stimulation with subpools V8.2 and H8.2 revealed clearly higher numbers of activated T cells with 83-176 IFN-γ SFC/10^6 splenocytes. Comparable numbers of IFN-γ producing cells were again induced by i.p. or i.m. immunization. Of note, the stimulations with the 15-mer peptides #89 ($N_{353-367}$ = QNIDAYKTFPKKEKK) or #90 ($N_{357-371}$ = AYKTFPKKEKKQKAP) alone resulted in detection of substantial quantities of IFN-γ producing cells (mean levels about 71–107 SFC/10^6 splenocytes) in mice that had been vaccinated with MVA-MERS-N by both immunization routes (Figure 4b). CD8+ T cells purified from mice receiving non-recombinant MVA or mock vaccine (PBS) did not produce IFN-γ following stimulation with peptides from subpools V8.1-H8.2 and with peptides #89 and #90. When checking for the specific peptides contained within the subpools, we observed that the strongly stimulatory peptides #89 and #90 were part of the subpools V8.2 and H8.2, whereas these peptides were absent in V8.1 and H8.1. This data suggested that the overlapping 15-mer peptides #89 and #90 contained a valuable antigen epitope for the activation of MERS-CoV N-specific CD8+ T cell responses.

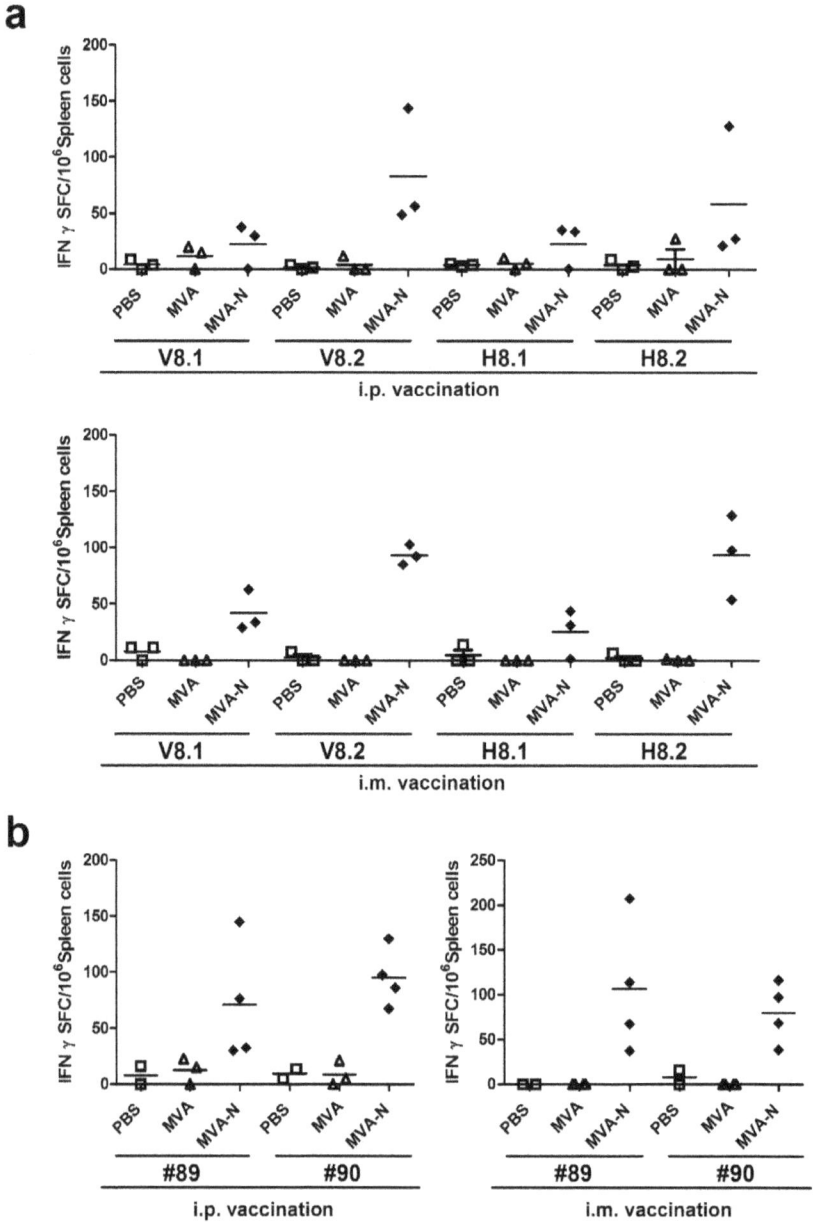

Figure 4. Mapping of H2-d restricted T cell epitopes in MERS-CoV N protein; (**a–b**) BALB/c mice (n = 2 to 4) were immunized twice (21-day interval) i.p. or i.m. with 10^8 PFU of recombinant MVA-MERS-N (MVA-N), non-recombinant MVA (MVA) or PBS. Splenocytes from vaccinated mice were incubated in the presence of subpools (V8.1, V8.2, H8.1, H8.2) from positive matrix pools (**a**) or individual 15-mers peptides #89 or #90 (**b**). IFN-γ spot-forming CD8+ T cells (IFN-γ SFC) were quantified by ELISPOT. The lines represent means.

3.2.3. Identification of MERS-CoV N-specific T Cell Epitope

To map more precisely this specific epitope within the MERS-CoV N protein, we concentrated on the overlapping 11-mer peptide shared between peptides #89 and #90 and obtained nine 8-10-mer peptides (Table 1).

Table 1. Peptide information.[1]

Peptide-ID	Sequence	Position
#89	QNID**AYKTFPKKEKK**	$N_{353-367}$
#90	**AYKTFPKKEKK**QKAP	$N_{357-371}$
11	AYKTFPKKEKK	$N_{357-367}$
10.1	AYKTFPKKEK	$N_{357-366}$
10.2	YKTFPKKEKK	$N_{358-367}$
9.1	AYKTFPKKE	$N_{357-365}$
9.2	YKTFPKKEK	$N_{358-366}$
9.3	KTFPKKEKK	$N_{359-367}$
8.1	AYKTFPKK	$N_{357-364}$
8.2	YKTFPKKE	$N_{358-365}$
8.3	KTFPKKEK	$N_{359-366}$
8.4	TFPKKEKK	$N_{360-367}$

[1] The common 11 amino acid sequence between positive 15-mers #89 and #90 was truncated into 8-10-mer peptides and tested by ELISPOT and ICS assay.

Using these peptides for the stimulation of splenocytes from MVA-MERS-N vaccinated mice, we obtained the highest numbers of IFN-γ producing T cells with peptide 10.2 (mean levels of 94 to 97 SFC/10^6 splenocytes), while the other peptides (10.1, 9.1, 9.2, 9.3, 8.1, 8.2, 8.3, 8.4) induced weaker responses with a mean of 5–76 IFN-γ SFC/10^6 splenocytes (Figure 5a,b).

Furthermore, we tested the 10.2 peptide to monitor N-specific T cell responses by IFN-γ ICS and fluorescence activated cell sorting (FACS) analysis. Indeed, we could detect significant numbers of 10.2 peptide-specific CD8+ T cells being induced and activated by the MVA-MERS-N prime-boost vaccination. In comparison, the CD4+ T cell populations from splenocytes of immunized animals demonstrated only background levels of IFN-γ producing cells (Figure 5c). The vaccinia virus-specific immunodominant CD8+ T cell determinant F2L [49] served as control peptide for the detection of MVA-specific CD8+ T cells (Figure 5d).

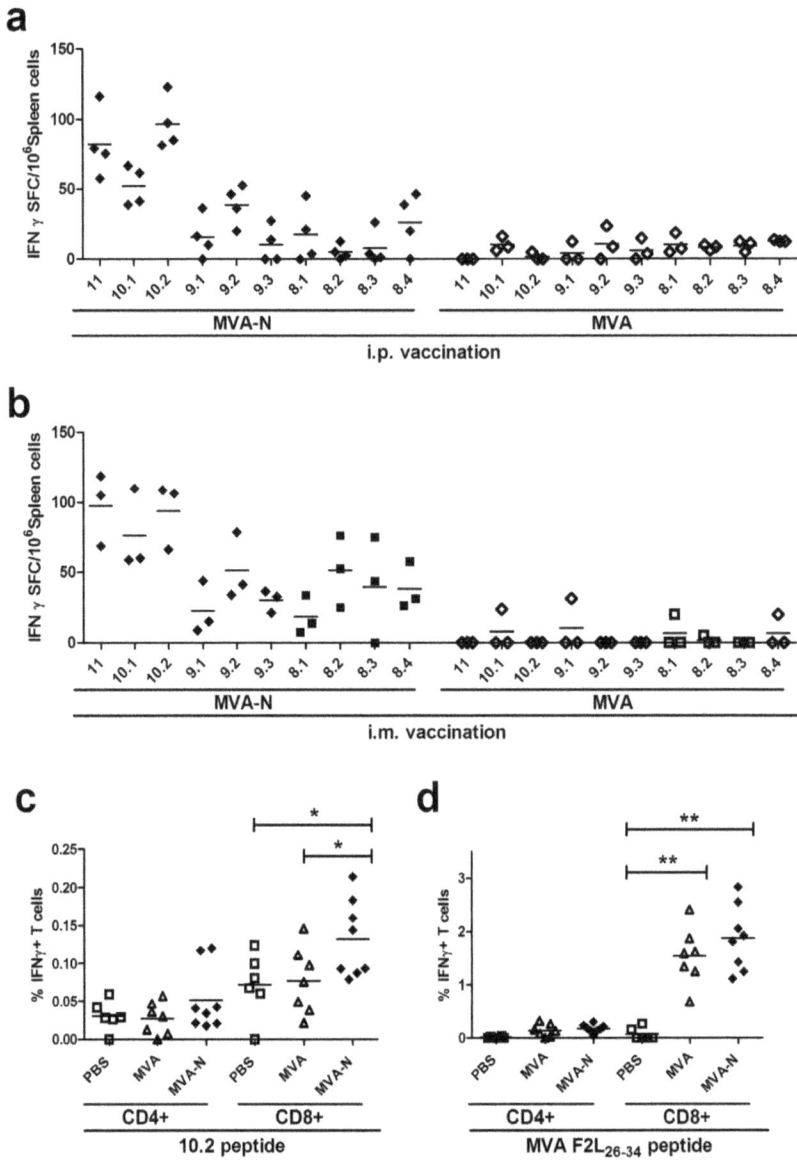

Figure 5. Identification of an H2-d restricted T cell epitope in MERS-CoV N protein; (**a–d**) Groups of BALB/c mice (*n* = 3 to 8) were vaccinated in a prime-boost regime with 10^8 PFU of MVA-MERS-N via i.p. (**a**) or i.m. (**b–d**) application. Mice immunized with non-recombinant MVA (MVA) and PBS served as negative controls. (**a-b**) Splenocytes were stimulated with individual 8-11-mer peptides and IFN-γ spot-forming CD8+ T cells (IFN-γ SFC) were measured by ELISPOT. (**c–d**) Splenocytes were stimulated with positive MERS-CoV N 10.2 peptide (**c**) or F2L$_{26-34}$ peptide (**d**) and IFN-γ producing CD8+ or CD4+ T cells were measured using intracellular cytokine staining assay and FACS analysis. The lines represent means. *< 0.05, **< 0.005.

4. Discussion

The availability of appropriate MERS-CoV-specific immune monitoring tools is a prerequisite for the successful development of vaccines and therapeutic approaches. The development of these tools is hampered by the fact that we still know little about the relevant viral antigens and the overall pathogenesis of the MERS-CoV infection. Recent studies describe a number of cases with asymptomatic MERS-CoV infection in humans and raise questions as to which factors influence the clinical manifestation of MERS [52,53]. Since asymptomatic or mild clinical manifestations of MERS-CoV infection are often associated with low levels of seropositivity, the analysis of MERS-CoV-specific cellular immune responses may facilitate further insight into the immune correlates of disease prevention. Moreover, in order to characterize the pathogenesis of MERS-CoV infection, it will be indispensable to monitor the role of virus-specific T cells in animal models of MERS and to precisely identify the antigen specificities of these T cell responses.

In the present study, we identified a major histocompatibility complex (MHC) haplotype H2-d restricted peptide epitope in the MERS-CoV N protein by stimulating T cells from MVA-MERS-N vaccinated BALB/c mice with a 2-D matrix pool of overlapping peptides. These mice have already been used in various preclinical studies to establish the MERS-CoV S protein as an important vaccine antigen for induction of virus neutralizing antibodies [23,40–42,54]. Moreover, BALB/c mice transduced with the human cell surface receptor dipeptidyl peptidase 4 (hDPP4) using an adenovirus vector are susceptible to productive MERS-CoV lung infection, which allows for the testing of the protective efficacy of MERS-CoV-specific immunization using the MERS-CoV S protein [24,40,41,43,44]. Here, we wished to specifically assess the suitability of an MVA-delivered MERS-CoV N antigen for the activation of cellular immune responses in mice. In general, the N protein is a well conserved internal protein and the major structural component incorporating the viral RNA within the viral nucleocapsid [55]. In previous studies, the SARS-CoV N protein has also been used as candidate antigen for vaccine development [56], and an experimental DNA vaccine efficiently induced SARS-CoV N-specific cellular immunity [57,58]. In line with this data, recent studies in MERS patients demonstrated that both antibody and T cell responses are associated with recovery from MERS-CoV infection [59].

The recombinant virus MVA-MERS-N produced stable amounts of MERS-CoV N antigen upon in vitro infection of human cells indicating the unimpaired expression of the target gene at the level of viral late transcription using the synthetic vaccinia virus-specific promoter PmH5 [46]. Moreover, the MERS-CoV N antigen produced in MVA-MERS-N infected cells was strongly recognized by antibodies from experimentally infected laboratory animals suggesting that N-specific immune responses were potently activated upon MERS-CoV infection. It seems noteworthy that MERS-CoV productively replicates in rabbits, but viral loads are low and the animals develop no overt disease symptoms. However, Haagmans et al. found infectious virus in the lung tissues of the rabbits and revealed the presence of the MERS-CoV N antigen in bronchiolar epithelial cells and in the epithelial cells of the nose [60]. The localization of N in these respiratory epithelial cells may result in an efficient recognition by innate and also adaptive immune cells similar to those described for other viruses inducing robust protective immunity [61]. This might be a possible explanation for efficient activation of MERS-CoV N-specific antibodies despite a barely productive MERS-CoV infection. Similar outcomes of infection were observed upon MERS-CoV infection in cynomolgus macaques and other relevant non-human primate models [62,63]. Thus, the induction of N-specific immune responses in these animals emphasizes the potential usefulness of the MERS-CoV N protein to serve as vaccine antigen.

Nevertheless, the immunogenicity of N requires further characterization in preclinical models for MERS-CoV infection. In addition to the induction of MERS-CoV-specific antibodies, the MERS-CoV N protein holds promise to efficiently activate virus-specific CD8+ T cell responses. For more detailed studies characterizing the possible role of these T cell specificities in MERS-CoV-associated immunity or pathogenesis it is highly relevant to determine the N peptide epitopes allowing for the appropriate MHC-restricted antigen presentation and the activation of virus-specific T cells. In this study, we identified a new H2-d restricted CD8+ T cell epitope in the MERS-CoV N protein using a 2-D matrix and pools of 84 overlapping 15-mer N peptides. First, we have identified two MERS-CoV N derived peptides ($N_{353-367}$ = QNIDAYKTFPKKEKK and $N_{357-371}$ = AYKTFPKKEKKQKAP) and further mapped these 15-mer peptides to the minimal aa sequence of $N_{358-367}$ = YKTFPKKEKK representing a decamer peptide epitope (Figure S1) [64]. Analysis of MERS-CoV sequences reveals that $N_{358-367}$ is conserved among different strains of MERS-CoV (Table S2) [65]. The availability of such an epitope may allow for more detailed experimental monitoring of cellular immune responses induced by a MVA based candidate vaccine against MERS-CoV in the mouse model and potentially also in other preclinical models. Of note, the H2-d restricted CD8+ T cell epitope enables characterization of T cell responses in BALB/c mice that serve as a well-established MERS-CoV infection model following adenovirus vector mediated transduction with hDPP4 [24,40,41,43,44]. A particular feature of MERS in humans, as observed upon the investigation of cluster outbreaks in hospitals, is the lack of detectable MERS-CoV neutralizing antibodies in patients with confirmed disease [66,67]. This observation is attributed to the emergence of specific virus mutants evading the neutralizing antibody response, as already described for SARS-CoV [68,69]. Thus, future use of a T cell-specific immune monitoring might contribute to a more detailed understanding of MERS pathogenesis. Here, studies on the function of MERS-CoV-specific CD8+ T cells in this BALB/c mouse MERS-CoV lung infection model will be helpful to better estimate the role of cellular immunity in vaccine mediated protection in MERS-CoV infection.

Finally, the MVA-MERS-N vector virus generated for this study proved to be a stable recombinant virus that can be readily amplified to obtain vaccine preparations technically fulfilling all requirements for further preclinical or even clinical development. Future work with MVA-MERS-N candidate vaccines should help to elucidate the potential protective capacity of N-specific immune responses in MERS-CoV infections models and contribute to our better understanding of MERS vaccine-induced protection.

Supplementary Materials: The following are available online at http://www.mdpi.com/1999-4915/10/12/718/s1, Table S1: The design of matrix peptide pools for systematically screening of H2-d restricted T cell epitopes in MERS-CoV N protein protein, Figure S1: Sequence analysis and modular organization of MERS-CoV N protein, Table S2: Comparative analysis of MERS-CoV $N_{358-367}$ epitope in different MERS-CoV strains.

Author Contributions: S.V., R.F., A.V. and G.S. conceived and designed the experiments; S.V. and S.J. performed the experiments; S.V. and A.V. analyzed the data; S.V., A.V. and G.S. wrote the paper.

Funding: This work was supported by the Federal Ministry of Education and Research (BMBF), grant numbers DZIF TTU 01.802 and RAPID 01KI1723C.

Acknowledgments: We thank Fei Song for expert help with the construction of first generation recombinant MVA-MERS viruses and Georgia Kalodimou for proficient proofreading of the manuscript. Ursula Klostermeier, Patrizia Bonert, Johannes Döring and Axel Groß provided valuable help in animal studies.

Conflicts of Interest: The authors declare no conflict of interest.

References

1. Zaki, A.M.; van Boheemen, S.; Bestebroer, T.M.; Osterhaus, A.D.; Fouchier, R.A. Isolation of a novel coronavirus from a man with pneumonia in Saudi Arabia. *N. Engl. J. Med.* **2012**, *367*, 1814–1820. [CrossRef] [PubMed]

2. WHO. Middle East respiratory Syndrome Coronavirus (MERS-CoV). Available online: http://www.who.int/emergencies/mers-cov/en/ (accessed on 22 November 2018).

3. Park, J.E.; Jung, S.; Kim, A.; Park, J.E. MERS transmission and risk factors: A systematic review. *BMC Public Health* **2018**, *18*, 574. [CrossRef] [PubMed]

4. Alraddadi, B.M.; Watson, J.T.; Almarashi, A.; Abedi, G.R.; Turkistani, A.; Sadran, M.; Housa, A.; Almazroa, M.A.; Alraihan, N.; Banjar, A.; et al. Risk Factors for Primary Middle East Respiratory Syndrome Coronavirus Illness in Humans, Saudi Arabia, 2014. *Emerg. Infect. Dis.* **2016**, *22*, 49–55. [CrossRef] [PubMed]

5. Alshukairi, A.N.; Zheng, J.; Zhao, J.; Nehdi, A.; Baharoon, S.A.; Layqah, L.; Bokhari, A.; Al Johani, S.M.; Samman, N.; Boudjelal, M.; et al. High Prevalence of MERS-CoV Infection in Camel Workers in Saudi Arabia. *mBio* **2018**, *9*, e01985-18. [CrossRef] [PubMed]

6. Assiri, A.; McGeer, A.; Perl, T.M.; Price, C.S.; Al Rabeeah, A.A.; Cummings, D.A.; Alabdullatif, Z.N.; Assad, M.; Almulhim, A.; Makhdoom, H.; et al. Hospital outbreak of Middle East respiratory syndrome coronavirus. *N. Engl. J. Med.* **2013**, *369*, 407–416. [CrossRef] [PubMed]

7. Ki, M. 2015 MERS outbreak in Korea: Hospital-to-hospital transmission. *Epidemiol. Health* **2015**, *37*, e2015033. [CrossRef] [PubMed]

8. Al-Tawfiq, J.A.; Auwaerter, P.G. Healthcare-associated infections: The hallmark of Middle East respiratory syndrome coronavirus with review of the literature. *J. Hosp. Infect.* **2018**. [CrossRef]

9. Oh, M.-D.; Park, W.B.; Park, S.-W.; Choe, P.G.; Bang, J.H.; Song, K.-H.; Kim, E.S.; Kim, H.B.; Kim, N.J. Middle East respiratory syndrome: What we learned from the 2015 outbreak in the Republic of Korea. *Korean J. Int. Med.* **2018**, *33*, 233–246. [CrossRef]

10. Kraaij-Dirkzwager, M.; Timen, A.; Dirksen, K.; Gelinck, L.; Leyten, E.; Groeneveld, P.; Jansen, C.; Jonges, M.; Raj, S.; Thurkow, I.; et al. Middle East respiratory syndrome coronavirus (MERS-CoV) infections in two returning travellers in the Netherlands, May 2014. *Eurosurveillance* **2014**, *19*, 20817. [CrossRef]

11. Assiri, A.; Al-Tawfiq, J.A.; Al-Rabeeah, A.A.; Al-Rabiah, F.A.; Al-Hajjar, S.; Al-Barrak, A.; Flemban, H.; Al-Nassir, W.N.; Balkhy, H.H.; Al-Hakeem, R.F.; et al. Epidemiological, demographic, and clinical characteristics of 47 cases of Middle East respiratory syndrome coronavirus disease from Saudi Arabia: A descriptive study. *Lancet Infect. Dis.* **2013**, *13*, 752–761. [CrossRef]

12. Memish, Z.A.; Cotten, M.; Meyer, B.; Watson, S.J.; Alsahafi, A.J.; Al Rabeeah, A.A.; Corman, V.M.; Sieberg, A.; Makhdoom, H.Q.; Assiri, A.; et al. Human infection with MERS coronavirus after exposure to infected camels, Saudi Arabia, 2013. *Emerg. Infect. Dis.* **2014**, *20*, 1012–1015. [CrossRef] [PubMed]

13. Al-Tawfiq, J.A.; Memish, Z.A. Middle East respiratory syndrome coronavirus: Epidemiology and disease control measures. *Infect. Drug Resist.* **2014**, *7*, 281–287. [CrossRef] [PubMed]

14. Rottingen, J.A.; Gouglas, D.; Feinberg, M.; Plotkin, S.; Raghavan, K.V.; Witty, A.; Draghia-Akli, R.; Stoffels, P.; Piot, P. New Vaccines against Epidemic Infectious Diseases. *N. Engl. J. Med.* **2017**, *376*, 610–613. [CrossRef] [PubMed]

15. Cho, H.; Excler, J.L.; Kim, J.H.; Yoon, I.K. Development of Middle East Respiratory Syndrome Coronavirus vaccines—Advances and challenges. *Hum. Vaccin. Immunother.* **2018**, *14*, 304–313. [CrossRef] [PubMed]

16. Volz, A.; Sutter, G. Modified Vaccinia Virus Ankara: History, Value in Basic Research, and Current Perspectives for Vaccine Development. *Adv. Virus Res.* **2017**, *97*, 187–243. [CrossRef] [PubMed]

17. Kolb, A.F.; Pewe, L.; Webster, J.; Perlman, S.; Whitelaw, C.B.; Siddell, S.G. Virus-neutralizing monoclonal antibody expressed in milk of transgenic mice provides full protection against virus-induced encephalitis. *J. Virol.* **2001**, *75*, 2803–2809. [CrossRef] [PubMed]

18. Sune, C.; Smerdou, C.; Anton, I.M.; Abril, P.; Plana, J.; Enjuanes, L. A conserved coronavirus epitope, critical in virus neutralization, mimicked by internal-image monoclonal anti-idiotypic antibodies. *J. Virol.* **1991**, *65*, 6979–6984.

19. Bisht, H.; Roberts, A.; Vogel, L.; Bukreyev, A.; Collins, P.L.; Murphy, B.R.; Subbarao, K.; Moss, B. Severe acute respiratory syndrome coronavirus spike protein expressed by attenuated vaccinia virus protectively immunizes mice. *Proc. Natl. Acad. Sci. USA* **2004**, *101*, 6641–6646. [CrossRef]

20. Yang, Z.Y.; Kong, W.P.; Huang, Y.; Roberts, A.; Murphy, B.R.; Subbarao, K.; Nabel, G.J. A DNA vaccine induces SARS coronavirus neutralization and protective immunity in mice. *Nature* **2004**, *428*, 561–564. [CrossRef]

21. Malczyk, A.H.; Kupke, A.; Prufer, S.; Scheuplein, V.A.; Hutzler, S.; Kreuz, D.; Beissert, T.; Bauer, S.; Hubich-Rau, S.; Tondera, C.; et al. A Highly Immunogenic and Protective Middle East Respiratory Syndrome Coronavirus Vaccine Based on a Recombinant Measles Virus Vaccine Platform. *J. Virol.* **2015**, *89*, 11654–11667. [CrossRef]

22. Muthumani, K.; Falzarano, D.; Reuschel, E.L.; Tingey, C.; Flingai, S.; Villarreal, D.O.; Wise, M.; Patel, A.; Izmirly, A.; Aljuaid, A.; et al. A synthetic consensus anti-spike protein DNA vaccine induces protective immunity against Middle East respiratory syndrome coronavirus in nonhuman primates. *Sci. Transl. Med.* **2015**, *7*, 301ra132. [CrossRef]

23. Song, F.; Fux, R.; Provacia, L.B.; Volz, A.; Eickmann, M.; Becker, S.; Osterhaus, A.D.; Haagmans, B.L.; Sutter, G. Middle East respiratory syndrome coronavirus spike protein delivered by modified vaccinia virus Ankara efficiently induces virus-neutralizing antibodies. *J. Virol.* **2013**, *87*, 11950–11954. [CrossRef] [PubMed]

24. Volz, A.; Kupke, A.; Song, F.; Jany, S.; Fux, R.; Shams-Eldin, H.; Schmidt, J.; Becker, C.; Eickmann, M.; Becker, S.; et al. Protective Efficacy of Recombinant Modified Vaccinia Virus Ankara Delivering Middle East Respiratory Syndrome Coronavirus Spike Glycoprotein. *J. Virol.* **2015**, *89*, 8651–8656. [CrossRef]

25. Wirblich, C.; Coleman, C.M.; Kurup, D.; Abraham, T.S.; Bernbaum, J.G.; Jahrling, P.B.; Hensley, L.E.; Johnson, R.F.; Frieman, M.B.; Schnell, M.J. One-Health: A Safe, Efficient, Dual-Use Vaccine for Humans and Animals against Middle East Respiratory Syndrome Coronavirus and Rabies Virus. *J. Virol.* **2017**, *91*, e02040-16. [CrossRef]

26. Lau, S.K.; Woo, P.C.; Wong, B.H.; Tsoi, H.W.; Woo, G.K.; Poon, R.W.; Chan, K.H.; Wei, W.I.; Peiris, J.S.; Yuen, K.Y. Detection of severe acute respiratory syndrome (SARS) coronavirus nucleocapsid protein in sars patients by enzyme-linked immunosorbent assay. *J. Clin. Microbiol.* **2004**, *42*, 2884–2889. [CrossRef] [PubMed]

27. Timani, K.A.; Ye, L.; Ye, L.; Zhu, Y.; Wu, Z.; Gong, Z. Cloning, sequencing, expression, and purification of SARS-associated coronavirus nucleocapsid protein for serodiagnosis of SARS. *J. Clin. Virol.* **2004**, *30*, 309–312. [CrossRef] [PubMed]

28. Chen, Y.; Chan, K.H.; Kang, Y.; Chen, H.; Luk, H.K.; Poon, R.W.; Chan, J.F.; Yuen, K.Y.; Xia, N.; Lau, S.K.; et al. A sensitive and specific antigen detection assay for Middle East respiratory syndrome coronavirus. *Emerg. Microbes Infect.* **2015**, *4*, e26. [CrossRef]

29. He, Q.; Du, Q.; Lau, S.; Manopo, I.; Lu, L.; Chan, S.W.; Fenner, B.J.; Kwang, J. Characterization of monoclonal antibody against SARS coronavirus nucleocapsid antigen and development of an antigen capture ELISA. *J. Virol. Methods* **2005**, *127*, 46–53. [CrossRef]

30. Yamaoka, Y.; Matsuyama, S.; Fukushi, S.; Matsunaga, S.; Matsushima, Y.; Kuroyama, H.; Kimura, H.; Takeda, M.; Chimuro, T.; Ryo, A. Development of Monoclonal Antibody and Diagnostic Test for Middle East Respiratory Syndrome Coronavirus Using Cell-Free Synthesized Nucleocapsid Antigen. *Front. Microbiol.* **2016**, *7*, 509. [CrossRef]

31. McBride, R.; van Zyl, M.; Fielding, B.C. The coronavirus nucleocapsid is a multifunctional protein. *Viruses* **2014**, *6*, 2991–3018. [CrossRef]

32. Hsin, W.-C.; Chang, C.-H.; Chang, C.-Y.; Peng, W.-H.; Chien, C.-L.; Chang, M.-F.; Chang, S.C. Nucleocapsid protein-dependent assembly of the RNA packaging signal of Middle East respiratory syndrome coronavirus. *J. Biomed. Sci.* **2018**, *25*, 47. [CrossRef] [PubMed]

33. Almazan, F.; Galan, C.; Enjuanes, L. The nucleoprotein is required for efficient coronavirus genome replication. *J. Virol.* **2004**, *78*, 12683–12688. [CrossRef] [PubMed]

34. Zuniga, S.; Cruz, J.L.; Sola, I.; Mateos-Gomez, P.A.; Palacio, L.; Enjuanes, L. Coronavirus nucleocapsid protein facilitates template switching and is required for efficient transcription. *J. Virol.* **2010**, *84*, 2169–2175. [CrossRef]

35. Spiegel, M.; Pichlmair, A.; Martinez-Sobrido, L.; Cros, J.; Garcia-Sastre, A.; Haller, O.; Weber, F. Inhibition of Beta interferon induction by severe acute respiratory syndrome coronavirus suggests a two-step model for activation of interferon regulatory factor 3. *J. Virol.* **2005**, *79*, 2079–2086. [CrossRef] [PubMed]

36. Kopecky-Bromberg, S.A.; Martinez-Sobrido, L.; Frieman, M.; Baric, R.A.; Palese, P. Severe acute respiratory syndrome coronavirus open reading frame (ORF) 3b, ORF 6, and nucleocapsid proteins function as interferon antagonists. *J. Virol.* **2007**, *81*, 548–557. [CrossRef] [PubMed]

37. Lu, X.; Pan, J.; Tao, J.; Guo, D. SARS-CoV nucleocapsid protein antagonizes IFN-beta response by targeting initial step of IFN-beta induction pathway, and its C-terminal region is critical for the antagonism. *Virus Genes* **2011**, *42*, 37–45. [CrossRef] [PubMed]

38. Aboagye, J.O.; Yew, C.W.; Ng, O.W.; Monteil, V.M.; Mirazimi, A.; Tan, Y.J. Overexpression of the nucleocapsid protein of Middle East respiratory syndrome coronavirus up-regulates CXCL10. *Biosci. Rep.* **2018**, *38*, BSR20181059. [CrossRef]

39. Tang, F.; Quan, Y.; Xin, Z.T.; Wrammert, J.; Ma, M.J.; Lv, H.; Wang, T.B.; Yang, H.; Richardus, J.H.; Liu, W.; et al. Lack of peripheral memory B cell responses in recovered patients with severe acute respiratory syndrome: A six-year follow-up study. *J. Immunol.* **2011**, *186*, 7264–7268. [CrossRef]

40. Chi, H.; Zheng, X.; Wang, X.; Wang, C.; Wang, H.; Gai, W.; Perlman, S.; Yang, S.; Zhao, J.; Xia, X. DNA vaccine encoding Middle East respiratory syndrome coronavirus S1 protein induces protective immune responses in mice. *Vaccine* **2017**, *35*, 2069–2075. [CrossRef]

41. Coleman, C.M.; Venkataraman, T.; Liu, Y.V.; Glenn, G.M.; Smith, G.E.; Flyer, D.C.; Frieman, M.B. MERS-CoV spike nanoparticles protect mice from MERS-CoV infection. *Vaccine* **2017**, *35*, 1586–1589. [CrossRef]

42. Jung, S.Y.; Kang, K.W.; Lee, E.Y.; Seo, D.W.; Kim, H.L.; Kim, H.; Kwon, T.; Park, H.L.; Kim, H.; Lee, S.M.; et al. Heterologous prime-boost vaccination with adenoviral vector and protein nanoparticles induces both Th1 and Th2 responses against Middle East respiratory syndrome coronavirus. *Vaccine* **2018**, *36*, 3468–3476. [CrossRef] [PubMed]

43. Liu, W.J.; Lan, J.; Liu, K.; Deng, Y.; Yao, Y.; Wu, S.; Chen, H.; Bao, L.; Zhang, H.; Zhao, M.; et al. Protective T Cell Responses Featured by Concordant Recognition of Middle East Respiratory Syndrome Coronavirus-Derived CD8+ T Cell Epitopes and Host MHC. *J. Immunol.* **2017**, *198*, 873–882. [CrossRef] [PubMed]

44. Zhao, J.; Li, K.; Wohlford-Lenane, C.; Agnihothram, S.S.; Fett, C.; Zhao, J.; Gale, M.J., Jr.; Baric, R.S.; Enjuanes, L.; Gallagher, T.; et al. Rapid generation of a mouse model for Middle East respiratory syndrome. *Proc. Natl. Acad. Sci. USA* **2014**, *111*, 4970–4975. [CrossRef] [PubMed]

45. Kremer, M.; Volz, A.; Kreijtz, J.H.; Fux, R.; Lehmann, M.H.; Sutter, G. Easy and efficient protocols for working with recombinant vaccinia virus MVA. *Methods Mol. Biol.* **2012**, *890*, 59–92. [CrossRef]

46. Wyatt, L.S.; Shors, S.T.; Murphy, B.R.; Moss, B. Development of a replication-deficient recombinant vaccinia virus vaccine effective against parainfluenza virus 3 infection in an animal model. *Vaccine* **1996**, *14*, 1451–1458. [CrossRef]

47. Fiore-Gartland, A.; Manso, B.A.; Friedrich, D.P.; Gabriel, E.E.; Finak, G.; Moodie, Z.; Hertz, T.; De Rosa, S.C.; Frahm, N.; Gilbert, P.B.; et al. Pooled-Peptide Epitope Mapping Strategies Are Efficient and Highly Sensitive: An Evaluation of Methods for Identifying Human T Cell Epitope Specificities in Large-Scale HIV Vaccine Efficacy Trials. *PLoS ONE* **2016**, *11*, e0147812. [CrossRef] [PubMed]

48. Malm, M.; Tamminen, K.; Vesikari, T.; Blazevic, V. Norovirus-Specific Memory T Cell Responses in Adult Human Donors. *Front. Microbiol.* **2016**, *7*, 1570. [CrossRef]

49. Tscharke, D.C.; Woo, W.P.; Sakala, I.G.; Sidney, J.; Sette, A.; Moss, D.J.; Bennink, J.R.; Karupiah, G.; Yewdell, J.W. Poxvirus CD8+ T-cell determinants and cross-reactivity in BALB/c mice. *J. Virol.* **2006**, *80*, 6318–6323. [CrossRef]

50. Gierer, S.; Bertram, S.; Kaup, F.; Wrensch, F.; Heurich, A.; Kramer-Kuhl, A.; Welsch, K.; Winkler, M.; Meyer, B.; Drosten, C.; et al. The spike protein of the emerging betacoronavirus EMC uses a novel coronavirus receptor for entry, can be activated by TMPRSS2, and is targeted by neutralizing antibodies. *J. Virol.* **2013**, *87*, 5502–5511. [CrossRef]

51. Millet, J.K.; Whittaker, G.R. Host cell entry of Middle East respiratory syndrome coronavirus after two-step, furin-mediated activation of the spike protein. *Proc. Natl. Acad. Sci. USA* **2014**, *111*, 15214–15219. [CrossRef]

52. Song, Y.-J.; Yang, J.-S.; Yoon, H.J.; Nam, H.-S.; Lee, S.Y.; Cheong, H.-K.; Park, W.-J.; Park, S.H.; Choi, B.Y.; Kim, S.S.; et al. Asymptomatic Middle East Respiratory Syndrome coronavirus infection using a serologic survey in Korea. *Epidemiol. Health* **2018**, *40*, e2018014. [CrossRef] [PubMed]

53. Al Hammadi, Z.M.; Chu, D.K.; Eltahir, Y.M.; Al Hosani, F.; Al Mulla, M.; Tarnini, W.; Hall, A.J.; Perera, R.A.; Abdelkhalek, M.M.; Peiris, J.S.; et al. Asymptomatic MERS-CoV Infection in Humans Possibly Linked to Infected Dromedaries Imported from Oman to United Arab Emirates, May 2015. *Emerg. Infect. Dis.* **2015**, *21*, 2197–2200. [CrossRef] [PubMed]

54. Jiaming, L.; Yanfeng, Y.; Yao, D.; Yawei, H.; Linlin, B.; Baoying, H.; Jinghua, Y.; Gao, G.F.; Chuan, Q.; Wenjie, T. The recombinant N-terminal domain of spike proteins is a potential vaccine against Middle East respiratory syndrome coronavirus (MERS-CoV) infection. *Vaccine* **2017**, *35*, 10–18. [CrossRef] [PubMed]

55. Narayanan, K.; Maeda, A.; Maeda, J.; Makino, S. Characterization of the coronavirus M protein and nucleocapsid interaction in infected cells. *J. Virol.* **2000**, *74*, 8127–8134. [CrossRef] [PubMed]

56. Zhao, P.; Cao, J.; Zhao, L.J.; Qin, Z.L.; Ke, J.S.; Pan, W.; Ren, H.; Yu, J.G.; Qi, Z.T. Immune responses against SARS-coronavirus nucleocapsid protein induced by DNA vaccine. *Virology* **2005**, *331*, 128–135. [CrossRef] [PubMed]

57. Zhao, J.; Zhao, J.; Mangalam, A.K.; Channappanavar, R.; Fett, C.; Meyerholz, D.K.; Agnihothram, S.; Baric, R.S.; David, C.S.; Perlman, S. Airway Memory CD4(+) T Cells Mediate Protective Immunity against Emerging Respiratory Coronaviruses. *Immunity* **2016**, *44*, 1379–1391. [CrossRef] [PubMed]

58. Zhao, J.; Zhao, J.; Perlman, S. T cell responses are required for protection from clinical disease and for virus clearance in severe acute respiratory syndrome coronavirus-infected mice. *J. Virol.* **2010**, *84*, 9318–9325. [CrossRef]

59. Zhao, J.; Alshukairi, A.N.; Baharoon, S.A.; Ahmed, W.A.; Bokhari, A.A.; Nehdi, A.M.; Layqah, L.A.; Alghamdi, M.G.; Al Gethamy, M.M.; Dada, A.M.; et al. Recovery from the Middle East respiratory syndrome is associated with antibody and T-cell responses. *Sci. Immunol.* **2017**, *2*, 14. [CrossRef]

60. Haagmans, B.L.; van den Brand, J.M.; Provacia, L.B.; Raj, V.S.; Stittelaar, K.J.; Getu, S.; de Waal, L.; Bestebroer, T.M.; van Amerongen, G.; Verjans, G.M.; et al. Asymptomatic Middle East respiratory syndrome coronavirus infection in rabbits. *J. Virol.* **2015**, *89*, 6131–6135. [CrossRef]

61. Ascough, S.; Paterson, S.; Chiu, C. Induction and Subversion of Human Protective Immunity: Contrasting Influenza and Respiratory Syncytial Virus. *Front. Immunol.* **2018**, *9*, 323. [CrossRef]

62. Falzarano, D.; de Wit, E.; Feldmann, F.; Rasmussen, A.L.; Okumura, A.; Peng, X.; Thomas, M.J.; van Doremalen, N.; Haddock, E.; Nagy, L.; et al. Infection with MERS-CoV causes lethal pneumonia in the common marmoset. *PLoS Pathog.* **2014**, *10*, e1004250. [CrossRef] [PubMed]

63. De Wit, E.; Rasmussen, A.L.; Falzarano, D.; Bushmaker, T.; Feldmann, F.; Brining, D.L.; Fischer, E.R.; Martellaro, C.; Okumura, A.; Chang, J.; et al. Middle East respiratory syndrome coronavirus (MERS-CoV) causes transient lower respiratory tract infection in rhesus macaques. *Proc. Natl. Acad. Sci. USA* **2013**, *110*, 16598–16603. [CrossRef] [PubMed]

64. Papageorgiou, N.; Lichiere, J.; Baklouti, A.; Ferron, F.; Sevajol, M.; Canard, B.; Coutard, B. Structural characterization of the N-terminal part of the MERS-CoV nucleocapsid by X-ray diffraction and small-angle X-ray scattering. *Acta Crystallogr. Sect. D Struct. Biol.* **2016**, *72*, 192–202. [CrossRef]

65. Lee, J.Y.; Kim, Y.J.; Chung, E.H.; Kim, D.W.; Jeong, I.; Kim, Y.; Yun, M.R.; Kim, S.S.; Kim, G.; Joh, J.S. The clinical and virological features of the first imported case causing MERS-CoV outbreak in South Korea, 2015. *BMC Infect. Dis.* **2017**, *17*, 498. [CrossRef] [PubMed]

66. Wang, L.; Shi, W.; Joyce, M.G.; Modjarrad, K.; Zhang, Y.; Leung, K.; Lees, C.R.; Zhou, T.; Yassine, H.M.; Kanekiyo, M.; et al. Evaluation of candidate vaccine approaches for MERS-CoV. *Nat. Commun.* **2015**, *6*, 7712. [CrossRef] [PubMed]

67. Pallesen, J.; Wang, N.; Corbett, K.S.; Wrapp, D.; Kirchdoerfer, R.N.; Turner, H.L.; Cottrell, C.A.; Becker, M.M.; Wang, L.; Shi, W.; et al. Immunogenicity and structures of a rationally designed prefusion MERS-CoV spike antigen. *Proc. Natl. Acad. Sci. USA* **2017**, *114*, E7348–E7357. [CrossRef] [PubMed]

68. Sui, J.; Deming, M.; Rockx, B.; Liddington, R.C.; Zhu, Q.K.; Baric, R.S.; Marasco, W.A. Effects of human anti-spike protein receptor binding domain antibodies on severe acute respiratory syndrome coronavirus neutralization escape and fitness. *J. Virol.* **2014**, *88*, 13769–13780. [CrossRef]

69. Tai, W.; Wang, Y.; Fett, C.A.; Zhao, G.; Li, F.; Perlman, S.; Jiang, S.; Zhou, Y.; Du, L. Recombinant Receptor-Binding Domains of Multiple Middle East Respiratory Syndrome Coronaviruses (MERS-CoVs) Induce Cross-Neutralizing Antibodies against Divergent Human and Camel MERS-CoVs and Antibody Escape Mutants. *J. Virol.* **2017**, *91*, e01651-16. [CrossRef]

Article

Enhanced Ability of Oligomeric Nanobodies Targeting MERS Coronavirus Receptor-Binding Domain

Lei He [1,†], Wanbo Tai [2,†], Jiangfan Li [1,†], Yuehong Chen [1], Yaning Gao [2], Junfeng Li [1], Shihui Sun [1], Yusen Zhou [1,3,*], Lanying Du [2,*] and Guangyu Zhao [1,*]

[1]   State Key Laboratory of Pathogen and Biosecurity, Beijing Institute of Microbiology and Epidemiology, Beijing 100071, China; helei_happy@126.com (L.H.); anatee@163.com (J.L.); chenyuehong.happy@163.com (Y.C.); lijunfeng2113@126.com (J.L.); sunsh01@163.com (S.S.)
[2]   Lindsley F. Kimball Research Institute, New York Blood Center, New York, NY 10065, USA; wtai@nybc.org (W.T.); ygao@nybc.org (Y.G.)
[3]   Institute of Medical and Pharmaceutical Sciences, Zhengzhou University, Zhengzhou 450052, China
*   Correspondence: yszhou@bmi.ac.cn (Y.Z.); ldu@nybc.org (L.D.); zhaogy@bmi.ac.cn (G.Z.); Tel.: +1-212-570-3459 (L.D.); +86-10-6385-8045 (Y.Z. & G.Z.)
†   These authors contributed equally to this work.

Received: 28 January 2019; Accepted: 15 February 2019; Published: 19 February 2019

Abstract: Middle East respiratory syndrome (MERS) coronavirus (MERS-CoV), an infectious coronavirus first reported in 2012, has a mortality rate greater than 35%. Therapeutic antibodies are key tools for preventing and treating MERS-CoV infection, but to date no such agents have been approved for treatment of this virus. Nanobodies (Nbs) are camelid heavy chain variable domains with properties distinct from those of conventional antibodies and antibody fragments. We generated two oligomeric Nbs by linking two or three monomeric Nbs (Mono-Nbs) targeting the MERS-CoV receptor-binding domain (RBD), and compared their RBD-binding affinity, RBD–receptor binding inhibition, stability, and neutralizing and cross-neutralizing activity against MERS-CoV. Relative to Mono-Nb, dimeric Nb (Di-Nb) and trimeric Nb (Tri-Nb) had significantly greater ability to bind MERS-CoV RBD proteins with or without mutations in the RBD, thereby potently blocking RBD–MERS-CoV receptor binding. The engineered oligomeric Nbs were very stable under extreme conditions, including low or high pH, protease (pepsin), chaotropic denaturant (urea), and high temperature. Importantly, Di-Nb and Tri-Nb exerted significantly elevated broad-spectrum neutralizing activity against at least 19 human and camel MERS-CoV strains isolated in different countries and years. Overall, the engineered Nbs could be developed into effective therapeutic agents for prevention and treatment of MERS-CoV infection.

Keywords: Coronavirus; MERS-CoV; receptor-binding domain; therapeutic antibodies; nanobodies; cross-neutralization

---

1. Introduction

Middle East respiratory syndrome (MERS) coronavirus (MERS-CoV), an emerging infectious coronavirus, was first reported in humans in Saudi Arabia in 2012 [1]. Bats are a likely natural reservoir of this virus, and dromedary camels are an important intermediate [2–6]. Camels are an important mode of transportation, particularly in the Middle East, and this application of these animals contributes significantly to camel-to-camel and camel-to-human transmission of MERS-CoV [7,8]. In addition, MERS-CoV may also be transmitted between humans in community or hospital settings [9–13]. Since its first emergence, MERS-CoV has continued to infect humans with a high mortality rate (>35%) (http://www.who.int/emergencies/mers-cov/en/). This situation calls for a consistent effort to develop effective countermeasures, including therapeutic antibodies and vaccines, to prevent and treat MERS-CoV infection.

MERS-CoV spike (S) protein, an enveloped glycoprotein, plays a key role in viral infection, viral attachment, and viral entry [14,15]. It is composed of S1 and S2 subunits: the receptor-binding domain (RBD) in the S1 subunit mediates MERS-CoV binding to its cellular receptor, dipeptidyl peptidase 4 (DPP4), and the S2 subunit subsequently mediates viral and cell membrane fusion, leading to viral entry into target cells [16–20]. The RBD of MERS-CoV S protein contains a critical neutralizing domain fragment capable of inducing strong neutralizing antibodies, and it is therefore considered to be an important therapeutic and vaccine target [21–26].

Several monoclonal antibodies (mAbs) have been developed to prevent and treat MERS-CoV infection, and most of these agents are based on the RBD [26–31]. However, conventional IgG mAbs and antibody fragments often have complex structures and unstable behavior [32–34]. Consequently, anti-MERS-CoV therapeutic antibodies with strong stability and simplified structures would be clinically valuable.

Camelid heavy chain variable domains (VHHs), also termed nanobodies (Nbs), are derived from the variable domains of the camelid heavy chain-only antibodies (HcAbs). These antibodies have distinctive properties, including high binding affinity, strong specificity for target antigens, good tissue penetration, and intrinsic stability under harsh conditions, such as extreme pH values, proteases, chemicals, and high temperature. Accordingly, they represent promising therapeutic tools for the treatment of human diseases [35–37]. Moreover, because Nbs do not require paired light and heavy chain domains to maintain antigen-binding activity, they can be easily modified by protein engineering techniques without loss of functionality [38]. Several monomeric Nbs can be readily fused to form multivalent or multispecific constructs, thereby improving their binding affinity and functionality. For example, monomeric proteins have been engineered into dimeric or trimeric proteins [38,39]. Also, multidomain Nbs have been generated by linking different Nbs targeting influenza virus hemagglutinin protein; the resultant agents have greater breadth, avidity, potency, and cross-neutralizing activity against divergent influenza viruses than the parent molecules [40], demonstrating the feasibility of engineering Nbs targeting multiple epitopes to increase their activity. It is therefore worthwhile to attempt to generate multidomain Nbs with improved activity against other emerging and re-emerging infectious viruses.

Previously, by immunizing llamas with a recombinant MERS-CoV RBD protein, we generated a monomeric Nb (Mono-Nb, NbMS10) that targets the RBD of MERS-CoV S protein [41]. In this study, we constructed two oligomeric Nbs, including dimeric Nb (Di-Nb) and trimeric Nb (Tri-Nb), and compared them to the Mono-Nb in terms of their ability to bind RBD proteins, inhibit RBD–DPP4 receptor binding, and cross-neutralize MERS-CoV infection. In addition, to demonstrate the advantages of the oligomeric Nbs relative to conventional antibodies, we evaluated the stability of these Nbs under the extreme conditions mentioned above. Overall, our data show that the engineered oligomeric Nbs have been significantly improved from the standpoint of binding affinity to the RBD, inhibition of the RBD-DPP4 binding, and cross-neutralizing activity against divergent strains of MERS-CoV. They also maintained greater pH, protease, chemical, and thermal stability than their mAb counterparts.

2. Materials and Methods

2.1. Construction and Expression of MERS-CoV RBD-Specific Dimeric and Trimeric Nbs

Dimeric and trimeric Nbs specific for MERS-CoV RBD were constructed by linking two or three monomeric Nb (Mono-Nb: NbMS10) [41] with a GGGGS linker and a C-terminal His$_6$ tag followed by insertion into the *Pichia pastoris* yeast secretory expression vector pPICZαA (Invitrogen, Carlsbad, CA, USA). The recombinant Nbs were expressed in *Pichia pastoris* GS115 cells and purified using Ni-NTA columns (GE Healthcare, Cincinnati, OH, USA).

2.2. SDS-PAGE and Western Blot

MERS-CoV RBD-specific Nbs were detected by SDS-PAGE and Western blot, as previously described [42,43]. Briefly, Nbs (3 μg) were resolved on 10% Tris-Glycine SDS-PAGE gels, followed by staining with Coomassie Brilliant Blue or transferring to nitrocellulose membranes. The membranes were further blocked overnight at 4 °C with PBST containing 5% non-fat milk, and incubated for 1 h at room temperature with goat anti-llama IgG antibody (1:3000, Abcam, Cambridge, MA, USA) and horseradish peroxidase (HRP)-conjugated anti-goat IgG antibody (1:1000, R&D Systems, Minneapolis, MN, USA). The treated membranes were further incubated with ECL Western blot substrate reagents (Abcam) and visualized using Amersham Hyperfilm (GE Healthcare). A MERS-CoV RBD-specific mouse mAb (MERS mAb) and a SARS-CoV RBD-specific mouse mAb (SARS mAb) [44] were included as controls.

2.3. ELISA

Binding between Nbs and MERS-CoV RBD proteins was detected by ELISA as previously described [42,45]. Briefly, ELISA plates were coated overnight at 4 °C with recombinant wild-type or mutant MERS-CoV RBDs containing a C-terminal human Fc tag. The plates were blocked with 2% PBST at 37 °C for 2 h, and sequentially incubated at 37 °C with serially diluted Nbs, goat anti-llama antibody (1:5000, Abcam), and HRP-conjugated anti-goat IgG antibody (1:3000, Abcam) for 1 h each. After washing, the plates were further incubated with substrate (3,3′,5,5′-tetramethylbenzidine, Sigma, St. Louis, MO, USA), and the reactions were stopped with 1 N H_2SO_4. Absorbance at 450 nm (A450) was measured by ELISA microplate reader (Tecan, Morrisville, NC, USA). To compare binding activity, the median effective concentration (EC_{50}) was calculated as previously described [46].

2.4. Surface Plasmon Resonance (SPR)

Binding between Nbs and MERS-CoV RBD protein was detected using a BiacoreS200 instrument (GE Healthcare) as previously described [41]. Briefly, recombinant Fc-fused MERS-CoV RBD protein (5 μg/mL) was captured on a Sensor Chip Protein A (GE Healthcare), and recombinant His_6-tagged NbMS10 Nb at various concentrations was flowed over the chip surface in 10 mM HEPES (pH 7.4), 150 mM NaCl, 3 mM EDTA, and 0.05% surfactant P20 buffer. The sensorgram was analyzed using the Biacore S200 software (GE Healthcare). A 1:1 binding model was fitted to the data.

2.5. Flow Cytometry

Inhibition of binding between MERS-CoV RBD and cell-surface hDPP4 receptor by Nbs was analyzed by flow cytometry as previously described [24]. Briefly, hDPP4-expressing Huh-7 cells were incubated at room temperature for 30 min with MERS-CoV RBD-Fc protein (20 μg/mL), with or without serially diluted Nbs. The cells were incubated for 30 min with FITC-labeled anti-human IgG antibody (1:50, Sigma), and then analyzed by flow cytometry. Percentage inhibition was calculated based on the fluorescence intensity of RBD–Huh-7 binding in the presence vs. absence of Nbs.

2.6. MERS-CoV Micro-Neutralization Assay

The neutralizing activity of MERS-CoV RBD-specific Nbs was initially measured by a live MERS-CoV-based neutralization assay, as previously described [28,45]. Briefly, MERS-CoV (EMC2012 strain, 100 $TCID_{50:}$ median tissue culture infective dose) was incubated with Nbs at 37 °C for 1 h. The Nb/virus mixture was added to Vero E6 cells, which were then cultured for 72 h at 37 °C. The cytopathic effect (CPE) was observed daily. The neutralizing activity of the Nbs was reported as 50% neutralization dose (ND_{50}). The Reed–Muench method was used to calculate the values of ND_{50} for each Nb [47].

2.7. MERS Pseudovirus Neutralization Assay

The cross-neutralizing activity of MERS-CoV RBD-specific Nbs was measured by pseudotyped MERS-CoV neutralization assay as previously described [24,45]. Briefly, 293T cells were cotransfected with a plasmid encoding Env-defective, luciferase-expressing HIV-1 genome (pNL4-3.luc.RE) and a plasmid encoding the MERS-CoV S protein. Pseudotyped MERS-CoV was harvested from culture supernatants 72 h after transfection, incubated with serially diluted Nbs at 37 °C for 1 h, and added to Huh-7 cells. After 72 h, the cells were lysed in cell lysis buffer (Promega, Madison, WI, USA), incubated with luciferase substrate (Promega), and assayed for relative luciferase activity by Tecan Infinite 200 PRO Luminator (Tecan). The ND_{50} of Nbs was calculated as previously described [46].

2.8. Detection of Nb Stability

The stability of Nbs with respect to changes in pH was evaluated by incubation in PBS at various pH values (5.0, 7.0, or 8.0) for 24 h at room temperature [48]. The stability of Nbs in the presence of chaotropic denaturants was evaluated by incubation in PBS containing a gradient of concentrations of urea (Sigma) for 24 h at 25 °C [49]. The stability of Nbs with respect to proteolysis was evaluated by incubation in 10 mM HCl buffer (pH 2.0) containing various concentrations of pepsin (Sigma) for 1 h at 37 °C [49,50]. The thermal stability of Nbs was evaluated by incubation in PBS at various temperatures (4 °C, 37 °C, and 60 °C) for 24 h [49]. Treated and non-treated Nbs were subjected to the MERS pseudovirus neutralization assay. MERS mAb and SARS mAb were included as controls.

2.9. Statistical Analysis

Statistical analysis was performed by Student's two-tailed t-test using the GraphPad Prism statistical software (San Diego, CA, USA). p values lower than 0.05 were considered statistically significant. *, **, and *** indicate $p < 0.05$, $p < 0.01$, and $p < 0.001$, respectively.

3. Results

3.1. Construction and Characterization of MERS-CoV RBD-Targeting Dimeric and Trimeric Nbs

HcAbs, presented in camelids and sharks, contain heavy chains but no light chains. The antigen-binding fragments of camelid HcAbs are also called VHHs (Figure 1A, left). Previously, using PCR to amplify the MERS-CoV RBD-specific VHH gene, we constructed a Mono-Nb targeting the MERS-CoV RBD. We linked this construct to a C-terminal His_6 tag for easy purification [41], resulting in a total molecular weight of about 16 kDa. We generated the Di-Nb and Tri-Nb specific for MERS-CoV RBD by linearly linking two and three Mono-Nbs, respectively, with a flexible GGGGS linker between each Mono-Nb and a C-terminal His_6 tag (Figure 1A, right). As with Mono-Nb, Di-Nb and Tri-Nb were expressed in culture supernatants of yeast expression cells at high yield and purity, and formed dimers and trimers with molecular weights of about 32 and 48 kDa, respectively (Figure 1B, left). These MERS-CoV RBD-targeting Nbs reacted strongly with an anti-llama antibody (Figure 1B, right). These data suggest that, like Mono-Nb, MERS-CoV RBD-specific Di-Nb and Tri-Nb maintained their native conformations and strong antigenicity.

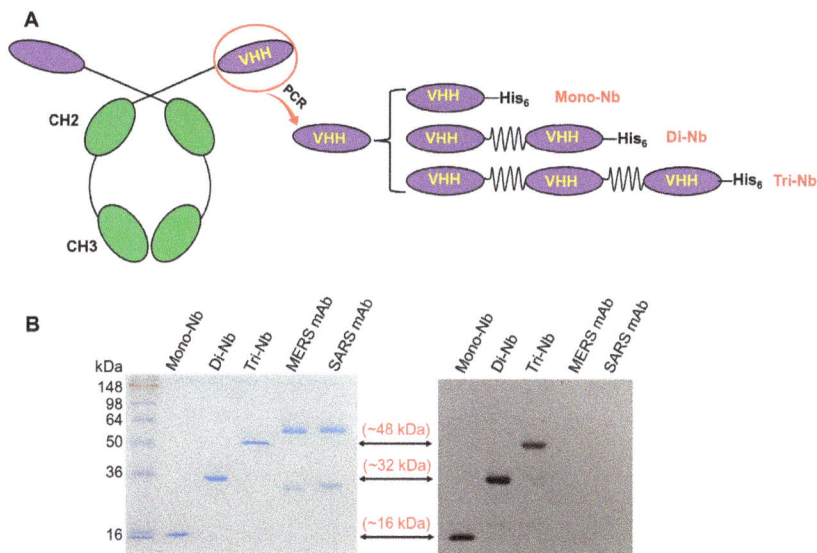

Figure 1. Construction and characterization of dimeric and trimeric nanobodies (Nbs) targeting the Middle East respiratory syndrome (MERS) coronavirus (MERS-CoV). (**A**) Heavy chain-only antibody (HcAb) consists of two constant heavy domains (CH2 and CH3) and heavy chain variable domains (VHHs). Monomeric Nb (Mono-Nb) was constructed previously by linking a MERS-CoV receptor-binding domain (RBD)-specific VHH and a C-terminal His_6, and dimeric Nb (Di-Nb) and trimeric Nb (Tri-Nb) were constructed by linking two or three Mono-Nbs with GGGGS linkers and a C-terminal His_6 tag for easy purification. (**B**) SDS-PAGE (left) and Western blot (right) analysis of MERS-CoV RBD-specific Nbs. The molecular weight marker (in kDa) is shown on the left. MERS-CoV RBD-targeting Mono-Nb was included as comparison, and MERS-CoV RBD-specific mAb (MERS mAb) and SARS-CoV RBD-specific mAb (SARS mAb) were used as controls. Anti-llama antibody was used for Western blot analysis.

3.2. MERS-CoV RBD-Targeting Dimeric and Trimeric Nbs Exhibited Superior Binding toward MERS-CoV RBD, Neutralization of MERS-CoV Infection, and Inhibition of RBD–hDPP4 Binding

To determine whether the engineered Nbs had stronger binding affinity to MERS-CoV RBD proteins, we performed ELISA to test their binding to wild-type MERS-CoV RBD protein fused to C-terminal hIgG1-Fc (RBD-WT), as well as Fc-fused RBD proteins containing mutations from MERS-CoV strains isolated from human and camel in 2012, 2013, 2014, and 2015 [24]. The results revealed that, relative to Mono-Nb, Di-Nb and especially Tri-Nb bound significantly more strongly to all RBDs tested; as expected, binding was dose-dependent (Figure 2A). In addition, the binding affinity of the MERS mAb control was similar to that of Mono-Nb, whereas the binding of the SARS mAb control was indistinguishable from background (Figure 2A). We then performed a SPR assay to test the binding affinity of these Nbs for RBD-WT. The results revealed that Mono-Nb, Di-Nb, and Tri-Nb had antibody binding affinity (K_d) values of 0.87 nM, 5.9 pM, and 7 pM, respectively, toward RBD-WT (Figure 2B).

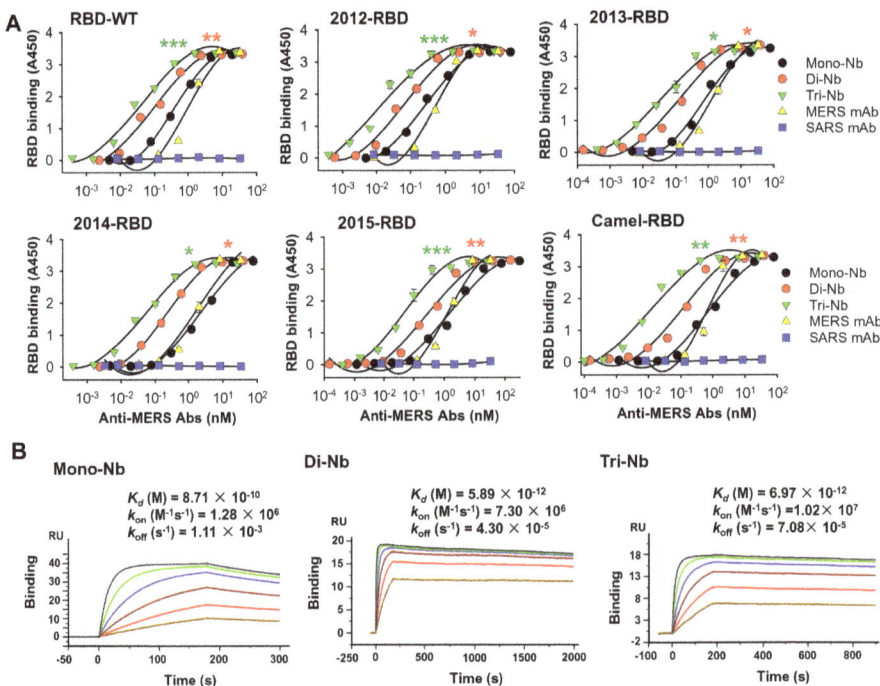

Figure 2. Detection of binding between MERS-CoV RBD-specific Nbs and MERS-CoV RBD proteins. (**A**) ELISA for binding between Di-Nb or Tri-Nb and RBD wild-type (RBD-WT) protein of the EMC2012 strain and mutant proteins containing RBD mutations from strains isolated from human and camel in 2012, 2013, 2014, and 2015. MERS-CoV RBD-targeting Mono-Nb was used for comparison, and MERS-CoV RBD-specific mAb (MERS mAb) and SARS-CoV RBD-specific mAb (SARS mAb) were included as controls. Data are presented as mean A450 $\pm$ standard error (s.e.m.) ($n = 2$). Experiments were repeated twice, yielding similar results. Significant differences in median effective concentration (EC$_{50}$) $\pm$ s.e.m. were observed between Di-Nb and Mono-Nb, as well as between Tri-Nb and Mono-Nb, indicated by red and green asterisk (*, **, and ***), respectively. Concentration (in nM) was calculated based on predicted molecular weights of 16, 32, 48, and 150 kDa for Mono-Nb, Di-Nb, Tri-Nb, and mAb, respectively. (**B**) Surface Plasmon Resonance (SPR) analysis of binding between Di-Nb or Tri-Nb and RBD protein (i.e., RBD-WT). MERS-CoV RBD-targeting Mono-Nb was used for comparison. Binding parameters are shown in each figure.

We then performed a micro-neutralization assay to investigate the neutralizing activity of the engineered MERS-CoV RBD-specific Nbs against live MERS-CoV (EMC2012 strain) infection. The results revealed that both Di-Nb and Tri-Nb potently neutralized MERS-CoV infection with a significantly lower ND$_{50}$ than Mono-Nb (Figure 3A). Previously, we demonstrated that the molecular mechanism of MERS-CoV RBD-specific Mono-Nb suppression of MERS-CoV involves inhibition of RBD–DPP4 receptor binding [41]. Here, we performed ELISA and flow cytometry assays to investigate whether the engineered Di-Nb and Tri-Nb could inhibit RBD-DPP4 binding to a greater extent than Mono-Nb. The ELISA result revealed that Di-Nb and Tri-Nb blocked the binding of MERS-CoV RBD (i.e., RBD-WT) to hDPP4 more strongly than Mono-Nb; the inhibition was dose-dependent (Figure 3B). The flow cytometry assay revealed that both Di-Nb and Tri-Nb had significantly greater ability than Mono-Nb to block the binding of RBD (i.e., RBD-WT) to Huh-7 cell-associated hDPP4; again, the inhibition was dose-dependent (Figure 3C). In both cases, the MERS mAb control inhibited

RBD–DPP4 binding as well as or better than Mono-Nb, whereas inhibition by the SARS mAb control was indistinguishable from background (Figure 3B,C).

Taken together, the data described above suggest that, relative to monomeric Nb, MERS-CoV RBD-specific dimeric and trimer Nbs exhibited significantly improved binding affinity toward MERS-CoV RBD proteins, elevated neutralizing activity toward MERS-CoV infection, and more potent inhibition of MERS-CoV RBD binding to the DPP4 receptor.

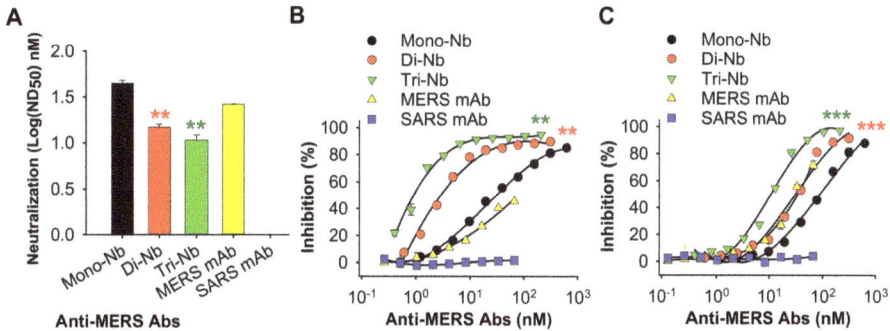

Figure 3. Detection of the neutralizing activity of MERS-CoV RBD-specific Nbs and their inhibition of RBD–DPP4 binding. (**A**) Neutralizing activity of Di-Nb and Tri-Nb against the prototypic MERS-CoV (EMC2012 strain). The neutralizing activity of the Nbs is expressed as the Nb concentration (nM) that completely inhibited the cytopathic effect (CPE) of MERS-CoV in at least 50% of the wells (50% neutralization dose: ND_{50}). Data are expressed as mean $ND_{50} \pm$ s.e.m. ($n = 3$). (**B**) Inhibition of binding between Di-Nb or Tri-Nb and hDPP4 protein, as determined by ELISA. Percentage inhibition is expressed as RBD–hDPP4 binding in the presence or absence of Nbs based on the formula (1 − [RBD–hDPP4-Nb]/[RBD–hDPP4]) × 100. Data are presented as mean percentage inhibition $\pm$ s.e.m. ($n = 2$). (**C**) Inhibition of binding between Di-Nb or Tri-Nb and hDPP4-expressing Huh-7 cells, as determined by flow cytometry analysis. Percentage inhibition is expressed as RBD–Huh-7 binding in the presence or absence of Nbs, which is calculated based on the formula (1−[RBD–Huh-7–Nb]/[RBD–Huh-7]) × 100. Data are presented as mean percentage inhibition $\pm$ s.e.m. ($n = 2$). For (**A**)–(**C**), MERS-CoV RBD-targeting Mono-Nb was used for comparison, and MERS mAb and SARS mAb were included as controls. Experiments were repeated twice, yielding similar results. Significant differences among groups were compared by $ND_{50} \pm$ s.e.m. (**A**) or median inhibitory concentration (IC_{50}) $\pm$ s.e.m. (**A**,**C**). Significant differences between Di-Nb and Mono-Nb are shown as red asterisk, and those between Tri-Nb and Mono-Nb are shown as green asterisk (** and ***).

3.3. MERS-CoV RBD-Targeting Nbs Maintain Strong pH, Protease, Chemical, and Thermal Stability

Nbs generally have intrinsic stability under a variety of extreme conditions, including low or high pH and temperatures, exposure to proteases (such as pepsin), and chaotropic agents (such as urea) [36,37]. To investigate the stability of MERS-CoV RBD-targeting Nbs, we subjected them to these extreme conditions, and then tested their neutralizing activity against pseudotyped MERS-CoV expressing the S protein of strain EMC2012. As shown in Table 1, all Nbs, including Mono-Nb, Di-Nb, and Tri-Nb, were still able to neutralize pseudotyped MERS-CoV infection after treatment at three different pH values (pH 5.0, 7.0, and 8.0), various concentrations of pepsin and urea, and three temperatures (4 °C, 37 °C, and 60 °C). All samples maintained neutralizing activity similar to their respective untreated counterparts. Although MERS mAb control, treated or not treated at the three pH values, maintained similar neutralizing activity against pseudotyped MERS-CoV infection, it significantly lost neutralizing ability after incubation with urea or pepsin, or after pre-treatment at 37 °C or 60 °C. As expected, the SARS mAb control had no cross-neutralizing activity against MERS-CoV after any of these treatments. These data indicate that, relative to traditional mAbs targeting MERS-CoV RBD, MERS-CoV RBD-specific Nbs maintain greater stability under all extreme conditions tested.

Table 1. Stability of MERS-CoV RBD-specific Nbs against extreme conditions.

Conditions	Treatment	ND_{50} (nM, Mean Value)				
		Mono-Nb	Di-Nb	Tri-Nb	MERS mAb	SARS mAb
pH	pH 5.0	1.94	0.21	0.03	0.53	ND
	pH 7.0	2.00	0.21	0.03	0.57	ND
	pH 8.0	1.93	0.21	0.03	0.53	ND
	No treatment	2.00	0.23	0.03	0.56	ND
Pepsin	0 µg/mL	2.23	0.29	0.04	0.56	ND
	25 µg/mL	2.24	0.27	0.04	1.19 **	ND
	625 µg/mL	2.20	0.26	0.04	1.86 **	ND
	No treatment	2.15	0.24	0.04	0.55	ND
Urea	0 mM	2.27	0.25	0.04	0.56	ND
	50 mM	2.01	0.29	0.04	0.85 **	ND
	400 mM	2.18	0.25	0.03	1.46 **	ND
	No treatment	2.19	0.25	0.04	0.54	ND
Temperature	4 °C	2.06	0.25	0.03	0.56	ND
	37 °C	1.80	0.22	0.04	1.32 **	ND
	60 °C	2.21	0.26	0.03	2.06 ***	ND
	No treatment	2.17	0.26	0.04	0.58	ND

Note: Nbs were tested for stability under extreme conditions, including pH, protease (pepsin), chaotropic agent (urea), and temperature. Nbs were treated at different pH values (pH 5.0, 7.0, and 8.0) for 24 h at room temperature, the indicated concentrations of pepsin for 1 h at 37 °C or urea for 24 h at 25 °C, and different temperatures (4 °C, 37 °C, and 60 °C) for 24 h, followed by measurement of their neutralizing activity against pseudotyped MERS-CoV (EMC2012 strain) infection. Neutralizing activity of Nbs is expressed as mean 50% neutralization dose (ND_{50}) ($n = 2$). Experiments were repeated twice, yielding similar results. MERS mAb and SARS mAb were used as controls. Significant differences between treatment and no-treatment groups under each condition were compared by mean $ND_{50} \pm$ s.e.m. *, **, and *** indicate the level of significance of the differences between MERS mAb, with or without treatment under the indicated conditions. ND, not detectable.

3.4. MERS-CoV RBD-Targeting Dimeric and Trimeric Nbs Had Significantly Elevated Cross-Neutralizing Activity Against Multiple Heterologous MERS-CoV Isolates

MERS-CoV has undergone a number of mutations, including those in the RBD [24]. Hence, it is critical that MERS-CoV RBD-specific Nbs maintain potent cross-neutralizing activity against MERS-CoV of divergent strains. In addition to pseudotyped MERS-CoV expressing the S protein of the prototypic MERS-CoV strain (EMC2012), we constructed 18 additional pseudotyped MERS-CoVs containing RBD mutations from MERS-CoVs isolated from seven countries (Saudi Arabia, UK, Qatar, Oman, Jordan, South Korea, and UAE), different time periods (2012–2016), and two hosts (human and camel) (Table 2). In addition, we tested the neutralizing activity of MERS-CoV RBD-specific Nbs. As shown in Table 2, relative to Mono-Nb, Di-Nb, and especially Tri-Nb, had significantly elevated neutralizing activity against all these 18 viruses tested. ND_{50} ranged from 0.81 to 27.1 nM for Mono-Nb, from 0.07 to 3.29 nM for Di-Nb, and from 0.01 to 0.61 nM for Tri-Nb. The neutralizing activity of MERS mAb was much lower than that of Di-Nb and Tri-Nb against all MERS-CoV strains tested, and SARS mAb had no neutralizing activity against these viruses. These data indicate that MERS-CoV RBD-specific oligomeric Nbs exhibited higher levels of cross-neutralization activity against divergent MERS-CoV strains.

Table 2. Source of divergent MERS-CoV strains and cross-neutralizing activity of MERS-CoV RBD-specific Nbs against these strains.

Accession No.	Isolate Year	Host	Country	S Protein RBD Mutation(s)	ND50 (nM, Mean Value)				
					Mono-Nb	Di-Nb	Tri-Nb	MERS mAb	SARS mAb
AFS88936	2012	Human	Saudi Arabia	—	2.14	0.24 **	0.03 **	0.57	ND
AGV08379	2012	Human	Saudi Arabia	D509G	3.39	0.19 ***	0.06 ***	122	ND
AGV08584	2012	Human	Saudi Arabia	V534A	6.64	0.38 *	0.08 *	2.02	ND
AFY13307	2012	Human	UK	L506F	27.1	3.29 ***	0.37 ***	67.5	ND
AHI48528	2013	Human	Saudi Arabia	A431P, A482V	1.10	0.10 ***	0.02 ***	0.52	ND
AHI48733	2013	Human	Saudi Arabia	A434V	6.67	0.11 *	0.05 *	2.11	ND
AHC74088	2013	Human	Qatar	S460F	2.57	0.26 **	0.04 **	0.50	ND
AKM76239	2013	Human	Oman	V514L	9.07	0.95 *	0.09 **	3.65	ND
AID55090	2014	Human	Saudi Arabia	T424I	0.81	0.07 **	0.01 **	0.80	ND
AID55087	2014	Human	Saudi Arabia	Q522H	1.49	0.09 ***	0.02 ***	0.28	ND
ALX27228	2014	Human	Jordan	E536K	14.7	2.96 **	0.61 **	5.54	ND
ALJ76277	2014	Human	Saudi Arabia	D537E	7.30	1.49 *	0.41 *	3.02	ND
ALJ54518	2015	Human	Saudi Arabia	L507P	13.1	2.76 ***	0.43 ***	186	ND
ALB08322	2015	Human	South Korea	D510G	2.28	0.23 *	0.04 **	8.92	ND
ALB08289	2015	Human	South Korea	I529T	3.77	0.24 ***	0.08 ***	865	ND
ATC84888	2016	Human	Saudi Arabia	S426R	12.6	1.60 **	0.34 ***	15.6	ND
AHY22545	2013	Camel	Saudi Arabia	K400N	1.77	0.20 ***	0.02 ***	1.12	ND
AHY22555	2013	Camel	Saudi Arabia	A520S	1.11	0.17 **	0.08 **	1.10	ND
ASU90076	2015	Camel	UAE	S460T	6.84	1.26 ***	0.26 ***	3.36	ND

Note: MERS-CoV strains were isolated in human and camel from 2012 to 2016 in different countries. EMC2012 (accession no. AFS88936) is the prototypic MERS-CoV strain. RBD mutations indicate mutant residues in the RBD of S protein of the indicated MERS-CoV isolates. A MERS-CoV neutralization assay was performed to test cross-neutralizing activity of MERS-CoV RBD-specific Nbs against pseudotyped MERS-CoV expressing the S protein of these strains. Neutralizing activity of Nbs is expressed as mean ND$_{50}$ ($n = 2$). Experiments were repeated twice, yielding similar results. MERS mAb and SARS mAb were used as controls. Significant differences between Di-Nb and Mono-Nb, as well as between Tri-Nb and Mono-Nb, were compared by mean ND$_{50}$ $\pm$ s.e.m. *, **, and *** indicate the level of significance of the differences between Di-Nb or Tri-Nb and Mono-Nb for the indicated MERS-CoV strains. ND, not detectable.

4. Discussion

MERS-CoV continues to pose a severe threat to public health worldwide due to its high mortality rate and the steady increase in clinical cases, particularly in Saudi Arabia. Currently, no MERS vaccines or therapeutics have been approved for use in humans, creating an urgent demand for efficacious vaccines and therapeutic agents capable of preventing MERS-CoV transmission and infection, as well as treating MERS-CoV-infected humans and camels. In terms of antibody therapy, antibodies with high productivity, good antigen-binding affinity, and potent neutralizing activity against divergent strains of MERS-CoV infection would be of the greatest practical use.

Unlike conventional IgG antibodies (~150 kDa), or antibody fragments such as antigen-binding fragment (Fab, ~55 kDa) and single-chain variable fragment (scFv, 28 kDa), Nb monomers generally exhibit excellent solubility, strong stability, and good tissue penetration, mainly due to their small size (~15–16 kDa) [32,33,51]. In some cases, to increase efficacy against viral infection, bispecific, multispecific, or multivalent Nbs can be constructed by tandemly linking two or more Nb monomers recognizing either the same or different epitopes [52]. In contrast to mAb fragments, which might exhibit reduced expression, stability, or affinity after engineering or recombination, engineered Nbs exhibit elevated antigen-binding affinity, thus extending the time period during which Nbs are bound to their targets, while maintaining their beneficial characteristics, without negatively affecting production yields, solubility, or stability [53–55].

Taking advantage of the properties of Nbs, especially their ability to form functional bi- or multi-specific Nbs with elevated activity, we constructed two MERS-CoV RBD-targeting oligomeric Nbs, Di-Nb and Tri-Nb, based on the previously developed Mono-Nb [41], with the goal of identifying anti-MERS-CoV Nbs with improved binding ability, superior inhibition, and broad-spectrum neutralizing activity against MERS-CoV infection without negatively affecting expression level or stability.

Di-Nb and Tri-Nb had molecular weights of about 32 and 48 kDa that were double and triple the size of Mono-Nb (~16 kDa), respectively. As expected, engineering of the RBD-specific Mono-Nb had no impact on antibody expression. As with Mono-Nb, Di-Nb and Tri-Nb could be expressed at high levels in a yeast cell expression system and purified with high purity, retaining their conformation and antigenicity. In addition, the larger size of the oligomeric Nbs did not affect their stability. Like Mono-Nb, the oligomerized Di-Nb and Tri-Nb maintained stability under all extreme conditions tested, including acidic or alkaline pH, protease (pepsin), chaotropic denaturant (urea), and high temperature. By contrast, the protease, chemical, and thermal stability of MERS mAb significantly decreased after these treatments. Thus, unlike the mAb control, the engineered MERS-CoV RBD-specific Nbs developed in this study maintained the key characteristics of Nbs, including intrinsic stability, high expression, and intact conformation. Because these Nbs are stable under above extreme conditions, they can be transported and stored without the need for refrigeration or special care, a marked advantage relative to traditional antibodies. This will significantly simplify the transportation and therapeutic processes of antibodies, particularly in Middle Eastern countries that generally have inadequate transport services and high temperatures during summer.

The MERS-CoV RBD-specific dimeric and trimeric Nbs developed in this study also have several important features related to interference with three critical steps of MERS-CoV infection. First, they exhibited significantly greater ability than the Mono-Nb to bind MERS-CoV RBD proteins with or without RBD mutations from divergent MERS-CoV strains isolated from different years and hosts, facilitating the Nb–RBD interaction and increasing the time that the Nbs are bound to MERS-CoV. We have previously demonstrated that the Mono-Nb recognizes an epitope at residue around D539 of MERS-CoV RBD [41]. Second, relative to Mono-Nb, the engineered oligomeric Nbs strongly blocked RBD binding to MERS-CoV receptor DPP4, which is involved in a key step of viral entry and infection [14,56], thereby more effectively blocking entry of MERS-CoV into its target cells. Third, Di-Nb and Tri-Nb had significantly greater capacities to neutralize homologous MERS-CoV (EMC2012 strain), as well as potently cross-neutralize dozens of heterologous MERS-CoV strains harboring one or two mutations in their RBDs, all of which were isolated from different countries, hosts (human and

Viruses **2019**, *11*, 166

camel), and time periods. Overall, the engineered oligomeric Nbs have great potential to neutralize new MERS-CoV strains with mutations in the RBD, and could therefore play a key role in preventing camel-to-human and human-to-human transmissions of MERS-CoV.

Despite the significant advantages of Nbs, it should be noted that the engineered dimeric (~32 kDa) and trimeric (~48 kDa) Nbs, as well as their monomeric counterpart (~16 kDa), are smaller than IgG antibodies (~150 kDa), and their molecular weights are far below the kidney filtration threshold (~60 kDa). Therefore, these small Nbs may be rapidly eliminated from the bloodstream by renal clearance, thus they might have shorter half-lives than mAbs [41,55]. One approach of increasing the in vivo half-life of Nbs is to fuse them with albumin-binding domain (ABD) or human Fc (hFc) [55,57,58]. We have also successfully fused monomeric Nb with a C-terminal hFc, increasing its half-life as well as its therapeutic and prophylactic efficacy against MERS-CoV infection [41]. In future studies, we plan to extend the half-lives of the constructed oligomeric Nbs by fusing them with ABD or hFc, and then compare their in vivo efficacy with that of Mono-Nb with or without hFc against MERS-CoV.

To summarize, in this study we developed oligomeric MERS-CoV RBD-specific Nbs and demonstrated that their in vitro activities were superior to those of their monomeric Nb counterparts, including antigen-binding affinity, inhibition of virus–receptor binding, and enhanced neutralizing and cross-neutralizing activity against variant strains of MERS-CoV infection, without reducing their stability under harsh conditions. These Nbs have the potential to be developed as therapeutics to prevent and treat MERS-CoV infection. Similar strategies could be applied to developing therapeutic agents against other emerging and re-emerging infectious viruses with pandemic potential.

Author Contributions: L.H., W.T., Y.Z., L.D., and G.Z. designed the study. L.H., W.T., J.L. (Jiangfan Li), Y.C., Y.G., J.L. (Junfeng Li), and S.S. performed the experiments. L.H., W.T., J.L. (Jiangfan Li), Y.C., Y.G., Y.Z., L.D., and G.Z. summarized and analyzed the data. Y.Z., L.D., and G.Z. wrote and revised the manuscript.

Funding: This study was supported by SKLPBS1805 (to G.Z.), National Project of Infectious Diseases (2017ZX10304402-003), National Natural Science Foundation of China 81571983 (to Y.Z.), and NIH grants R21AI128311, R01AI137472, and R01AI139092 (to L.D.).

Conflicts of Interest: The authors declare no competing interests.

References

1. Zaki, A.M.; Van Boheemen, S.; Bestebroer, T.M.; Osterhaus, A.D.; Fouchier, R.A. Isolation of a novel coronavirus from a man with pneumonia in Saudi Arabia. *N. Engl. J. Med.* **2012**, *367*, 1814–1820. [CrossRef] [PubMed]

2. Yang, Y.; Du, L.; Liu, C.; Wang, L.; Ma, C.; Tang, J.; Baric, R.S.; Jiang, S.; Li, F. Receptor usage and cell entry of bat coronavirus HKU4 provide insight into bat-to-human transmission of MERS coronavirus. *Proc. Natl. Acad. Sci. USA* **2014**, *111*, 12516–12521. [CrossRef] [PubMed]

3. Munster, V.J.; Adney, D.R.; Van Doremalen, N.; Brown, V.R.; Miazgowicz, K.L.; Milne-Price, S.; Bushmaker, T.; Rosenke, R.; Scott, D.; Hawkinson, A.; et al. Replication and shedding of MERS-CoV in Jamaican fruit bats (Artibeus jamaicensis). *Sci. Rep.* **2016**, *6*, 21878. [CrossRef]

4. Anthony, S.J.; Gilardi, K.; Menachery, V.D.; Goldstein, T.; Ssebide, B.; Mbabazi, R.; Navarrete-Macias, I.; Liang, E.; Wells, H.; Hicks, A.; et al. Further evidence for bats as the evolutionary source of Middle East respiratory syndrome coronavirus. *MBio* **2017**, *8*, e00373-17. [CrossRef] [PubMed]

5. Lau, S.K.P.; Wong, A.C.P.; Lau, T.C.K.; Woo, P.C.Y. Molecular evolution of MERS Coronavirus: Dromedaries as a recent intermediate host or long-time animal reservoir? *Int. J. Mol. Sci.* **2017**, *18*, 2138. [CrossRef] [PubMed]

6. van Doremalen, N.; Hijazeen, Z.S.; Holloway, P.; Al, O.B.; McDowell, C.; Adney, D.; Talafha, H.A.; Guitian, J.; Steel, J.; Amarin, N.; et al. High prevalence of Middle East respiratory coronavirus in young dromedary camels in Jordan. *Vector Borne Zoonotic Dis.* **2017**, *17*, 155–159. [CrossRef] [PubMed]

7. Madani, T.A.; Azhar, E.I.; Hashem, A.M. Evidence for camel-to-human transmission of MERS coronavirus. *N. Engl. J. Med.* **2014**, *371*, 1360. [PubMed]

8. Hemida, M.G.; Elmoslemany, A.; Al-Hizab, F.; Alnaeem, A.; Almathen, F.; Faye, B.; Chu, D.K.; Perera, R.A.; Peiris, M. Dromedary camels and the transmission of Middle East respiratory syndrome coronavirus (MERS-CoV). *Transbound. Emerg. Dis.* **2015**, *64*, 344–353. [CrossRef]

9. Hunter, J.C.; Nguyen, D.; Aden, B.; Al, B.Z.; Al, D.W.; Abu, E.K.; Khudair, A.; Al, M.M.; El, S.F.; Imambaccus, H.; et al. Transmission of Middle East respiratory syndrome coronavirus infections in healthcare settings, Abu Dhabi. *Emerg. Infect. Dis.* **2016**, *22*, 647–656. [CrossRef]

10. Alhakeem, R.F.; Midgley, C.M.; Assiri, A.M.; Alessa, M.; Al, H.H.; Saeed, A.B.; Almasri, M.M.; Lu, X.; Abedi, G.R.; Abdalla, O.; et al. Exposures among MERS case-patients, Saudi Arabia, January-February 2016. *Emerg. Infect. Dis.* **2016**, *22*, 2020–2022. [CrossRef]

11. Chen, X.; Chughtai, A.A.; Dyda, A.; MacIntyre, C.R. Comparative epidemiology of Middle East respiratory syndrome coronavirus (MERS-CoV) in Saudi Arabia and South Korea. *Emerg. Microbes Infect.* **2017**, *6*, e51. [CrossRef] [PubMed]

12. Choi, J.Y. An outbreak of Middle East respiratory syndrome coronavirus infection in South Korea, 2015. *Yonsei Med. J.* **2015**, *56*, 1174–1176. [CrossRef] [PubMed]

13. Lee, S.S.; Wong, N.S. Probable transmission chains of Middle East respiratory syndrome coronavirus and the multiple generations of secondary infection in South Korea. *Int. J. Infect. Dis.* **2015**, *38*, 65–67. [CrossRef] [PubMed]

14. Zhou, Y.; Jiang, S.; Du, L. Prospects for a MERS-CoV spike vaccine. *Expert Rev. Vaccines* **2018**, *17*, 677–686. [CrossRef] [PubMed]

15. Du, L.; Tai, W.; Zhou, Y.; Jiang, S. Vaccines for the prevention against the threat of MERS-CoV. *Expert Rev. Vaccines* **2016**, *15*, 1123–1134. [CrossRef] [PubMed]

16. Lu, G.; Hu, Y.; Wang, Q.; Qi, J.; Gao, F.; Li, Y.; Zhang, Y.; Zhang, W.; Yuan, Y.; Bao, J.; et al. Molecular basis of binding between novel human coronavirus MERS-CoV and its receptor CD26. *Nature* **2013**, *500*, 227–231. [CrossRef] [PubMed]

17. Raj, V.S.; Mou, H.; Smits, S.L.; Dekkers, D.H.; Muller, M.A.; Dijkman, R.; Muth, D.; Demmers, J.A.; Zaki, A.; Fouchier, R.A.; et al. Dipeptidyl peptidase 4 is a functional receptor for the emerging human coronavirus-EMC. *Nature* **2013**, *495*, 251–254. [CrossRef]

18. Lu, L.; Liu, Q.; Zhu, Y.; Chan, K.H.; Qin, L.; Li, Y.; Wang, Q.; Chan, J.F.; Du, L.; Yu, F.; et al. Structure-based discovery of Middle East respiratory syndrome coronavirus fusion inhibitor. *Nat. Commun.* **2014**, *5*, 3067. [CrossRef]

19. Li, F. Receptor recognition mechanisms of coronaviruses: A decade of structural studies. *J. Virol.* **2015**, *89*, 1954–1964. [CrossRef]

20. Chen, Y.; Rajashankar, K.R.; Yang, Y.; Agnihothram, S.S.; Liu, C.; Lin, Y.L.; Baric, R.S.; Li, F. Crystal structure of the receptor-binding domain from newly emerged Middle East respiratory syndrome coronavirus. *J. Virol.* **2013**, *87*, 10777–10783. [CrossRef]

21. Ma, C.; Wang, L.; Tao, X.; Zhang, N.; Yang, Y.; Tseng, C.T.; Li, F.; Zhou, Y.; Jiang, S.; Du, L. Searching for an ideal vaccine candidate among different MERS coronavirus receptor-binding fragments—The importance of immunofocusing in subunit vaccine design. *Vaccine* **2014**, *32*, 6170–6176. [CrossRef]

22. Zhang, N.; Tang, J.; Lu, L.; Jiang, S.; Du, L. Receptor-binding domain-based subunit vaccines against MERS-CoV. *Virus Res.* **2015**, *202*, 151–159. [CrossRef] [PubMed]

23. Du, L.; Tai, W.; Yang, Y.; Zhao, G.; Zhu, Q.; Sun, S.; Liu, C.; Tao, X.; Tseng, C.K.; Perlman, S.; et al. Introduction of neutralizing immunogenicity index to the rational design of MERS coronavirus subunit vaccines. *Nat. Commun.* **2016**, *7*, 13473. [CrossRef] [PubMed]

24. Tai, W.; Wang, Y.; Fett, C.A.; Zhao, G.; Li, F.; Perlman, S.; Jiang, S.; Zhou, Y.; Du, L. Recombinant receptor-binding domains of multiple Middle East respiratory syndrome coronaviruses (MERS-CoVs) induce cross-neutralizing antibodies against divergent human and camel MERS-CoVs and antibody escape mutants. *J. Virol.* **2017**, *91*, e01651-16. [CrossRef] [PubMed]

25. Zhang, N.; Channappanavar, R.; Ma, C.; Wang, L.; Tang, J.; Garron, T.; Tao, X.; Tasneem, S.; Lu, L.; Tseng, C.T.; et al. Identification of an ideal adjuvant for receptor-binding domain-based subunit vaccines against Middle East respiratory syndrome coronavirus. *Cell. Mol. Immunol.* **2016**, *13*, 180–190. [CrossRef] [PubMed]

26. Du, L.; Yang, Y.; Zhou, Y.; Lu, L.; Li, F.; Jiang, S. MERS-CoV spike protein: A key target for antivirals. *Expert Opin. Ther. Targets* **2017**, *21*, 131–143. [CrossRef] [PubMed]

27. Qiu, H.; Sun, S.; Xiao, H.; Feng, J.; Guo, Y.; Tai, W.; Wang, Y.; Du, L.; Zhao, G.; Zhou, Y. Single-dose treatment with a humanized neutralizing antibody affords full protection of a human transgenic mouse model from lethal Middle East respiratory syndrome (MERS)-coronavirus infection. *Antiviral Res.* **2016**, *132*, 141–148. [CrossRef] [PubMed]

28. Du, L.; Zhao, G.; Yang, Y.; Qiu, H.; Wang, L.; Kou, Z.; Tao, X.; Yu, H.; Sun, S.; Tseng, C.T.; et al. A conformation-dependent neutralizing monoclonal antibody specifically targeting receptor-binding domain in Middle East respiratory syndrome coronavirus spike protein. *J. Virol.* **2014**, *88*, 7045–7053. [CrossRef]

29. Corti, D.; Zhao, J.; Pedotti, M.; Simonelli, L.; Agnihothram, S.; Fett, C.; Fernandez-Rodriguez, B.; Foglierini, M.; Agatic, G.; Vanzetta, F.; et al. Prophylactic and postexposure efficacy of a potent human monoclonal antibody against MERS coronavirus. *Proc. Natl Acad. Sci. USA* **2015**, *112*, 10473–10478. [CrossRef]

30. Ying, T.; Prabakaran, P.; Du, L.; Shi, W.; Feng, Y.; Wang, Y.; Wang, L.; Li, W.; Jiang, S.; Dimitrov, D.S.; et al. Junctional and allele-specific residues are critical for MERS-CoV neutralization by an exceptionally potent germline-like antibody. *Nat. Commun.* **2015**, *6*, 8223. [CrossRef]

31. Jiang, L.; Wang, N.; Zuo, T.; Shi, X.; Poon, K.M.; Wu, Y.; Gao, F.; Li, D.; Wang, R.; Guo, J.; et al. Potent neutralization of MERS-CoV by human neutralizing monoclonal antibodies to the viral spike glycoprotein. *Sci. Transl. Med.* **2014**, *6*, 234ra59. [CrossRef] [PubMed]

32. Lowe, D.; Dudgeon, K.; Rouet, R.; Schofield, P.; Jermutus, L.; Christ, D. Aggregation, stability, and formulation of human antibody therapeutics. *Adv. Protein Chem. Struct. Biol.* **2011**, *84*, 41–61. [PubMed]

33. Rouet, R.; Lowe, D.; Christ, D. Stability engineering of the human antibody repertoire. *FEBS Lett.* **2014**, *588*, 269–277. [CrossRef] [PubMed]

34. Chames, P.; Van, R.M.; Weiss, E.; Baty, D. Therapeutic antibodies: Successes, limitations and hopes for the future. *Br. J. Pharmacol.* **2009**, *157*, 220–233. [CrossRef] [PubMed]

35. Muyldermans, S. Nanobodies: Natural single-domain antibodies. *Annu. Rev. Biochem.* **2013**, *82*, 775–797. [CrossRef] [PubMed]

36. Siontorou, C.G. Nanobodies as novel agents for disease diagnosis and therapy. *Int. J. Nanomed.* **2013**, *8*, 4215–4227. [CrossRef] [PubMed]

37. Wesolowski, J.; Alzogaray, V.; Reyelt, J.; Unger, M.; Juarez, K.; Urrutia, M.; Cauerhff, A.; Danquah, W.; Rissiek, B.; Scheuplein, F.; et al. Single domain antibodies: Promising experimental and therapeutic tools in infection and immunity. *Med. Microbiol. Immunol.* **2009**, *198*, 157–174. [CrossRef] [PubMed]

38. Hmila, I.; Abdallah, R.B.; Saerens, D.; Benlasfar, Z.; Conrath, K.; Ayeb, M.E.; Muyldermans, S.; Bouhaouala-Zahar, B. VHH, bivalent domains and chimeric Heavy chain-only antibodies with high neutralizing efficacy for scorpion toxin AahI'. *Mol. Immunol.* **2008**, *45*, 3847–3856. [CrossRef]

39. Els, C.K.; Lauwereys, M.; Wyns, L.; Muyldermans, S. Camel single-domain antibodies as modular building units in bispecific and bivalent antibody constructs. *J. Biol. Chem.* **2001**, *276*, 7346–7350.

40. Laursen, N.S.; Friesen, R.H.E.; Zhu, X.; Jongeneelen, M.; Blokland, S.; Vermond, J.; van Eijgen, A.; Tang, C.; van Diepen, H.; Obmolova, G.; et al. Universal protection against influenza infection by a multidomain antibody to influenza hemagglutinin. *Science* **2018**, *362*, 598–602. [CrossRef]

41. Zhao, G.; He, L.; Sun, S.; Qiu, H.; Tai, W.; Chen, J.; Li, J.; Chen, Y.; Guo, Y.; Wang, Y.; et al. A novel nanobody targeting Middle East respiratory syndrome coronavirus (MERS-CoV) receptor-binding domain has potent cross-neutralizing activity and protective efficacy against MERS-CoV. *J. Virol.* **2018**, *92*, e00837-18. [CrossRef] [PubMed]

42. Tai, W.; Zhao, G.; Sun, S.; Guo, Y.; Wang, Y.; Tao, X.; Tseng, C.K.; Li, F.; Jiang, S.; Du, L.; et al. A recombinant receptor-binding domain of MERS-CoV in trimeric form protects human dipeptidyl peptidase 4 (hDPP4) transgenic mice from MERS-CoV infection. *Virology* **2016**, *499*, 375–382. [CrossRef] [PubMed]

43. Tai, W.; He, L.; Wang, Y.; Sun, S.; Zhao, G.; Luo, C.; Li, P.; Zhao, H.; Fremont, D.H.; Li, F.; et al. Critical neutralizing fragment of Zika virus EDIII elicits cross-neutralization and protection against divergent Zika viruses. *Emerg. Microbes Infect.* **2018**, *7*, 7. [CrossRef] [PubMed]

44. He, Y.; Lu, H.; Siddiqui, P.; Zhou, Y.; Jiang, S. Receptor-binding domain of severe acute respiratory syndrome coronavirus spike protein contains multiple conformation-dependent epitopes that induce highly potent neutralizing antibodies. *J. Immunol.* **2005**, *174*, 4908–4915. [CrossRef] [PubMed]

45. Wang, Y.; Tai, W.; Yang, J.; Zhao, G.; Sun, S.; Tseng, C.K.; Jiang, S.; Zhou, Y.; Du, L.; Gao, J. Receptor-binding domain of MERS-CoV with optimal immunogen dosage and immunization interval protects human transgenic mice from MERS-CoV infection. *Hum. Vaccin. Immunother.* **2017**, *13*, 1615–1624. [CrossRef] [PubMed]

46. Chou, T.C. Theoretical basis, experimental design, and computerized simulation of synergism and antagonism in drug combination studies. *Pharmacol. Rev.* **2006**, *58*, 621–681. [CrossRef] [PubMed]

47. Biacchesi, S.; Skiadopoulos, M.H.; Yang, L.; Murphy, B.R.; Collins, P.L.; Buchholz, U.J. Rapid human metapneumovirus microneutralization assay based on green fluorescent protein expression. *J. Virol. Methods* **2005**, *128*, 192–197. [CrossRef] [PubMed]

48. Gaiotto, T.; Hufton, S.E. Cross-neutralising nanobodies bind to a conserved pocket in the hemagglutinin stem region identified using yeast display and deep mutational scanning. *PLoS ONE* **2016**, *11*, e0164296. [CrossRef]

49. Ardekani, L.S.; Gargari, S.L.; Rasooli, I.; Bazl, M.R.; Mohammadi, M.; Ebrahimizadeh, W.; Bakherad, H.; Zare, H. A novel nanobody against urease activity of Helicobacter pylori. *Int. J. Infect. Dis.* **2013**, *17*, e723–e728. [CrossRef] [PubMed]

50. Hussack, G.; Hirama, T.; Ding, W.; Mackenzie, R.; Tanha, J. Engineered single-domain antibodies with high protease resistance and thermal stability. *PLoS ONE* **2011**, *6*, e28218. [CrossRef] [PubMed]

51. AlDeghaither, D.; Smaglo, B.G.; Weiner, L.M. Beyond peptides and mAbs–current status and future perspectives for biotherapeutics with novel constructs. *J. Clin. Pharmacol.* **2015**, S4–S20. [CrossRef] [PubMed]

52. Ibanez, L.I.; De, F.M.; Hultberg, A.; Verrips, T.; Temperton, N.; Weiss, R.A.; Vandevelde, W.; Schepens, B.; Vanlandschoot, P.; Saelens, X. Nanobodies with in vitro neutralizing activity protect mice against H5N1 influenza virus infection. *J. Infect. Dis.* **2011**, *203*, 1063–1072. [CrossRef] [PubMed]

53. Li, T.; Bourgeois, J.P.; Celli, S.; Glacial, F.; Le Sourd, A.M.; Mecheri, S.; Weksler, B.; Romero, I.; Couraud, P.O.; Rougeon, F.; et al. Cell-penetrating anti-GFAP VHH and corresponding fluorescent fusion protein VHH-GFP spontaneously cross the blood-brain barrier and specifically recognize astrocytes: Application to brain imaging. *FASEB J.* **2012**, *26*, 3969–3979. [CrossRef] [PubMed]

54. Liu, A.; Yin, K.; Mi, L.; Ma, M.; Liu, Y.; Li, Y.; Wei, W.; Zhang, Y.; Liu, S. A novel photoelectrochemical immunosensor by integration of nanobody and ZnO nanorods for sensitive detection of nucleoside diphosphatase kinase-A. *Anal. Chim. Acta* **2017**, *973*, 82–90. [CrossRef] [PubMed]

55. Bannas, P.; Hambach, J.; Koch-Nolte, F. Nanobodies and nanobody-based human heavy chain antibodies as antitumor therapeutics. *Front. Immunol.* **2017**, *8*, 1603. [CrossRef] [PubMed]

56. Wang, N.; Shi, X.; Jiang, L.; Zhang, S.; Wang, D.; Tong, P.; Guo, D.; Fu, L.; Cui, Y.; Liu, X.; et al. Structure of MERS-CoV spike receptor-binding domain complexed with human receptor DPP4. *Cell Res.* **2013**, *23*, 986–993. [CrossRef]

57. Van Roy, M.; Ververken, C.; Beirnaert, E.; Hoefman, S.; Kolkman, J.; Vierboom, M.; Breedveld, E.; Hart, B.; Poelmans, S.; Bontinck, L.; et al. The preclinical pharmacology of the high affinity anti-IL-6R Nanobody(R) ALX-0061 supports its clinical development in rheumatoid arthritis. *Arthritis Res. Ther.* **2015**, *17*, 135. [CrossRef]

58. Raj, V.S.; Okba, N.M.; Gutierrez-Alvarez, J.; Drabek, D.; van Dieren, B.; Widagdo, W.; Lamers, M.M.; Widjaja, I.; Fernandez-Delgado, R.; Sola, I.; et al. Chimeric camel/human heavy-chain antibodies protect against MERS-CoV infection. *Sci. Adv.* **2018**, *4*, eaas9667.

viruses

MDPI

Review

Neutralizing Monoclonal Antibodies as Promising Therapeutics against Middle East Respiratory Syndrome Coronavirus Infection

Hui-Ju Han [1], Jian-Wei Liu [1], Hao Yu [2] and Xue-Jie Yu [1,*]

[1] School of Health Sciences, and State Key Laboratory of Virology, Wuhan University, Wuhan 430071, China; nikihuijuhan@163.com (H.-J.H.); liujw_2012@163.com (J.-W.L.)
[2] Fudan University School of Medicine, Shanghai 200032, China; howardyu89@163.com
* Correspondence: yuxuejie@whu.edu.cn

Received: 19 November 2018; Accepted: 29 November 2018; Published: 30 November 2018

Abstract: Since emerging in 2012, Middle East Respiratory Syndrome Coronavirus (MERS-CoV) has been a global public health threat with a high fatality rate and worldwide distribution. There are no approved vaccines or therapies for MERS until now. Passive immunotherapy with neutralizing monoclonal antibodies (mAbs) is an effective prophylactic and therapeutic reagent against emerging viruses. In this article, we review current advances in neutralizing mAbs against MERS-CoV. The receptor-binding domain (RBD) in the spike protein of MERS-CoV is a major target, and mouse, camel, or human-derived neutralizing mAbs targeting RBD have been developed. A major problem with neutralizing mAb therapy is mutant escape under selective pressure, which can be solved by combination of neutralizing mAbs targeting different epitopes. Neutralizing mAbs are currently under preclinical evaluation, and they are promising candidate therapeutic agents against MERS-CoV infection.

Keywords: Middle East Respiratory Syndrome Virus; MERS-CoV; neutralizing monoclonal antibodies

1. Introduction

Middle East Respiratory Syndrome (MERS) emerged in 2012 in Saudi Arabia with the death of a man with pneumonia; the causative agent was subsequently identified as MERS-CoV, which belonged to lineage C betacoronaviruses [1]. With dromedary camels (*Camelus dromedarius*, also known as Arabian camel) as direct sources and bats as potential reservoirs [2], MERS-CoV has been frequently introduced into human populations. Once MERS-CoV is introduced into a person, person-to-person transmission might occur, and is responsible for approximately 40% of MERS cases globally [3]. MERS-CoV has been a consistent threat to humans. As of October 2018, MERS-CoV has caused 2254 laboratory-confirmed human cases, including 800 deaths in 27 countries, with the fatality rate as high as 35% (http://www.who.int/emergencies/mers-cov/en/). Although MERS cases are primarily reported in the Middle East, facilitated by international travelling, MERS-CoV can also be a worldwide threat, which is well illustrated by the MERS outbreak in South Korea in 2015 [4]. Given the potential risk of causing worldwide public health emergencies and the absence of licensed vaccines and antiviral therapeutics, the World Health Organization has listed MERS-CoV in the "List of Blueprint priority diseases" (http://www.who.int/blueprint/priority-diseases/en/).

Vaccines are the most important approach against viral infections, but usually take a long time to develop. They are also unable to provide either immediate prophylactic protection or treat ongoing viral infections. Neutralizing monoclonal antibodies (mAbs) have recently emerged as a powerful tool to provide prophylactic and therapeutic protection against emerging viruses [5]. Potent neutralizing mAbs can be achieved by various technologies, such as hybridoma technology, humanized mouse, phage or yeast display, and single B cell isolation [5].

2. Spike (S) Protein of MERS-CoV as Target for Neutralizing mAbs

MERS-CoV is a single, positive-stranded RNA virus of about 30 kb, which encodes four major viral structural proteins—including spike (S), envelope (E), membrane (M) and nucleocapsid (N)—as well as several accessory proteins [6]. The S protein (1353 aa) plays an important role in virus infection and consists of a receptor-binding subunit S1 (aa 18–751) and a membrane-fusion subunit S2 (aa 752–1353). S1 mediates viral attachment to host cells and S2 mediates virus-cell membrane fusion [7]. The S1 subunit contains a receptor-binding domain (RBD) (aa 367–606) [8] that can bind to cell receptor dipeptidyl peptidase 4 (DPP4, also known as CD26), and mediates viral attachment target cells [9]. The RBD consists of a core subdomain and a receptor-binding motif (RBM) (aa 484–567). The schematic representation of MERS-CoV S protein is shown in Figure 1A.

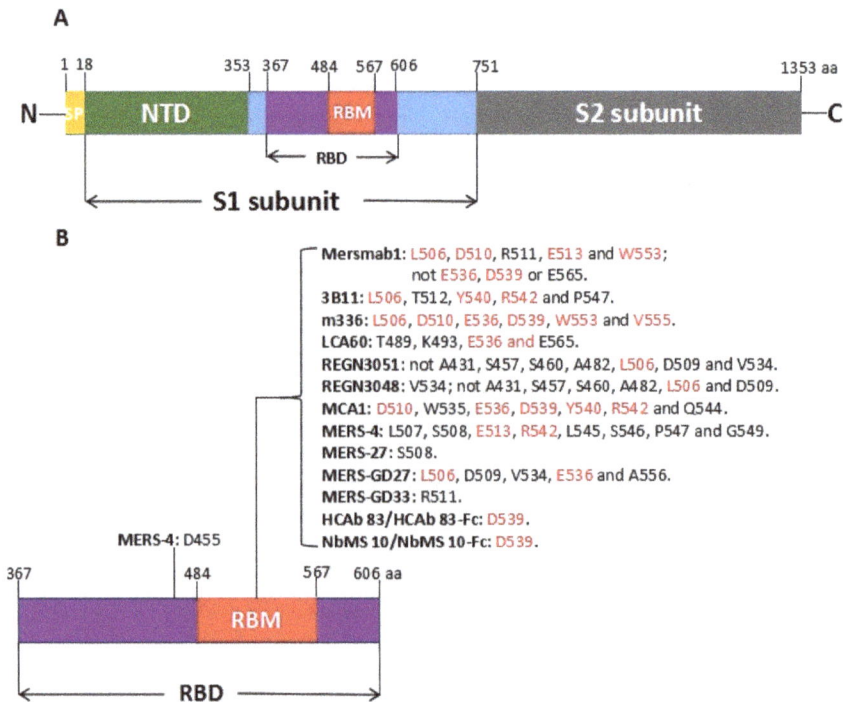

Figure 1. (**A**) Schematic representation of MERS-CoV S protein. (**B**) Residues on RBD critical for mAb neutralization. SP: signal peptide; NTD: N-terminal domain; RBD: receptor-binding domain; RBM: receptor-binding motif. Conserved residues on RBM critical for hDPP4 binding are shown in red.

Neutralizing mAbs binding to the S protein of MERS-CoV can prevent viral attachment to the cell receptor and inhibit viral entry [7]. The S protein of MERS-CoV is a key target for antivirals, and RBD is the most popular focus. In this study, we review the current knowledge on neutralizing mAbs targeting the RBD of MERS-CoV.

3. Mouse Neutralizing mAbs

3.1. 4C2 and 2E6

Stable hybridoma cell lines were generated by fusing myeloma cells with splenocytes of mice that were immunized with MERS-RBD protein. Two neutralizing mAbs, 4C2 and 2E6, had high affinity for the RBD of MERS-CoV and blocked both pseudovirus and live MERS-CoV entry into cells with high efficacy [10]. Humanized 4C2 showed similar neutralizing activity in cell entry tests. In vivo tests indicated that 4C2 could significantly reduce the virus titers in the lungs of Ad5-hCD26-transduced mice which were infected with MERS-CoV, highlighting its potential application in humans not only for preventing but also treating MERS-CoV infection. Crystallization of the 4C2 Fab/MERS-RBD complex showed that the 4C2 recognized conformational epitopes (Y397-N398, K400, L495-K496, P525, V527-S532, W535-E536, and D539-Q544), which were partially overlapped the receptor-binding footprint in the RBD of MERS-CoV. The 4C2 complex interfered with MERS-CoV binding to DPP4 by both steric hindrance and interface-residue competition. 2E6 competed with 4C2 to bind to MERS-RBD, indicating that they recognized proximate or overlapping epitopes [10].

3.2. Mersmab1

Neutralizing mAb Mersmab1 was obtained by fusing myeloma cells with splenocytes of a mouse that was immunized with recombinant MERS-CoV S1 [11]. Mersmab1 effectively blocked the entry of pseudovirus and live MERS-CoV into cells. Structural analysis showed that Mersmab1 bound to the RBD of MERS-CoV through recognizing conformational epitopes, and all of the residues critical for Mersmab1 binding were located on the left ridge of RBM. Mersmab1 neutralized MERS-CoV by competitively blocking the binding of MERS-CoV RBD to DPP4. Based on escape mutant analysis of the key residues on the RBD, it was found that residue L506, D510, R511, E513, and W553 were critical for Mersmab1 binding to the RBD, while mutation of E536, D539, or E565 did not affect the interaction of Mersmab1 and the RBD at all [11].

4. Human Neutralizing mAbs

4.1. 3B11

An ultra-large nonimmune human antibody-phage display library was constructed with B cells of unimmunized donors. With a unique spanning strategy, seven human neutralizing mAbs with varying neutralization efficacy to MERS-CoV were identified [12]. Binding detection demonstrated that the epitopes of these mAbs lay within aa 349–590 of the S protein, which overlapped a large part of the RBD of MERS-CoV. Binding competition assays showed that these mAbs recognized at least three distinct epitope groups, which was further confirmed by escape studies. With no cross-epitope resistance, these mAbs neutralized MERS-CoV by competitively blocking the binding of the RBD of MERS-CoV to DPP4. Escape mutant assays showed that five residues were critical for neutralization of these mAbs, namely L506, T512, Y540, R542, and P547. Of the seven mAbs, 3B11 exhibited the best neutralization activity against both pseudovirus and live MERS-CoV infectivity in cells. Moreover, under the selective pressure of these mAbs, the IgG form of 3B11 was superior, since it did not induce neutralization escape [12]. In vivo tests demonstrated that 3B11 reduced lung pathology in rhesus monkeys infected with MERS-CoV [13]. With its high neutralizing activity and suppression of mutant escape, 3B11 in the IgG form is a promising therapeutic mAb against MERS-CoV.

4.2. m336

Three human mAbs—m336, m337, and m338—were identified from a large naïve human phage display antibody library, which was constructed with peripheral blood mononuclear cells from healthy volunteers [14]. The binding sties of the three mAbs were within the RBD of MERS-CoV (aa 377–588), therefore they neutralized MERS-CoV by competing with DPP4 binding to the RBD. The three mAbs also competed with each other to bind to the RBD of MERS-CoV, and mutant analysis showed that the three mAbs possessed overlapping but distinct epitopes. Of the three mAbs, m336 neutralized both pseudovirus and live MERS-CoV infectivity in cells with exceptional potency (m336 inhibited 90% MERS-CoV pseudovirus infection at a concentration of 0.039 g/mL, and neutralized live MERS-CoV with IC_{95} of 1 g/mL and IC_{50} of 0.07 g/mL). Residues in the RBD crucial for m336 binding were L506, D510, E536, D539, W553, and V555 [14]. In vivo study demonstrated that prophylaxis with m336 reduced virus titers in the lung of rabbits infected with MERS-CoV [15], and m336 also provided transgenic mice expressing human DPP4 with full prophylactic and therapeutic protection from MERS-CoV [16]. However, another study with a non-human primate, the common marmoset showed that m336 could only alleviate the severity of the disease, and did not provide complete protection against MERS-CoV [17].

4.3. LCA60

IgG+ memory B cells were isolated from a MERS patient, and were subsequently immortalized with Epstein–Barr virus. A neutralizing mAbs, LCA60, was identified, and was the first fully human neutralizing mAb with naïve heavy and light chain pairs [18]. LCA60 efficiently neutralize MERS-CoV infectivity in cells. In vivo study showed that LCA60 provided BALB/c mice transduced with adenoviral vectors expressing human DPP4 (hDPP4) with both prophylactic and postexposure protection against MERS-CoV. Furthermore, the neutralizing efficacy of LCA60 was evaluated in IFN-α/β receptor-knockout mice that were more stringent models of MERS-CoV infection. After transducing with hDPP4, these mice showed more profound clinical symptoms when challenged with MERS-CoV [19]; administration of LCA60 reduced MERS-CoV titer in the lungs of these mice more effectively (lung viral titer reduced by three logs in one day for IFN-α/β receptor-knockout mice vs. three days for BALB/c mice) [18]. With naïve heavy and light chain pairs, LCA60 was more potent than 3B11 and comparable to m336. Cross-competition experiment demonstrated that LCA60 competed with 3B11 to bind to the RBD. LCA60 interacted with RBD residues around K493, and the LCA60 footprint on the RBD was partially overlapped with that of DPP4. Four residues in the RBD affected the binding of LCA60—namely T489, K493, E536, and E565—which were conserved in all MERS-CoV isolates. Moreover, compared with DPP4, the binding affinity of LCA60 to RBD was significantly higher (~500-fold). Therefore, one major neutralization mechanism of LCA60 was to competitively inhibit the interaction of the RBD with DPP4. Interestingly, virus escape studies demonstrated that under the selective pressure of LCA60, a mutant variant (V33A) in the N-terminal domain (NTD) of MERS-CoV S1 subunit was also generated [18]. A GMP-approved cell line (LCA60.273.1) that expresses LCA60 in high concentrations has been established, highlighting its application as promising therapeutics against MERS-CoV infection [20].

4.4. REGN3051 and REGN3048

Hybridoma B cells producing neutralizing mAbs against the S protein of MERS-CoV were generated by immunizing humanized transgenic mice (VelocImmune mice) with DNA encoding the MERS-CoV S protein. Two fully human neutralizing mAbs, REGN3051 and REGN3048, were obtained [21]. The two mAbs bound with high affinity to distinct epitopes on the RBD of MERS-CoV, which were conserved during the natural evolution of MERS-CoV. Mutation as a result of selective pressure by one mAb should not affect the binding of the other mAb. REGN3051 neutralized a broad range of MERS-CoV isolates, the prototype EMC/2012 strain and all clinical mutants including A431P, S457G, S460F, A482V, L506F, D509G, and V534A. With the exception of V534A variant, REGN3048 achieved similar neutralizing activity. In vivo study demonstrated that REGN3051 and REGN3048 reduced MERS-CoV replication in humanized DPP4 mice in both prophylactic and therapeutic settings [21]. When evaluated in the common marmoset, both mAbs seemed to be more effective for prophylaxis rather than for treatment of MERS-CoV infection [22].

4.5. MCA1

An anti-MERS-CoV phage display antibody library was constructed with the peripheral B cells of a MERS survivor, and a human neutralizing mAb against MERS-CoV, MCA1, was identified [23]. MCA1 showed potent neutralizing activity against MERS-CoV in cell entry tests. In vivo, MCA1 completely inhibited the replication of MERS-CoV in common marmosets when administrated prophylactically or therapeutically. Structure analysis of the MCA1 Fab-RBD complex showed that MCA1 formed direct contacts with the receptor-binding site (RBS) subdomain on the RBD. Epitopes on the RBS critical for MCA1 binding were D510, W535, E536, D539, Y540, R542, and Q544. Superimposed structure analysis of MCA1-RBD and hDPP4-RBD complexes showed that the binding interface of MCA1 was largely overlapped with that of hDPP4. Therefore, the neutralizing mechanism of MCA1 was achieved by competing with DPP4 for binding to the RBD [23].

4.6. MERS-4 and MERS-27

Two potent human neutralizing mAbs, MERS-4 and MERS-27, were derived from a nonimmune human yeast display antibody library, which was constructed with spleen and lymph node polyadenylated RNA from normal humans [24]. MERS-4 and MERS-27 inhibited the entry of both pseudovirus and live MERS-CoV into cells. Mutant analysis suggested that MERS-4 and MERS-27 recognized distinct epitopes in the RBD of MERS-CoV, and the epitopes of MERS-27 might be located away from those that recognized by MERS-4. Mutant analysis also demonstrated that residues D455, L507, E513, R542, L545, S546, P547, G549, and S508 in the RBD were critical for MERS-4 binding, while only S508 was important for MERS-27 binding. Combined use of MERS-4 and MERS-27 demonstrated a synergistic effect against pseudotyped MERS-CoV. MERS-4 bound to the RBD with about 45-fold higher affinity than DPP4. The primary neutralizing mechanism of MERS-4 and MERS-27 was through blocking the binding of the RBD to DPP4 [24]. Further structural analysis showed that MERS-4 bound to unique epitopes and caused conformational changes in the RBD interface critical for accommodating DPP4, therefore indirectly disrupting the interaction between the two. Moreover, MERS-4 also demonstrated synergistic effects with m336 and 5F9 (a NTD-specific mAb). The special neutralizing mechanism made MERS-4 a valuable addition for the combined use of mAbs against MERS-CoV infection [25].

4.7. MERS-GD27 and MERS-GD33

Thirteen ultrapotent neutralizing mAbs, which all targeted the RBD of MERS-CoV were generated following a protocol for the rapid production of antigen-specific human mAbs [26]. Briefly, antibody-secreting B cells were isolated from the whole blood of a MERS patient, and the antibody genes were amplified and cloned into vectors to transfect human cell lines for mAb production. Of the 13 mAbs, MERS-GD27 and MERS-GD33 exhibited the strongest neutralizing activity against both pseudovirus and live MERS-CoV in cell infection tests. MERS-GD27 directly competed with DPP4 to bind to the RBD to DPP4, and the crystal structure of MERS-GD27 showed that its epitopes were almost completely overlapped with DPP4-binding sites. MERS-GD27 and MERS-GD33 recognized distinct epitopes on the RBD, and had a low level of competing activity. The combined use of the two mAbs demonstrated synergistic effects in neutralization against pseudotyped MERS-CoV. Mutant analysis demonstrated that residues L506, D509, V534, E536, and A556 on RBD were important for the neutralizing activity of MERS-GD27, and residue R511was critical for MERS-GD33 [26]. Moreover, In vivo study found that MERS-GD27 could provide both prophylactic and therapeutic protection for hDPP4-trangenic mice against MERS-CoV infection [27].

5. Camel Neutralizing mAbs

Dromedary camels exposed to MERS-CoV showed mild clinical signs but developed exceptionally potent neutralizing antibodies. Camelid species naturally produced heavy chain-only antibodies

(HCAbs) [28], which are dimeric and devoid of light chains, and their antigen recognition region is solely formed by the variable heavy chains (VHHs) (also called nanobodies, Nbs). VHHs or Nbs have long complementarity-determining region 3 (CDR3) loops and are capable of binding to unique epitopes not accessible to conventional antibodies [29]. Notably, camelid VHHs are relatively stable and can be produced with high yields in prokaryotic systems [30]. Because of their small size; good tissue permeability; and cost-effective production, storage, and transportation [31–33], VHHs or Nbs have been gaining acceptance as antiviral agents.

5.1. Chimeric Camel/Human HCAb-83

A VHH complementary DNA library was constructed with the bone marrow of dromedary camels infected with MERS-CoV. Four VHHs (VHH-1, VHH-4, VHH-83, and VHH-101) with high neutralizing activity were identified by direct cloning and screening of the phage display antibody library [34]. The four VHHs competed for a single epitope that partially overlapped with the RBD-DPP4 interface. Mutant analysis showed that the four VHHs did not bind to the D539N variant, which was a critical residue on the RBD for DPP4 binding [8,35]. Therefore, these VHHs most likely neutralized MERS-CoV by blocking its binding to DPP4. Of the 4 VHHs, VHH-83 showed the best neutralizing activity and epitope recognition. VHH-83 efficiently blocked the entry of MERS-CoV into cells, and it also prophylactically protected K18 transgenic mouse expressing hDPP4 from MERS-CoV infection. To extend the half-life of VHH-83 in serum, it was linked to a human Fc domain lacking the CH1 exon to construct the chimeric camel/human HCAb-83, which showed similar neutralizing activity as VHH-83. The chimeric camel/human HCAb-83 was highly stable in mice and provided K18 mice with fully prophylactic protection against MERS-CoV infection [34].

5.2. NbMS10 and NbMS10-Fc

Alpacas were immunized with recombinant MERS-CoV RBD-containing a C-terminal human IgG1 Fc tag, and VHHs were amplified from their peripheral blood mononuclear cells to construct a VHH phage display library. A neutralizing Nb, NbMS10, which bound with high affinity to the RBD of MERS-CoV and blocked the binding of RBD to DPP4, was identified [36]. To extend its in vivo half-life, the human-Fc-fused version, NbMS10-Fc, was constructed. NbMS10 competed with DPP4 to bind to RBD, indicating that the binding site of NbMS10 on RBD overlapped with that of DPP4. The binding site of the NbMS10 on the RBD was mapped to be around residue D539, which is part of a highly conserved conformational epitope at the receptor-binding interface in almost all the natural MERS-CoV published to date. NbMS10 did not neutralize psuedotyped MERS-CoV bearing a mutation in D539, confirming that residue D539 was critical for NbMS10 binding. NbMS10 efficiently neutralized the cell entry of live MERS-CoV. Moreover, NbMS10 showed potent prophylactic and therapeutic efficacy in protecting hDPP4-transgenic mice against MERS-CoV infection [36].

6. Discussion

For their exceptionally high neutralization activity in vitro and in vivo, these newly identified neutralizing mAbs are promising candidate therapeutics against the infection of MERS-CoV. However, the use of a single neutralizing antibody bears the risk of selecting escape mutants, a fact that has been observed for LCA60 and other described antibodies [12,18,37]. Notably, the majority of these escape mutations had little impact on viral fitness and the interaction of DPP4 with the RBD [12]. Moreover, mutants of MERS-CoV during natural infection have also been reported [38]. Escape from neutralization is a major concern with therapeutic neutralizing mAbs, however, this potential problem can be solved by combining mAbs that target distinct epitopes and show different neutralizing mechanisms [37]. This strategy can take advantage of the synergistic effects while decreasing the possibility of viral escape.

Currently, most of the MERS-CoV neutralizing mAbs compete with DPP4 binding to the RBD, and residues on the RBD critical for mAb neutralization are identified by mutant analysis. Almost all of the residues identified critical for mAb neutralization are located in RBM, and overlap with those critical for DPP4 binding (Figure 1B). With the availability of crystal structure of mAb Fab-RBD complex, the neutralization mechanism of these mAbs will be better illustrated.

Based on the crystal structure of RBD-DPP4, it was found that several conserved residues in the RBD are critical for the interaction of the RBD with DPP4 (Y499, L506, D510, E513, E536, D537, D539, Y540, R542, W553, and V555) [35,39]. Development of therapeutic neutralizing mAbs targeting those critically conserved residues might be important for combating MERS-CoV. Moreover, a study found a mouse-derived neutralizing mAb, 5F9, which bound to a possible linear epitope in the NTD of the MERS-CoV S1 subunit, exhibited efficient neutralizing activity against pseudovirus and live MERS-CoV in cell entry tests. This study highlighted the important role of NTD during the infection process of MERS-CoV. NTD might have significant implications for the development of prophylactic and therapeutic mAbs against MERS-CoV infection [40]. Although the in vitro neutralizing potency of 5F9 was approximately 10-fold lower than that of the RBD-targeting neutralizing mAbs [40], it may provide an alternative for the immunotherapy against MERS-CoV, once the virus mutates and is no longer susceptible to RBD-specific mAbs.

So far, there is a lack of appropriate animal models to mimic the pathology of MERD-CoV in humans. Commonly-used laboratory animals—such as wild-type mouse, ferret, hamster, and guinea pig—are not susceptible to MERS-CoV infection due to differences in critical amino acids in the S-binding domain of their DPP4 [41–43]. New Zealand rabbits, hDPP4-transduced/transgenic mice, camelids and non-human primates (rhesus macaque and common marmoset) are susceptible to MERS-CoV infection, however, rabbits showed asymptomatic infection [44]; dromedary camels displayed different clinical manifestations to that of humans [45]; rhesus macaque only showed transient lower respiratory infection [46], while common marmoset developed progressive pneumonia [47]; hDPP4-trangenic mouse expressed hDPP4 extensively, and resulted in multiple organ damage [48]; hDPP4-transduced mouse only exhibited mild transient clinical diseases [19]. With robust animal models, the protective effects of these neutralizing mAbs will be better evaluated. Furthermore, ongoing efforts on developing therapeutic neutralizing mAbs against MERS-CoV should also consider the different target populations (dromedary camels and humans) and their protective efficacy.

Author Contributions: H.J.H. conceived the ideas; H.J.H., J.W.L., H.Y., and X.J.Y. wrote the paper.

Funding: This study was supported by a grant from National Natural Science Funds of China (nos. 31570167).

Conflicts of Interest: The authors declare that they have no conflict of interest.

References

1. Zaki, A.M.; van Boheemen, S.; Bestebroer, T.M.; Osterhaus, A.D.M.E.; Fouchier, R.A.M. Isolation of a Novel Coronavirus from a Man with Pneumonia in Saudi Arabia. *N. Engl. J. Med.* **2012**, *367*, 1814–1820. [CrossRef] [PubMed]

2. Han, H.J.; Yu, H.; Yu, X.J. Evidence for zoonotic origins of Middle East respiratory syndrome coronavirus. *J. Gen. Virol.* **2016**, *97*, 274–280. [CrossRef] [PubMed]

3. Hui, D.S.; Azhar, E.I.; Kim, Y.J.; Memish, Z.A.; Oh, M.D.; Zumla, A. Middle East respiratory syndrome coronavirus: Risk factors and determinants of primary, household, and nosocomial transmission. *Lancet Infect. Dis.* **2018**, *18*, E217–E227. [CrossRef]

4. Lee, S.S.; Wong, N.S. Probable transmission chains of Middle East respiratory syndrome coronavirus and the multiple generations of secondary infection in South Korea. *Int. J. Infect. Dis.* **2015**, *38*, 65–67. [CrossRef] [PubMed]

5. Jin, Y.J.; Lei, C.; Hu, D.; Dimitrov, D.S.; Ying, T.L. Human monoclonal antibodies as candidate therapeutics against emerging viruses. *Front. Med.* **2017**, *11*, 462–470. [CrossRef] [PubMed]

6. Van Boheemen, S.; de Graaf, M.; Lauber, C.; Bestebroer, T.M.; Raj, V.S.; Zaki, A.M.; Osterhaus, A.D.M.E.; Haagmans, B.L.; Gorbalenya, A.E.; Snijder, E.J.; et al. Genomic Characterization of a Newly Discovered Coronavirus Associated with Acute Respiratory Distress Syndrome in Humans. *mBio* **2012**, *3*, e00473-12. [CrossRef] [PubMed]

7. Du, L.Y.; Yang, Y.; Zhou, Y.S.; Lu, L.; Li, F.; Jiang, S.B. MERS-CoV spike protein: A key target for antivirals. *Expert Opin. Ther. Targets* **2017**, *21*, 131–143. [CrossRef] [PubMed]
8. Wang, N.S.; Shi, X.L.; Jiang, L.W.; Zhang, S.Y.; Wang, D.L.; Tong, P.; Guo, D.X.; Fu, L.L.; Cui, Y.; Liu, X.; et al. Structure of MERS-CoV spike receptor-binding domain complexed with human receptor DPP4. *Cell Res.* **2013**, *23*, 986–993. [CrossRef] [PubMed]
9. Raj, V.S.; Mou, H.H.; Smits, S.L.; Dekkers, D.H.W.; Muller, M.A.; Dijkman, R.; Muth, D.; Demmers, J.A.A.; Zaki, A.; Fouchier, R.A.M.; et al. Dipeptidyl peptidase 4 is a functional receptor for the emerging human coronavirus-EMC. *Nature* **2013**, *495*, 251–254. [CrossRef] [PubMed]
10. Li, Y.; Wan, Y.H.; Liu, P.P.; Zhao, J.C.; Lu, G.W.; Qi, J.X.; Wang, Q.H.; Lu, X.C.; Wu, Y.; Liu, W.J.; et al. A humanized neutralizing antibody against MERS-CoV targeting the receptor-binding domain of the spike protein. *Cell Res.* **2015**, *25*, 1237–1249. [CrossRef] [PubMed]
11. Du, L.Y.; Zhao, G.Y.; Yang, Y.; Qiu, H.J.; Wang, L.L.; Kou, Z.H.; Tao, X.R.; Yu, H.; Sun, S.H.; Tseng, C.T.K.; et al. A Conformation-Dependent Neutralizing Monoclonal Antibody Specifically Targeting Receptor-Binding Domain in Middle East Respiratory Syndrome Coronavirus Spike Protein. *J. Virol.* **2014**, *88*, 7045–7053. [CrossRef] [PubMed]
12. Tang, X.C.; Agnihothram, S.S.; Jiao, Y.J.; Stanhope, J.; Graham, R.L.; Peterson, E.C.; Avnir, Y.; Tallarico, A.S.; Sheehan, J.; Zhu, Q.; et al. Identification of human neutralizing antibodies against MERS-CoV and their role in virus adaptive evolution. *Proc. Natl. Acad. Sci. USA* **2014**, *111*, 6863. [CrossRef] [PubMed]
13. Johnson, R.F.; Bagci, U.; Keith, L.; Tang, X.C.; Mollura, D.J.; Zeitlin, L.; Qin, J.; Huzella, L.; Bartos, C.J.; Bohorova, N.; et al. 3B11-N, a monoclonal antibody against MERS-CoV, reduces lung pathology in rhesus monkeys following intratracheal inoculation of MERS-CoV Jordan-n3/2012. *Virology* **2016**, *490*, 49–58. [CrossRef] [PubMed]
14. Ying, T.L.; Du, L.Y.; Ju, T.W.; Prabakaran, P.; Lau, C.C.Y.; Lu, L.; Liu, Q.; Wang, L.L.; Feng, Y.; Wang, Y.P.; et al. Exceptionally Potent Neutralization of Middle East Respiratory Syndrome Coronavirus by Human Monoclonal Antibodies. *J. Virol.* **2014**, *88*, 7796–7805. [CrossRef] [PubMed]
15. Houser, K.V.; Gretebeck, L.; Ying, T.L.; Wang, Y.P.; Vogel, L.; Lamirande, E.W.; Bock, K.W.; Moore, I.N.; Dimitrov, D.S.; Subbarao, K. Prophylaxis with a Middle East Respiratory Syndrome Coronavirus (MERS-CoV)-Specific Human Monoclonal Antibody Protects Rabbits From MERS-CoV Infection. *J. Infect. Dis.* **2016**, *213*, 1557–1561. [CrossRef] [PubMed]
16. Agrawal, A.S.; Ying, T.L.; Tao, X.R.; Garron, T.; Algaissi, A.; Wang, Y.P.; Wang, L.L.; Peng, B.H.; Jiang, S.B.; Dimitrov, D.S.; et al. Passive Transfer of a Germline-like Neutralizing Human Monoclonal Antibody Protects Transgenic Mice Against Lethal Middle East Respiratory Syndrome Coronavirus Infection. *Sci. Rep.* **2016**, *6*, 31629. [CrossRef] [PubMed]
17. Van Doremalen, N.; Falzarano, D.; Ying, T.L.; de Wit, E.; Bushmaker, T.; Feldmann, F.; Okumura, A.; Wang, Y.; Scott, D.P.; Hanley, P.W.; et al. Efficacy of antibody-based therapies against Middle East respiratory syndrome coronavirus (MERS-CoV) in common marmosets. *Antivir. Res.* **2017**, *143*, 30–37. [CrossRef] [PubMed]
18. Corti, D.; Zhao, J.C.; Pedotti, M.; Simonelli, L.; Agnihothram, S.; Fett, C.; Fernandez-Rodriguez, B.; Foglierini, M.; Agatic, G.; Vanzetta, F.; et al. Prophylactic and postexposure efficacy of a potent human monoclonal antibody against MERS coronavirus. *Proc. Natl. Acad. Sci. USA* **2015**, *112*, 10473–10478. [CrossRef] [PubMed]
19. Zhao, J.C.; Li, K.; Wohlford-Lenane, C.; Agnihothram, S.S.; Fett, C.; Zhao, J.X.; Gale, M.J.; Baric, R.S.; Enjuanes, L.; Gallagher, T.; et al. Rapid generation of a mouse model for Middle East respiratory syndrome. *Proc. Natl. Acad. Sci. USA* **2014**, *111*, 4970–4975. [CrossRef] [PubMed]
20. Corti, D.; Passini, N.; Lanzavecchia, A.; Zambon, M. Rapid generation of a human monoclonal antibody to combat Middle East respiratory syndrome. *J. Infect. Public Health* **2016**, *9*, 231–235. [CrossRef] [PubMed]
21. Pascal, K.E.; Coleman, C.M.; Mujica, A.O.; Kamat, V.; Badithe, A.; Fairhurst, J.; Hunt, C.; Strein, J.; Berrebi, A.; Sisk, J.M.; et al. Pre- and postexposure efficacy of fully human antibodies against Spike protein in a novel humanized mouse model of MERS-CoV infection. *Proc. Natl. Acad. Sci. USA* **2015**, *112*, 8738–8743. [CrossRef] [PubMed]
22. De Wit, E.; Feldmann, F.; Okumura, A.; Horne, E.; Haddock, E.; Saturday, G.; Scott, D.; Erlandson, K.J.; Stahl, N.; Lipsich, L.; et al. Prophylactic and therapeutic efficacy of mAb treatment against MERS-CoV in common marmosets. *Antivir. Res.* **2018**, *156*, 64–71. [CrossRef] [PubMed]

23. Chen, Z.; Bao, L.L.; Chen, C.; Zou, T.T.; Xue, Y.; Li, F.D.; Lv, Q.; Gu, S.Z.; Gao, X.P.; Cui, S.; et al. Human Neutralizing Monoclonal Antibody Inhibition of Middle East Respiratory Syndrome Coronavirus Replication in the Common Marmoset. *J. Infect. Dis.* **2017**, *215*, 1807–1815. [CrossRef] [PubMed]
24. Jiang, L.W.; Wang, N.S.; Zuo, T.; Shi, X.L.; Poon, K.M.V.; Wu, Y.K.; Gao, F.; Li, D.Y.; Wang, R.K.; Guo, J.Y.; et al. Potent Neutralization of MERS-CoV by Human Neutralizing Monoclonal Antibodies to the Viral Spike Glycoprotein. *Sci. Transl. Med.* **2014**, *6*, 234ra59. [CrossRef] [PubMed]
25. Zhang, S.Y.; Zhou, P.P.; Wang, P.F.; Li, Y.Y.; Jiang, L.W.; Jia, W.X.; Wang, H.; Fan, A.; Wang, D.L.; Shi, X.L.; et al. Structural Definition of a Unique Neutralization Epitope on the Receptor-Binding Domain of MERS-CoV Spike Glycoprotein. *Cell Rep.* **2018**, *24*, 441–452. [CrossRef] [PubMed]
26. Niu, P.; Zhang, S.; Zhou, P.; Huang, B.; Deng, Y.; Qin, K.; Wang, P.; Wang, W.; Wang, X.; Zhou, J.; et al. Ultrapotent Human Neutralizing Antibody Repertoires Against Middle East Respiratory Syndrome Coronavirus from a Recovered Patient. *J. Infect. Dis.* **2018**, *218*, 1249–1260. [CrossRef] [PubMed]
27. Niu, P.; Zhao, G.; Deng, Y.; Sun, S.; Wang, W.; Zhou, Y.; Tan, W. A novel human mAb (MERS-GD27) provides prophylactic and postexposure efficacy in MERS-CoV susceptible mice. *Sci. China Life Sci.* **2018**, *61*, 1280–1282. [CrossRef] [PubMed]
28. Hamers-casterman, C.; Atarhouch, T.; Muyldermans, S.; Robinson, G.; Hamers, C.; Songa, E.B.; Bendahman, N.; Hamers, R. Naturally-Occurring Antibodies Devoid of Light-Chains. *Nature* **1993**, *363*, 446–448. [CrossRef] [PubMed]
29. De Genst, E.; Silence, K.; Decanniere, K.; Conrath, K.; Loris, R.; Kinne, J.; Muyldermans, S.; Wyns, L. Molecular basis for the preferential cleft recognition by dromedary heavy-chain antibodies. *Proc. Natl. Acad. Sci. USA* **2016**, *103*, 4586–4591. [CrossRef] [PubMed]
30. Muyldermans, S. Nanobodies: Natural Single-Domain Antibodies. *Annu. Rev. Biochem.* **2013**, *82*, 775–797. [CrossRef] [PubMed]
31. Wilken, L.; McPherson, A. Application of camelid heavy-chain variable domains (VHHs) in prevention and treatment of bacterial and viral infections. *Int. Rev. Immunol.* **2018**, *37*, 69–76. [CrossRef] [PubMed]
32. Van Heeke, G.; Allosery, K.; De Brabandere, V.; De Smedt, T.; Detalle, L.; de Fougerolles, A. Nanobodies(R) as inhaled biotherapeutics for lung diseases. *Pharmacol. Ther.* **2017**, *169*, 47–56. [CrossRef] [PubMed]
33. Detalle, L.; Stohr, T.; Palomo, C.; Piedra, P.A.; Gilbert, B.E.; Mas, V.; Millar, A.; Power, U.F.; Stortelers, C.; Allosery, K.; et al. Generation and Characterization of ALX-0171, a Potent Novel Therapeutic Nanobody for the Treatment of Respiratory Syncytial Virus Infection. *Antimicrob. Agents Chemother.* **2016**, *60*, 6–13. [CrossRef] [PubMed]
34. Stalin Raj, V.; Okba, N.M.A.; Gutierrez-Alvarez, J.; Drabek, D.; van Dieren, B.; Widagdo, W.; Lamers, M.M.; Widjaja, I.; Fernandez-Delgado, R.; Sola, I.; et al. Chimeric camel/human heavy-chain antibodies protect against MERS-CoV infection. *Sci. Adv.* **2018**, *4*, eaas9667. [CrossRef] [PubMed]
35. Lu, G.W.; Hu, Y.W.; Wang, Q.H.; Qi, J.X.; Gao, F.; Li, Y.; Zhang, Y.F.; Zhang, W.; Yuan, Y.; Bao, J.K.; et al. Molecular basis of binding between novel human coronavirus MERS-CoV and its receptor CD26. *Nature* **2013**, *500*, 227. [CrossRef] [PubMed]
36. Zhao, G.; He, L.; Sun, S.; Qiu, H.; Tai, W.; Chen, J.; Li, J.; Chen, Y.; Guo, Y.; Wang, Y.; et al. A Novel Nanobody Targeting Middle East Respiratory Syndrome Coronavirus (MERS-CoV) Receptor-Binding Domain Has Potent Cross-Neutralizing Activity and Protective Efficacy against MERS-CoV. *J. Virol.* **2018**, *92*, e00837-18. [CrossRef] [PubMed]
37. Wang, L.; Shi, W.; Chappell, J.D.; Joyce, M.G.; Zhang, Y.; Kanekiyo, M.; Becker, M.M.; van Doremalen, N.; Fischer, R.; Wang, N.; et al. Importance of neutralizing monoclonal antibodies targeting multiple antigenic sites on MERS-CoV Spike to avoid neutralization escape. *J. Virol.* **2018**. [CrossRef]
38. Bermingham, A.; Chand, M.A.; Brown, C.S.; Aarons, E.; Tong, C.; Langrish, C.; Hoschler, K.; Brown, K.; Galiano, M.; Myers, R.; et al. Severe respiratory illness caused by a novel coronavirus, in a patient transferred to the United Kingdom from the Middle East, September 2012. *Eurosurveillance* **2012**, *17*, 20290. [PubMed]
39. Song, W.F.; Wang, Y.; Wang, N.S.; Wang, D.L.; Guo, J.Y.; Fu, L.L.; Shi, X.L. Identification of residues on human receptor DPP4 critical for MERS-CoV binding and entry. *Virology* **2014**, *471*, 49–53. [CrossRef] [PubMed]
40. Chen, Y.Z.; Lu, S.; Jia, H.; Deng, Y.; Zhou, J.F.; Huang, B.Y.; Yu, Y.Y.; Lan, J.M.; Wang, W.L.; Lou, Y.L.; et al. A novel neutralizing monoclonal antibody targeting the N-terminal domain of the MERS-CoV spike protein. *Emerg. Microbes Infect.* **2017**, *6*, e37. [CrossRef] [PubMed]

41. Coleman, C.M.; Matthews, K.L.; Goicochea, L.; Frieman, M.B. Wild-type and innate immune-deficient mice are not susceptible to the Middle East respiratory syndrome coronavirus. *J. Gen. Virol.* **2014**, *95*, 408–412. [CrossRef] [PubMed]

42. De Wit, E.; Prescott, J.; Baseler, L.; Bushmaker, T.; Thomas, T.; Lackemeyer, M.G.; Martellaro, C.; Milne-Price, S.; Haddock, E.; Haagmans, B.L.; et al. The Middle East respiratory syndrome coronavirus (MERS-CoV) does not replicate in Syrian hamsters. *PLoS ONE* **2013**, *8*, e69127. [CrossRef] [PubMed]

43. Raj, V.S.; Smits, S.L.; Provacia, L.B.; van den Brand, J.M.; Wiersma, L.; Ouwendijk, W.J.; Bestebroer, T.M.; Spronken, M.I.; van Amerongen, G.; Rottier, P.J.; et al. Adenosine deaminase acts as a natural antagonist for dipeptidyl peptidase 4-mediated entry of the Middle East respiratory syndrome coronavirus. *J. Virol.* **2014**, *88*, 1834–1838. [CrossRef] [PubMed]

44. Haagmans, B.L.; van den Brand, J.M.A.; Provacia, L.B.; Raj, V.S.; Stittelaar, K.J.; Getu, S.; de Waal, L.; Bestebroer, T.M.; van Amerongen, G.; Verjans, G.M.G.M.; et al. Asymptomatic Middle East Respiratory Syndrome Coronavirus Infection in Rabbits. *J. Virol.* **2015**, *89*, 6131–6135. [CrossRef] [PubMed]

45. Haverkamp, A.K.; Lehmbecker, A.; Spitzbarth, I.; Widagdo, W.; Haagmans, B.L.; Segales, J.; Vergara-Alert, J.; Bensaid, A.; van den Brand, J.M.A.; Osterhaus, A.; et al. Experimental infection of dromedaries with Middle East respiratory syndrome-Coronavirus is accompanied by massive ciliary loss and depletion of the cell surface receptor dipeptidyl peptidase 4. *Sci. Rep.* **2018**, *8*, 9778. [CrossRef] [PubMed]

46. Yao, Y.; Bao, L.; Deng, W.; Xu, L.; Li, F.; Lv, Q.; Yu, P.; Chen, T.; Xu, Y.; Zhu, H.; et al. An animal model of MERS produced by infection of rhesus macaques with MERS coronavirus. *J. Infect. Dis.* **2014**, *209*, 236–242. [CrossRef] [PubMed]

47. Falzarano, D.; de Wit, E.; Feldmann, F.; Rasmussen, A.L.; Okumura, A.; Peng, X.X.; Thomas, M.J.; van Doremalen, N.; Haddock, E.; Nagy, L.; et al. Infection with MERS-CoV Causes Lethal Pneumonia in the Common Marmoset. *PLoS Pathog.* **2014**, *10*, e1004250. [CrossRef] [PubMed]

48. Zhao, G.; Jiang, Y.; Qiu, H.; Gao, T.; Zeng, Y.; Guo, Y.; Yu, H.; Li, J.; Kou, Z.; Du, L.; et al. Multi-Organ Damage in Human Dipeptidyl Peptidase 4 Transgenic Mice Infected with Middle East Respiratory Syndrome-Coronavirus. *PLoS ONE* **2015**, *10*, e0145561. [CrossRef] [PubMed]

![viruses logo] *viruses*

MDPI

Article

Potent MERS-CoV Fusion Inhibitory Peptides Identified from HR2 Domain in Spike Protein of Bat Coronavirus HKU4

Shuai Xia [1], Qiaoshuai Lan [1], Jing Pu [1], Cong Wang [1], Zezhong Liu [1], Wei Xu [1], Qian Wang [1], Huan Liu [2,*], Shibo Jiang [1,3,*] and Lu Lu [1,*]

[1] Key Laboratory of Medical Molecular Virology of MOE/MOH, School of Basic Medical Sciences and Shanghai Public Health Clinical Center, Fudan University, Shanghai 200032, China; 15111010053@fudan.edu.cn (S.X.); 18111010010@fudan.edu.cn (Q.L.); 17111010015@fudan.edu.cn (J.P.); 16111010068@fudan.edu.cn (C.W.); 17111010065@fudan.edu.cn (Z.L.); xuwei11@fudan.edu.cn (W.X.); wang_qian@fudan.edu.cn (Q.W.)

[2] State Key Laboratory of Virology, Wuhan Institute of Virology, Chinese Academy of Sciences, Wuhan 430071, China

[3] Lindsley F. Kimball Research Institute, New York Blood Center, New York, NY 10065, USA

* Correspondence: liuhuan@wh.iov.cn (H.L.); shibojiang@fudan.edu.cn (S.J.); lul@fudan.edu.cn (L.L.); Tel.: +86-027-8719-9810 (H.L.); +1-212-570-3058 (S.J.); +86-21-5423-7671 (L.L.)

Received: 7 December 2018; Accepted: 10 January 2019; Published: 14 January 2019

Abstract: The Middle East respiratory syndrome coronavirus (MERS-CoV) emerged in 2012 and caused continual outbreaks worldwide with high mortality. However, no effective anti-MERS-CoV drug is currently available. Recently, numerous evolutionary studies have suggested that MERS-CoV originated from bat coronavirus (BatCoV). We herein reported that three peptides derived from the HR2 region in spike protein of BatCoV HKU4, including HKU4-HR2P1, HKU4-HR2P2 and HKU4-HR2P3, could bind the MERS-CoV HR1-derived peptide to form a six-helix bundle (6-HB) with high stability. Moreover, these peptides, particularly HKU4-HR2P2 and HKU4-HR2P3, exhibited potent inhibitory activity against MERS-CoV S-mediated cell–cell fusion and viral infection, suggesting that these HKU4 HR2-derived peptides could be candidates for futher development as antiviral agents against MERS-CoV infection.

Keywords: MERS-CoV; fusion inhibitor; peptide; cell–cell fusion; HKU4

1. Introduction

Middle East respiratory syndrome coronavirus (MERS-CoV), a novel human coronavirus (Figure 1A), emerged in Saudi Arabia in 2012 [1], rapidly spread wordwide, and caused continuous outbreaks with significant mortality and morbidity [2,3]. As of 27 November 2018, the World Health Organization (WHO) reported 2266 laboratory-confirmed cases of infection with MERS-CoV and 804 MERS-associated deaths (about 36% fatality rate) in 27 countries (available online: http://www.who.int/emergencies/mers-cov/en/). MERS-CoV uses dipeptidyl peptidase-4 (DPP4, also named CD26), a type-II transmembrane glycoprotein, as the cellular receptor to infect humans and cause severe respiratory disease and other severe complications, including renal failure and even multiorgan failure [4–8]. However, no licensed vaccines or effective therapeutics against MERS-CoV infection have been approved by the U.S. Food and Drug Administration (FDA). This calls for the urgent development of effective therapeutics against MERS-CoV infection for clinical use.

Similar to other coronaviruses [9], MERS-CoV is an enveloped positive-sense single-stranded RNA virus and belongs to the lineage C betacoronavirus in the family coronaviridae [10]. MERS-CoV displays its spike protein (S), a type I transmembrane glycoprotein, on its enveloped membrane surface in a trimer state. S protein contains two subunits: S1 subunit (S1) and S2 subunit (S2). S1 recognizes and binds the cellular receptor, while S2 mediates viral and cellular membrane fusion. The interaction between the receptor-binding domain (RBD) in S1 and the receptor (DPP4) on the host cell surface induces S2 structural changes, including the insertion of N-terminal fusion peptide into the host cell membrane, the formation of a homotrimer by heptad repeat 1 (HR1) helices, exposing three hydrophobic grooves on the surface, and the binding of three heptad repeat 2 (HR2) molecules to the hydrophobic grooves of HR1 trimer to form a six-helix bundle (6-HB), which brings the viral and cell membranes close together for fusion. Finally, MERS-CoV releases its genetic materials into target cells (Figure 1B) [11,12]. Disrupting the process of MERS-6HB formation can inhibit virus–cell membrane fusion and abolish viral infection [13,14]. Our previous studies found that a peptide, MERS-HR2P, derived from the MERS-CoV HR2 region, could prevent the formation of the fusion core by competitively binding to HR1 and blocking the native interaction between HR1 and HR2 (Figure 1B). MERS-HR2P has no effect on SARS-CoV infection, and SARS-CoV HR2-derived peptide, CP-1, failed to inhibit MERS-CoV infection [14,15], suggesting that MERS-HR2P is specific for inibiting MERS-CoV infection. Whether other novel and potent peptide fusion inhibitors can be developed against MERS-CoV from other coronavirus HR2 regions is still unclear.

Although camels are the intermediate host for the transmission of MERS-CoV, bats are likely to be its original source and serve as the natural host [10,16–18]. Phylogenetically, MERS-CoV is very closely related to *Tylonycteris sp. Bat CoV* HKU4 (Ty-BatCoV HKU4) (Figure 1A). Similar to MERS-CoV, HKU4 also harbors the ability to use human DPP4 as the receptor for infecting the target cell [19,20]. We herein identified that HKU4-HR2P1, HKU4-HR2P2 and HKU4-HR2P3 peptides, derived from the HKU4 HR2 region, could bind MERS-HR1P, a MERS-CoV HR1-derived peptide, and mimic the interaction of MERS-CoV HR1 and HR2 regions to form stable 6-HB. Moreover, these peptides exhibited significant inhibitory activity in both MERS-CoV S-mediated cell–cell fusion assay and pseudovirus infection assay, with even more potency than that of the reported peptide, MERS-HR2P. Taken together, these HKU4 HR2-derived peptides could be considered as candidates for the development of anti-MERS-CoV agents for clinical use.

2. Materials and Methods

2.1. Cells, Viruses and Peptides

Huh-7 and 293T cells were obtained from the Chinese Academy of Sciences Cell Bank (Shanghai, China). All cell lines were cultured in Dulbecco's Modified Eagle's Medium (DMEM) with 10% fetal bovine serum (FBS). Plasmids, including pcDNA3.1-MERS-S, pcDNA3.1-MERS-S with Q1020H or Q1020R mutation, pAAV-IRES-S-GFP and pNL4-3.Luc.R-E, were constructed in our laboratory. All peptides were synthesized by KareBay Biochem (Monmouth Junction, NJ, USA) with HPLC purity >90% [21,22]. The average hydrophilicity of peptides was predicted by using an online peptide-calculator program (available online: http://www.bachem.com/service-support/peptide-calculator/).

2.2. Coronavirus Phylogenetic Analysis

Phylogenetic tree was constructed using MEGA6.06. Accession numbers utilized for phylogenetic analysis are as follows: BtCoV-HKU4 (NC_009019.1), BtCoV-HKU5 (NC_009020), MERS-CoV (AID55097.1), BtSCoV-WIV1 (KF367457), BtCoV-HKU9 (EF065516), and IBV (KY421672).

2.3. Circular Dichroism Spectroscopic Analysis

Circular dichroism spectra (198-260 nm) were collected on a J-815 spectropolarimeter (Jasco, Inc., Tokyo, Japan) to evaluate the secondary structure of the individual peptide or the complexes dissolved in phosphate-buffered saline (PBS) with the final concentration at 10 μM [15,23]. The ellipticity value of $-33,000 \deg \text{cm}^2 \text{dmol}^{-1}$ at 222 nm was taken as 100% α-helicity [24,25]. Thermal denaturation of peptide complexes was monitored from 20 °C to 100 °C at 222 nm with a thermal gradient of 5°C/ min. The midpoint of the thermal unfolding transition (Tm) values was acquired by Jasco software utilities.

2.4. Native Polyacrylamide Gel Electrophoresis (N-PAGE)

Native polyacrylamide gel electrophoresis (N-PAGE) was conducted as described elsewhere [15]. Briefly, each HKU4 HR2-derived peptide (40 μM) in PBS was incubated with HKU4-HR1P or MERS-HR1Ps (20, 40, 80 μM), respectively, at 37°C for 30 min, using PBS as control, and then loaded on a tris-glycine gel (12%) with tricine glycine running buffer (pH 8.6). Finally, staining was performed with Coomassie blue, and the images were visualized on the FluorChem Imaging System (Alpha Innotech/ProteinSimple).

2.5. Cell–Cell Fusion Assay

The assay for MERS-CoV S protein-mediated cell–cell fusion was performed as described elsewhere [14,26–28]. Briefly, 293T cells transiently co-expressing MERS-CoV S protein and GFP protein on cell surface and in cytoplasm, respectively, were used as the effector cells (293T/S/GFP), and Huh-7 cells were used as the target cells. Then the effector cells (1×10^4 cells per well) and target cells (5×10^4 cells per well) were cocultured in the wells of a flat-bottom 96-well plate at 37 °C for 2 h in the presence or absence of peptides at the indicated concentrations. Finally, the fused and unfused cells were visualized, photographed and counted under an inverted fluorescence microscope (Nikon Eclipse Ti-S). PBS without peptides was used as no inhibition control, and the median inhibitory concentration (IC_{50}) was calculated using the CalcuSyn software [29].

2.6. Generation and Packaging of Middle East Respiratory Syndrome Coronavirus (MERS-CoV) Pseudovirus

The package of MERS-CoV pseudovirus was described in our previous studies [14,27]. Briefly, we cotransfected 293T cells with plasmid pcDNA3.1-MERS-S with or without Q1020H/R mutation and plasmid pNL4-3.luc.R-E encoding Env-defective, luciferase-expressing HIV-1 [30]. Then, the pseudoviruses in the supernatant were collected 48–72 h post-transfection and quantified by the level of lentivirus p24. Target cells, Huh-7, were preplated in 96-well plates (1×10^4 cells per well). The MERS-CoV pseudovirus was premixed with peptide at indicated concentration and incubated for 30 min at 37 °C. PBS was used as no inhibition control. Then, the mixture was added to the Huh-7 cells, replaced by fresh medium 12 h post-infection, and then incubated for an additional 48 h. Transduced Huh-7 cells were lysed for detection of relative light units (RLU) according to the luciferase assay system manual (Promega, Madison, WI, USA) [31].

2.7. Time-of-Addition Assay

The time-of-addition assay was conducted as described in our previous study [14]. Briefly, Huh-7 cells plated in 96-well plates were incubated with MERS-CoV PsV, while each peptide at a final concentration of 10 μM was added 0, 0.5, 1, 2, 4 or 6 h post-infection. Cells were lysed 72 h later to determine the entry inhibition ratio.

2.8. Time-of-Removal Assay

Each peptide (10 μM) was added to Huh-7 cells, followed by an incubation at 37 °C for 30 min, respectively. The cells were then washed with PBS to remove the unbound peptide and the cells that were not washed were used as control, followed by addition of the MERS-CoV pseudovirus. After co-culture at 37 °C for 3 days, cells were lysed to test the inhibitory activity of peptides.

2.9. Cytotoxicity Assay

The cytotoxicity of peptides to Huh-7 cells was measured following the instructions in the manual provided in the Cell Counting Kit-8 (CCK-8; Dojindo, Kumamoto, Kyushu, Japan) [22]. Briefly, each peptide with indicated concentrations was mixed with 10^4 target cells in each well of 96-well plates to co-incubate for 48 h. Then culture medium was replaced by fresh cell medium with 4% CCK-8 solution for an additional incubation 2 h at 37 °C. Finally, the absorbance was detected at 450 nm (A450).

3. Results

3.1. Design of HKU4 HR2-Peptides

Our previous studies crystallized the fusion core structure of the MERS-CoV S2 subunit and designed peptides MERS-HR1P and MERS-HR2P based on the fusion structure [14]. These peptides can mimic the interaction of HR1 and HR2 regions to form 6-HB [14]. Through multiple sequence alignment with the MERS-CoV S protein S2 subunit, we found that the S2 subunit of HKU4 also harbors a fusion peptide (FP, residues 942–981), an HR1 domain (residues 983–1103), an HR2 domain (residues 1247–1296), a transmembrane domain (residues 1296–1319) and an intracellular domain (residues1319–1352). The designated HKU4-HR1P spans residues 997–1038 in the HR1, while peptides HKU4-HR2P1, HKU4-HR2P2, HKU4-HR2P3 span residues 1247–1282, 1252–1287, and 1261–1296, respectively (Figure 1C). Similar to MERS-HR2P, the three HKU4 HR2-derived peptides all contain the helical region (residues1263–1281), which contains some different amino acids, such as 1264S, 1265D, 1268A, 1269M, 1272E, 1276Q and1279D (Figure 1C). However, the *a* and *d* residues in the HKU4 HR2 helical region, which are fully the same as those of the MERS-CoV HR2 helical region, are responsible for interaction with the *g* and *e* residues in the HR1 region, correspondingly [14], suggesting that the HKU4 HR2-derived peptides might cross-target the MERS-CoV HR1 region, as does MERS-HR2P.

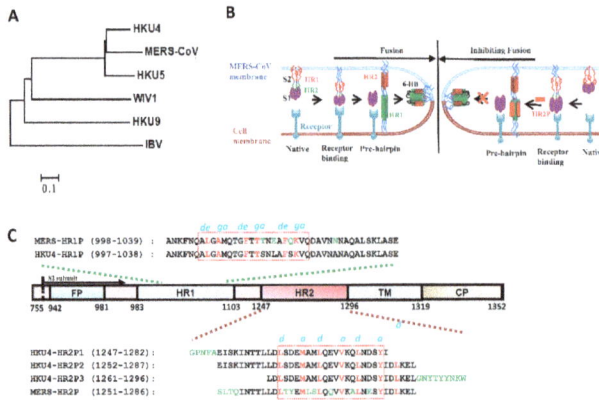

Figure 1. The design of HKU4 HR2-derived peptides. (**A**) Neighbor-joining tree created with the spike sequence of representative bat-CoVs and Infectious Bronchitis Virus (IBV). (**B**) The antiviral mechanism of HR2-derived peptides. (**C**) Schematic representation of HKU4 S protein S2 subunit. FP, fusion peptide; HR, heptad repeat domain; TM, transmembrane domain; CP, cytoplasmic domain. Corresponding sequences of the designed peptides (HKU4-HR1P and HKU4-HR2Ps) are shown in the diagram. The helical region in the HR1- and HR2-derived peptides is highlighted in the red boxes, according to the sequences of MERS-HR1P and MERS-HR2P. The identical amino acids at the *e* and *g* positions in HR1 and those at *a* and *d* positions in HR2 of MERS-CoV and HKU4 are highlighted in red. The different amino acids in HR1Ps and HR2Ps between MERS-CoV and HKU4 are highlighted in green.

3.2. Interaction between MERS-HR1P and HKU4 HR2-Derived Peptides

To investigate whether the HKU4-HR2Ps could interact with HKU4-HR1P to mimic the HR1 and HR2 regions, we determined their interaction by using N-PAGE. After coincubation of HKU4-HR1P at

20, 40, and 80 µM and HKU-HR2P1 (or HKU1-HR2P2 or HKU4-HR2P3) in PBS at 40 µM, respectively, the samples were loaded into the tris-glycine gel for electrophoresis. As shown in Figure 2A–C, individual HKU4 HR2-derived peptides moved down to the lowest part of the gel under native electrophoresis, mainly by their carried negative charges. On the contrary, HKU4-HR1P with positive charges could not move into the gel. Meanwhile, the mixtures of HKU4-HR1P and each HKU4 HR2-derived peptide showed new bands at the upper part in the gel (lanes 3, 4 and 5), confirming that all three HKU4-HR2 peptides could, indeed, bind to HKU4-HR1P.

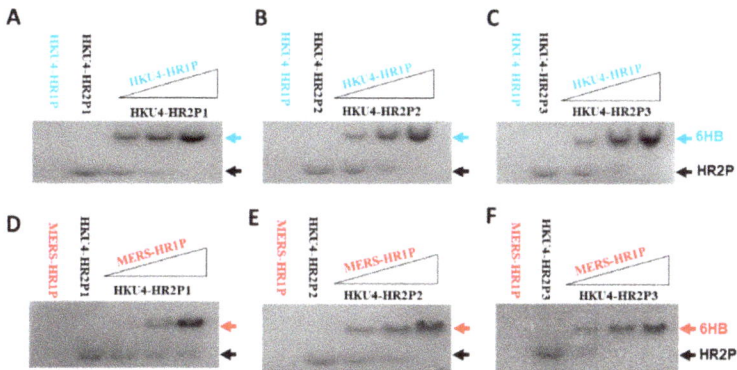

Figure 2. Determination of the interactions between HKU4-HR1P with HKU4-HR2P1 (**A**), HKU4-HR2P2 (**B**) and HKU4-HR2P3 (**C**), as well as the interaction between MERS-HR1P with HKU4-HR2P1 (**D**), HKU4-HR2P2 (**E**) and HKU4-HR2P3 (**F**) by native polyacrylamide gel electrophoresis (N-PAGE).

We subsequently assessed the ability of HKU4 HR2-derived peptides to bind the MERS-CoV HR1 region, thereby allowing us to detect the interaction between HKU4-HR2Ps and MERS-HR1P, which was derived from the MERS-CoV HR1 region. We saw that MERS-HR1P did interact with all three HKU4-HR2 peptides and form new bands on N-PAGE, as well as HKU4-HR1P (Figure 2D,E), suggesting that the HKU4-HR2 peptides could cross-target the MERS-CoV HR1 region.

To characterize the interaction between MERS-HR1P and those HR2-derived peptides, we used circular dichroism (CD) to determine the secondary structures of each peptide and their complexes. As shown in Figure 3A–D, each peptide (MERS-HR2P, HKU4-HR2P1, HKU4-HR2P2, HKU4-HR2P3 or MERS-HR1P) exhibited a low α-helicity structure (10.5%-13.5%), whereas the CD spectrum of HKU4-HR2P1/MERS-HR1P, HKU4-HR2P1/MERS-HR1P, HKU4-HR2P2/MERS-HR1P and HKU4-HR2P3/MERS-HR1P complexes display the characteristic α-helical structure, i.e., double negative peak at 208 and 222 nm, and possess helicity at 70.9%, 66.4%, 80.3% and 66.8%, respectively. Consistent with our previous studies [14], the complex of MERS-HR1P/MERS-HR2P showed high thermostability with the Tm value of 87.1 °C. Similarly, the complexes formed by MERS-HR1P and HKU4-HR2P peptides also possess strong thermostability with Tm values ranging from 83.5 to 90.5 °C (Figure 3E). These results show that all three HKU4-HR2 peptides can bind MERS-HR1P and form 6-HB formation with good stability. Among them, HKU4-HR2P2 exhibited the highest binding affinity to MERS-HR1P and formed the most α-helicity, suggesting that it alone might have the most potent anti-MERS-CoV effect.

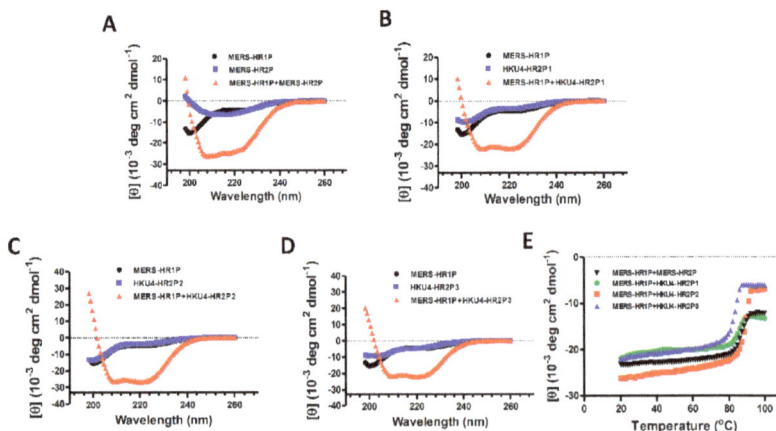

Figure 3. Circular dichroism (CD) spectra of MERS-HR1P, MERS-HR2P and MERS-HR1P/MERS-HR2P complex (**A**), MERS-HR1P, HKU4-HR2P1 and MERS-HR1P/ HKU4-HR2P1 complex (**B**), MERS-HR1P, HKU4-HR2P2 and MERS-HR1P/ HKU4-HR2P2 complex (**C**), MERS-HR1P, HKU4-HR2P3 and MERS-HR1P/ HKU4-HR2P3 complex (**D**) in phosphate buffer (pH 7.2). (**E**) Melting curves of the complexes formed by MERS-HR1P and MERS-HR2P, HKU4-HR2P1, HKU4-HR2P2, or HKU4-HR2P3, respectively.

3.3. Inhibition of MERS-CoV S Protein-Mediated Cell–Cell Fusion

To determine whether the peptides could inhibit MERS-CoV S-mediated cell–cell fusion, we tested them with a MERS-CoV S-mediated cell–cell fusion assay, in which the GFP of effector cells diffused into the target cells such that the fused cells showed larger size and weaker fluorescence than unfused effector cells. Similar to MERS-HR2P, HKU4-HR2P1, HKU4-HR2P2 and HKU4-HR2P3 peptides could effectively inhibit MERS-CoV S-mediated cell–cell fusion with IC_{50}s at concentrations of 1.09, 0.38 and 0.55 μM, respectively (Figure 4A,B, and Table 1). However, HKU4-HR1P did not show fusion-inhibitory activity comparable to that of MERS-HR1P, whereas HKU4-HR2P2 and HKU4-HR2P3 showed more potent inhibitory activity than that of MERS-HR2P (IC50 = 1.07 μM), which was, in fact, derived from the MERS-CoV HR2 region, indicating that HKU4-HR2P2 and HKU4-HR2P3 are good candidates for development as potent anti-MERS-CoV agents.

3.4. Inhibition of Pseudotyped MERS-CoVs Infection

To further assess the anti-MERS-CoV activities of the HKU4 HR2-derived peptides, we tested them using a MERS-CoV pseudovirus infection assay, which is able to mimic the entry process of authentic MERS-CoV and has been widely used in basic and antiviral research [14]. HKU4-HR2P1, HKU4-HR2P2 and HKU4-HR2P3 peptides inhibited entry of MERS-CoV in a dose-dependent manner with IC_{50}s of 2.15, 0.34 and 0.48 μM (Figure 5A and Table 1), respectively, and did so in a manner comparable with that of MERS-HR2P (IC_{50} = 1.14 μM), whereas neither MERS-HR1P nor HKU4-HR1P exhibited inhibitory activity. The results from the time-of-addition assay indicated that the inhibitory activity of HKU4-HR2P2 and HKU4-HR2P3 on pseudotyped MERS-CoV infection was significantly decreased when the peptides were added to cells 4 h post-infection (Figure S1), suggesting that these peptides target the viral fusion and entry stages.

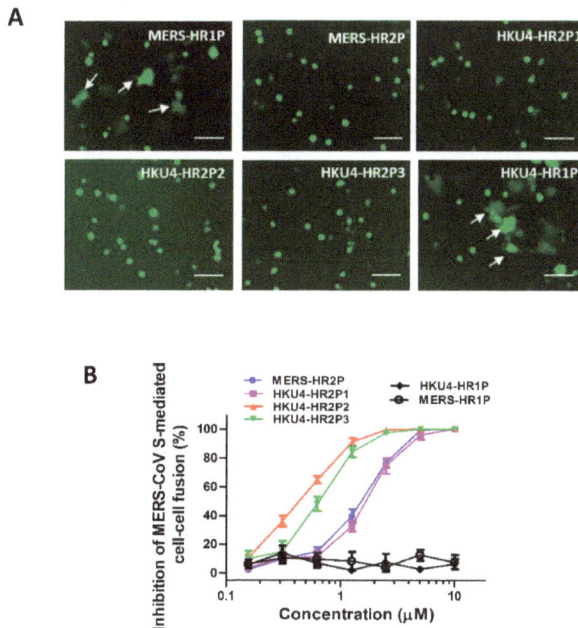

Figure 4. Inhibitory activity of peptides as demonstrated by MERS-CoV S-mediated cell–cell fusion assay. (**A**) Images of MERS-CoV S protein-mediated cell–cell fusion in the presence of MERS-HR1P, MERS-HR2P, HKU4-HR2P1, HKU4-HR2P2, HKU4-HR2P3 and HKU4-HR1P, respectively, all at the concentration of 10 μM. Arrows show the fused cells. Scale bars, 800 μm. (**B**) Inhibitory activities of MERS-HR2P, HKU4-HR2P1, HKU4-HR2P2, HKU4-HR2P3, HKU4-HR1P and MERS-HR1P against MERS-CoV S-mediated cell–cell fusion. Each sample was tested in triplicate and the data are expressed as means ± standard deviation (SD). Each experiment was repeated twice and similar results were obtained.

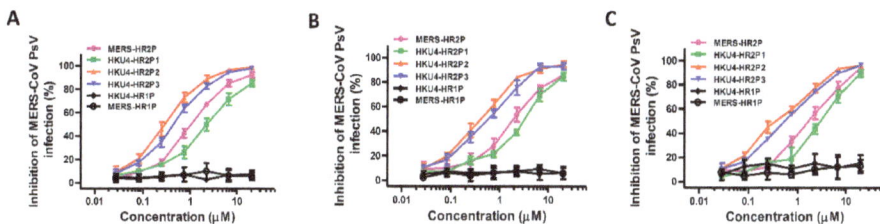

Figure 5. Inhibitory activity of peptides as demonstrated by pseudotyped MERS-CoV infection mediated by S protein (**A**) or S protein with Q1020H (**B**) or Q1020R (**C**). Each sample was tested in triplicate and the data are expressed as means ± SD. Each experiment was repeated twice and similar results were obtained.

Based on numerous epidemiological studies, researchers have found some mutant sites in the spike protein of MERS-CoV, in particular Q1020, which is under strong positive selection in the S2 subunit [32,33]. More importantly, Q1020 is located at the MERS-CoV HR1 region, which is the target site for peptide fusion inhibitors, including HKU4 HR2-derived peptides, and might affect the antiviral activity of these peptides. Therefore, to further assess whether these new peptides would possess broad-spectrum anti-MERS-CoV activity, we used the pseudotyped S-mediated MERS-CoVs with Q1020H and Q1020R mutant sites to infect target cells in the presence of each peptide. As shown in

Figure 5B,C, similar to MERS-HR2P, the three HKU4-HR2 peptides could effectively inhibit infection caused by the two pseudotyped MERS-CoV mutants. Again, HKU4-HR2P2 and HKU4-HR2P3 exhibited more potent inhibitory activity than either that of MERS-HR2P or HKU4-HR2P1. However, these peptides exhibited no obvious inhibition on vesicular stomatitis virus (VSV) G protein-mediated pseudovirus infection at the concentration up to 10 µM (Figure S2A). The results from a time-of-removal assay indicated that the inhibitory activity of the peptides on MERS-CoV pseudovirus is not associated with the binding of the peptides to the target cells (Figure S2B). Besides, these peptide at high concentration (100 µM) showed no cytotoxicity (Figure S2C). All these results suggest that the fusion inhibitory activity of these peptides is not because of their non-specific interaction with the viral or cellular membranes or proteins, nor their cytotoxicity to the target cells.

Table 1. Solubility and inhibitory activities of peptides.

Peptide	Average Hydro-Philicity	Solubility (µM) in				Inhibitory Activity of Peptides, IC_{50} (µM)							
		PBS (pH 7.2)		Water		Cell–Cell Fusion		Pseudovirus Infection					
								Q1020		Q1020H		Q1020R	
		Mean	SD	Mean	SD	Mean	SD	Mean	SD	Mean	SD	Mean	SD
HKU4-HR2P1	0.0	1569	110	651	40	1.09	0.21	2.15	0.17	2.72	0.59	2.42	0.93
HKU4-HR2P2	0.2	2924	131	150	25	0.38	0.01	0.34	0.06	0.44	0.1	0.3	0.04
HKU4-HR2P3	0.0	118	13	1674	73	0.55	0.06	0.48	0.08	0.52	0.1	0.4	0.03
MERS-HR2P	−0.2	105	12	29	14	1.07	0.22	1.14	0.02	1.71	0.02	1.31	0.07

IC_{50}: Concentration of peptide at which 50% of HCoV S-mediated cell–cell fusion was blocked.

4. Discussion

In the early 1990s, Jiang et al and Wild et al identified potent HIV-1 fusion inhibitory peptides derived from the HIV-1 gp41 CHR (or HR2) domain, SJ-2176 and DP-178 (it was later renamed as T-20) [34,35]. Subsequently, there have been numerous reports showing that peptides derived from the HR2 domain in the class I membrane fusion proteins of some enveloped viruses, such as respiratory syncytial virus (RSV) [36], paramyxoviruses simian virus 5 (SV5) [37], Nipah virus [38], and mouse hepatitis virus (MHV) [39], exhibit inhibitory activity against the corresponding viruses. Similarly, in our previous work, we found that both MERS-HR2P, derived from MERS-CoV HR2 region, and CP-1, derived from SARS-CoV HR2 region, exhibited effective fusion-inhibitory activity against MERS-CoV and SARS-CoV, respectively [14,15]. However, MERS-HR2P had no inhibitory activity against SARS-CoV, and SARS-HR2P (CP-1) had no inhibitory activity on MERS-CoV, suggesting that the HR2-derived peptides lacked cross-inhibitory activity, thus limiting the development of broad HCoV fusion inhibitors. To the best of our knowledge, we herein report, for the first time, that BatCoV HR2-derived peptides, namely HKU4-HR2P1, HKU4-HR2P2 and HKU4-HR2P3, could successfully cross-bind MERS-HR1P to form stable 6-HB. These peptides, especially HKU4-HR2P2 and HKU4-HR2P3, exhibited potent inhibitory activity against MERS-CoV S-mediated cell–cell fusion and effectively blocked pseudotyped MERS-CoV infection, indicating the existence of a potent fusion inhibitor against MERS-CoV infection in the natural BatCoV HR2 region.

We previously reported that the fusion core region of HR1 and HR2 is important for the formation of 6-HB [14]; therefore, we herein designed HKU4 HR2-derived peptides containing the helical region. All peptides exhibited antiviral activity against MERS-CoV infection. Apart from HR2 helices, previous studies also suggested that the N- and C-terminal tail regions of HR2 could be important to the inhibitory activity of peptides [40]. Indeed, the three HKU4 HR2-derived peptides possess different terminal tail regions and exhibited various antiviral activity against MERS-CoV infection. On the other hand, the location of HKU4-HR2P2 in the HKU4 HR2 region is the same as that of MERS-HR2P in the MERS-CoV HR2 region. However, the higher helicity (80.3%) and Tm value (90.5 °C) of the HKU4-HR2P2/MERS-HR2 complex than those of the MERS-HR2P2/MERS-HR2 complex may contribute to the higher anti-MERS-CoV activity of HKU4-HR2P2 than that of MERS-HR2P.

More interestingly, the values of helicity and Tm of the HKU4-HR2P3/MERS-HR1P complex are equal to those of HKU4-HR2P1/MERS-HR1P, but lower than those of MERS-HR2P/MERS-HR1P, while HKU4-HR2P3 showed more potent antiviral activity compared to either HKU4-HR2P1 or MERS-HR2P. The precise antiviral mechanism is worth elucidating in future studies, but it can be speculated here that the answer might lie in the C-terminal tail of HKU4-HR2P3, which belongs to the tryptophan (Trp)-rich membrane proximal external region (MPER). This is best exemplified by SARS-CoV MPER-derived peptide, which can interact with the internal fusion peptide (IFP23) to inhibit the formation of MPER-IFP heteromer, a putative quaternary structure that extends from the 6-HB and functions in membrane fusion [41,42]. Moreover, the MPER-derived peptide could directly inhibit the SARS-CoV entry. HKU4-HR2P3 exhibited similar potent inhibitory activity likely from the blocking of MERS-CoV 6-HB formation and disturbance of the interaction between MPER and IFP in MERS-CoV infection. This clearly constitutes a novel strategy for the development of MERS-CoV peptide fusion inhibitors with bispecific targets: HR1 and IFP.

Consistent with the reports that unlike the HR2 (or CHR) peptides, the HR1 (or NHR) peptides derived from the enveloped viruses with class I membrane fusion proteins (e.g., HIV-1, SARS-CoV, and MERS-CoV) have no or low activity to inhibit 6-HB formation, cell–cell fusion, and viral infection, mainly because these HR1 peptides tend to aggregate in physiological solutions [14,15,43–45], we found that MERS-HR1P and HKU4-HR1P also exhibited no fusion inhibitory activity, possibly for the same reason described above. Nonetheless, we have previously reported that conjugation of the T4 fibritin trimerization domain, Foldon (Fd), to the HR1 (or NHR) peptides derived from the HIV-1 gp41, such as N36Fd and N28Fd, make the NHR peptides fold into stable and soluble trimeric coiled-coils with potent HIV-1 fusion inhibitory activity [46]. We will use a similar approach to construct MERS-HR1P-Fd and HKU4-HR1P-Fd with MERS-CoV and HKU4 fusion inhibitory activity, respectively.

The HKU4 coronavirus is closely related to MERS-CoV, and it can recognize and bind the same receptor, human DPP4 [47]. More importantly, the risk of infecting humans could be sharply increased through the simple induction of two mutants in S protein [48]. However, only sparse information about the HKU4 fusion core is currently available, and no effective fusion inhibitors have been reported. Here, based on the structure of the MERS-CoV 6-HB fusion core, we designed HKU4-HR1P and HKU4-HR2P peptides, which can interact to mimic the formation process of 6-HB by their own HR1 and HR2 regions. Hence, this study contributes to a better understanding about the mechanism of HKU4 infection and provides more information for the development of HKU4 peptide fusion inhibitors in the event that HKU4 crosses the species barrier to infect humans.

In summary, we have developed a series of bat-CoV-derived peptide fusion inhibitors, including HKU4-HR2P1, HKU4-HR2P2 and HKU4-HR2P3, which can interact with MERS- HR1P to form highly stable 6-HB, mimicking the formation of MERS-CoV 6-HB structure. Additionally, these peptides, especially HKU4-HR2P2 and HKU4-HR2P3, exhibited potent inhibitory activity against MERS-CoV infection, indicating that these bat-CoV peptides are promising as candidates for development as clinical agents against MERS-CoV infection.

Supplementary Materials: Supplementary materials can be found at http://www.mdpi.com/1999-4915/11/1/56/s1.

Author Contributions: L.L., S.J., and H.L. conceived and designed the experiments; S.X., Q.L., J.P., C.W., Z.L., W.X. and Q.W. performed the experiments; S.X., Q.L., C.W., Z.L., and W.X. analyzed the data; S.X., L.L., S.J., and H.L. wrote the paper. All authors revised the paper. All authors discussed the results and contributed to the final manuscript.

Funding: This work was supported by the National Megaprojects of China for Major Infectious Diseases (2018ZX10301403 to L.L.), Ministry of Science and Technology of the People's Republic of China (2016YFC1201000 and 2016YFC1200405 to S.J. and 2016YFC1202901 to L.L.), Wuhan Science and Technology Project (2016060101010066 to H.L.), and China Postdoctoral Science Foundation (2018M640341 to S.X.).

Conflicts of Interest: The authors declare no conflicts of interest. The funding sponsors had no role in the writing of the manuscript or the decision to publish this article.

References

1. Zaki, A.M.; van Boheemen, S.; Bestebroer, T.M.; Osterhaus, A.D.; Fouchier, R.A. Isolation of a novel coronavirus from a man with pneumonia in Saudi Arabia. *N. Engl. J. Med.* **2012**, *367*, 1814–1820. [CrossRef] [PubMed]

2. Bermingham, A.; Chand, M.A.; Brown, C.S.; Aarons, E.; Tong, C.; Langrish, C.; Hoschler, K.; Brown, K.; Galiano, M.; Myers, R.; et al. Severe respiratory illness caused by a novel coronavirus, in a patient transferred to the United Kingdom from the Middle East, September 2012. *Eurosurveillance* **2012**, *17*, 20290. [PubMed]

3. Muller, M.A.; Raj, V.S.; Muth, D.; Meyer, B.; Kallies, S.; Smits, S.L.; Wollny, R.; Bestebroer, T.M.; Specht, S.; Suliman, T.; et al. Human coronavirus EMC does not require the SARS-coronavirus receptor and maintains broad replicative capability in mammalian cell lines. *MBio* **2012**, *3*, e00515-12. [CrossRef]

4. Letko, M.; Miazgowicz, K.; McMinn, R.; Seifert, S.N.; Sola, I.; Enjuanes, L.; Carmody, A.; van Doremalen, N.; Munster, V. Adaptive evolution of MERS-CoV to species variation in DPP4. *Cell Rep.* **2018**, *24*, 1730–1737. [CrossRef] [PubMed]

5. Lu, G.W.; Hu, Y.W.; Wang, Q.H.; Qi, J.X.; Gao, F.; Li, Y.; Zhang, Y.F.; Zhang, W.; Yuan, Y.; Bao, J.K.; et al. Molecular basis of binding between novel human coronavirus MERS-CoV and its receptor CD26. *Nature* **2013**, *500*, 227–231. [CrossRef] [PubMed]

6. Chan, J.F.; Lau, S.K.; To, K.K.; Cheng, V.C.; Woo, P.C.; Yuen, K.Y. Middle East respiratory syndrome coronavirus: Another zoonotic betacoronavirus causing SARS-like disease. *Clin. Microbiol. Rev.* **2015**, *28*, 465–522. [CrossRef]

7. Chan, J.F.; Li, K.S.; To, K.K.; Cheng, V.C.; Chen, H.; Yuen, K.Y. Is the discovery of the novel human betacoronavirus 2c EMC/2012 (HCoV-EMC) the beginning of another SARS-like pandemic? *J. Infect.* **2012**, *65*, 477–489. [CrossRef]

8. Lu, L.; Xia, S.; Ying, T.L.; Jiang, S.B. Urgent development of effective therapeutic and prophylactic agents to control the emerging threat of Middle East respiratory syndrome (MERS). *Emerg. Microbes Infect.* **2015**, *4*, e37. [CrossRef]

9. Yuan, Y.; Cao, D.; Zhang, Y.; Ma, J.; Qi, J.; Wang, Q.; Lu, G.; Wu, Y.; Yan, J.; Shi, Y.; et al. Cryo-EM structures of MERS-CoV and SARS-CoV spike glycoproteins reveal the dynamic receptor binding domains. *Nat. Commun.* **2017**, *8*, 15092. [CrossRef]

10. Woo, P.C.; Lau, S.K.; Huang, Y.; Yuen, K.Y. Coronavirus diversity, phylogeny and interspecies jumping. *Exp. Biol. Med.* **2009**, *234*, 1117–1127. [CrossRef]

11. Du, L.Y.; Zhao, G.Y.; Kou, Z.H.; Ma, C.Q.; Sun, S.H.; Poon, V.K.M.; Lu, L.; Wang, L.L.; Debnath, A.K.; Zheng, B.J.; et al. Identification of a receptor-binding domain in the s protein of the novel human coronavirus Middle East respiratory syndrome coronavirus as an essential target for vaccine development. *J. Virol.* **2013**, *87*, 11963. [CrossRef]

12. Xia, S.; Liu, Q.; Wang, Q.; Sun, Z.; Su, S.; Du, L.; Ying, T.; Lu, L.; Jiang, S. Middle East respiratory syndrome coronavirus (MERS-CoV) entry inhibitors targeting spike protein. *Virus. Res.* **2014**, *194*, 200–210. [CrossRef] [PubMed]

13. Gao, J.; Lu, G.W.; Qi, J.X.; Li, Y.; Wu, Y.; Deng, Y.; Geng, H.Y.; Li, H.B.; Wang, Q.H.; Xiao, H.X.; et al. Structure of the fusion core and inhibition of fusion by a heptad repeat peptide derived from the s protein of Middle East respiratory syndrome coronavirus. *J. Virol.* **2013**, *87*, 13134–13140. [CrossRef] [PubMed]

14. Lu, L.; Liu, Q.; Zhu, Y.; Chan, K.H.; Qin, L.; Li, Y.; Wang, Q.; Chan, J.F.; Du, L.; Yu, F.; et al. Structure-based discovery of Middle East respiratory syndrome coronavirus fusion inhibitor. *Nat. Commun.* **2014**, *5*, 3067. [CrossRef] [PubMed]

15. Liu, S.W.; Xiao, G.F.; Chen, Y.B.; He, Y.X.; Niu, J.K.; Escalante, C.R.; Xiong, H.B.; Farmar, J.; Debnath, A.K.; Tien, P.; et al. Interaction between heptad repeat 1 and 2 regions in spike protein of SARS-associated coronavirus: Implications for virus fusogenic mechanism and identification of fusion inhibitors. *Lancet* **2004**, *363*, 938–947. [CrossRef]

16. Anthony, S.J.; Gilardi, K.; Menachery, V.D.; Goldstein, T.; Ssebide, B.; Mbabazi, R.; Navarrete-Macias, I.; Liang, E.; Wells, H.; Hicks, A.; et al. Further evidence for bats as the evolutionary source of Middle East respiratory syndrome coronavirus. *Mbio* **2017**, *8*, e00373-17. [CrossRef] [PubMed]

17. Mohd, H.A.; Al-Tawfiq, J.A.; Memish, Z.A. Middle East respiratory syndrome coronavirus (MERS-CoV) origin and animal reservoir. *Virol. J.* **2016**, *13*, 87. [CrossRef]
18. Dudas, G.; Carvalho, L.M.; Rambaut, A.; Bedford, T. MERS-CoV spillover at the camel-human interface. *Elife* **2018**, *7*, e31257. [CrossRef]
19. Lau, S.K.P.; Zhang, L.B.; Luk, H.K.H.; Xiong, L.F.; Peng, X.W.; Li, K.S.M.; He, X.Y.; Zhao, P.S.H.; Fan, R.Y.Y.; Wong, A.C.P.; et al. Receptor usage of a novel bat lineage c betacoronavirus reveals evolution of Middle East respiratory syndrome-related coronavirus spike proteins for human dipeptidyl peptidase 4 binding. *J. Infect. Dis.* **2018**, *218*, 197–207. [CrossRef]
20. Wang, Q.H.; Qi, J.X.; Yuan, Y.; Xuan, Y.F.; Han, P.C.; Wan, Y.H.; Ji, W.; Li, Y.; Wu, Y.; Wang, J.W.; et al. Bat origins of MERS-CoV supported by bat coronavirus HKU4 usage of human receptor CD26. *Cell Host Microbe* **2014**, *16*, 328–337. [CrossRef]
21. Su, S.; Zhu, Y.; Ye, S.; Qi, Q.Q.; Xia, S.; Ma, Z.X.; Yu, F.; Wang, Q.; Zhang, R.G.; Jiang, S.B.; et al. Creating an artificial tail anchor as a novel strategy to enhance the potency of peptide-based HIV fusion inhibitors. *J. Virol.* **2017**, *91*, e01445-16. [CrossRef] [PubMed]
22. Xia, S.; Xu, W.; Wang, Q.; Wang, C.; Hua, C.; Li, W.H.; Lu, L.; Jiang, S.B. Peptide-based membrane fusion inhibitors targeting HCoV-229E spike protein HR1 and HR2 domains. *Int. J. Mol. Sci.* **2018**, *19*, 487. [CrossRef] [PubMed]
23. Chan, D.C.; Fass, D.; Berger, J.M.; Kim, P.S. Core structure of gp41 from the HIV envelope glycoprotein. *Cell* **1997**, *89*, 263–273. [CrossRef]
24. Weng, Y.K.; Weiss, C.D. Mutational analysis of residues in the coiled-coil domain of human immunodeficiency virus type 1 transmembrane protein gp41. *J. Virol.* **1998**, *72*, 9676–9682. [PubMed]
25. Takeda, K.; Sasa, K.; Nagao, M.; Batra, P.P. Secondary structural-changes of non-reduced and reduced ribonuclease-a in solutions of urea, guanidine-hydrochloride and sodium dodecyl-sulfate. *Biochim. Biophys. Acta* **1988**, *957*, 340–344. [CrossRef]
26. Wang, C.; Zhao, L.; Xia, S.; Zhang, T.; Cao, R.; Liang, G.; Li, Y.; Meng, G.; Wang, W.; Shi, W.; et al. De novo design of alpha-helical lipopeptides targeting viral fusion proteins: A promising strategy for relatively broad-spectrum antiviral drug discovery. *J. Med. Chem.* **2018**, *61*, 8734–8745. [CrossRef] [PubMed]
27. Channappanavar, R.; Lu, L.; Xia, S.; Du, L.; Meyerholz, D.K.; Perlman, S.; Jiang, S. Protective Effect of Intranasal Regimens Containing Peptidic Middle East Respiratory Syndrome Coronavirus Fusion Inhibitor Against MERS-CoV Infection. *J. Infect. Dis.* **2015**, *212*, 1894–1903. [CrossRef]
28. Liu, Q.; Xia, S.; Sun, Z.; Wang, Q.; Du, L.; Lu, L.; Jiang, S. Testing of Middle East respiratory syndrome coronavirus replication inhibitors for the ability to block viral entry. *Antimicrob. Agents Chemother.* **2015**, *59*, 742–744. [CrossRef]
29. Chou, T.C. Theoretical basis, experimental design, and computerized simulation of synergism and antagonism in drug combination studies. *Pharmacol. Rev.* **2006**, *58*, 621–681. [CrossRef]
30. Zhao, G.Y.; Du, L.Y.; Ma, C.Q.; Li, Y.; Li, L.; Poon, V.K.M.; Wang, L.L.; Yu, F.; Zheng, B.J.; Jiang, S.B.; et al. A safe and convenient pseudovirus-based inhibition assay to detect neutralizing antibodies and screen for viral entry inhibitors against the novel human coronavirus MERS-CoV. *Virol. J.* **2013**, *10*, 266. [CrossRef]
31. Jiang, S.B.; Lu, H.; Liu, S.W.; Zhao, Q.; He, Y.X.; Debnath, A.K. N-substituted pyrrole derivatives as novel human immunodeficiency virus type 1 entry inhibitors that interfere with the gp41 six-helix bundle formation and block virus fusion. *Antimicrob. Agents Chemother.* **2004**, *48*, 4349–4359. [CrossRef] [PubMed]
32. Cotten, M.; Watson, S.J.; Zumla, A.I.; Makhdoom, H.Q.; Palser, A.L.; Ong, S.H.; Al Rabeeah, A.A.; Alhakeem, R.F.; Assiri, A.; Al-Tawfiq, J.A.; et al. Spread, circulation, and evolution of the Middle East respiratory syndrome coronavirus. *Mbio* **2014**, *5*, e01062-13. [CrossRef] [PubMed]
33. Forni, D.; Filippi, G.; Cagliani, R.; De Gioia, L.; Pozzoli, U.; Al-Daghri, N.; Clerici, M.; Sironi, M. The heptad repeat region is a major selection target in MERS-CoV and related coronaviruses. *Sci. Rep.* **2015**, *5*, 14480. [CrossRef] [PubMed]
34. Jiang, S.; Lin, K.; Strick, N.; Neurath, A.R. HIV-1 inhibition by a peptide. *Nature* **1993**, *365*, 113. [CrossRef] [PubMed]
35. Wild, C.; Greenwell, T.; Matthews, T. A Synthetic peptide from HIV-1 gp41 Is a potent inhibitor of virus-mediated cell-cell fusion. *Aids Res. Hum. Retrovir.* **1993**, *9*, 1051–1053. [CrossRef] [PubMed]

36. Lawless-Delmedico, M.K.; Sista, P.; Sen, R.; Moore, N.C.; Antczak, J.B.; White, J.M.; Greene, R.J.; Leanza, K.C.; Matthews, T.J.; Lambert, D.M. Heptad-repeat regions of respiratory syncytial virus F1 protein form a six-membered coiled-coil complex. *Biochemistry* **2000**, *39*, 11684–11695. [CrossRef] [PubMed]

37. Russell, C.J.; Jardetzky, T.S.; Lamb, R.A. Membrane fusion machines of paramyxoviruses: Capture of intermediates of fusion. *EMBO J.* **2001**, *20*, 4024–4034. [CrossRef]

38. Bossart, K.N.; Mungall, B.A.; Crameri, G.; Wang, L.F.; Eaton, B.T.; Broder, C.C. Inhibition of Henipavirus fusion and infection by heptad-derived peptides of the Nipah virus fusion glycoprotein. *Virol. J.* **2005**, *2*, 57. [CrossRef]

39. Bosch, B.J.; van der Zee, R.; de Haan, C.A.M.; Rottier, P.J.M. The coronavirus spike protein is a class I virus fusion protein: Structural and functional characterization of the fusion core complex. *J. Virol.* **2003**, *77*, 8801–8811. [CrossRef]

40. Wang, C.; Xia, S.; Zhang, P.; Zhang, T.; Wang, W.; Tian, Y.; Meng, G.; Jiang, S.; Liu, K. Discovery of hydrocarbon-stapled short alpha-helical peptides as promising Middle East respiratory syndrome coronavirus (MERS-CoV) fusion inhibitors. *J. Med. Chem.* **2018**, *61*, 2018–2026. [CrossRef]

41. Liao, Y.; Zhang, S.M.; Neo, T.L.; Tam, J.P. Tryptophan-dependent membrane interaction and heteromerization with the internal fusion peptide by the membrane proximal external region of SARS-CoV spike protein. *Biochemistry* **2015**, *54*, 1819–1830. [CrossRef] [PubMed]

42. Lu, Y.N.; Neo, T.L.; Liu, D.X.; Tam, J.P. Importance of SARS-CoV spike protein Trp-rich region in viral infectivity. *Biochem. Biophys. Res. Commun.* **2008**, *371*, 356–360. [CrossRef] [PubMed]

43. Lu, M.; Blacklow, S.C.; Kim, P.S. A trimeric structural domain of the HIV-1 transmembrane glycoprotein. *Nat. Struct. Biol.* **1995**, *2*, 1075–1082. [CrossRef]

44. Lu, M.; Kim, P.S. A trimeric structural subdomain of the HIV-1 transmembrane glycoprotein. *J. Biomol. Struct. Dyn.* **1997**, *15*, 465–471. [CrossRef] [PubMed]

45. Liu, S.; Lu, H.; Niu, J.; Xu, Y.; Wu, S.; Jiang, S. Different from the HIV fusion inhibitor C34, the anti-HIV drug Fuzeon (T-20) inhibits HIV-1 entry by targeting multiple sites in gp41 and gp120. *J. Biol. Chem.* **2005**, *280*, 11259–11273. [CrossRef] [PubMed]

46. Chen, X.; Lu, L.; Qi, Z.; Lu, H.; Wang, J.; Yu, X.; Chen, Y.; Jiang, S. Novel recombinant engineered gp41 N-terminal heptad repeat trimers and their potential as anti-HIV-1 therapeutics or microbicides. *J. Biol. Chem.* **2010**, *285*, 25506–25515. [CrossRef] [PubMed]

47. Yang, Y.; Du, L.Y.; Liu, C.; Wang, L.L.; Ma, C.Q.; Tang, J.; Baric, R.S.; Jiang, S.; Li, F. Receptor usage and cell entry of bat coronavirus HKU4 provide insight into bat-to-human transmission of MERS coronavirus. *Proc. Natl. Acad. Sci. USA* **2014**, *111*, 12516–12521. [CrossRef]

48. Yang, Y.; Liu, C.; Du, L.Y.; Jiang, S.B.; Shi, Z.L.; Baric, R.S.; Li, F. Two mutations were critical for bat-to-human transmission of Middle East respiratory syndrome coronavirus. *J. Virol.* **2015**, *89*, 9119–9123. [CrossRef]

viruses

MDPI

Article

Combining a Fusion Inhibitory Peptide Targeting the MERS-CoV S2 Protein HR1 Domain and a Neutralizing Antibody Specific for the S1 Protein Receptor-Binding Domain (RBD) Showed Potent Synergism against Pseudotyped MERS-CoV with or without Mutations in RBD

Cong Wang [1,†], Chen Hua [1,†], Shuai Xia [1], Weihua Li [2], Lu Lu [1,*] and Shibo Jiang [1,2,*]

1   Key Laboratory of Medical Molecular Virology of MOE/MOH, School of Basic Medical Sciences and Shanghai Public Health Clinical Center, Fudan University, Shanghai 200032, China; 16111010068@fudan.edu.cn (C.W.); 16211010047@fudan.edu.cn (C.H.); 15111010053@fudan.edu.cn (S.X.)

2   NHC Key Laboratory of Reproduction Regulation (Shanghai Institute of Planned Parenthood Research), Fudan University, Shanghai 200032, China; weihua.li@sippr.org.cn

*   Correspondence: lul@fudan.edu.cn (L.L.); shibojiang@fudan.edu.cn (S.J.); Tel.: +86-21-5423-7673 (L.L.); +86-21-5423-7673 (S.J.)

†   These authors contributed equally to this work.

Received: 9 December 2018; Accepted: 2 January 2019; Published: 6 January 2019

Abstract: Middle East respiratory syndrome coronavirus (MERS-CoV) has continuously posed a threat to public health worldwide, yet no therapeutics or vaccines are currently available to prevent or treat MERS-CoV infection. We previously identified a fusion inhibitory peptide (HR2P-M2) targeting the MERS-CoV S2 protein HR1 domain and a highly potent neutralizing monoclonal antibody (m336) specific to the S1 spike protein receptor-binding domain (RBD). However, m336 was found to have reduced efficacy against MERS-CoV strains with mutations in RBD, and HR2P-M2 showed low potency, thus limiting the clinical application of each when administered separately. However, we herein report that the combination of m336 and HR2P-M2 exhibited potent synergism in inhibiting MERS-CoV S protein-mediated cell–cell fusion and infection by MERS-CoV pseudoviruses with or without mutations in the RBD, resulting in the enhancement of antiviral activity in contrast to either one administered alone. Thus, this combinatorial strategy could be used in clinics for the urgent treatment of MERS-CoV-infected patients.

Keywords: MERS-CoV; RBD; mutation; peptide; neutralizing antibody; combination

1. Introduction

Middle East respiratory syndrome (MERS) coronavirus (MERS-CoV), a lineage C beta-coronavirus, was reported to cause severe respiratory tract infection [1,2]. To date, 2266 laboratory-confirmed cases of infection with MERS-CoV, including 804 MERS-CoV associated deaths, have been reported to the World Health Organization (WHO) from 27 countries. Currently, no effective therapeutics or vaccines are available to treat or prevent MERS-CoV infection.

The spike (S) protein of MERS-CoV plays important roles in virus attachment, fusion, and entry into the target cell [3–5]. Similar to other coronaviruses, the S protein of MERS-CoV consists of S1 and S2 subunits. The S1 subunit is responsible for the binding of the virion by its receptor binding domain (RBD) to the cellular receptor, dipeptidyl peptidase-4 (DPP4), while the S2 subunit mediates the fusion between viral and cellular membranes through the interaction between its HR1 and HR2 domains and entry of the viral genetic materials into the host cell [5–8]. Thus, both the RBD in S1 subunit and HR1 domain in S2 subunit can serve as important targets for development of antiviral agents against MERS-CoV infection.

Recently, we identified a peptide derived from the HR2 domain of MERS-CoV S protein S2 subunit, designated HR2P, which could interact with the HR1 domain of S protein S2 subunit to form a six-helix bundle (6-HB) complex and block viral fusion and replication with IC_{50}s ranging from 0.6 to 1 μM [5]. By replacing amino acid residues at the *i* to *i* + 4 positions with a negatively charged amino acid (e.g., E) and a positively charged amino acid (e.g., K) in HR2P for introduction of intramolecular salt-bridges, the resultant peptide HR2P-M2 exhibited improved solubility, stability, and anti-MERS-CoV activity [5,9]. However, its potency is still not strong enough to warrant clinical development.

By screening an extra-large phage-displayed antibody Fab library, Ying et al. identified a human neutralizing monoclonal antibody (hmAb), m336, which is specific for the RBD in the S protein S1 subunit. It exhibited highly potent neutralizing activity against MERS-CoV infection, both in vitro and in vivo [10–13]. X-ray crystallography has shown that the binding epitope of m336 on MERS-CoV S protein almost completely overlaps with the binding site of DPP4 [14]. However, the future clinical application of m336 could be limited by its inability to neutralize MERS-CoV strains with mutations in RBD, like the mouse neutralizing mAb Mersmab1 with similar weakness [15,16].

In this study, we compared the sensitivity of a pseudotyped MERS-CoV wild-type strain with that of strains with key mutations, including D509G, D510G, Q522H, and I529T, which were detected in the RBD of some MERS-CoV strains isolated from different regions and at different times throughout the course of the MERS outbreak from 2012 to 2015 [17,18]. We found that these strains with mutations in RBD were significantly less sensitive than the wild-type strain to the neutralizing activity of m336, while the pseudoviruses with or without mutations showed equal sensitivity to the fusion inhibitory activity of HR2P-M2. Interestingly, when m336 was combined with HR2P-M2, a strong synergism emerged against MERS-CoV S-mediated cell–cell fusion and infection by pseudotyped MERS-CoV strains with or without mutations in RBD, suggesting that this combinational therapy could be further developed for clinical use to treat patients infected by the MERS-CoV strains with or without mutations in RBD.

2. Materials and Methods

2.1. Cells, Peptides, Human mAb m336, and Plasmids

The 293T cell line was obtained from ATCC (Manassas, VA, USA), and the Huh-7 cell line was from the Cell Bank of the Chinese Academy of Sciences (Shanghai, China). These two cell lines were propagated in Dulbecco's Modified Eagle's Medium (DMEM) supplemented with 10% fetal bovine serum (FBS). Peptide HR2P-M2 was synthesized by solid phase peptide synthesis at SYN Inc. (Shanghai, China), and human mAb m336 was provided by Prof. Tianlei Ying at Fudan University, Shanghai, China. Recombinant plasmids encoding the MERS-CoV S protein with D509G, D510G, Q522H, or I529T mutations were kindly provided by Dr. Lanying Du at the New York Blood Center, NY, USA.

2.2. Production of Pseudoviruses

MERS-CoV pseudoviruses were constructed as described previously [19,20]. Briefly, 293T cells were plated in a T175 tissue culture flask and incubated at 37 °C for 16 h. Cells were cotransfected with plasmids pNL4-3.luc.RE encoding Env-defective, luciferase-expressing HIV-1 and pcDNA3.1-MERS-CoV-S encoding S protein with or without mutation in RBD at mass ratio of 1:1 using VigoFect (Vigorous Biotechnology, Beijing, China), according to the manufacturer's recommendation. The supernatant was replaced with fresh DMEM at 8–10 h post-transfection and harvested after incubation for an additional 72 h. In order to remove cell debris, the supernatant was centrifuged at

3000 rpm for 10 min, followed by filtration through a 0.45 μm filter. MERS-CoV pseudovirus in the supernatant was quantified by testing p24 content in the product of MERS-CoV pseudovirus.

2.3. Inhibition of Pseudotyped MERS-CoV Infection

A MERS-CoV pseudovirus inhibition assay was performed as previously described [5,15,21]. Briefly, Huh-7 cells were seeded (10^4 cells/well) into a 96-well plate and incubated overnight at 37 °C. MERS-CoV pseudovirus was incubated with a serially diluted inhibitor for 30 min at 37 °C, followed by the addition of Huh-7 cells. The cells were incubated with or without pseudovirus as virus control and cell control, respectively. The culture was replaced with fresh medium 12 h post-infection and incubated for an additional 72 h. Cells were lysed using lysis reagent (Promega, Madison, WI, USA), and cell lysates were transferred to a 96-well Costar flat-bottom luminometer plate (Corning Costar, New York, NY, USA), followed by the addition of luciferase substrate (Promega) to measure luminescence using an Infinite M200 PRO (Tecan, GröDig, Austria).

2.4. Inhibition of MERS-CoV S Protein-Mediated Cell–Cell Fusion

MERS-CoV S protein-mediated cell–cell fusion was performed as previously described [5]. Briefly, plasmid pAAV-IRES-MERS-EGFP encoding the MERS-CoV S protein was transfected into 293T cells (293T/MERS/EGFP) using the transfection reagent, VigoFect (Vigorous Biotechnology, Beijing, China). The target Huh-7 cells expressing DPP4were incubated at 2×10^4 cells/well in wells of a 96-well plate for 12 h. The effector 293T/MERS/EGFP cells that express MERS-CoV S protein and EGFP or the control 293T/EGFP cells that express EGFP only were preincubated at 10^4 cells/well with an inhibitor at the indicated concentration or phosphate buffered saline (PBS) as control at 37 °C for 30 min. The mixture of 293T/MERS/EGFP cells and an inhibitor or PBS were added to Huh-7 cells in the wells, followed by a co-culture at 37 °C for 2 h. The 293T/MERS/EGFP cells fused or unfused with Huh-7 cells were fixed with 4% PFA and counted under an inverted fluorescence microscope (Nikon, Tokyo, Japan). The fused cell showed much larger size and weaker fluorescence intensity than the unfused cell because of the diffusion of EGFP from one cell to more cells (Figure 1). Almost no fused cells could be observed in the groups of negative control (PBS+293T/EGFP+Huh-7) or peptide treatment (HR2P-M2+293T/MERS/EGFP+Huh-7) (Figure 1). The concentration for 50% inhibition (IC_{50}) was calculated using CalcuSyn software kindly provided by Dr. T.C. Chou [22].

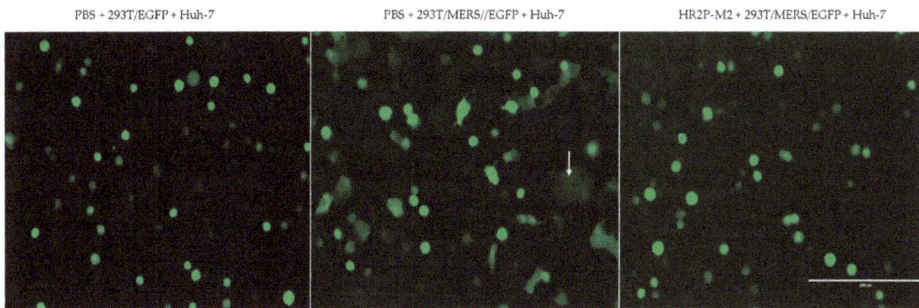

Figure 1. Images of Middle East respiratory syndrome coronavirus (MERS-CoV) S protein-mediated cell–cell fusion. Huh-7 cells were co-cultured with 293T/EGFP cells (**left**) or 293T/MERS/EGFP cells in the presence of PBS (**middle**) or 10 μM HR2P-M2 peptide (**right**) at 37 °C for 2h. The fused cells (one of the fused cell is indicated by an arrow) and the unfused cells were counted under a fluorescence microscope. Scale bars, 200 μm.

2.5. Inhibitory Activity of Sera from Mice Treated with m336 Alone, HR2P-M2 Alone, or m336/HR2P-M2 Combined

The animal experiment was performed under ethical guidelines for the care and use of laboratory animals of Fudan University, and the protocol was approved by the Institutional Laboratory Animal

Care and Use Committee at Fudan University (approval number 20160927-1, 27 September 2016). Six-week-old female specific-pathogen-free (SPF) BALB/c mice (bodyweight about 20 g) were divided into 3 groups of 3 mice each. Mice in group 1, 2, and 3 were intraperitoneally (i.p.) injected with m336 (0.01 mg in 100 μL PBS) alone, HR2P-M2 (1 mg in 100 μL PBS) alone, and the combination of m336 (0.01 mg in 100 μL PBS) and HR2P-M2 (1 mg in 100 μL PBS), respectively. Mice were sedated with Nembutal (100 mg/kg body weight) before and 2 h after injection of the inhibitors, respectively, and bled retro-orbitally. The blood was centrifuged at 6000 rpm for 10 min after standing at room temperature for 3 h. The sera were collected and heat-inactivated at 56 °C for 30 min. Inhibitory activity of the inhibitors on MERS-CoV pseudovirus was evaluated in serum as described above.

2.6. Inhibitor Combination Assay

To assess the potential synergistic effect, HR2P-M2 and m336 were mixed at the indicated molar concentration ratio, while HR2P-M2 alone and m336 alone were included as controls. The mixtures were serially diluted and tested for their inhibitory activity on MERS-CoV pseudovirus infection as described above. Each sample was tested in triplicate, and data were analyzed for synergistic effect by calculating the combination index (CI), using the CalcuSyn program. CI values of <1 and >1 indicate synergy and antagonism, respectively, and synergy was divided into different strengths, according to CI values, as follows: <0.1 indicates very strong synergism; 0.1–0.3 indicates strong synergism; 0.3–0.7 indicates synergism; 0.7–0.85 indicates moderate synergism; and 0.85–0.90 indicates slight synergism [23,24]. Fold of potency enhancement was calculated with the ratio of concentrations of inhibitor testing alone and in combination.

2.7. Statistical Analysis

To determine the significance of difference in sensitivity between wild-type and mutant viruses to inhibitors and the inhibitory activity detected in sera from BALB/c mice treated with inhibitors alone or combination, statistical analyses were performed using a two-tailed unpaired Student's *t*-test, using GraphPad Prism, version 5.0. Values with $p < 0.05$ and $p < 0.01$ were considered statistically significant and very significant, respectively.

3. Results

3.1. Combining HR2P-M2 with m336 Exhibited Strong Synergism against MERS-CoV Pseudovirus Infection

We first investigated the potential cooperative effects of combining HR2P-M2 with m336 on MERS-CoV pseudovirus infection. In our preliminary study, we found that IC_{50} values of HR2P-M2 and m336 for inhibiting MERS-CoV pseudovirus infection were about 600 nM and 0.06 nM, respectively. Therefore, we tested the inhibitory activity of HR2P-M2 alone, m336 alone, and HR2P-M2/m336 in combination at a molar concentration ratio of 10,000:1, respectively. As shown in Figure 2 and Table 1, combining HR2P-M2 and m336 resulted in strong synergistic inhibitory activity against MERS-CoV pseudovirus infection with CI values of 0.13–0.20 for 50–90% inhibition, including potency enhancement of 12.9- to 18.9-fold for m336 and 8.4- to 12.9-fold for HR2P-M2. This result suggested that the MERS-CoV fusion inhibitory peptide HR2P-M2 and the MERS-CoV neutralizing mAb m336 could be used in combination to enhance anti-MERS-CoV activity.

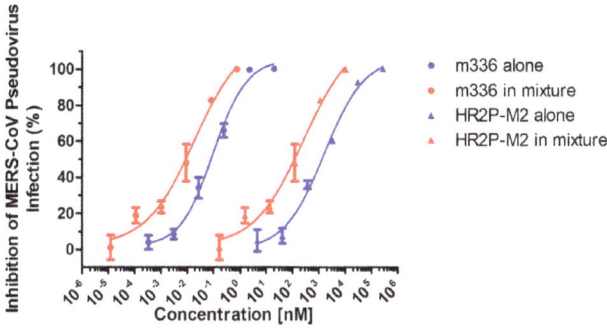

Figure 2. Strong synergism of HR2P-M2 combined with m336 against MERS-CoV pseudovirus infection. The effective concentrations for inhibiting MERS-CoV pseudovirus infection are plotted in two curves. The blue curves represent inhibitors used alone, and the red curves represent each inhibitor used in combination. The width between two curves represents the fold of enhancement between an inhibitor used alone and in combination.

Table 1. Combination index (CI) and fold of enhancement for inhibiting MERS-CoV pseudovirus infection by HR2P-M2 (μM) and m336 (nM) tested in combination.

% Inhibition	CI	HR2P-M2			m336		
		Concentration (μM)		Fold of	Concentration (nM)		Fold of
		Alone	in Mixture	Enhancement	Alone	in Mixture	Enhancement
50	0.197	0.574	0.069	8.36	0.066	0.005	12.94
60	0.183	0.874	0.097	9.05	0.099	0.007	13.88
70	0.168	1.381	0.140	9.87	0.155	0.010	14.98
80	0.152	2.415	0.220	10.96	0.268	0.016	16.44
90	0.131	5.598	0.436	12.85	0.610	0.032	18.92

Note: Each sample was tested in triplicate, and the mean values are presented. Ratio of molar concentration of HR2P-M2 and m336 in combination is 10,000:1.

3.2. Combining HR2P-M2 with m336 Displayed Strong Synergism against MERS-CoV S Protein-Mediated Cell–Cell Fusion

Next, we tested the potential synergistic activity of the HR2P-M2/m336 combination on MERS-CoV S protein-mediated cell–cell fusion. We adjusted the molar concentration ratio of HR2P-M2 and m336 in the combination to 4500:1, since the IC_{50} values of HR2P-M2 and m336 for inhibiting MERS-CoV S protein-mediated cell–cell fusion in our preliminary studies were about 700 nM and 0.15 nM, respectively. As shown in Figure 3 and Table 2, the combination also exhibited strong synergism against MERS-CoV S protein-mediated cell–cell fusion (CI = 0.27) with enhancement of 18-fold for m336 and 4-fold for HR2P-M2. This result confirms that combining HR2P-M2, a MERS-CoV fusion inhibitor, with m336, a human neutralizing mAb, results in strong synergism on S protein-mediated membrane fusion because they target the different stages of MERS-CoV fusion and entry processes.

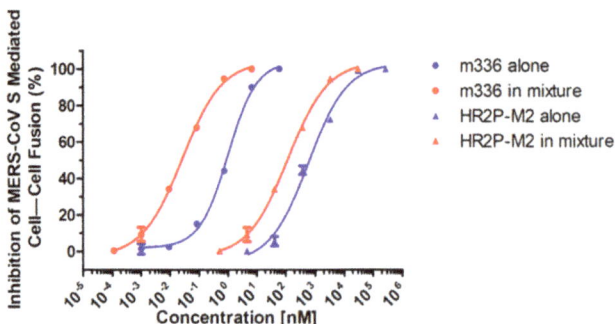

Figure 3. Strong synergism resulting from the HR2P-M2/m336 combination against MERS-CoV S protein-mediated cell–cell fusion.

Table 2. Combination index and fold of enhancement for inhibiting MERS-CoV S protein-mediated cell–cell fusion by the HR2P-M2/m336 combination.

% Inhibition	CI	HR2P-M2			m336		
		Concentration (μM)		Fold of	Concentration (nM)		Fold of
		Alone	in Mixture	Enhancement	Alone	in Mixture	Enhancement
50	0.271	0.511	0.110	4.64	0.440	0.024	17.96
60	0.274	0.713	0.156	4.57	0.625	0.035	18.04
70	0.278	1.025	0.228	4.50	0.918	0.051	18.12
80	0.282	1.596	0.362	4.41	1.466	0.080	18.22
90	0.288	3.106	0.726	4.28	2.965	0.161	18.38

Note: Each sample was tested in triplicate, and the mean values are presented. The molar concentration ratio of HR2P-M2 and m336 in combination is 4500:1.

3.3. MERS-CoV Pseudoviruses with Mutations in RBD Mutant of MERS-CoV Were Resistant to RBD-Specific mAb m336, While They Were Equally Sensitive to the HR1-Targeting Peptide HR2P-M2

Du et al. have previously shown that MERS-CoV pseudoviruses with mutations in RBD, such as D509G and D510G detected in some MERS-CoV strains isolated from different regions and at different times [17,18], are resistant to the neutralizing activity of an RBD-specific mouse mAb Mersmab1 [16]. In the present study, the sensitivity of pseudotyped MERS-CoV strains with key mutations in RBD, as identified in some MERS-CoV mutants isolated during the 2012–2015 outbreaks [17], including D509G, D510G, Q522H, and I529T, along with wild-type MERS-CoV, was compared between the inhibitory activity of HR2P-M2 peptide alone and m336 neutralizing mAb alone. As shown in Table 3, the resistance of MERS-CoV mutants to the neutralizing activity of m336 is about 2- to 8-fold, whereas the pseudoviruses with or without mutations were equally sensitive to fusion inhibitory activity of HR2P-M2. This result suggested that use of mAb m336 alone is unable to control the infection by MERS-CoV strains with mutations in RBD.

Table 3. Sensitivity of MERS-CoV pseudoviruses with or without mutations in the receptor-binding domain (RBD) to the inhibitory activity of m336 (nM) and HR2P-M2 (μM) separately.

MERS-CoV Pseudovirus	IC_{50} (nM) of m336	RR_{50} (Fold of Resistance)	*p*	IC_{50} (μM) of HR2P-M2	RR_{50} (Fold of Resistance)	*p*
Wild-type	0.055 ± 0.009	—	—	0.553 ± 0.056	—	—
D509G	0.116 ± 0.020	2.11	<0.01	0.619 ± 0.079	1.12	>0.05
D510G	0.450 ± 0.085	8.18	<0.05	0.679 ± 0.144	1.23	>0.05
Q522H	0.148 ± 0.051	2.69	<0.01	0.677 ± 0.071	1.22	>0.05
I529T	0.215 ± 0.055	3.91	<0.01	0.574 ± 0.209	1.04	>0.05

Note: Each sample was tested in triplicate. Data are presented as means $\pm$ SD. Resistance ratio (RR_{50}) values are based on IC_{50} of a mutant strain divided by IC_{50} of the wild-type stain. The significance of the difference between a mutant strain and the wild-type strain was statistically analyzed by a two-tailed unpaired Student's *t*-test using GraphPad Prism, version 5.0. Values with $p < 0.05$ and $p < 0.01$ were considered statistically significant and very significant, respectively.

3.4. Combining m336 with HR2P-M2 Exhibited Potent Synergism against MERS-CoV Pseudoviruses with or without Mutations in RBD or Those in the HR1 Domain

To determine whether the combination of HR2P-M2 and m336 also exhibited synergistic antiviral activity against infection of MERS-CoV strains with mutations in RBD or in the HR1 domain, we constructed pseudoviruses bearing MERS-CoV S protein with mutations in RBD, including D509G, D510G, Q522H, or I529T, and those in the HR1 domain, including Q1020H and Q1020R [25,26]. We then tested their sensitivity to the inhibition of HR2P-M2 alone, m336 alone, and the HR2P-M2/m336 combination. As shown in Table 4, combining m336 with HR2P-M2 exhibited strong synergism against infection by pseudotyped MERS-CoV strains with or without mutations in the RBD or HR1 domain with CI value less than 0.3 and potency enhancement in the range of 6- to 25-fold, suggesting that this combinational therapy has potential to be further developed for treatment of patients infected by different MERS-CoV strains, including those with resistance to RBD-specific neutralizing antibodies.

Table 4. Combination index and fold of enhancement for inhibiting MERS-CoV pseudoviruses with or without mutations in RBD in S1 subunit and HR1 in S2 subunit of MERS-CoV S protein by HR2P-M2 and m336.

% Inhibition	CI	HR2P-M2			m336		
		Concentration (μM)		Fold of	Concentration (μM)		Fold of
		Alone	in Mixture	Enhancement	Alone	in Mixture	Enhancement
Wild type							
50	0.197	0.574	0.069	8.36	0.066	0.005	12.94
60	0.183	0.874	0.097	9.05	0.099	0.007	13.88
70	0.168	1.381	0.140	9.87	0.155	0.010	14.98
80	0.152	2.415	0.220	10.96	0.268	0.016	16.44
90	0.131	5.598	0.436	12.85	0.610	0.032	18.92
D509G in RBD							
50	0.296	0.912	0.155	5.88	0.273	0.034	7.91
60	0.29	1.335	0.217	6.15	0.379	0.048	7.87
70	0.282	2.023	0.313	6.47	0.544	0.070	7.83
80	0.274	3.358	0.489	6.87	0.844	0.109	7.77
90	0.263	7.199	0.956	7.53	1.635	0.213	7.69
D510G in RBD							
50	0.137	0.962	0.088	10.96	0.429	0.020	21.96
60	0.145	1.523	0.153	9.96	0.763	0.034	22.43
70	0.155	2.512	0.280	8.97	1.429	0.062	22.97
80	0.169	4.625	0.586	7.90	3.075	0.130	23.63
90	0.194	11.59	1.777	6.52	9.74	0.395	24.67

Table 4. *Cont.*

% Inhibition	CI	HR2P-M2			m336		
		Concentration (μM)		Fold of	Concentration (μM)		Fold of
		Alone	in Mixture	Enhancement	Alone	in Mixture	Enhancement
Q522H in RBD							
50	0.135	0.799	0.045	17.59	0.129	0.010	12.76
60	0.143	1.243	0.067	18.67	0.166	0.015	11.21
70	0.153	2.012	0.101	19.93	0.219	0.022	9.74
80	0.168	3.619	0.168	21.57	0.306	0.037	8.21
90	0.199	8.757	0.36	24.30	0.508	0.08	6.34
I529T in RBD							
50	0.256	0.864	0.098	8.81	0.153	0.022	7.03
60	0.242	1.295	0.139	9.32	0.23	0.031	7.45
70	0.227	2.013	0.203	9.90	0.358	0.045	7.92
80	0.211	3.449	0.323	10.67	0.614	0.072	8.55
90	0.188	7.752	0.649	11.94	1.383	0.144	9.58
Q1020H in HR1							
50	0.186	0.762	0.086	8.83	0.088	0.006	13.82
60	0.189	0.954	0.106	9.04	0.100	0.008	12.84
70	0.192	1.218	0.131	9.27	0.115	0.010	11.85
80	0.198	1.642	0.172	9.56	0.137	0.013	10.74
90	0.208	2.571	0.257	10.01	0.176	0.019	9.27
Q1020R in HR1							
50	0.293	0.69	0.119	5.79	0.073	0.009	8.30
60	0.28	0.905	0.143	6.32	0.087	0.011	8.19
70	0.268	1.216	0.175	6.95	0.104	0.013	8.06
80	0.254	1.743	0.223	7.80	0.131	0.017	7.92
90	0.238	2.997	0.323	9.29	0.184	0.024	7.70

Note: The molar concentration ratio of HR2P-M2 and m336 in combination against wildtype virus, viruses with mutations in RBD, and those in HR1 is 10,000:1, 4500:1, and 10,000:1, respectively.

3.5. Sera from Mice Treated with the m336/HR2P-M2 Combination Showed More Efficacy in Inhibiting MERS-CoV Pseudovirus Infection than Either HR2P-M2 or m336 Alone

To determine whether the HR2P-M2/m336 combination could sustain its efficacy in vivo compared to HR2P-M2 or m336 alone, we tested the anti-MERS-CoV pseudovirus activity of the inhibitors in sera of mice treated with i.p. injection of HR2P-M2, m336, and the HR2P-M2/m336 combination, respectively. As shown in Figure 4, the inhibitory activity detected in sera from mice treated with HR2P-M2 or m336 alone was significantly higher than that detected in sera from mice before inhibition of any inhibitor. On the other hand, the anti-MERS-CoV activity detected in sera from mice treated with the HR2P-M2/m336 combination was significantly more potent than that detected in sera of mice administered with HR2P-M2 or m336 alone. This result confirms that combining HR2P-M2 with m336 affords synergism against MERS-CoV S infection, both in vitro and in vivo.

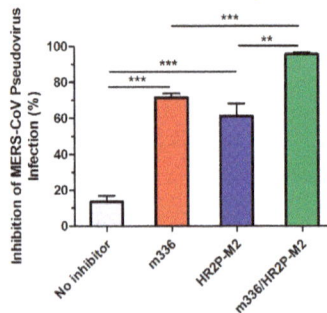

Figure 4. MERS-CoV pseudovirus inhibitory activity as determined from sera of BALB/c mice treated with m336 (0.01 mg) alone, HR2P-M2 (1 mg) alone, or m336 (0.01 mg)/HR2P-M2 (1 mg) in combination. Data are presented as means ± SD. **, and *** represent $p < 0.01$, and $p < 0.001$, respectively.

4. Discussion

The high mortality of MERS-CoV-infected patients [27–29] calls for the development of highly effective anti-MERS-CoV therapeutics. Although we and others have previously identified a MERS-CoV fusion inhibitory peptide (HR2P-M2) targeting the MERS-CoV S2 protein HR1 domain and a highly potent human neutralizing mAb (m336) targeting the MERS-CoV S1 protein RBD [10,26,30], their further development is limited by low potency in the case of HR2P-M2 and low efficacy to neutralize MERS-CoV strains with RBD mutations in the case of m336 [10,31–33].

The combinatorial use of drugs with different mechanisms of action, i.e., cocktail regimen, has been widely applied in clinics [22]. For example, the combinatorial use of HIV reverse transcriptase (RT) inhibitors and protease inhibitors, known as highly active anti-retrovirus therapy (HAART), has shown significant synergism in inhibiting HIV-1 infection, reducing adverse effects and delaying the emergence of drug resistance, thus extending the lifespan of millions of HIV/AIDS patients [34–36]. Moreover, we previously showed that combining HIV-1 attachment inhibitors with RT inhibitors, or combining the 1st, 2nd, and/or 3rd generation HIV fusion inhibitors that target different sites in the HIV-1 gp41 HR1 domain, exhibited synergistic and complementary effect against infection by a broad spectrum of HIV-1 strains, including those resistant to HIV attachment inhibitors, fusion inhibitors, and RT inhibitors [37–39].

In this study, we compared the anti-MERS-CoV activity of HR2P-M2 alone and m336 alone with that of HR2P-M2/m336 in combination and found that the inhibitory activity of the HR2P-M2/m336 combination was significantly more potent than either one administered alone against MERS-CoV S protein-mediated cell–cell fusion and MERS-CoV pseudovirus infection, suggesting synergistic activity based on the dual mechanisms of action whereby HR2P-M2 targets the S2 subunit HR1 domain for inhibiting S2-mediated virus–cell or cell–cell fusion [5] and m336 targets the S1 subunit RBD for inhibiting virus–cell binding or virus attachment [10]. It has been well known that drug synergism can be expected when drugs that act by different mechanisms of action are mixed together [22]. While MERS-CoV pseudoviruses with mutations in RBD were resistant to the RBD-specific mAb m336, they were equally sensitive to HR1-targeting peptide HR2P-M2. Notably, however, the HR2P-M2/m336 combination exhibited strong synergistic antiviral activity against all pseudotyped MERS-CoV strains, including those with mutations in RBD of S protein, which are even resistant to an RBD-specific mouse mAb Mersmab1 [16]. We also demonstrated that sera from mice treated with the HR2P-M2/m336 combination revealed significant efficacy in inhibiting MERS-CoV pseudovirus infection compared to HR2P-M2 or m336 alone. Collectively, these results suggest that the combinatorial strategy overcomes the weaknesses of HR2P-M2 peptide and m336 mAb, while, at the same time, takes advantage of the unique mechanism of action of each to provide, by the sum of both, much more effective inhibitory activity against MERS-CoV infection than either peptide or mAb used alone. The strong synergy of the combination is expected to reduce the dosage of the individual inhibitor in such combinational therapy, resulting in decreased cost and toxicity, thus making the final product more affordable and safer. Therefore, this combinational therapy shows promise for further clinical development.

Author Contributions: S.J. and L.L. conceived and designed the experiments; C.W. and C.H. performed the experiments; C.W., C.H., S.X., W.L. and S.J. analyzed the data; C.W., W.L., L.L. and S.J. wrote the paper. All authors discussed the results and contributed to the final manuscript.

Funding: This work was supported by the National Natural Science Foundation of China (81630090 to Shibo Jiang; 81672019 and 8161101485 to Lu Lu), the National Key Research and Development Program of China (2016YFC1201000 and 2016YFC1200405 to Shibo Jiang, 2016YFC1202901 to Lu Lu), and the National Megaprojects of China for Major Infectious Diseases (2018ZX10301403 to Lu Lu).

Acknowledgments: We thank Lanying Du at the New York Blood Center for providing the plasmid encoding MERS-CoV S protein with mutation in RBD and Tianlei Ying at Fudan University for providing mAb m336.

Conflicts of Interest: The authors declare no competing financial interests.

References

1. Zaki, A.M.; van Boheemen, S.; Bestebroer, T.M.; Osterhaus, A.D.; Fouchier, R.A. Isolation of a novel coronavirus from a man with pneumonia in Saudi Arabia. *N. Engl. J. Med.* **2012**, *367*, 1814–1820. [CrossRef] [PubMed]

2. de Groot, R.J.; Baker, S.C.; Baric, R.S.; Brown, C.S.; Drosten, C.; Enjuanes, L.; Fouchier, R.A.; Galiano, M.; Gorbalenya, A.E.; Memish, Z.A.; et al. Middle East respiratory syndrome coronavirus (MERS-CoV): Announcement of the Coronavirus Study Group. *J. Virol.* **2013**, *87*, 7790–7792. [CrossRef] [PubMed]

3. Belouzard, S.; Millet, J.K.; Licitra, B.N.; Whittaker, G.R. Mechanisms of coronavirus cell entry mediated by the viral spike protein. *Viruses* **2012**, *4*, 1011–1033. [CrossRef] [PubMed]

4. Wang, N.; Shi, X.; Jiang, L.; Zhang, S.; Wang, D.; Tong, P.; Guo, D.; Fu, L.; Cui, Y.; Liu, X.; et al. Structure of MERS-CoV spike receptor-binding domain complexed with human receptor DPP4. *Cell Res.* **2013**, *23*, 986–993. [CrossRef] [PubMed]

5. Lu, L.; Liu, Q.; Zhu, Y.; Chan, K.H.; Qin, L.; Li, Y.; Wang, Q.; Chan, J.F.; Du, L.; Yu, F.; et al. Structure-based discovery of Middle East respiratory syndrome coronavirus fusion inhibitor. *Nat. Commun.* **2014**, *5*, 3067. [CrossRef] [PubMed]

6. Du, L.; Zhao, G.; Kou, Z.; Ma, C.; Sun, S.; Poon, V.K.; Lu, L.; Wang, L.; Debnath, A.K.; Zheng, B.J.; et al. Identification of a receptor-binding domain in the S protein of the novel human coronavirus Middle East respiratory syndrome coronavirus as an essential target for vaccine development. *J. Virol.* **2013**, *87*, 9939–9942. [CrossRef]

7. Raj, V.S.; Mou, H.; Smits, S.L.; Dekkers, D.H.; Muller, M.A.; Dijkman, R.; Muth, D.; Demmers, J.A.; Zaki, A.; Fouchier, R.A.; et al. Dipeptidyl peptidase 4 is a functional receptor for the emerging human coronavirus-EMC. *Nature* **2013**, *495*, 251–254. [CrossRef]

8. Gao, J.; Lu, G.; Qi, J.; Li, Y.; Wu, Y.; Deng, Y.; Geng, H.; Li, H.; Wang, Q.; Xiao, H.; et al. Structure of the fusion core and inhibition of fusion by a heptad repeat peptide derived from the S protein of Middle East respiratory syndrome coronavirus. *J. Virol.* **2013**, *87*, 13134–13140. [CrossRef]

9. Tao, X.; Garron, T.; Agrawal, A.S.; Algaissi, A.; Peng, B.H.; Wakamiya, M.; Chan, T.S.; Lu, L.; Du, L.; Jiang, S.; et al. Characterization and Demonstration of the Value of a Lethal Mouse Model of Middle East Respiratory Syndrome Coronavirus Infection and Disease. *J. Virol.* **2016**, *90*, 57–67. [CrossRef]

10. Ying, T.; Du, L.; Ju, T.W.; Prabakaran, P.; Lau, C.C.; Lu, L.; Liu, Q.; Wang, L.; Feng, Y.; Wang, Y.; et al. Exceptionally potent neutralization of Middle East respiratory syndrome coronavirus by human monoclonal antibodies. *J. Virol.* **2014**, *88*, 7796–7805. [CrossRef]

11. Houser, K.V.; Gretebeck, L.; Ying, T.; Wang, Y.; Vogel, L.; Lamirande, E.W.; Bock, K.W.; Moore, I.N.; Dimitrov, D.S.; Subbarao, K. Prophylaxis With a Middle East Respiratory Syndrome Coronavirus (MERS-CoV)-Specific Human Monoclonal Antibody Protects Rabbits From MERS-CoV Infection. *J. Infect. Dis.* **2016**, *213*, 1557–1561. [CrossRef] [PubMed]

12. Agrawal, A.S.; Ying, T.; Tao, X.; Garron, T.; Algaissi, A.; Wang, Y.; Wang, L.; Peng, B.H.; Jiang, S.; Dimitrov, D.S.; et al. Passive Transfer of A Germline-like Neutralizing Human Monoclonal Antibody Protects Transgenic Mice Against Lethal Middle East Respiratory Syndrome Coronavirus Infection. *Sci. Rep.* **2016**, *6*, 31629. [CrossRef] [PubMed]

13. van Doremalen, N.; Falzarano, D.; Ying, T.; de Wit, E.; Bushmaker, T.; Feldmann, F.; Okumura, A.; Wang, Y.; Scott, D.P.; Hanley, P.W.; et al. Efficacy of antibody-based therapies against Middle East respiratory syndrome coronavirus (MERS-CoV) in common marmosets. *Antiviral Res.* **2017**, *143*, 30–37. [CrossRef] [PubMed]

14. Ying, T.; Prabakaran, P.; Du, L.; Shi, W.; Feng, Y.; Wang, Y.; Wang, L.; Li, W.; Jiang, S.; Dimitrov, D.S.; et al. Junctional and allele-specific residues are critical for MERS-CoV neutralization by an exceptionally potent germline-like antibody. *Nat. Commun.* **2015**, *6*, 8223. [CrossRef] [PubMed]

15. Du, L.; Zhao, G.; Yang, Y.; Qiu, H.; Wang, L.; Kou, Z.; Tao, X.; Yu, H.; Sun, S.; Tseng, C.T.; et al. A conformation-dependent neutralizing monoclonal antibody specifically targeting receptor-binding domain in Middle East respiratory syndrome coronavirus spike protein. *J. Virol.* **2014**, *88*, 7045–7053. [CrossRef] [PubMed]

16. Tai, W.; Wang, Y.; Fett, C.A.; Zhao, G.; Li, F.; Perlman, S.; Jiang, S.; Zhou, Y.; Du, L. Recombinant Receptor-Binding Domains of Multiple Middle East Respiratory Syndrome Coronaviruses (MERS-CoVs) Induce Cross-Neutralizing Antibodies against Divergent Human and Camel MERS-CoVs and Antibody Escape Mutants. *J. Virol.* **2017**, *91*. [CrossRef] [PubMed]

17. Kim, Y.; Cheon, S.; Min, C.K.; Sohn, K.M.; Kang, Y.J.; Cha, Y.J.; Kang, J.I.; Han, S.K.; Ha, N.Y.; Kim, G.; et al. Spread of Mutant Middle East Respiratory Syndrome Coronavirus with Reduced Affinity to Human CD26 during the South Korean Outbreak. *mBio* **2016**, *7*, e00019. [CrossRef]

18. Kim, D.W.; Kim, Y.J.; Park, S.H.; Yun, M.R.; Yang, J.S.; Kang, H.J.; Han, Y.W.; Lee, H.S.; Kim, H.M.; Kim, H.; et al. Variations in Spike Glycoprotein Gene of MERS-CoV, South Korea, 2015. *Emerg. Infect. Dis.* **2016**, *22*, 100–104. [CrossRef]

19. Zhao, G.; Du, L.; Ma, C.; Li, Y.; Li, L.; Poon, V.K.; Wang, L.; Yu, F.; Zheng, B.J.; Jiang, S.; et al. A safe and convenient pseudovirus-based inhibition assay to detect neutralizing antibodies and screen for viral entry inhibitors against the novel human coronavirus MERS-CoV. *Virol. J.* **2013**, *10*, 266. [CrossRef]

20. Gunaratne, G.S.; Yang, Y.; Li, F.; Walseth, T.F.; Marchant, J.S. NAADP-dependent Ca(2+) signaling regulates Middle East respiratory syndrome-coronavirus pseudovirus translocation through the endolysosomal system. *Cell Calcium* **2018**, *75*, 30–41. [CrossRef]

21. Wang, L.; Shi, W.; Chappell, J.D.; Joyce, M.G.; Zhang, Y.; Kanekiyo, M.; Becker, M.M.; van Doremalen, N.; Fischer, R.; Wang, N.; et al. Importance of neutralizing monoclonal antibodies targeting multiple antigenic sites on MERS-CoV Spike to avoid neutralization escape. *J. Virol.* **2018**. [CrossRef]

22. Chou, T.C. Theoretical basis, experimental design, and computerized simulation of synergism and antagonism in drug combination studies. *Pharmacol. Rev.* **2006**, *58*, 621–681. [CrossRef] [PubMed]

23. Qi, Q.; Wang, Q.; Chen, W.; Yu, F.; Du, L.; Dimitrov, D.S.; Lu, L.; Jiang, S. Anti-HIV antibody and drug combinations exhibit synergistic activity against drug-resistant HIV-1 strains. *J. Infect.* **2017**, *75*, 68–71. [CrossRef] [PubMed]

24. Xu, W.; Wang, Q.; Yu, F.; Lu, L.; Jiang, S. Synergistic effect resulting from combinations of a bifunctional HIV-1 antagonist with antiretroviral drugs. *J. Acquir. Immune Defic. Syndr.* **2014**, *67*, 1–6. [CrossRef] [PubMed]

25. Cotten, M.; Watson, S.J.; Zumla, A.I.; Makhdoom, H.Q.; Palser, A.L.; Ong, S.H.; Al Rabeeah, A.A.; Alhakeem, R.F.; Assiri, A.; Al-Tawfiq, J.A.; et al. Spread, circulation, and evolution of the Middle East respiratory syndrome coronavirus. *mBio* **2014**, *5*. [CrossRef] [PubMed]

26. Channappanavar, R.; Lu, L.; Xia, S.; Du, L.; Meyerholz, D.K.; Perlman, S.; Jiang, S. Protective Effect of Intranasal Regimens Containing Peptidic Middle East Respiratory Syndrome Coronavirus Fusion Inhibitor Against MERS-CoV Infection. *J. Infect. Dis.* **2015**, *212*, 1894–1903. [CrossRef] [PubMed]

27. Zumla, A.; Hui, D.S.; Perlman, S. Middle East respiratory syndrome. *Lancet* **2015**, *386*, 995–1007. [CrossRef]

28. van Boheemen, S.; de Graaf, M.; Lauber, C.; Bestebroer, T.M.; Raj, V.S.; Zaki, A.M.; Osterhaus, A.D.; Haagmans, B.L.; Gorbalenya, A.E.; Snijder, E.J.; et al. Genomic characterization of a newly discovered coronavirus associated with acute respiratory distress syndrome in humans. *mBio* **2012**, *3*. [CrossRef]

29. Assiri, A.; McGeer, A.; Perl, T.M.; Price, C.S.; Al Rabeeah, A.A.; Cummings, D.A.; Alabdullatif, Z.N.; Assad, M.; Almulhim, A.; Makhdoom, H.; et al. Hospital outbreak of Middle East respiratory syndrome coronavirus. *N. Engl. J. Med.* **2013**, *369*, 407–416. [CrossRef]

30. Lu, L.; Xia, S.; Ying, T.; Jiang, S. Urgent development of effective therapeutic and prophylactic agents to control the emerging threat of Middle East respiratory syndrome (MERS). *Emerg. Microbes Infect.* **2015**, *4*, e37. [CrossRef]

31. Xia, S.; Liu, Q.; Wang, Q.; Sun, Z.; Su, S.; Du, L.; Ying, T.; Lu, L.; Jiang, S. Middle East respiratory syndrome coronavirus (MERS-CoV) entry inhibitors targeting spike protein. *Virus Res.* **2014**, *194*, 200–210. [CrossRef] [PubMed]

32. Zhao, G.; He, L.; Sun, S.; Qiu, H.; Tai, W.; Chen, J.; Li, J.; Chen, Y.; Guo, Y.; Wang, Y.; et al. A Novel Nanobody Targeting Middle East Respiratory Syndrome Coronavirus (MERS-CoV) Receptor-Binding Domain Has Potent Cross-Neutralizing Activity and Protective Efficacy against MERS-CoV. *J. Virol.* **2018**, *92*. [CrossRef] [PubMed]

33. Kleine-Weber, H.; Elzayat, M.T.; Wang, L.; Graham, B.S.; Muller, M.A.; Drosten, C.; Pohlmann, S.; Hoffmann, M. Mutations in the spike protein of MERS-CoV transmitted in Korea increase resistance towards antibody-mediated neutralization. *J. Virol.* **2018**. [CrossRef] [PubMed]

34. Zhang, L.; Ramratnam, B.; Tenner-Racz, K.; He, Y.; Vesanen, M.; Lewin, S.; Talal, A.; Racz, P.; Perelson, A.S.; Korber, B.T.; et al. Quantifying residual HIV-1 replication in patients receiving combination antiretroviral therapy. *N. Engl. J. Med.* **1999**, *340*, 1605–1613. [CrossRef] [PubMed]
35. Hogg, R.S.; Rhone, S.A.; Yip, B.; Sherlock, C.; Conway, B.; Schechter, M.T.; O'Shaughnessy, M.V.; Montaner, J.S. Antiviral effect of double and triple drug combinations amongst HIV-infected adults: Lessons from the implementation of viral load-driven antiretroviral therapy. *AIDS* **1998**, *12*, 279–284. [CrossRef] [PubMed]
36. Richman, D.D.; Margolis, D.M.; Delaney, M.; Greene, W.C.; Hazuda, D.; Pomerantz, R.J. The challenge of finding a cure for HIV infection. *Science* **2009**, *323*, 1304–1307. [CrossRef] [PubMed]
37. Liu, S.; Lu, H.; Neurath, A.R.; Jiang, S. Combination of candidate microbicides cellulose acetate 1,2-benzenedicarboxylate and UC781 has synergistic and complementary effects against human immunodeficiency virus type 1 infection. *Antimicrob. Agents Chemother.* **2005**, *49*, 1830–1836. [CrossRef]
38. Pan, C.; Cai, L.; Lu, H.; Qi, Z.; Jiang, S. Combinations of the first and next generations of human immunodeficiency virus (HIV) fusion inhibitors exhibit a highly potent synergistic effect against enfuvirtide-sensitive and -resistant HIV type 1 strains. *J. Virol.* **2009**, *83*, 7862–7872. [CrossRef]
39. Pan, C.; Lu, H.; Qi, Z.; Jiang, S. Synergistic efficacy of combination of enfuvirtide and sifuvirtide, the first- and next-generation HIV-fusion inhibitors. *AIDS* **2009**, *23*, 639–641. [CrossRef]

Article

Complement Receptor C5aR1 Inhibition Reduces Pyroptosis in hDPP4-Transgenic Mice Infected with MERS-CoV

Yuting Jiang [1,†], Junfeng Li [1,†], Yue Teng [1,†], Hong Sun [2], Guang Tian [1], Lei He [1], Pei Li [1], Yuehong Chen [1], Yan Guo [1], Jiangfan Li [1], Guangyu Zhao [1], Yusen Zhou [1,3,*] and Shihui Sun [1,*]

[1] State Key Laboratory of Pathogen and Biosecurity, Beijing Institute of Microbiology and Epidemiology, Beijing 100071, China; captain99@126.com (Y.J.); lijunfeng2113@126.com (J.L.); yueteng@me.com (Y.T.); tglnx05@tom.com (G.T.); helei_happy@126.com (L.H.); lwhisperer@163.com (P.L.); chenyuehong.happy@163.com (Y.C.); muhan0425@126.com (Y.G.); anatee@163.com (J.L.); guangyu0525@163.com (G.Z.)

[2] Department of Basic Medical Sciences, North China University of Science and Technology, Tangshan 063210, China; xun1@hotmail.com

[3] Institute of Medical and Pharmaceutical Sciences, Zhengzhou University, Zhengzhou 450052, China

* Correspondence: yszhou@bmi.ac.cn (Y.Z.); sunsh01@163.com (S.S.); Tel./Fax: +86-10-63858045 (Y.Z.)

† These authors contributed equally to this work.

Received: 28 November 2018; Accepted: 2 January 2019; Published: 9 January 2019

Abstract: Middle East respiratory syndrome coronavirus (MERS-CoV) is a highly pathogenic virus with a crude mortality rate of ~35%. Previously, we established a human DPP4 transgenic (hDPP4-Tg) mouse model in which we studied complement overactivation-induced immunopathogenesis. Here, to better understand the pathogenesis of MERS-CoV, we studied the role of pyroptosis in THP-1 cells and hDPP4 Tg mice with MERS-CoV infection. We found that MERS-CoV infection induced pyroptosis and over-activation of complement in human macrophages. The hDPP4-Tg mice infected with MERS-CoV overexpressed caspase-1 in the spleen and showed high IL-1β levels in serum, suggesting that pyroptosis occurred after infection. However, when the C5a-C5aR1 axis was blocked by an anti-C5aR1 antibody (Ab), expression of caspase-1 and IL-1β fell. These data indicate that MERS-CoV infection induces overactivation of complement, which may contribute to pyroptosis and inflammation. Pyroptosis and inflammation were suppressed by inhibiting C5aR1. These results will further our understanding of the pathogenesis of MERS-CoV infection.

Keywords: MERS-CoV; inflammation; pyroptosis; complement

1. Introduction

Middle East respiratory syndrome coronavirus (MERS-CoV), the second highly pathogenic coronavirus to emerge after severe acute respiratory syndrome coronavirus (SARS-CoV), causes severe acute respiratory failure and extra-pulmonary multi-organ damage accompanied by severe systemic inflammation [1–3]. However, the pathogenesis of MERS-CoV still needs to be explored. Complement activation and pyroptosis are two proteolytic cascades that defend the host against dangerous pathogens. They are important parts of the innate immune system and have some similar characteristics, including pore-formation and proinflammatory characteristics.

Pyroptosis is a lytic and inflammatory mode of regulated cell death catalyzed by the caspase family [4]. Activation of caspase-1 relies on assembly of inflammasome complexes, which contain NLRP1b, NLRC4, NLRP3, and AIM2. Different inflammasomes are activated by different pathogen-associated molecular patterns (PAMPs) or danger-associated molecular patterns (DAMPs) via particular pattern recognition receptors (PRRs). The best-characterized inflammasome is the

NLRP3 inflammasome, which responds to a variety of bacterial, viral, and fungal agents [5,6], DAMPs (e.g., ATP, monosodium urate crystals, and amyloid-β aggregates) [7,8], and even environmental and industrial particles such as silica and asbestos [9]. The NLRP3 inflammasome comprises the NLRP3 scaffold, the ASC (PYCARD) adaptor, and pro-caspase-1. The activated NLRP3 inflammasome promotes transformation of pro-caspase-1 to its active form, which proteolytically cleaves gasdermin D, pro-IL-1β, and pro-IL-18 to yield their bioactive forms. The N-terminal domain of cleaved gasdermin D perforates the cell membrane, resulting in osmotic lysis [10], whereas mature IL-1β and IL-18 act as proinflammatory cytokines [11].

The complement system is an ancient molecular cascade; indeed, homologs have been found in sea urchin [12] and mosquitoes [13]. Complement is activated via three pathways: the classical, lectin, and alternative pathways. During the process of activation, an enzyme named C3 convertase cleaves C3 to C3a and C3b, which are recruited to the C3 convertase to form the C5 convertase. C5 convertase catalyzes cleavage of C5 to C5a and C5b to initiate the terminal complement pathway, resulting in formation of the membrane attack complex, which has pore-forming properties. During this process, two split products, C3a and C5a (known as anaphylatoxins), promote inflammation or serve as chemoattractants by engaging their cognate receptors [14,15].

In a previous study we demonstrated that aberrant complement activation contributes to severe outcomes in hDPP4 transgenic mice infected with MERS-CoV, and that preventing over-activation of the complement system may be an effective clinical therapy for MERS [16]. Here, we examined the role of pyroptosis in the pathogenesis of MERS, along with the relationship between pyroptosis and complement. The results may help us to better understand the mechanism underlying severe outcomes after MERS-CoV infection.

2. Materials and Methods

2.1. Ethics Statement

All animal experiments were approved by the Institutional Animal Care and Use Committee (IACUC) of the Beijing Institute of Microbiology and Epidemiology (IACUC Permit No: BIME 2017-0011; Permit Date: 8 March 2017). Animal studies were carried out in strict accordance with the recommendations set out in the Guide for the Care and Use of Laboratory Animals.

2.2. Cells

Human monocytic cells (THP-1) were purchased from the American Type Culture Collection (Manassas, VA, USA, ATCC Number: TIB-202) and cultured in RPMI 1640 medium supplemented with 10% heat-inactivated fetal bovine serum (FBS), 100 U/mL penicillin, 100 µg/mL streptomycin sulfate, 1× Glutamax-I (L-glutamine alternate), and 0.05 mM 2-mercaptoethanol. THP-1 (differentiated) macrophages were obtained by exposing the cells to 60 nM phorbol-12-myristate-13-acetate (PMA) for 12 h, followed by culture for a further 24 h in complete growth medium without PMA.

2.3. Virus Infection

MERS-CoV (HCoV-EMC/2012 strain) was propagated and titrated on Vero cells in an approved biosafety level 3 laboratory. THP-1 differentiated macrophages cultured in 75-cm^2 flasks were infected with MERS-CoV at a multiplicity of infection of 0.1. The virus was adsorbed at 37 °C for 1 h and unbound virus was washed away.

HDPP4-Tg mice (6 weeks old, female) [17] were maintained in a pathogen-free facility and housed in cages containing sterilized feed and drinking water. Following intraperitoneal anesthetization with sodium pentobarbital (5 mg/kg body weight), mice were inoculated intranasally with MERS-CoV (103.3 50% tissue culture infectious dose (TCID50)) in 20 µL Dulbecco's modified Eagle's medium (DMEM). Mice in the sham group received the same volume of DMEM. For the experiments of C5aR1 inhibition, mice were received an intravenous (i.v.) injection (600 µg/kg) of a monoclonal Ab (mAb) specific for mouse C5aR1 (Hycult Biotech, Uden, the Netherlands) to block the interaction of C5a to C5aR1 or an injection of phosphate-buffered saline (PBS) (sham treatment control) at the same time as MERS-CoV inoculation. All infectious experiments related to MERS-CoV were performed in an approved biosafety level 3 facility.

2.4. Isolation of RNA and Proteins

THP-1 differentiated macrophages were lysed in TRIzol™ Reagent (Life Technologies, Carlsbad, CA, USA) at 24 h post-infection with MERS-CoV. Total RNA and proteins were isolated according to the reagent user guide.

Mice were euthanized by overdose inhalation of carbon dioxide at different time points after infection with MERS-CoV. Lungs were harvested and total RNA was extracted and purified using an RNeasy Extraction Kit (Qiagen, Hilden, Germany).

2.5. Quantitative Reverse Transcription-PCR

To detect expression of inflammasomes and complement components in MERS-CoV-infected THP-1 differentiated macrophages and hDPP4-Tg mice, 2 μg of total RNA from cells or the lung of mice we used as template for first-strand cDNA synthesis. The resulting cDNA was subjected to quantitative PCR using Power SYBR® Green PCR Master Mix (Life Technologies, Carlsbad, CA, USA) to determine the relative abundance of inflammasome and complement components. The forward and reverse primers used for each component are listed in the Table 1. The relative amount of each gene was obtained by normalization against an endogenous control gene (GAPDH) and calculated using the comparative $2^{-\Delta\Delta CT}$ method. Sham-infected monocytes and sham-infected hDPP4-Tg mice were used as respective calibrators. An identical amplification reaction comprising (i) polymerase activation and DNA denaturation at 95 °C for 10 min, (ii) 40 cycles each of the denaturation at 95 °C for 10 s, and (iii) an annealing/extension step at 60 °C for 30 s, was used for each gene analyzed.

Table 1. Primers used to amplify inflammasome and complement components.

Primer	Species	Gene	Orientation	Sequence (5'-3')
1	Human	NLRP3	F	ATTCGGAGATTGTGGTTGGG
			R	AGGGCGTTGTCACTCAGGTC
2		pro-caspase-1	F	CTCAGGCTCAGAAGGGAATGTC
			R	TGTGCGGCTTGACTTGTCC
3		pro-IL-1b	F	GCTCGCCAGTGAAATGATGG
			R	CAGAGGGCAGAGGTCCAGG
4		C3	F	CACTATGATCCTTGAGATCTGTACCA
			R	GGAGCAAAGCCAGTCATCA
5		C3aR	F	GACATCCAGGTGCTGAAGCC
			R	ACTGGGGGCTCATTCCATG
6		C5aR1	F	GCTGACCATACCCTCCTTCCT
			R	CCGTTTGTCGTGGCTGTAGTC
7		C5aR2	F	TGCTGTTTGTCTCTGCCCATC
			R	GTCAGCAGGATGATGGAGGG
8	Mouse	pro-caspase-1	F	AATGAAGTTGCTGCTGGAGGA
			R	CAGAAGTCTTGTGCTCTGGGC
9		pro-IL-1b	F	TGGACCTTCCAGGATGAGGACA
			R	GTTCATCTCGGAGCCTGTAGTG

2.6. Western Blot Analysis

Proteins isolated from THP-1 monocytic cells and THP-1 differentiated macrophages were electrophoresed in a 12% SDS-PAGE gel and transferred to a PVDF membrane (GE Healthcare, Dassel, Germany). PVDF membranes were then blocked for 1 h at room temperature in 5% non-fat milk. Membranes were incubated overnight at 4 °C with primary antibodies, followed by secondary antibodies for 1 h at room temperature. The primary antibodies used for analysis were: mouse anti-caspase 1 (1:200; Santa Cruz Biotechnology, Cat: sc-392736, Santa Cruz, CA, USA), mouse anti-IL-1β (1:200; Santa Cruz Biotechnology, Cat: sc-32294), rabbit anti-cleaved IL-1β (1:2000; Cell Signaling, Cat: #83186, Danvers, MA, USA), rabbit anti-MERS-CoV Nucleocapsid Protein (1:2000; Sino Biological Inc., Cat: 40068-RP02, Beijing, China), and rabbit anti-β-actin (1:2000; Cell Signaling, Cat: #4970). The secondary antibody was anti-rabbit secondary (1:5000; TransGen, Cat: HS101-01, Beijing, China) or anti-mouse secondary (1:5000; TransGen Cat: HS201-01). Blots were developed using a Pierce ECL Western Blotting Substrate (Thermo Scientific, Waltham, MA, USA) and protein bands were detected using an Amersham Imager 600 (GE Healthcare, Chicago, IL, USA).

2.7. Analysis of Inflammatory Cytokines

Cytokines in mouse serum were measured using a Milliplex Mouse Cytokine/Chemokine Magnetic Panel Kit (Merck Millipore, Burlington, MA, USA). A panel of inflammatory cytokines (IL-1β, IL-6, TNF-α, and IFN-γ) was detected according to the manufacturer's protocol.

2.8. Immunohistochemistry (IHC)

Sections of paraffin-embedded spleen and lung tissues (4 μm thick) were prepared and stained to detect antigen expression. Briefly, retrieved sections were incubated overnight at 4 °C with the following antibodies: mouse anti-caspase-1 mAb (AdipoGen, San Diego, CA, USA), polyclonal rabbit anti-CD68 (Abcam, Cambridge, MA, USA), and polyclonal anti-IFN-γRα (Santa Cruz Biotechnology). Biotinylated immunoglobulin G was then added, followed by an avidin–biotin–peroxidase conjugate (Beijing Zhongshan Biotechnology Co., Ltd., Beijing, China). Immunoreactivity was detected using 3,3' diamino benzidine (DAB). Slides were counter-stained with hematoxylin.

2.9. Statistical Analysis

Statistical analyses were performed using GraphPad Prism software, version 5.01 (GraphPad Software, San Diego, CA, USA). Student's *t* test was used to compare two groups with respect to relative expression of mRNA and cytokine levels in serum. *p* values < 0.05 were considered significant.

3. Results

3.1. MERS-CoV Infection Induced Pyroptosis in THP-1 Macrophages

Unlike abortive infection of SARS-CoV in human macrophages, MERS-CoV can establish a productive infection in macrophages and induce production of proinflammatory cytokines and chemokines [18]. Many RNA viruses, such as EV71, H1N1, H7N9 influenza A virus, and Zika virus, can infect macrophages and trigger IL-1β secretion via the NLRP3 inflammasome [19–22]. To evaluate the response of macrophages to MERS-CoV infection, we inoculated THP-1 monocytic cells and THP-1 differentiated macrophages with MERS-CoV or RPMI 1640 medium (sham-infection). We then examined expression of NLRP3, pro-caspase-1, and pro-IL-1β 24 h later by RT-qPCR. As shown in Figure 1, MERS-CoV infection induced relatively higher expression of pro-caspase-1 (Figure 1A) and pro-IL-1β (Figure 1B), but not NLRP3 (Figure 1C), in both THP-1 monocytes and macrophages. Expression of pro-IL-1β in monocytes increased by 170-fold, whereas that in macrophages increased by 26-fold (on average).

We verified expression of caspase-1, IL-1β, and MERS nucleocapsid protein (NP) by Western blotting (Figure 1D). MERS-CoV-infected THP-1 macrophages expressed higher levels of pro-caspase-1, pro-IL-1β, and activated IL-1β (p17) than sham-infected THP-1 macrophages or MERS-CoV-infected THP-1 monocytes. MERS NP was detected in both MERS-CoV-infected THP-1 monocytes and macrophages. These results indicate that MERS-CoV infection induces high levels of proinflammatory IL-1β secretion and THP-1 macrophage pyroptosis.

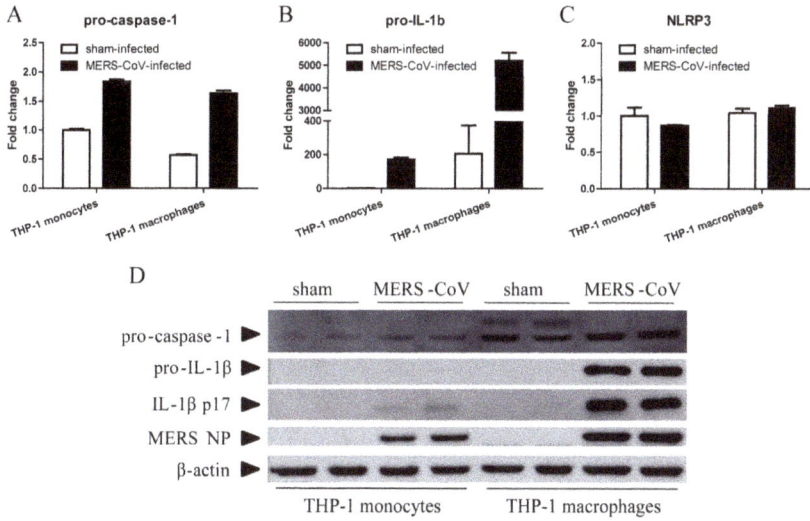

Figure 1. MERS-CoV infection induces pyroptosis in THP-1 macrophages. THP-1 monocytes and macrophages were infected with MERS-CoV for 24 h. Total RNA and protein was then extracted from the cells using TRIzol Reagent. (**A–C**) Total RNA was used for RT-qPCR to detect transcription of pro-caspase-1, pro-IL-1β, and NLRP3. Data are expressed as means ± SEM (*n* = 2 per group). (**D**) Samples of total protein were subjected to Western blotting to detect pro-caspase-1, pro-IL-1β, activated IL-1β, and MERS NP.

3.2. Pyroptosis in Mice Infected with MERS-CoV

To determine whether MERS-CoV infection induces pyroptosis in mice, we used RT-qPCR to detect mRNA encoding NLRP3, pro-caspase-1, and pro-IL-1β in lung tissue from hDPP4 transgenic mice at Day 3 post-MERS-CoV infection. Although there was no significant difference in expression of NLRP3 and pro-caspase-1 between the sham-infected and MERS-CoV-infected groups (Figure 2A,B), expression of pro-IL-1β mRNA was significantly higher after MERS-CoV infection (Figure 2C). In addition, we measured the concentration of IL-1β in serum. The results showed that MERS-CoV infection induced production of IL-1β (Figure 2D). Furthermore, we examined expression of caspase-1 in the lung and spleen at Day 7 post-MERS-CoV infection by IHC. In line with the mRNA results, there was no significant difference in expression of caspase-1 in the lung of sham-infected and MERS-CoV-infected mice. However, the spleens of mice infected with MERS-CoV showed higher expression of caspase-1 than those of mice in the sham group (Figure 2E). The results indicated that MERS-CoV infection could induce pyroptosis in mice.

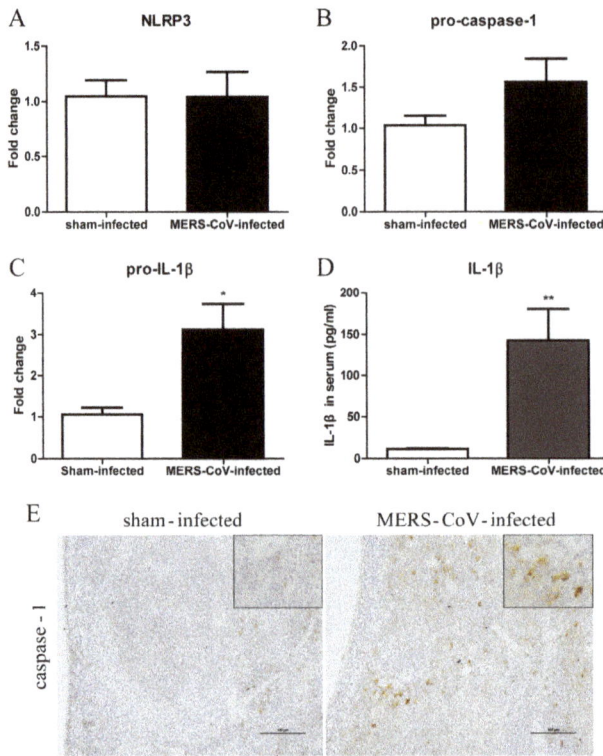

Figure 2. MERS-CoV infection induces pyroptosis in hDPP4-transgenic mice. (**A–C**) Transcription of NLRP3, pro-caspase-1, and pro-IL-1β in lung tissue at Day 3 post-MERS-CoV infection (n = 5–6 per group). (**D**) Concentration of IL-1β in serum at Day 3 post-MERS-CoV infection. Data are expressed as means ± SEM (n = 5–6 per group). * $p < 0.05$, ** $p < 0.01$ (Student's t test with Welch's correction). (**E**) Representative images of immunohistochemical staining of caspase-1 in lung tissue on Day 7 post-challenge of sham-infected and MERS-CoV-infected mice (scale bars = 100 μm).

3.3. Inflammatory Responses in Mice Infected with MERS-CoV

IL-1β plays an important role in mediating autoinflammatory diseases and in generating inflammatory responses to infection [23]. Therefore, to assess the inflammatory responses in mice, we measured TNF-α, IFN-γ, and IL-6 in serum at Day 3 post-MERS-CoV infection. As shown in Figure 3A–C, serum from mice in the MERS-CoV-infected group contained more TNF-α, IFN-γ, and IL-6 than that from sham-infected mice. IHC examination of CD68 and IFN-γ receptor expression also suggested greater macrophage infiltration and activation in the lung and spleen of mice at 7 days post-MERS-CoV infection (Figure 3D). These results indicate that MERS-CoV infection causes systemic inflammation, as reported in clinical MERS patients and MERS-CoV infected animal models [17,24].

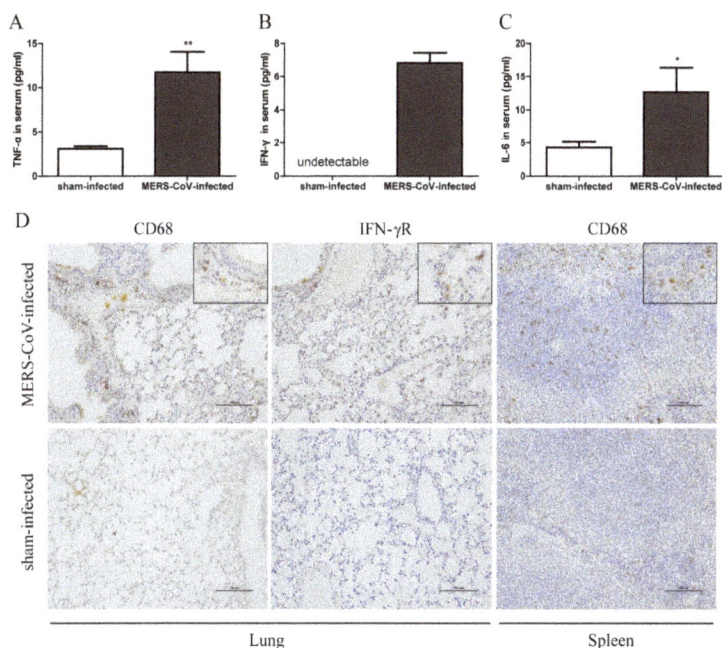

Figure 3. MERS-CoV infection induces systemic inflammation in hDPP4-transgenic mice. (**A–C**) Concentration of TNF-α, IFN-γ, and IL-6 in serum at Day 3 post-MERS-CoV infection. Data are expressed as means ± SEM (*n* = 5–6 per group). * *p* < 0.05, ** *p* < 0.01 (Student's *t* test with Welch's correction). (**D**) Macrophage infiltration and expression of IFN-γ receptor were assessed by immunohistochemical staining of lung and spleen at 7 days post-challenge (scale bars = 100 μm).

3.4. MERS-CoV Infection Alters Expression of Complement in THP-1 Monocytes and Macrophages

The complement system links activation of Toll-like receptors to transcription of IL-1β mRNA [25,26]. It has been studied that intracellular C3 is converted to biologically active C3a and C3b by the protease cathepsin L [27], and C3a activates NLRP3 and triggers IL-1β production in human monocytes by regulating efflux of ATP [28]. In addition, C5a is believed to induce a proinflammatory or anti-inflammatory response when ligated to C5aR1 or C5aR2 respectively [29–31]. Thus, we used RT-qPCR to examine expression of complement components and their receptors C3aR, C5aR1, and C5aR2. As shown in Figure 4, C3 and C3aR expression by both THP-1 monocytes and macrophages was highly upregulated (by 5–20-fold) after MERS-CoV infection. C5aR1 was upregulated, whereas C5aR2 was downregulated, after MERS-CoV infection.

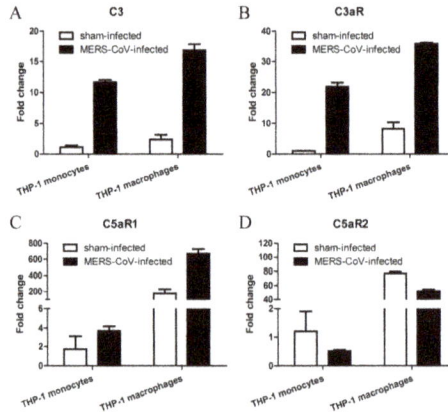

Figure 4. MERS-CoV infection alters complement expression in monocytes and macrophages. (**A–D**) Transcription of C3, C3aR, C5aR1, and C5aR2 in THP-1 monocytes and macrophages at 24 h post-MERS-CoV infection. Data are expressed as means ± SEM (*n* = 2 per group).

3.5. Inhibiting C5aR1 Reduces Pyroptosis in Mice Infected with MERS-CoV

Our previous study demonstrated that MERS-CoV infection results in dysregulated host immune responses and severe tissue damage [17] and inhibiting C5aR1 alleviates MERS-CoV infection-induced tissue damage by regulating host immune responses [16]. Here, we used an anti-C5aR1 Ab to block the C5a-C5aR1 axis. The antibody was administered at the same time as MERS-CoV infection. We then measured expression of caspase-1 in the spleen and IL-1β in the lung and serum. Compared with the PBS-treated group, mice receiving the anti-C5aR1 Ab expressed less caspase-1 in the spleen at Day 7 post-MERS-CoV infection (Figure 5A). Although there was no significant difference between the two groups with respect to pro-IL-1β mRNA expression (Figure 5B), serum levels of IL-1β were lower in the anti-C5aR1 Ab-treated group than in the PBS-treated group at Day 1 (Figure 5C) and Day 3 [16] post-MERS-CoV infection. These results suggest that complement inhibition decreased the expression of pyroptosis indicators, IL-1β and caspase-1, in mice infected with MERS-CoV.

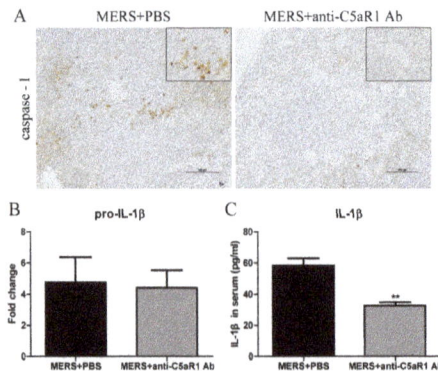

Figure 5. Blocking the C5a-C5aR1 axis reduces pyroptosis in hDPP4-transgenic mice. (**A**) Representative images of immunohistochemical staining of caspase-1 in lungs from PBS-treated and anti-C5aR1 Ab-treated groups on Day 7 post-challenge (scale bars = 100 μm). (**B**) Transcription of pro-IL-1β in lung tissues from PBS-treated and anti-C5aR1 Ab-treated groups at Day 7 post-infection (*n* = 4–5 per group). (**C**) Concentration of IL-1β in serum at Day 1 post-MERS-CoV infection. Data are expressed as means ± SEM (*n* = 4–5 per group). ** *p* < 0.01 (Student's *t* test with Welch's correction).

3.6. Inhibiting C5aR1 Reduces Inflammation in Mice Infected with MERS-CoV

At 1 day after MERS-CoV infection, we measured proinflammatory cytokines (IFN-γ, TNF-α, and IL-6) in serum. IFN-γ levels in mice treated with the anti-C5aR1 Ab were much lower than those in the PBS-treated group (Figure 6A). To further evaluate the effect of complement inhibition on the local inflammation at later time after MERS-CoV infection, we examined expression of CD68 and IFN-γ receptor in lung and spleen at 7 days post-MERS-CoV infection. IHC revealed that macrophage infiltration and activation were lower in the anti-C5aR1 Ab-treated group (Figure 6D). Taken together, these results suggest that inhibiting complement dampens the over-activated inflammatory response in mice infected with MERS-CoV.

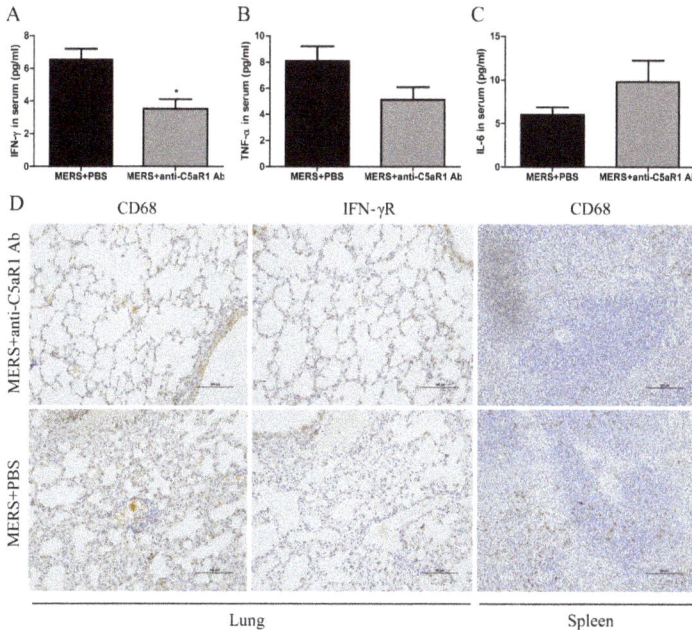

Figure 6. Blockade of the C5a-C5aR1 axis reduces pyroptosis in hDPP4-transgenic mice. (**A–C**) Concentration of IFN-γ, TNF-α, and IL-6 in serum at Day 1 post-MERS-CoV infection. Data are expressed as means ± SEM (*n* = 4–5 per group). * *p* < 0.05 (Student's *t* test with Welch's correction). (**D**) Infiltration of the lungs and spleen by macrophages and expression of IFN-γ receptor were assessed by immunohistochemical staining 7 days after challenge (scale bars = 100 μm).

4. Discussion

Macrophages play important roles in host defense by clearing dead cells, ingesting and destroying microbes, and presenting antigens to T lymphocytes. In addition, macrophages produce the full array of complement components [32] and PRRs, which are closely associated with inflammasome activation and pyroptosis. The accumulated studies indicate that macrophages play an important role in the pathogenesis of SARS and MERS [18,33]. Macrophages infected with MERS-CoV secrete proinflammatory cytokines and chemokines [18]. Widespread distribution of these macrophages throughout many organs is one of the reasons underlying multi-organ damage and systemic inflammation after virus infection. Pyroptosis or inflammasome activation plays an important role in virus-mediated pathogenesis. For example, abortive HIV-1 infection in quiescent lymphoid CD4 T cells leads to CD4 T cell pyroptosis independent of the NLRP3 inflammasome [34]. EV71 3D and ZIKV NS5 activate the NLRP3 inflammasome by interacting with NACHT and the LRR domain of

NLRP3 [19,22]. The PB1-F2 protein of avian influenza A virus H1N1 or H7N9 induces inflammation by activating the NLRP3 inflammasome [35,36], and deficiency of NLRP3 or caspase-1 protects mice against H7N9 infection-associated morbidity and mortality [21]. Here, we demonstrate that the dysregulated macrophage-mediated immune responses after MERS-CoV infection may contribute to a severe outcome.

Several studies demonstrate relationships between complement and inflammasomes. For example, C3-/- mice display reduced inflammasome activation in an intracerebral hemorrhage (ICH) model [37]. Engagement of C5a and C5aR1 on CD4+ T cells generates reactive oxygen species, which are a classical DAMP, thereby triggering inflammasome assembly [38]. In monocytes, C3a alters metabolic programming and increases ATP efflux, leading to NLRP3 activation and IL-1β generation via the receptor P2 × 7 [28]. Thus, the complement system is considered to be an essential regulatory component of the cellular alarm system that controls inflammasome activation [39]. Here, we inhibit MERS-CoV infection-induced inflammation and pyroptosis by blocking the C5a-C5aR1 axis with an anti-C5aR1 antibody.

In our study, we first demonstrated that virus infection leads to pyroptosis by measuring activation of caspase-1 and IL-1β in human macrophages infected with MERS-CoV (Figure 1). Next, we found that MERS-CoV infection induced transcription of pro-IL-1β in the lung and expression of caspase-1 in the spleen. Activated caspase-1 could be released into the extracellular space and delivered to the lung via exosomes [40] or the systemic circulation; it then cleaves pro-IL-1β into its bio-activated form, which is secreted into the serum (Figure 2D). We cannot exclude the possibility that pyroptosis occurs in the spleen because we did not detect expression of pro-IL-1β. Meanwhile, examination of CD68 and IFN-γ receptor expression revealed macrophage infiltration and activation in the lungs and spleen (Figure 3D). Thus, pyroptosis could occur in both organs.

In our previous study, we showed that inhibiting C5aR1 increased splenic cell regeneration and decreased splenic cell apoptosis, thereby alleviating MERS-CoV infection-induced tissue damage [16]. The spleen happens to be a large reservoir for myeloid lineage cells such as macrophages, dendritic cells and CD4+ T cells in which pyroptosis mainly occurred [4,34,41]. Here, we studied and demonstrated that inhibiting C5aR1 suppressed caspase-1 activation (Figure 5A) and macrophage infiltration and activation in the spleen (Figure 6D), and thereby reduced the secretion of IL-1β into the serum (Figure 5C) and the systemic inflammation (Figure 6). However, in the lung tissue, there was no significant difference of pro-IL-1β transcription between the Ab and sham-treated groups (Figure 5B), which may due to the limited macrophages, the main cell type in which pyroptosis occurred, when compared to that in spleen.

Although many studies have focused on the link between complement and pyroptosis, the pathways that link them remain unclear. Here, our results showed that MERS-CoV infection induces pro-IL-1β transcription, and complement activation, which leads to pyroptosis in macrophages. Induction of pyroptosis was related with complement activation and may be promoted by ligation of C5a and C5aR1, which was confirmed by the blockade of anti-C5aR1 antibody (Figure 7).

In summary, these data indicate that MERS-CoV infection induces overactivation of complement, which may contribute to pyroptosis and inflammation. Pyroptosis and inflammation were suppressed by inhibiting C5aR1. These research will further our understanding of the pathogenesis of MERS-CoV infection.

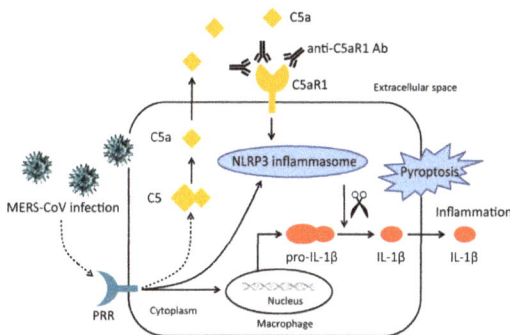

Figure 7. Diagram illustrating the relationship between complement and pyroptosis during MERS-CoV infection. MERS-CoV infection induces activation of complement and transcription of pro-IL-1β. Activated complement component C5a is released into the extracellular space where it interacts with C5aR1, which triggers assembly of the NLRP3 inflammasome. Activated caspase-1 cleaves pro-IL-1β into its active form thereby triggering pyroptosis. An anti-C5aR1 antibody prevents interaction between C5a and C5aR1, thereby ameliorating pyroptosis.

Author Contributions: Y.J., J.L. (Junfeng Li), Y.T., H.S., G.Z., S.S., G.T. performed the experiments. L.H., P.L., Y.C., Y.G., J.L. (Jiangfan Li) contributed reagents/materials/analysis tools. Y.J., J.L. (Junfeng Li), H.S., S.S. analyzed the data. Y.J., S.S., Y.Z. wrote and revised the manuscript. Y.J., Y.Z., S.S. conceived and designed the experiments. All authors read and approved the final manuscript.

Funding: This study was supported by the National Key Plan for Scientific Research and Development of China (2016YFD0500306), the National Natural Science Foundation of China (81571983, 31570158), the National Key Research and Development Program of China (2016YFC1202903, 2016YFC1202402), the National Project of Infectious Diseases (2017ZX10304402-003), BWS14J058, and 16CXZ039.

Conflicts of Interest: The authors declare no conflict of interest.

References

1. Arabi, Y.M.; Arifi, A.A.; Balkhy, H.H.; Najm, H.; Aldawood, A.S.; Ghabashi, A.; Hawa, H.; Alothman, A.; Khaldi, A.; Al Raiy, B. Clinical course and outcomes of critically ill patients with Middle East respiratory syndrome coronavirus infection. *Ann. Intern. Med.* **2014**, *160*, 389–397. [CrossRef]

2. Assiri, A.; Al-Tawfiq, J.A.; Al-Rabeeah, A.A.; Al-Rabiah, F.A.; Al-Hajjar, S.; Al-Barrak, A.; Flemban, H.; Al-Nassir, W.N.; Balkhy, H.H.; Al-Hakeem, R.F.; et al. Epidemiological, demographic, and clinical characteristics of 47 cases of Middle East respiratory syndrome coronavirus disease from Saudi Arabia: A descriptive study. *Lancet. Infect. Dis.* **2013**, *13*, 752–761. [CrossRef]

3. Al-Abdallat, M.M.; Payne, D.C.; Alqasrawi, S.; Rha, B.; Tohme, R.A.; Abedi, G.R.; Al Nsour, M.; Iblan, I.; Jarour, N.; Farag, N.H.; et al. Hospital-associated outbreak of Middle East respiratory syndrome coronavirus: A serologic, epidemiologic, and clinical description. *Clin. Infect. Dis.* **2014**, *59*, 1225–1233. [CrossRef] [PubMed]

4. Vande Walle, L.; Lamkanfi, M. Pyroptosis. *Curr. Biol.* **2016**, *26*, R568–R572. [CrossRef] [PubMed]

5. Kanneganti, T.D.; Body-Malapel, M.; Amer, A.; Park, J.H.; Whitfield, J.; Franchi, L.; Taraporewala, Z.F.; Miller, D.; Patton, J.T.; Inohara, N.; et al. Critical role for Cryopyrin/Nalp3 in activation of caspase-1 in response to viral infection and double-stranded RNA. *J. Biol. Chem.* **2006**, *281*, 36560–36568. [CrossRef] [PubMed]

6. Joly, S.; Sutterwala, F.S. Fungal pathogen recognition by the NLRP3 inflammasome. *Virulence* **2010**, *1*, 276–280. [CrossRef]

7. Mariathasan, S.; Weiss, D.S.; Newton, K.; McBride, J.; O'Rourke, K.; Roose-Girma, M.; Lee, W.P.; Weinrauch, Y.; Monack, D.M.; Dixit, V.M. Cryopyrin activates the inflammasome in response to toxins and ATP. *Nature* **2006**, *440*, 228–232. [CrossRef] [PubMed]

8. Shaw, O.M.; Steiger, S.; Liu, X.; Hamilton, J.A.; Harper, J.L. Brief report: Granulocyte-macrophage colony-stimulating factor drives monosodium urate monohydrate crystal-induced inflammatory macrophage differentiation and NLRP3 inflammasome up-regulation in an in vivo mouse model. *Arthritis Rheumatol.* **2014**, *66*, 2423–2428. [CrossRef]

9. Dostert, C.; Petrilli, V.; Van Bruggen, R.; Steele, C.; Mossman, B.T.; Tschopp, J. Innate immune activation through Nalp3 inflammasome sensing of asbestos and silica. *Science* **2008**, *320*, 674–677. [CrossRef]

10. Ding, J.; Wang, K.; Liu, W.; She, Y.; Sun, Q.; Shi, J.; Sun, H.; Wang, D.C.; Shao, F. Pore-forming activity and structural autoinhibition of the gasdermin family. *Nature* **2016**, *535*, 111–116. [CrossRef]

11. Schroder, K.; Tschopp, J. The inflammasomes. *Cell* **2010**, *140*, 821–832. [CrossRef] [PubMed]

12. Al-Sharif, W.Z.; Sunyer, J.O.; Lambris, J.D.; Smith, L.C. Sea urchin coelomocytes specifically express a homologue of the complement component C3. *J. Immunol.* **1998**, *160*, 2983–2997. [PubMed]

13. Levashina, E.A.; Moita, L.F.; Blandin, S.; Vriend, G.; Lagueux, M.; Kafatos, F.C. Conserved role of a complement-like protein in phagocytosis revealed by dsRNA knockout in cultured cells of the mosquito, Anopheles gambiae. *Cell* **2001**, *104*, 709–718. [CrossRef]

14. Peng, Q.; Li, K.; Sacks, S.H.; Zhou, W. The role of anaphylatoxins C3a and C5a in regulating innate and adaptive immune responses. *Inflamm. Allergy Drug Targets* **2009**, *8*, 236–246. [CrossRef] [PubMed]

15. DiScipio, R.G.; Schraufstatter, I.U. The role of the complement anaphylatoxins in the recruitment of eosinophils. *Int. Immunopharmacol.* **2007**, *7*, 1909–1923. [CrossRef]

16. Jiang, Y.; Zhao, G.; Song, N.; Li, P.; Chen, Y.; Guo, Y.; Li, J.; Du, L.; Jiang, S.; Guo, R.; et al. Blockade of the C5a-C5aR axis alleviates lung damage in hDPP4-transgenic mice infected with MERS-CoV. *Emerg. Microbes Infect.* **2018**, *7*, 77. [CrossRef] [PubMed]

17. Zhao, G.; Jiang, Y.; Qiu, H.; Gao, T.; Zeng, Y.; Guo, Y.; Yu, H.; Li, J.; Kou, Z.; Du, L.; et al. Multi-Organ Damage in Human Dipeptidyl Peptidase 4 Transgenic Mice Infected with Middle East Respiratory Syndrome-Coronavirus. *PLoS ONE* **2015**, *10*, e0145561. [CrossRef]

18. Zhou, J.; Chu, H.; Li, C.; Wong, B.H.; Cheng, Z.S.; Poon, V.K.; Sun, T.; Lau, C.C.; Wong, K.K.; Chan, J.Y.; et al. Active replication of Middle East respiratory syndrome coronavirus and aberrant induction of inflammatory cytokines and chemokines in human macrophages: Implications for pathogenesis. *J. Infect. Dis.* **2014**, *209*, 1331–1342. [CrossRef]

19. Wang, W.; Xiao, F.; Wan, P.; Pan, P.; Zhang, Y.; Liu, F.; Wu, K.; Liu, Y.; Wu, J. EV71 3D Protein Binds with NLRP3 and Enhances the Assembly of Inflammasome Complex. *PLoS Pathog.* **2017**, *13*, e1006123. [CrossRef]

20. Allen, I.C.; Scull, M.A.; Moore, C.B.; Holl, E.K.; McElvania-TeKippe, E.; Taxman, D.J.; Guthrie, E.H.; Pickles, R.J.; Ting, J.P. The NLRP3 inflammasome mediates in vivo innate immunity to influenza A virus through recognition of viral RNA. *Immunity* **2009**, *30*, 556–565. [CrossRef]

21. Ren, R.; Wu, S.; Cai, J.; Yang, Y.; Ren, X.; Feng, Y.; Chen, L.; Qin, B.; Xu, C.; Yang, H.; et al. The H7N9 influenza A virus infection results in lethal inflammation in the mammalian host via the NLRP3-caspase-1 inflammasome. *Sci. Rep.* **2017**, *7*, 7625. [CrossRef] [PubMed]

22. Wang, W.; Li, G.; De, W.; Luo, Z.; Pan, P.; Tian, M.; Wang, Y.; Xiao, F.; Li, A.; Wu, K.; et al. Zika virus infection induces host inflammatory responses by facilitating NLRP3 inflammasome assembly and interleukin-1beta secretion. *Nat. Commun.* **2018**, *9*, 106. [CrossRef] [PubMed]

23. Dinarello, C.A. Immunological and inflammatory functions of the interleukin-1 family. *Annu. Rev. Immunol.* **2009**, *27*, 519–550. [CrossRef] [PubMed]

24. Guery, B.; Poissy, J.; el Mansouf, L.; Sejourne, C.; Ettahar, N.; Lemaire, X.; Vuotto, F.; Goffard, A.; Behillil, S.; Enouf, V.; et al. Clinical features and viral diagnosis of two cases of infection with Middle East Respiratory Syndrome coronavirus: A report of nosocomial transmission. *Lancet* **2013**, *381*, 2265–2272. [CrossRef]

25. Hajishengallis, G.; Lambris, J.D. Crosstalk pathways between Toll-like receptors and the complement system. *Trends Immunol.* **2010**, *31*, 154–163. [CrossRef] [PubMed]

26. Samstad, E.O.; Niyonzima, N.; Nymo, S.; Aune, M.H.; Ryan, L.; Bakke, S.S.; Lappegard, K.T.; Brekke, O.L.; Lambris, J.D.; Damas, J.K.; et al. Cholesterol crystals induce complement-dependent inflammasome activation and cytokine release. *J. Immunol.* **2014**, *192*, 2837–2845. [CrossRef] [PubMed]

27. Liszewski, M.K.; Kolev, M.; Le Friec, G.; Leung, M.; Bertram, P.G.; Fara, A.F.; Subias, M.; Pickering, M.C.; Drouet, C.; Meri, S.; et al. Intracellular complement activation sustains T cell homeostasis and mediates effector differentiation. *Immunity* **2013**, *39*, 1143–1157. [CrossRef]

28. Asgari, E.; Le Friec, G.; Yamamoto, H.; Perucha, E.; Sacks, S.S.; Kohl, J.; Cook, H.T.; Kemper, C. C3a modulates IL-1beta secretion in human monocytes by regulating ATP efflux and subsequent NLRP3 inflammasome activation. *Blood* **2013**, *122*, 3473–3481. [CrossRef]

29. Gerard, N.P.; Gerard, C. The chemotactic receptor for human C5a anaphylatoxin. *Nature* **1991**, *349*, 614–617. [CrossRef]

30. Gao, H.; Neff, T.A.; Guo, R.F.; Speyer, C.L.; Sarma, J.V.; Tomlins, S.; Man, Y.; Riedemann, N.C.; Hoesel, L.M.; Younkin, E.; et al. Evidence for a functional role of the second C5a receptor C5L2. *FASEB J.* **2005**, *19*, 1003–1005. [CrossRef]

31. Gerard, N.P.; Lu, B.; Liu, P.; Craig, S.; Fujiwara, Y.; Okinaga, S.; Gerard, C. An anti-inflammatory function for the complement anaphylatoxin C5a-binding protein, C5L2. *J. Biol. Chem.* **2005**, *280*, 39677–39680. [CrossRef] [PubMed]

32. Morgan, B.P.; Gasque, P. Extrahepatic complement biosynthesis: Where, when and why? *Clin. Exp. Immunol.* **1997**, *107*, 1–7. [CrossRef] [PubMed]

33. Gu, J.; Gong, E.; Zhang, B.; Zheng, J.; Gao, Z.; Zhong, Y.; Zou, W.; Zhan, J.; Wang, S.; Xie, Z.; et al. Multiple organ infection and the pathogenesis of SARS. *J. Exp. Med.* **2005**, *202*, 415–424. [CrossRef] [PubMed]

34. Doitsh, G.; Galloway, N.L.; Geng, X.; Yang, Z.; Monroe, K.M.; Zepeda, O.; Hunt, P.W.; Hatano, H.; Sowinski, S.; Munoz-Arias, I.; et al. Cell death by pyroptosis drives CD4 T-cell depletion in HIV-1 infection. *Nature* **2014**, *505*, 509–514. [CrossRef] [PubMed]

35. McAuley, J.L.; Tate, M.D.; MacKenzie-Kludas, C.J.; Pinar, A.; Zeng, W.; Stutz, A.; Latz, E.; Brown, L.E.; Mansell, A. Activation of the NLRP3 inflammasome by IAV virulence protein PB1-F2 contributes to severe pathophysiology and disease. *PLoS Pathog.* **2013**, *9*, e1003392. [CrossRef] [PubMed]

36. Pinar, A.; Dowling, J.K.; Bitto, N.J.; Robertson, A.A.; Latz, E.; Stewart, C.R.; Drummond, G.R.; Cooper, M.A.; McAuley, J.L.; Tate, M.D.; et al. PB1-F2 Peptide Derived from Avian Influenza A Virus H7N9 Induces Inflammation via Activation of the NLRP3 Inflammasome. *J. Biol. Chem.* **2017**, *292*, 826–836. [CrossRef]

37. Yao, S.T.; Cao, F.; Chen, J.L.; Chen, W.; Fan, R.M.; Li, G.; Zeng, Y.C.; Jiao, S.; Xia, X.P.; Han, C.; et al. NLRP3 is Required for Complement-Mediated Caspase-1 and IL-1beta Activation in ICH. *J. Mol. Neurosci.* **2017**, *61*, 385–395. [CrossRef]

38. Arbore, G.; West, E.E.; Spolski, R.; Robertson, A.A.B.; Klos, A.; Rheinheimer, C.; Dutow, P.; Woodruff, T.M.; Yu, Z.X.; O'Neill, L.A.; et al. T helper 1 immunity requires complement-driven NLRP3 inflammasome activity in CD4(+) T cells. *Science* **2016**, *352*. [CrossRef]

39. Reichhardt, M.P.; Meri, S. Intracellular complement activation-An alarm raising mechanism? *Semin. Immunol.* **2018**. [CrossRef]

40. Zhang, Y.; Liu, F.; Yuan, Y.; Jin, C.; Chang, C.; Zhu, Y.; Zhang, X.; Tian, C.; He, F.; Wang, J. Inflammasome-Derived Exosomes Activate NF-kappaB Signaling in Macrophages. *J. Proteome Res.* **2017**, *16*, 170–178. [CrossRef]

41. Mebius, R.E.; Kraal, G. Structure and function of the spleen. *Nat. Rev. Immunol.* **2005**, *5*, 606–616. [CrossRef] [PubMed]

viruses

MDPI

Review

Development of Small-Molecule MERS-CoV Inhibitors

Ruiying Liang [1,†], Lili Wang [2,†], Naru Zhang [3,†], Xiaoqian Deng [1], Meng Su [1], Yudan Su [1], Lanfang Hu [1], Chen He [1], Tianlei Ying [4,*], Shibo Jiang [4,*] and Fei Yu [1,*]

1   College of Life and Science, Hebei Agricultural University, Baoding 071001, China;
    ruiyingliang@outlook.com (R.L.); dengxiaoqian0926@hotmail.com (X.D.); sumeng123@hotmail.com (M.S.);
    suyudan123@hotmail.com (Y.S.); hlf0519@hotmail.com (L.H.); hechen285@hotmail.com (C.H.)
2   Research Center of Chinese Jujube, Hebei Agricultural University, Baoding 071001, China;
    yywll@hebau.edu.cn
3   Department of Clinical Medicine, Faculty of Medicine, Zhejiang University City College, Hangzhou 310015,
    China; zhangnr@zucc.edu.cn
4   Key Laboratory of Medical Molecular Virology of MOE/MOH, School of Basic Medical Sciences,
    Fudan University, Shanghai 200032, China
*   Correspondence: shmyf@hebau.edu.cn (F.Y.); shibojiang@fudan.edu.cn (S.J.); tlying@fudan.edu.cn (T.Y.);
    Tel.: +86-312-7528935 (F.Y.); +86-21-54237673 (S.J.); +86-21-54237761 (T.Y.)
†   These authors contributed equally to this work.

Received: 24 November 2018; Accepted: 12 December 2018; Published: 17 December 2018

Abstract: Middle East respiratory syndrome coronavirus (MERS-CoV) with potential to cause global pandemics remains a threat to the public health, security, and economy. In this review, we focus on advances in the research and development of small-molecule MERS-CoV inhibitors targeting different stages of the MERS-CoV life cycle, aiming to prevent or treat MERS-CoV infection.

Keywords: MERS-CoV; mechanism of action; small-molecule inhibitor

1. Introduction

Middle East respiratory syndrome coronavirus (MERS-CoV) has posed a serious threat to public health worldwide because it can cause severe respiratory disease in humans with high mortality (about 36%) [1]. As of 27 November 2018, a total of 2266 human MERS-CoV infections with 804 deaths had been reported from 27 countries in the Middle East, North Africa, Europe, Asia, and North America to the World Health Organization (WHO), with 83% reported by the Kingdom of Saudi Arabia (Figure 1) (https://www.who.int/emergencies/mers-cov/en/).

Phylogenetic and sequencing data strongly suggest that MERS-CoV belongs to the C-lineage of the genus betacoronavirus, the first known lineage C betacoronavirus associated with human infections [2]. The clinical features of MERS-CoV infection range from asymptomatic infection to rapidly progressive acute hypoxemic respiratory failure and extrapulmonary organ dysfunction [3–5]. At present, no effective vaccine or therapeutics are available for the prevention or treatment of MERS-CoV infection [6–8]. However, many basic and clinical studies on anti-MERS-CoV agents have been completed or are ongoing. In this review, we focus on current progress in the research and development of small-molecule MERS-CoV inhibitors, either peptides or compounds, targeting different stages of the MERS-CoV life cycle, aiming to prevent or treat MERS-CoV infection.

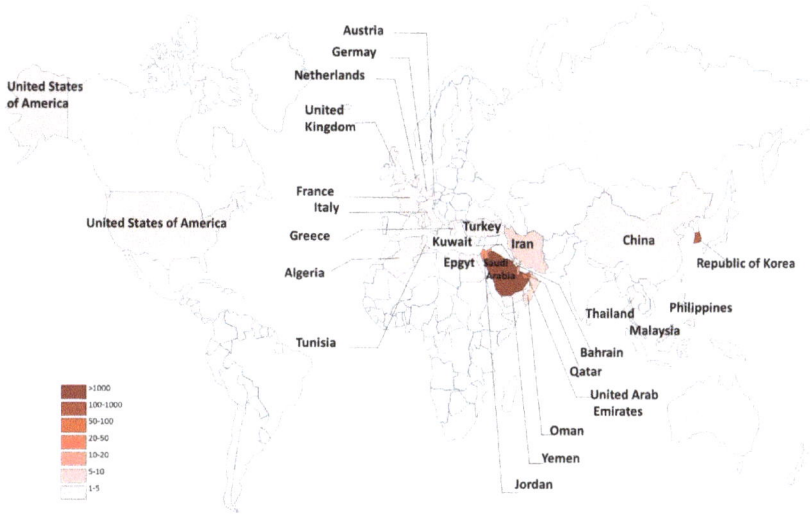

Figure 1. Summary of morbidity statistics with country- and quarter-level panel data.

2. MERS-CoV Life Cycle and Potential Targets for the Development of Small-Molecule Inhibitors Against MERS-CoV Infection

MERS-CoV enters host cells through two pathways. The first involves plasma membrane fusion, which relies on spike (S) protein activation by secreted or surface proteases, such as the transmembrane protease serine 2 (TMPRSS2) and the human airway trypsin-like protease (HAT). The second involves endosomal membrane fusion, in which spike protein activation is facilitated by the pH-dependent endosomal protease cathepsin L (CTSL) [9,10]. The spike protein plays a key role in MERS-CoV attachment to host cells and virus-cell membrane fusion [11]. It contains 1353 amino acids within the viral envelope in trimeric state [12]. Spike protein consists of S1 and S2 subunits. The S1 subunit contains the receptor binding domain (RBD), while the S2 subunit contains the fusion peptide (FP), a long heptad repeat 1 domain (HR1) and a short heptad repeat 2 domain (HR2) [13,14]. MERS-CoV enters the host cell by binding the viral particle via the RBD in spike protein to the cellular receptor dipeptidyl peptidase-4 (DPP4) on the surface of the host cell [12,15]. Then, S2 changes its conformation and inserts its FP into the plasma membrane, or the endosomal membrane if the virion is in the endosome. The HR2 binds to the HR1 to form a six-helix bundle (6-HB) fusion core, which brings viral and cell membranes into close apposition for fusion [14,16,17]. During this process, RBD, DPP4, HR1, HR2, and the related proteases, e.g., HAT and TMPRSS2, can all serve as targets for the development of MERS-CoV fusion/entry inhibitors.

After MERS-CoV entry into the host's cells, the positive RNA genome is translated in the cytoplasm. The genome can be translated into two polyproteins: ppla and pplb, which are cleaved into 16 nonstructural proteins by PL_{pro} (papain-like protease) and $3CL_{pro}$ (3-chymotrypsin-like protease). Hence, the proteases that are critically important for MERS-CoV replication can also be considered as targets for developing MERS-CoV replication inhibitors. However, information about the enzymes required for producing more genome copies and subgenomic mRNA for virus replication is limited. Then, the RNA genome and structural proteins are packaged into viral particles in host cells, and the progeny virus particles are finally released from host cells (Figure 2). Although these steps can also be used as targets for the development of MERS-CoV maturation-and-release inhibitors, no such inhibitors have been reported so far.

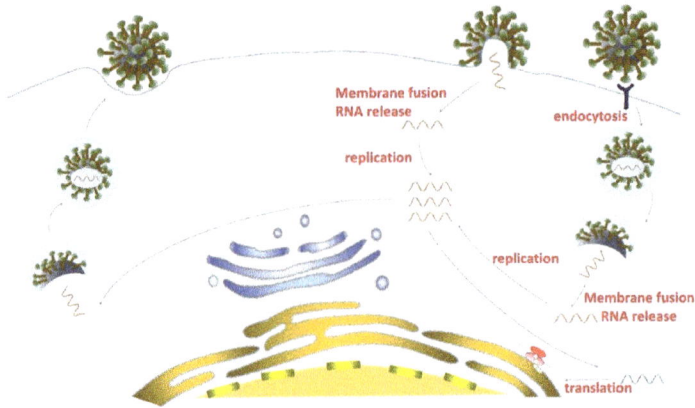

Figure 2. Schematic diagram of Middle East respiratory syndrome coronavirus (MERS-CoV) infection. MERS-CoV enters host cells by plasma membrane fusion (membrane fusion) or endosomal membrane fusion (endocytosis), and then releases the viral RNA into the cytoplasm. The RNA genome is replicated and viral proteins are produced. The progeny virus is generated and released from the infected cells.

3. Current Small-Molecule Inhibitors Against MERS-CoV Infection and Their Mechanisms of Action

3.1. MERS-CoV Entry Inhibitors

MERS-CoV S protein plays a key role in mediating virus entry into host target cells. This process includes binding to host receptors, viral fusion, and final entry into host cells. MERS-CoV pseudovirus expressing S protein, which allows for single-cycle infection in cells expressing receptor DPP4, can be used for screening MERS-CoV fusion/entry inhibitors.

HR2P, spanning residues 1251–1286 in the HR2 domain, with low or no toxic effect in vitro, can effectively inhibit MERS-CoV replication by interacting with the HR1 domain to block spike protein-mediated cell–cell fusion and MERS-CoV pseudovirus entry (Table 1; Figure 3) [16]. To increase its stability, solubility, and anti-MERS-CoV activity, Lu et al. introduced a Glu, Lys, or Arg residue into HR2P, generating a new peptide, HR2P-M2 (Table 1). HR2P-M2 was indeed found to be more stable and soluble than HR2P. It blocked fusion core formation between HR1 and HR2 peptides by binding to the viral S protein HR1 domain and inhibiting S protein-mediated membrane fusion with an EC_{50} of 0.55 µM (Figure 4) [16,23]. HR2P-M2 is highly effective in inhibiting MERS-CoV infection in both Calu-3 and Vero cells with an EC_{50} of about 0.6 µM. Intranasal application of HR2P-M2 could significantly reduce the titers of MERS-CoV in the lung of Ad5-hDPP4 (adenovirus serotype-5–human dipeptidyl peptidase 4)-transduced mice [16,18]. Furthermore, intranasal administration of HR2P-M2 before viral challenge fully protected hDPP4-transgenic mice from MERS-CoV infection, whereas all untreated mice died 8 days after viral challenge [24]. Furthermore, by combining HR2P-M2 with interferon β, protection was enhanced for Ad5-hDPP4-transduced mice against infection by MERS-CoV strains with or without mutations in the HR1 region of the S protein, with >1000-fold reduction of viral titers in lung [18].

Table 1. Peptide viral inhibitors against MERS-CoV.

Compound	Sequence	Testing Model	Cell Lines Tested	EC50 (μM)	CC50 (μM)	Ref.
	Peptide inhibitors disturbing membrane fusion					
HR2P	SLIQNTLLDLTYEMLSLQQVVKALNESYIDLKEL	In vitro	Vero cells Huh-7 cells	0.6 0.93 ± 0.15 b	>1000	[16]
HR2P-M2	SLIQNTLLDLEYEMKKLEVVKKLEESYIDLKEL	In vitro; in vivo: hDPP4 Tg mice	Calu-3 and Vero cells; Ad5-hDPP4 mice	0.55 ± 0.04 b	-	[16,18,19]
P21S10	LDLTYEM LSLQQVV K*LNE*Y	In vitro	Huh-7 cells	0.97 ± 0.08; 0.33 ± 0.04 b	>100	[20]
P21S2	L*LIY*M LSLQQVV KALNESY	In vitro	Huh-7 cells	3.90 ± 1.1 b	-	[20]
P21S4	LDLI*EM L*LQQVV KALNESY	In vitro	Huh-7 cells	7.14 ± 0.7 b	-	[20]
P21S5	LDLTYEM *SLQ*VV KALNESY	In vitro	Huh-7 cells	10.7 ± 2.6 b	-	[20]
P21S8	LDLTYEM LSLQ*VV K*LNESY	In vitro	Huh-7 cells	3.03 ± 0.29; 0.26 ± 0.05 b	>100	[20]
P21S9	LDLTYEM LSLQQVV *ALN*SY	In vitro	Huh-7 cells	14.1 ± 2.3 b	-	[20]
P21L2	LXLIYXM LSLQQVV KALNESY	In vitro	Huh-7 cells	10.9 ± 1.1 b	-	[20]
P21L4	LDLTXEM LXLQQVV KALNESY	In vitro	Huh-7 cells	8.21 ± 0.9 b	-	[20]
P21L5	LDLTYEM XSLQXVV KALNESY	In vitro	Huh-7 cells	4.49 ± 0.6 b	-	[20]
P21L8	LDLTYEM LSLQXVV KXLNESY	In vitro	Huh-7 cells	20.6 ± 3.3 b	-	[20]
P21L9	LDLTYEM LSLQQVV XALNXSY	In vitro	Huh-7 cells	10.9 ± 1.0 b	-	[20]
P21L10	LDLTYEM LSLQQVV KXLNEXY	In vitro	Huh-7 cells	3.55 ± 0.2 b	-	[20]
P21R8	LDLTYEM LSLQ*VV K*LNESY	In vitro	Huh-7 cells	16.3 ± 1.1 b	-	[20]
P21S8Z	LDLTYEZ LSLQ*VV K*LNESY	In vitro	Huh-7 cells	2.80 ± 0.74; 0.63 ± 0.05 b	>100	[20]
P21S8F	LDLTYEM LSLQ*VV K*LNESF	In vitro	Huh-7 cells	2.16 ± 1.1 b	-	[20]
P21S8ZF	LDLTYES LSLQ*VV K*LNESF	In vitro	Huh-7 cells	3.89 ± 0.8 b	-	[20]
P9 a	NGAICWGPCPTAFRQICNCGHFKVRCCKIR	In vitro	MDCK cells	5.00 μg/mL	380 μg/mL	[21]
LLS	LEEISKKLEELSKKLEELSKKLEELSKKLEELSKK-βA-K (C16)	In vitro	Huh-7 cells	0.24 ± 0.08 b	4.04 ± 0.4	[22]
IIS	IEEISKKIEEISKKIEEISKKIEEISKKIEEISKK-βA-K (C16)	In vitro	Huh-7 cells	0.10 ± 0.02 b	88.8 ± 28	[22]
AAS	AEEASKKAEEASKKAEEASKKAEEASKKAEEASKK-βA-K(C16)	In vitro	Huh-7 cells	4.47 ± 1.7 b	2.38 ± 0.9	[22]
FFS	FEEFSKKFEEFSKKFEEFSKKFEEFSKKFEEFSKK-βA-K (C16)	In vitro	Huh-7 cells	3.11 ± 0.9 b	>100	[22]
YYS	YEEYSKKYEEYSKKYEEYSKKYEEYSKKYEEYSKK-βA-K(C16)	In vitro	Huh-7 cells	6.26 ± 2.1 b	19.8 ± 1.6	[22]
IIY	IEEIYKKIEEIYKKIEEIYKKIEEIYKKIEEIYKK-βA-K (C16)	In vitro	Huh-7 cells	0.52 ± 0.4 b	>100	[22]
IIW	IEEIWKKIEEIWKKIEEIWKKIEEIWKKIEEIWKK-βA-K (C16)	In vitro	Huh-7 cells	10.6 ± 2.4 b	>100	[22]
IIH	IEEIHKKIEEIHKKIEEIHKKIEEIHKKIEEIHKK-βA-K (C16)	In vitro	Huh-7 cells	1.68 ± 0.47 b	>100	[22]
IIQ	IEEIQKKIEEIQKKIEEIQKKIEEIQKKIEEIQKK-βA-K (C16)	In vitro	Huh-7 cells	0.13 ± 0.1; 0.11 ± 0.02 b	>100	[22]
IIK	IEEIKKKIEEIKKKIEEIKKKIEEIKKKIEEIKKK-βA-K (C16)	In vitro	Huh-7 cells	0.45 ± 0.13 b	4.54 ± 0.6	[22]
IIE	IEEIEKKIEEIEKKIEEIEKKIEEIEKKIEEIEKK-βA-K (C16)	In vitro	Huh-7 cells	2.93 ± 0.95 b	>100	[22]

a P9-aci-1: three acidic amino acids D, E, and D were added to the C-terminus of P9. b Concentration of peptide that blocks MERS-CoV S-mediated cell–cell fusion. "-" indicates data not available. "*" indicates the position of the S5 residues, which react to form the all hydrocarbon staple. "X" indicates the positions of the R5 amino acids, which react to form staples. EC50: concentration for 50% of maximal effect. CC50: the 50% cytotoxicity concentrations.

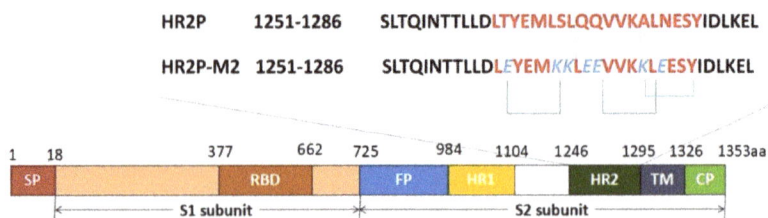

Figure 3. Schematic representation of MERS-CoV S (spike) protein S1 subunit and S2 subunit. RBD, receptor binding domain; FP, fusion peptide; HR1, heptad repeat 1 domain; HR2, heptad repeat 2 domain; TM, transmembrane domain; CP, cytoplasmic domain. The residue numbers of each region correspond to their positions in the S protein of MERS-CoV. HR2P, the peptide derived from the HR2 domain of MERS-CoV S protein S2 subunit; HR2P-M2, HR2P analogous peptide with mutations.

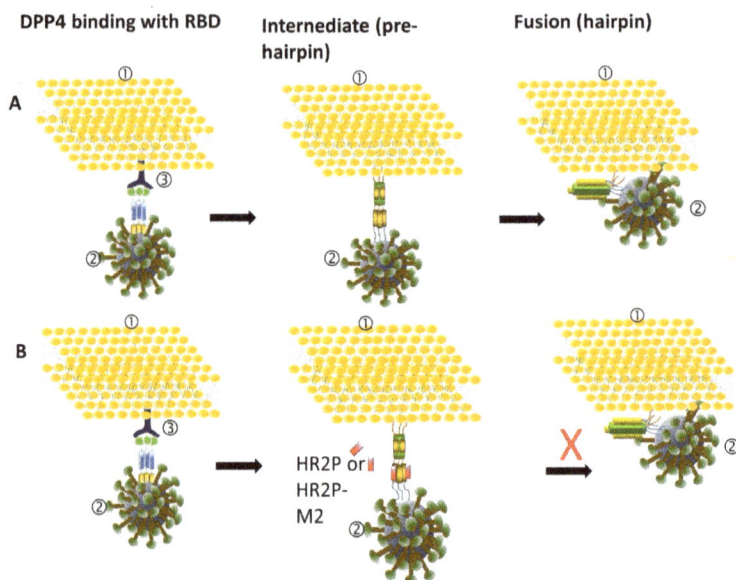

Figure 4. Schematic representation of the inhibition mechanism of HR2P and HR2P-M2. ① Target cell membrane; ② MERS-CoV; ③ dipeptidyl peptidase-4 (DPP4). (**A**) Mechanism of normal binding between a host cell and MERS-CoV. MERS-CoV enters the host cell by binding the viral particle via the RBD in spike protein to the cellular receptorDPP4 on the surface of the host cell. The HR2 binds to the HR1 to form a six-helix bundle (6-HB) fusion core, which brings viral and cell membranes into close apposition for fusion. (**B**) HR2P and HR2P-M2 block six-bundle fusion core formation between HR1 and HR2 peptides by binding to the viral S protein HR1 domain.

P21S10, the most effective fusion inhibitor of MERS-CoV, can inhibit MERS-CoV pseudovirus infection with an EC_{50} of about 1 µM in Huh-7 cells and a CC_{50} of >100 µM in Huh-7 cells by CCK8 (Cell Counting Kit-8) assay (Table 1) [20]. In addition, a series of synthesized stapled peptides, such as P21S10, P21S2, P21S4, P21S5, P21S8, P21S9, P21S8F, P21S8ZF, etc., could effectively inhibit infection by MERS-CoV pseudovirus and its spike protein-mediated cell fusion by blocking helix-mediated NHR (N-terminal heptad repeats) /CHR (C-terminal heptad repeats) interactions with a low EC_{50} and a high CC_{50} in Huh-7 cells [20].

P9, a short peptide, exhibited potent and broad spectrum antiviral effects against multiple respiratory viruses in vitro and in vivo [21,25]. P9 inhibited MERS-CoV with an EC_{50} of about 5 µg/mL in Madin-Darby canine kidney (MDCK) cells, obtained by plaque assay, and a CC_{50} of 380 µg/mL in MDCK cells obtained by MTT (3-(4,5-dimethyl-2-thiazolyl)-2,5-diphenyl-2-H-tetrazolium bromide) assay (Table 1) [21].

Lipopeptides are bioactive peptides that replicate the α-helical chain from the viral fusion machinery [22]. All 12 lipopeptides inhibit cell−cell fusion mediated by MERS-CoV S protein with EC_{50} values ranging from 0.1 to >10.0 µM in Huh-7 cells (Table 1) [22]. Among these lipopeptides, LLS and IIS were found to be the most potent MERS-CoV fusion inhibitors with EC_{50} values of 0.24 µM and 0.1 µM, respectively [22]. Other lipopeptides such as AAS, FFS, YYS, IIY, IIW, IIH, IIQ, IIK, and IIE can also inhibit cell−cell fusion mediated by MERS-CoV S protein with variable EC_{50} values [22].

Three neurotransmitter inhibitors, including chlorpromazine, fluphenazine, and promethazine, were moderate inhibitors of cell–cell fusion with EC_{50} values of about 23, 15, and 17 µM, respectively (Table 2; Figure 5(5), (45), (46)) [26]. They can also disrupt clathrin-mediated endocytosis to inhibit MERS-CoV [26].

A small-molecule HIV entry inhibitor targeting gp41 ADS-J1 (Figure 5(1)) at the concentration of 20 µM could inhibit >90% of MERS-CoV pseudovirus infection in NBL-7 and Huh-7 cells. ADS-J1 could interrupt the interactions between the HR1 and HR2 of MERS-CoV to form the six-helix bundle, thus inhibiting the entry of pseudotyped MERS-CoV with an EC_{50} of 0.6 µM in the DPP4-expressing cell line and with a CC_{50} of 26.9 µM in NBL-7 and Huh-7 cells by MTT assay (Table 2) [27].

The elucidation of MERS-CoV interaction with its host cell is critical to the development of antiviral interventions. In order to gain entry into host cells, MERS-CoV not only uses DPP4 as a functional virus receptor, but also utilizes certain cellular proteases, such as TMPRSS2 and members of the cathepsin family, as activators of the S glycoprotein [9]. TMPRSS2 is expressed in epithelial cells of the human respiratory and gastrointestinal tracts [28–31]. The respective enzymes from host cells are also excellent targets for the identification of small-molecule MERS-CoV inhibitors. The serine protease inhibitor camostat mesylate (camostat) could completely block syncytium formation, but only partially block virus entry into TMPRSS2-expressing Vero cells (Figure 5(2)) [31].

K11777, a compound known to inhibit cruzain, a cathepsin-like protease from the protozoan parasite *Trypanosoma cruzi*, can inhibit MERS-CoV with an EC_{50} of 46 nM (Figure 5(3)) [32,33].

Chloroquine inhibited MERS-CoV replication and blocked infection at an early step with an EC_{50} of 3 µM and a CC_{50} of 58 µM (Table 2; Figure 5(4)) [34]. Chlorpromazine inhibited MERS-CoV replication at both early and post-entry stages with an EC_{50} of about 5 µM and a CC_{50} of 21 µM (Table 2; Figure 5(5)) [34]. However, high cytotoxicity narrowed the therapeutic window in both monocyte-derived macrophages (MDMs) and dendritic cells (MDDCs) [34].

Ouabain and bufalin can inhibit MERS-CoV entry by blocking clathrin-mediated endocytosis (Figure 5(6), (7)) [25,35]. The addition of small amounts of ouabain (50 nM) or bufalin (10 to 15 nM) inhibited infection with MERS-CoV and VSV (vesicular stomatitis virus) (Table 2), but only when the drug was added prior to inoculation in Huh-7 cells [35].

Dihydrotanshinone, a lipophilic compound, showed a decimal reduction at 0.5 µg/mL and excellent antiviral effects at ≥ 2 µg/mL with a reduction in titer from 6.5 Log to 1.8 Log $TCID_{50}$/mL by using a pseudovirus expressing MERS-CoV spike protein (Figure 5(8)) [36].

During the biosynthesis of MERS-CoV S protein, the furin inhibitor decanoyl-RVKR-chloromethylketone (dec-RVKR-CMK) at 75 µM can lead to a decrease of the 85-kDa cleaved product in MERS-CoV S wt and S2′ mutant (Figure 5(9)) [37].

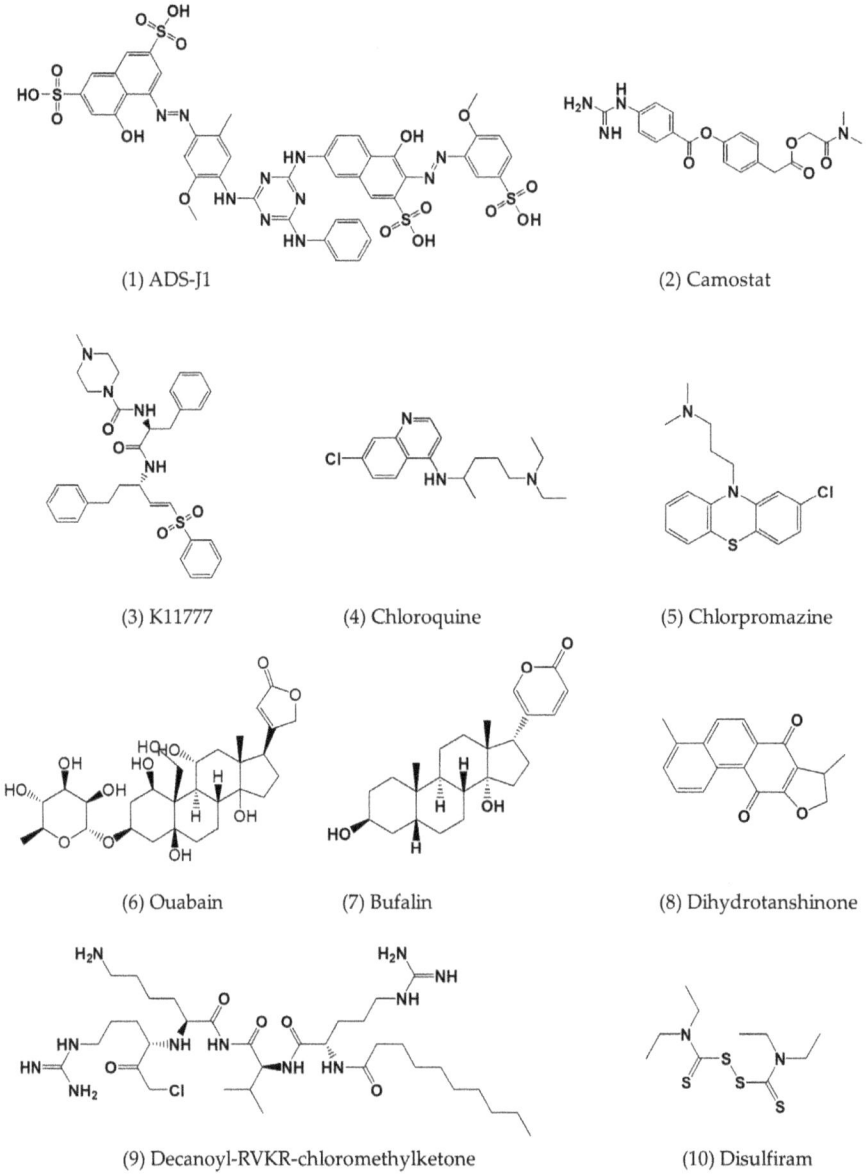

(1) ADS-J1

(2) Camostat

(3) K11777

(4) Chloroquine

(5) Chlorpromazine

(6) Ouabain

(7) Bufalin

(8) Dihydrotanshinone

(9) Decanoyl-RVKR-chloromethylketone

(10) Disulfiram

Figure 5. *Cont.*

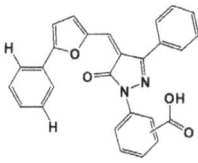

(11) **3k**

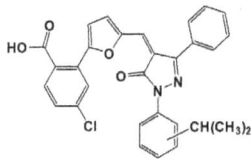

(12) **3h**

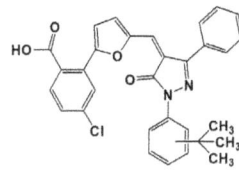

(13) **3i**

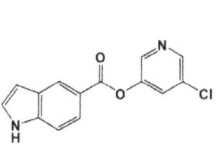

(14) CE-5

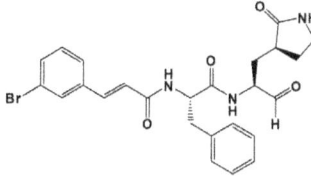

(15) **6b**

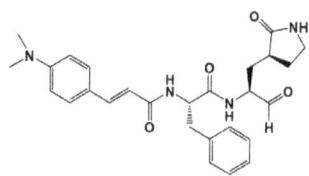

(16) **6c**

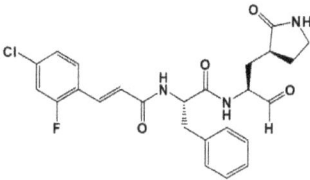

(17) **6d**

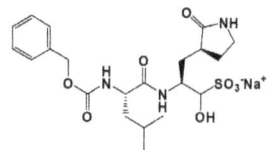

(18) GC376

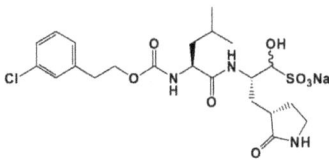

(19) GC813

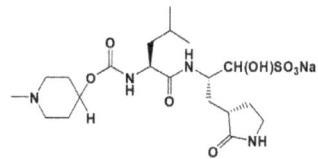

(20) **10a**

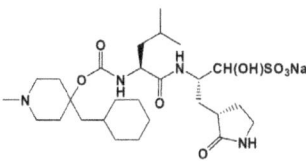

(21) **10c**

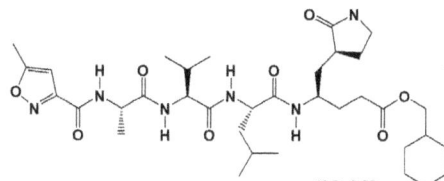

(22) N3

Figure 5. *Cont.*

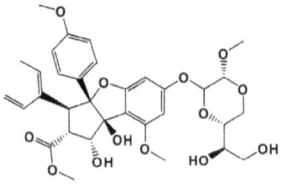

(23) Silvestrol

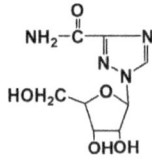

(24) Ribavirin

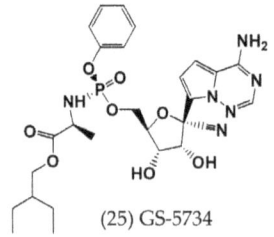

(25) GS-5734

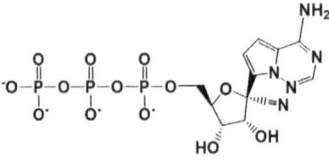

(26) GS-441524

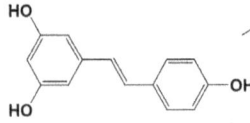

(27) Resveratrol

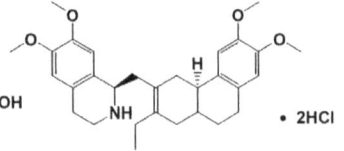

(28) Emetine dihydrochloride hydrate

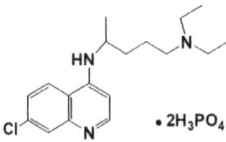

(29) Chloroquine diphosphate

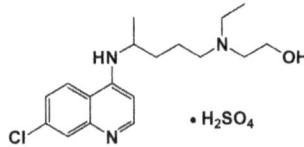

(30) Hydroxychloroquine sulfate

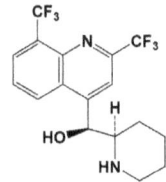

(31) Mefloquine

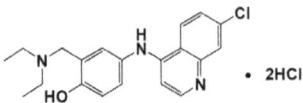

(32) Amodiaquine dihydrochloride dehydrate

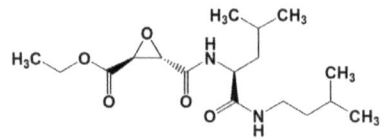

(33) E-64-D

Figure 5. *Cont.*

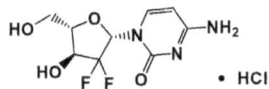

(34) Gemcitabine hydrochloride

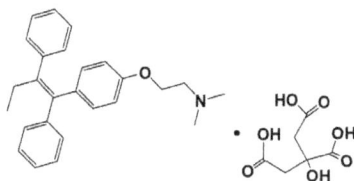

(35) Tamoxifen citrate

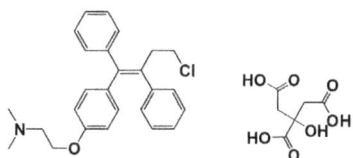

(36) Toremifene citrate

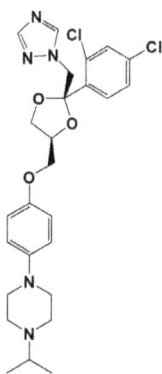

(37) Terconazole

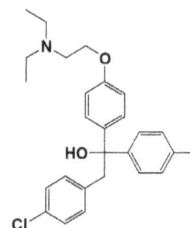

(38) Triparanol

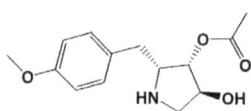

(39) Anisomycin

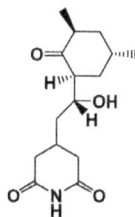

(40) Cycloheximide

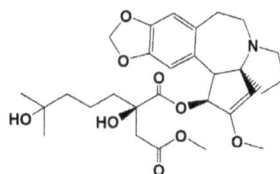

(41) Homoharringtonine

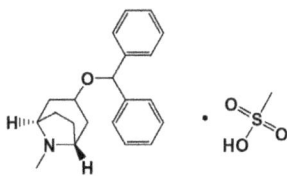

(42) Benztropine mesvlate

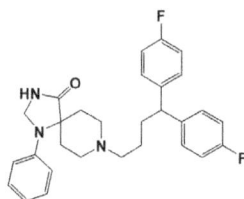

(43) Fluspirilene

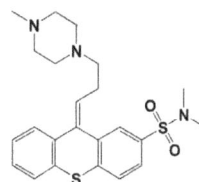

(44) Thiothixene

Figure 5. *Cont.*

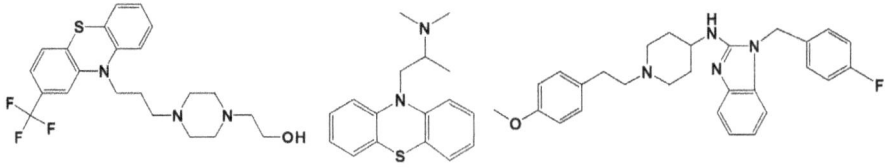

(45) Fluphenazine hydrochloride   (46) Promethazine hydrochloride   (47) Astemizole

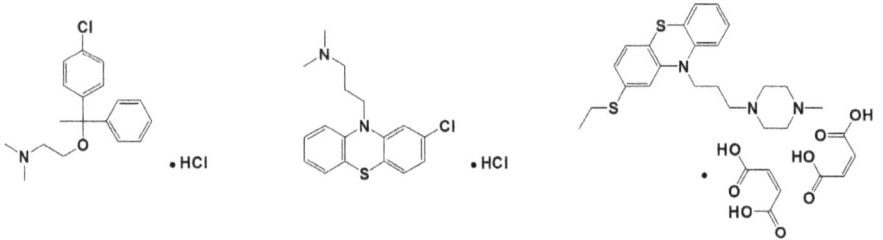

(48) Chlorophenoxamine
hydrochloride

(49) Chlorpromazine
hydrochloride

(50) Thiethyperazine
maleate

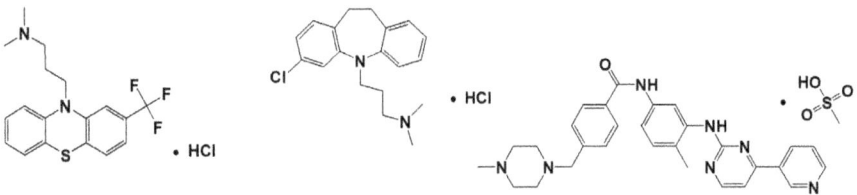

(51) Triflupromazine
hydrochloride

(52) Clomipramine
hydrochloride

(53) Imatinib mesylate

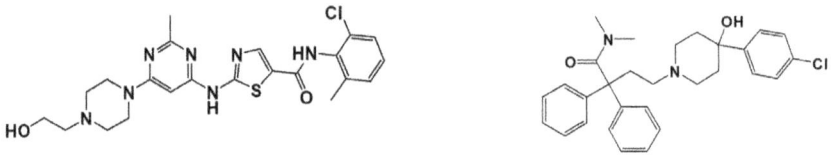

(54) Dasatinib

(55) Loperamide

Figure 5. *Cont.*

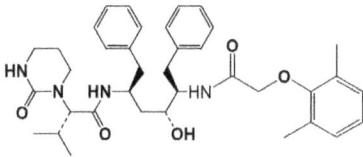

(56) Lopinavir

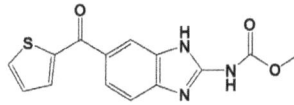

(57) Nocodazole

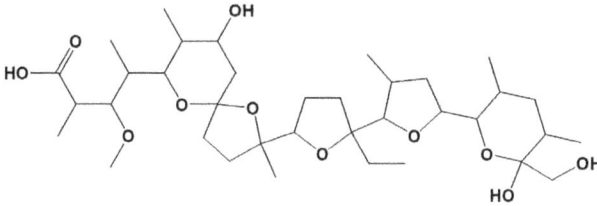

(58) Monensin

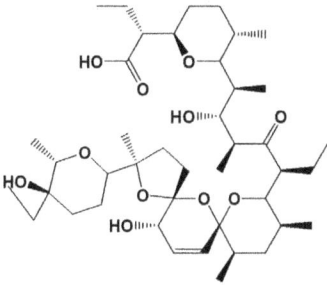

(59) Salinomycin sodium

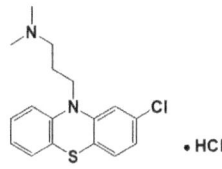

(60) Chlorpromazine
hydrochloride

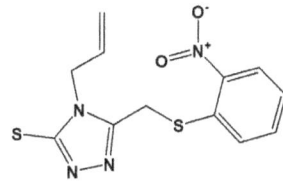

(61) SSYA10-001

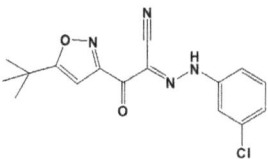

(62) ESI-09

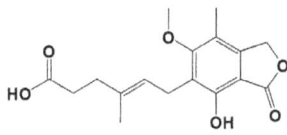

(63) Mycophenolic acid

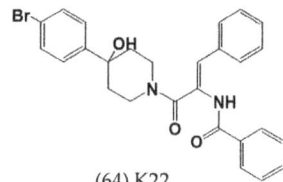

(64) K22

Figure 5. *Cont.*

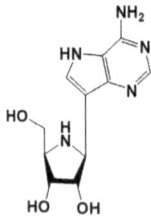

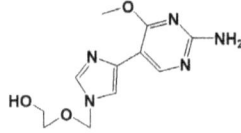

(65) BCX4430   (66) target fleximer analogue 2   (67) Cyclosporine

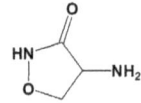

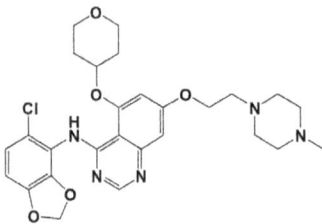

(68) Saracatinib   (69) Nutlin-3

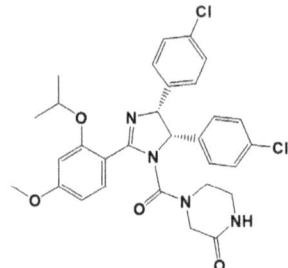

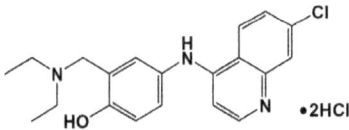

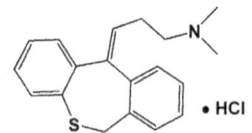

(70) Amodiaquine
dihydrochloride   (71) Sotrastaurin   (72) Dosulepin
hydrochloride

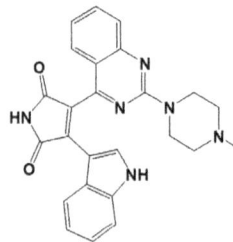

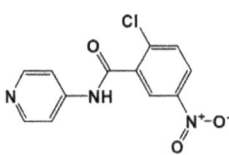

(73) N1-(4-pyridyl)-2-chloro-5-nitrobenzamide   (74) Acetophenazine maleate

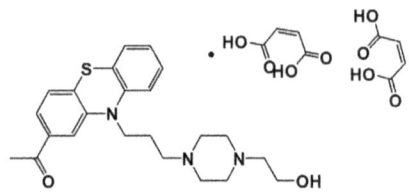

Figure 5. *Cont.*

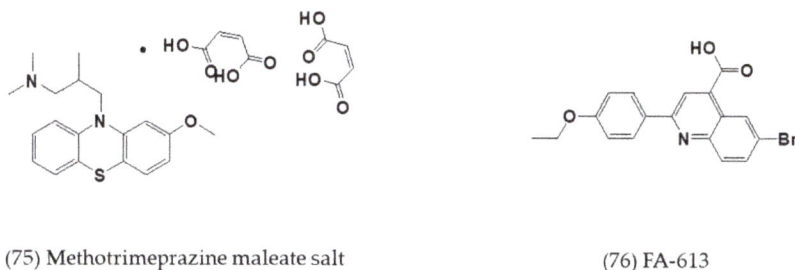

(75) Methotrimeprazine maleate salt   (76) FA-613

Figure 5. Chemical structure formulae of small-molecule inhibitors of MERS-CoV described in this review.

3.2. MERS-CoV Replication Inhibitors

3.2.1. MERS-CoV Inhibitors Targeting Papain-Like Protease

Papain-like protease is a cysteine protease that uses the thiol group of cysteine as a nucleophile to attack the carbonyl group of the scissile peptide bond [38,39]. The genome of MERS-CoV encodes two polyproteins, ppla and pplb, which are processed by papain-like protease (PL$_{pro}$) and 3C-like protease (3CL$_{pro}$) [40]. MERS-CoV has only one papain-like protease, as does SARS-CoV, while other coronaviruses have two enzymes [41,42]. MERS-PL$_{pro}$ is a part of the nonstructural protein nsp3, which includes three domains—namely, ubiquitin-like domain (UBL), a catalytic triad consisting of C1594–H1761–D1776, and the ubiquitin-binding domain (UBD) at the zinc finger—according to the homology model [40,43]. MERS-PL$_{pro}$ is a multifunctional enzyme with deISGylating and deubiquitinating (DUB) activities [43], but it can also block the interferon regulatory factor 3 (IRF3) pathway [43,44].

Disulfiram, a drug used in alcohol aversion therapy, has been approved by the U.S. Food and Drug Administration (FDA) since 1951 (Figure 5(10)). It can inhibit the activity of some enzymes, such as urease [45], methyltransferase [46], and kinase [45], all by reacting with cysteine residues, suggesting broad-spectrum characteristics [47]. Notably, disulfiram also acts as an allosteric inhibitor of MERS-CoV papain-like protease [47]. Multiple inhibition assays also support a kinetic mechanism by which disulfiram, together with 6TG (6-thioguanine) and/or MPA (mycophenolic acid), can synergistically inhibit MERS-CoV papain-like protease [47]. Hence, the recombination of three clinically available drugs could feasibly be used to treat MERS-CoV infection.

3.2.2. MERS-CoV Inhibitors Targeting 3C-Like Protease

The active site of MERS-3CL$_{pro}$ can be divided into subsites S1–S6 [48]. Subsite S1 consists of vital catalytic residue Cys145 with His41 to process polyproteins at 11 conserved Gln sites, followed by small amino acids like Ala, Ser, or Gly [49]. Another crucial component of the S1 subsite is the oxyanion hole formed by the interaction of a carboxylate anion of conserved Gln with Gly143, Ser144, and Cys145, which stabilizes the transition state during proteolysis [50,51]. Glu166 at the entrance of the pocket interacts via H-bond with the Nε2 of the conserved Gln [50]. The S2 and S4 subsites contain hydrophobic and bulky side chains such as Val, Leu, or Phe. Subsites S5 and S6 are near the surface of the active site and have little participation in substrate binding [48].

Table 2. Small molecule viral inhibitors against MERS-CoV.

Inhibitor	Testing Model	Cell Lines	EC$_{50}$ (µM)	CC$_{50}$ (µM)	Ref.
Inhibitors blocking the binding between virus and host cells					
ADS-J1	In vitro	NBL-7 and Huh-7 cells	0.6	26.9	[27]
Inhibitors disrupting endocytosis					
Chlorpromazine	In vitro	Huh-7 cells	23.33 ± 2.89 [a]; 49 ± 1.2; 9.514	>40; 21.3 ± 1.0	[5–7,26]
Promethazine	In vitro	Huh-7 cells	16.67 ± 7.22 [a]; 11.802	>40	[7,26]
Fluphenazine	In vitro	Huh-7 cells	15.00 ± 4.33 [a]; 5.868	~40	[7,26]
K11777	In vitro	Vero cells	0.046	>10	[32]
Camostat	In vitro	Vero-TMPRSS2 cells	~1	-	[31]
Ouabain	In vitro	Huh-7 cells	~0.05	-	[35]
Bufalin	In vitro	Huh-7 cells	0.01–0.015	-	[35]
Dihydrotanshinone	In vitro	-	0.5–1 µg/mL	-	[36]
Inhibitors interrupting MERS-CoV MERS-CoV RNA replication and translation					
Disulfiram	In vitro	-	22.7 ± 0.5	-	[47]
3k	In vitro	-	5.8 ± 1.6	-	[48]
3h	In vitro	-	7.3 ± 2.1	-	[48]
3i	In vitro	-	7.4 ± 2.2	-	[48]
CE-5	In vitro	HEK293T cells	~12.5	-	[53]
6b	In vitro	Huh-7 cells	1.4 ± 0.0	>100	[54]
6c	In vitro	Huh-7 cells	1.2 ± 0.6	>100	[54]
6d	In vitro	Huh-7 cells	0.6 ± 0.0	58.6 ± 1.2	[54]
GC376	In vitro	-	1.56 ± 0.09; 0.9	>150	[52,55]
GC813	In vitro	-	0.5	-	[52]
10a	In vitro	Vero81 cells	0.5	>100	[52]
10c	In vitro	Vero81 cells	0.8	>100	[52]
N3	In vitro	-	0.28 ± 0.02	-	[56]

Table 2. *Cont.*

Inhibitor	Testing Model	Cell Lines	EC$_{50}$ (μM)	CC$_{50}$ (μM)	Ref.
Silvestrol	In vitro	MRC-5 cells	0.0013	0.4	[57]
GS-5734	In vitro	HAE cells	0.074 ± 0.023	>10	[58]
GS-441524	In vitro	HAE cells	0.86 ± 0.78	>100	[58]
Chloroquine	In vitro	MDMs and MDDCs cells	3.0 ± 1.1; 6.275	58.1 ± 1.1	[7,59]
		Inhibitors with undefined mechanisms			
Emetine dihydrochloride hydrate	In vitro	Vero E6 cells	0.014	-	[7]
Hydroxychloroquine sulfate	In vitro	Vero E6 cells	8.279	-	[7]
Mefloquine	In vitro	Vero E6 cells	7.416	-	[7]
Amodiaquine dihydrochloride dehydrate	In vitro	Vero E6 cells	6.212	-	[7]
E-64-D	In vitro	Vero E6 cells	1.275	-	[7]
Gemcitabine hydrochloride	In vitro	Vero E6 cells	1.216	-	[7]
Tamoxifen citrate	In vitro	Vero E6 cells	10.117	-	[7]
Toremifene citrate	In vitro	Vero E6 cells	12.915	-	[7]
Terconazole	In vitro	Vero E6 cells	12.203	-	[7]
Triparanol	In vitro	Vero E6 cells	5.283	-	[7]
Anisomycin	In vitro	Vero E6 cells	0.003	-	[7]
Cycloheximide	In vitro	Vero E6 cells	0.189	-	[7]
Homoharringtonine	In vitro	Vero E6 cells	0.0718	-	[7]
Benztropine mesylate	In vitro	Vero E6 cells	16.627	-	[7]
Fluspirilene	In vitro	Vero E6 cells	7.477	-	[7]
Thiothixene	In vitro	Vero E6 cells	9.297	-	[7]
Astemizole	In vitro	Vero E6 cells	4.884	-	[7]
Chlorphenoxamine hydrochloride	In vitro	Vero E6 cells	12.646	-	[7]
Thiethylperazine maleate	In vitro	Vero E6 cells	7.865	-	[7]
Triflupromazine hydrochloride	In vitro	Vero E6 cells	5.758	-	[7]
Clomipramine hydrochloride	In vitro	Vero E6 cells	9.332	-	[7]
Imatinib mesylate	In vitro	Vero E6 cells	17.689	-	[7]
Dasatinib	In vitro	Vero E6 cells	5.468	-	[7]
Loperamide	In vitro	Vero E6 cells	4.8 ± 1.5	15.5 ± 1.0	[7]
Lopinavir	In vitro	Vero E6 cells	8.0 ± 1.5	24.4 ± 1.0	[7]
SSYA10-001	In vitro	Vero E6 cells	~25	>500	[60]
ESI-09	In vitro	Calu-3 and Vero E6 cells	5–10	>50	[61]
Mycophenolic acid	In vitro	Vero E6 cells	2.87	-	[60]

Table 2. *Cont.*

Inhibitor	Testing Model	Cell Lines	EC$_{50}$ (μM)	CC$_{50}$ (μM)	Ref.
Inhibitors with undefined mechanisms					
BCX4430	In vitro	-	68.4	>100	[62]
Fleximer analogues 2	In vitro	Vero cells	23 ± 0.6;	71 ± 14;	[63]
Nutlin-3	In vitro	Huh-7 cells	27 ± 0.0	149 ± 6.8	[64]
Amodiaquine dihydrochloride	In vitro	Huh-7 cells	6.9 ± 1.4	26.8 ± 1.6	[64]
Saracatinib	In vitro	Huh-7 cells	2.1 ± 0.7	12.3 ± 5.9	[64]
Sotrastaurin	In vitro	Huh-7 cells	2.9 ± 0.6	57 ± 5.5	[64]
Acetophenazine maleate	In vitro	Huh-7 cells	9.7 ± 3.3	>50	[64]
Dosulepin hydrochloride	In vitro	Huh-7 cells	11.2 ± 5.0	23.6 ± 3.8	[64]
Methotrimeprazine maleate salt	In vitro	Huh-7 cells	3.4 ± 0.0	28.9 ± 0.0	[64]
N1-(4-pyridyl)-2–chloro-5-nitrobenzamide	In vitro	Huh-7 cells	2.5 ± 0.0	24.5 ± 0.0	[64]
FA-613	In vitro	Huh-7 cells	10.5 ± 0.3	>50	[64]
	In vitro	Huh-7 cells	10.2 ± 0.2	-	[65]

[a] 50% effective concentration (EC$_{50}$) values of inhibiting cell–cell fusion. "–" indicates data not available.

Polyproteins pp1a and pp1b are processed by $3CL_{pro}$ (11 cleavage sites) and PL_{pro} (3 cleavage sites), resulting in 16 mature nonstructural proteins, including RNA-dependent RNA polymerase (RdRp) and helicase, which play important roles in the transcription and replication of coronaviruses [40,52]. Therefore, both proteases are essential for viral replication, making them attractive targets for drug development [52].

The analogues of hits of neuraminidase (NA) inhibitors on MERS-CoV $3CL_{pro}$ have been synthesized and showed average-to-good inhibition of MERS-$3CL_{pro}$. The better one is the compound **3k** with an EC_{50} of 5.8 µM (Table 2; Figure 5(11)) [48]. Another two are compounds **3h** (Figure 5(12)) and **3i** (Figure 5(13)) with EC_{50} values of 7.3 and 7.4 µM, repsectively (Table 2) [48]. Furthermore, researchers have concluded that pharmacophores phenyl at R3 and carboxylate, either at R1 or R4, are essential for the antiviral activity [48]. Since the modification of rings A and B is well tolerated, these rings can be further altered to enhance the activity of the compounds. The SARS-CoV $3CL_{pro}$ inhibitor CE-5 can block the function of the MERS-CoV $3CL_{pro}$ (Figure 5(14)) [53]. Treatment with CE-5 inhibited the activity of MERS-CoV $3CL_{pro}$ to 30% of that of DMSO-treated cells at a maximum dose of 50 µM [53]. The endpoint evaluation of CE-5 indicated an EC_{50} of ~12.5 µM in cell culture (Table 2) [53].

Peptidomimetic inhibitors of enterovirus (**6b**, **6c**, and **6d**) inhibit MERS-CoV with EC_{50} values ranging from 1.7 to 4.7 µM, as shown by enzymatic assay (Figure 5(15), (16), (17)) [54]. As shown in Table 1, compounds **6b**, **6c**, and **6d** efficiently suppressed viral replication with EC_{50} values of 1.4, 1.2, and 0.6 µM, respectively, after performing a cytopathic inhibition assay using MERS-CoV-infected Huh-7 cells (Table 2) [54].

GC376, a dipeptidyl transition state $3CL_{pro}$ inhibitor, can substantially inhibit the activity of MERS-CoV $3CL_{pro}$ with an EC_{50} of 1.6 µM by fluorescence resonance energy transfer (FRET) assay (Table 2; Figure 5(18)) [55].

GC813 as well as its synthesizing extended compounds **10a** and **10c** exhibit inhibition for MERS-CoV with EC_{50} values of 0.5 µM, 0.5 µM, and 0.8 µM in cell culture (Table 2; Figure 5(18), (19), (20), (21)) [52].

N3, a broad-spectrum anti-CoV inhibitor, can inhibit the proteolytic activity of MERS-CoV $3CL_{pro}$ by binding with the interface of domain I and II of MERS-CoV $3CL_{pro}$ with an EC_{50} of about 0.3 µM (Table 2; Figure 5(22)) [56].

3.3. Other Small-Molecule Inhibitors with Defined or Undefined Mechanisms of Action

Silvestrol, an eIF4A inhibitor, can inhibit MERS-CoV infection with an EC_{50} of 1.3 nM, as shown by plaque assay in MRC-5 cells and CC_{50} of 400 nM by MTT assay in peripheral blood mononuclear cells (PBMCs) (Table 2; Figure 5(23)) [57]. Silvestrol has broad-spectrum antiviral activity via the inhibition of the expression of CoV structural and nonstructural proteins (N, nsp8) and the formation of viral replication/transcription complexes [57].

The combination of interferon-α2b and ribavirin can effectively reduce MERS-CoV replication in vitro and in vivo (Table 2; Figure 5(24)) [6]. Rhesus macaques treated with IFN-α2b and ribavirin 8 h after MERS-CoV infection showed improved clinical parameters with no or very mild radiographic evidence of pneumonia compared with untreated macaques [6]. Moreover, treated macaques showed lower levels of systemic (serum) and local (lung) proinflammatory markers in addition to fewer viral genome copies, distinct gene expression, and less severe histopathological changes in the lungs [6].

GS-5734 (Remdesivir), the monophosphoramidate prodrug of the C-adenosine nucleoside analogue GS-441524, can inhibit the replication of the model β-coronavirus murine hepatitis virus (MHV) and RNA synthesis in wild-type (WT) virus, while an nsp14 ExoN (-) mutant lacking proofreading demonstrated increased susceptibility to GS-5734 (Figure 5(25)) [58]. GS-5734 also inhibits MERS-CoV infection with an EC_{50} of 0.074 $\pm$ 0.023 µM and a CC_{50} of >10 µM in human amniotic epithelial (HAE) cells (Table 2) [58]. Furthermore, GS-5734 acts at the early post-infection stage to decrease viral RNA levels, whereas delaying the addition of GS-5734 until 24 h post-infection resulted in decreased viral titer in HAE cell cultures at 48 and 72 h post-infection [58]. The nucleotide analogue GS-441524 also inhibits the infection of MERS-CoV with an EC_{50} of 0.9 µM and a CC_{50} of >100 µM in HAE cells (Table 2; Figure 5(26)) [58].

Resveratrol was found to significantly inhibit MERS-CoV infection as well as prolong cellular survival after virus infection (Figure 5. (27)) [66]. It was found that resveratrol could reduce RNA levels and infection titers in Vero cells [66]. Although resveratrol has minimal cytotoxicity, even at the high concentration of 250 µM, it can be ignored when compared to the much more severe toxicity of MERS-CoV infection [66].

A series of FDA-approved compounds were screened against MERS-CoV (Table 2) by cell-based ELISA assay (Figure 5(28–56)) [7]. Pharmaceuticals that inhibit MERS-CoV include neurotransmitter inhibitors, estrogen receptor antagonists, kinase signaling inhibitors, inhibitors of lipid or sterol metabolism, protein processing inhibitors, inhibitors of DNA synthesis/repair, as well as inhibitors of ion transport, cytoskeleton (specifically tubulin), and apoptosis [7]. Antiparasitics and antibacterials are two classes of pharmaceuticals, the functions of which are not obviously linked to coronaviruses, or viruses in general, but nonetheless show antiviral activity against MERS-CoV.

Nocodazole, targeting the cytoskeleton, specifically interferes with microtubule polymerization. It is an antimitotic drug developed for the treatment of cancer, but it was found to show high activity against MERS-CoV (Figure 5(57)) [67,68]. Monensin and salinomycin sodium, two of the nine ion channel inhibitors, have inhibitory activity against MERS-CoV, indicating that MERS-CoV may be susceptible to ionophore activities (Figure 5 (58), (59)). Chlorpromazine and chloroquine appear to target host factors, rather than viral proteins specifically, and the treatment of viral infections in patients aimed at host factors could reconfigure overt manifestations of viral pathogenesis into a less virulent subclinical infection and lower adverse disease outcome (Figure 5(60), (29)) [34,69].

Loperamide, an antidiarrheal opioid receptor agonist that reduces intestinal motility, also inhibits the replication of MERS-CoV at low-micromolar concentrations (3.3–6.3 µM) *in vitro* (Table 2; Figure 5(55)) [34]. Lopinavir, the HIV-1 protease inhibitor, inhibits MERS-CoV replication with an EC_{50} of 8 µM (Table 2; Figure 5(56)) [34].

SSYA10-001 inhibits MERS-CoV replication with an EC_{50} of ~25 µM in Vero E6 cells (Table 2; Figure 5(61)) [70]. Molecular modeling data suggest that SSYA10-001 can be docked with a comparable "Glide" score [70].

ESI-09 can reduce virus yield by inhibiting cAMP signaling in a cell type-independent manner (Figure 5(62)) [61]. The concentration of MERS-CoV inhibition by ESI-09 was found with an EC_{50} of 5 to 10 µM and a $CC_{50} > 50$ µM for both Calu-3 and Vero E6 cells by using the lactate dehydrogenase (LDH)-based cytotoxicity assay [62]. In addition, the undetectable cytopathic effect (CPE) and minimal expression of viral antigen indicated that Calu-3 cells treated with ESI-09 were almost fully protected [61].

Mycophenolic acid (MPA) can strongly reduce MERS-CoV replication by inhibiting inosine monophosphate dehydrogenase (IMPDH) and guanine monophosphate synthesis with an EC_{50} of 2.87 µM by cell-based ELISA in Vero E6 cells (Table 2; Figure 5(63)) [60].

K22 is a spectrum inhibitor which can inhibit MERS-CoV replication by reducing the formation of double membrane vesicles (DMVs) and by the near-complete inhibition of RNA synthesis (Figure 5(64)) [25,71].

BCX4430, an adenosine analogue that acts as a non-obligate RNA chain terminator to inhibit viral RNA polymerase function, can inhibit MERS-CoV infection with EC_{50} of 68.4 µM in Vero E6 cells by highly charged ions (HCIs)-based analysis and CC_{50} of >100 µM by neutral-red uptake (Table 2; Figure 5(65)) [25,62].

Fleximer nucleoside analogues of acyclovir are doubly flexible nucleoside analogues based on the acyclic sugar scaffold of acyclovir and the flex-base moiety in fleximers responsible for inhibiting RNA-dependent RNA polymerase (RdRp) [25,63]. The target fleximer analogue 2 can inhibit MERS-CoV infection with EC_{50} of 27 μM and CC_{50} of 149 μM in Huh-7 cells, but EC_{50} of 23 μM and CC_{50} of 71 μM in Vero cells (Table 2; Figure 5(66)) [63].

Interferon alpha1 (IFN-α1) and cyclosporine (CsA) have additive or synergistic effects in limiting MERS-CoV replication in ex vivo cultures of human bronchus (Figure 5(67)) [72]. In addition, the combined treatment of IFN-α1 and CsA has the most potent effect on inducing interferon-stimulated genes (ISGs) in both lung (24 hpi) and bronchial (56 hpi) tissues [72].

Saracatinib, a potent inhibitor of the Src-family of tyrosine kinases (SFK), potently inhibits MERS-CoV with an EC_{50} of about 3 μM in Huh-7 cells (Table 2; Figure 5(68)) [64]. It possibly inhibits MERS-CoV replication through the suppression of SFK signaling pathways at the early stages of the viral life cycle [64]. In addition, another seven compounds, primarily classified as antiprotozoal, anticancer, and antipsychotic, were also determined by complete dose-response analyses (Table 2; Figure 5(69–75)) [64].

A spectrum-inhibitor, FA-613, can inhibit MERS-CoV with an EC_{50} of ~10 μM in the interferon-competent cell line of Huh-7 cells, as shown by MTT assay (Table 2; Figure 5(76)) [65].

4. Strategies for Developing Small-Molecule MERS-CoV Inhibitors

The luciferase-based biosensor assay is a cell-based screening assay for selecting MERS-CoV-specific or broad-spectrum coronavirus PL_{pro} and $3CL_{pro}$ inhibitors [53]. HEK293T cells were transfected by two artificial plasmids: protease expression plasmids and biosensor expression plasmids [53]. Protease expression plasmids contain the sequence of MERS-CoV PL_{pro}, the nonstructural proteins nsp4 and nsp5, as well as the N-terminal 6 region. Biosensor expression plasmids contain a circularly permuted *Photuris pennsylvanica* luciferase and the amino sequence of cleavage site of PL_{pro} or $3CL_{pro}$ [53]. After cell transfection and coexpression of a MERS-CoV protease domain with a cleavage-activated luciferase substrate, transfected live cells allow for both endpoint evaluation and live cell imaging profiles of protease activity [53]. This novel method can be performed in a biosafety level 2 research laboratory to evaluate the ability to inhibit the CoV protease activity of existing and new drugs [53].

Pseudovirus-based screening assays have been developed for identifying antiviral compounds in the MERS-CoV life cycle without using infectious viruses. The MERS-CoV pseudovirus allows for single-cycle infection of a variety of cells expressing DPP4, and results are consistent with those from a live MERS-CoV-based inhibition assay. More importantly, the pseudovirus assay can be carried out in a BSL-2, rather than a BSL-3 facility [9]. VSV- and HIV-luciferase pseudotyped with the MERS-CoV S protein are two more approaches [27].

Structure-Guided Design and Optimization of Small Molecules is a strategy that involves embodying a piperidine moiety as a design element to attain optimal pharmacological activity and protein kinase property [52]. This strategy permits the resultant hybrid inhibitor to participate in favorable binding interactions with the S3 and S4 subsites of $3CL_{pro}$ by attaching the piperidine moiety to a dipeptidyl component [52].

Ubiquitin-like domain 2 (Ubl2) is immediately adjacent to the N-terminus of the PL_{pro} domain in coronavirus polyproteins. In the past, the role of Ubl2 in PL_{pro} has remained undefined. However, evidence indicates that removing the Ubl2 domain from MERS PL_{pro} has no effect on its ability to process the viral polyprotein or act as an interferon antagonist, which involves deubiquitinating and deISGylating cellular proteins [73].

Analyzing the transcriptome of hosts infected with MERS-CoV can provide insight into how MERS-CoV infection influences and interacts with host cells. Josset et al. [74] infected a lung epithelial cell line, Calu3, with MERS-CoV and analyzed the transcriptome to identify inhibitory compounds resident in host factors that could be exploited as antiviral therapeutics. This approach can be used to identify host factors beneficial for virus propagation, thus establishing appropriate targets for existing or new antiviral inhibitors.

5. Conclusions

As a positive-sense, single-stranded RNA virus, MERS-CoV utilizes host cellular components to accomplish various physiological processes, including viral entry, genomic replication, and the assembly and budding of virions, thereby resulting in pathological damage to the host. Therefore, various stages of virus life cycle could be potential targets for developing small-molecule antiviral inhibitors. Inhibitors blocking MERS-CoV entry into host cells, viral protease inhibitors, and inhibitors targeting host cells and many other small-molecule inhibitors with defined or undefined mechanisms of action are summarized in this review.

Any compounds that interfere with virus infection may be harmful to host cells. Therefore, the establishment of a safety profile is essential. Furthermore, an antiviral inhibitor should effectively inhibit the growth of the virus because a small amount of virion replication can lead to resistant mutations. The advantages of small-molecule inhibitors include low price, stability, and the convenience of oral administration. Three main approaches are currently used to develop MERS-CoV small-molecule inhibitors. The first is the de novo synthesis of inhibitors targeting the unique structure in the proteins of MERS-CoV appearing in its infection process. The second approach involves screening inhibitors against MERS-CoV infection from an existing drug database by various chemical synthesis strategies. The third approach involves changing the chemical group of a fully developed drug to enhance its pharmacological activity against MERS-CoV. More novel strategies in improving the efficacy of screening small-molecule inhibitors are anticipated to reduce the threat of future MERS-CoV infections.

Author Contributions: R.L., L.W., N.Z., X.D., M.S., Y.S., L.H., and C.H. drafted the manuscript. T.Y., S.J., and F.Y. revised and edited the manuscript.

Funding: This work was supported by grants from the National Natural Science Foundation of China (81501735 and 81601761), Hebei Province's Program for Talents Returning from Studying Overseas (CN201707), a starting grant from Hebei Agricultural University (ZD2016026), and the Program for Youth Talent of Higher Learning Institutions of Hebei Province (BJ2018045).

Conflicts of Interest: The authors declare no conflict of interest.

References

1. Cotten, M.; Watson, S.J.; Zumla, A.I.; Makhdoom, H.Q.; Palser, A.L.; Ong, S.H.; Al Rabeeah, A.A.; Alhakeem, R.F.; Assiri, A.; Al-Tawfiq, J.A.; et al. Spread, circulation, and evolution of the middle east respiratory syndrome coronavirus. *mBio* **2014**, *5*, e01062-13. [CrossRef] [PubMed]
2. Chan, J.F.; Lau, S.K.; Woo, P.C. The emerging novel middle east respiratory syndrome coronavirus: The "knowns" and "unknowns". *J. Formos. Med. Assoc.* **2013**, *112*, 372–381. [CrossRef] [PubMed]
3. Arabi, Y.M.; Arifi, A.A.; Balkhy, H.H.; Najm, H.; Aldawood, A.S.; Ghabashi, A.; Hawa, H.; Alothman, A.; Khaldi, A.; Al Raiy, B. Clinical course and outcomes of critically ill patients with middle east respiratory syndrome coronavirus infection. *Ann. Intern. Med.* **2014**, *160*, 389–397. [CrossRef] [PubMed]
4. Drosten, C.; Meyer, B.; Muller, M.A.; Corman, V.M.; Al-Masri, M.; Hossain, R.; Madani, H.; Sieberg, A.; Bosch, B.J.; Lattwein, E.; et al. Transmission of mers-coronavirus in household contacts. *N. Engl. J. Med.* **2014**, *371*, 828–835. [CrossRef] [PubMed]
5. Lu, L.; Liu, Q.; Du, L.; Jiang, S. Middle east respiratory syndrome coronavirus (mers-cov): Challenges in identifying its source and controlling its spread. *Microbes Infect.* **2013**, *15*, 625–629. [CrossRef]
6. Falzarano, D.; de Wit, E.; Rasmussen, A.L.; Feldmann, F.; Okumura, A.; Scott, D.P.; Brining, D.; Bushmaker, T.; Martellaro, C.; Baseler, L.; et al. Treatment with interferon-alpha2b and ribavirin improves outcome in mers-cov-infected rhesus macaques. *Nat. Med.* **2013**, *19*, 1313–1317. [CrossRef]
7. Dyall, J.; Coleman, C.M.; Hart, B.J.; Venkataraman, T.; Holbrook, M.R.; Kindrachuk, J.; Johnson, R.F.; Olinger, G.G., Jr.; Jahrling, P.B.; Laidlaw, M.; et al. Repurposing of clinically developed drugs for treatment of middle east respiratory syndrome coronavirus infection. *Antimicrob. Agents Chemother.* **2014**, *58*, 4885–4893. [CrossRef]

8. Lu, L.; Xia, S.; Ying, T.; Jiang, S. Urgent development of effective therapeutic and prophylactic agents to control the emerging threat of middle east respiratory syndrome (mers). *Emerg. Microbes Infect.* **2015**, *4*, e37. [CrossRef]

9. Gierer, S.; Bertram, S.; Kaup, F.; Wrensch, F.; Heurich, A.; Kramer-Kuhl, A.; Welsch, K.; Winkler, M.; Meyer, B.; Drosten, C.; et al. The spike protein of the emerging betacoronavirus emc uses a novel coronavirus receptor for entry, can be activated by tmprss2, and is targeted by neutralizing antibodies. *J. Virol.* **2013**, *87*, 5502–5511. [CrossRef]

10. Bertram, S.; Dijkman, R.; Habjan, M.; Heurich, A.; Gierer, S.; Glowacka, I.; Welsch, K.; Winkler, M.; Schneider, H.; Hofmann-Winkler, H.; et al. Tmprss2 activates the human coronavirus 229e for cathepsin-independent host cell entry and is expressed in viral target cells in the respiratory epithelium. *J. Virol.* **2013**, *87*, 6150–6160. [CrossRef]

11. Du, L.Y.; Yang, Y.; Zhou, Y.S.; Lu, L.; Li, F.; Jiang, S.B. Mers-cov spike protein: A key target for antivirals. *Expert Opin. Ther. Target* **2017**, *21*, 131–143. [CrossRef] [PubMed]

12. Xia, S.; Liu, Q.; Wang, Q.; Sun, Z.W.; Su, S.; Dub, L.Y.; Ying, T.L.; Lu, L.; Jiang, S.B. Middle east respiratory syndrome coronavirus (mers-cov) entry inhibitors targeting spike protein. *Virus Res.* **2014**, *194*, 200–210. [CrossRef] [PubMed]

13. Forni, D.; Filippi, G.; Cagliani, R.; De Gioia, L.; Pozzoli, U.; Al-Daghri, N.; Clerici, M.; Sironi, M. The heptad repeat region is a major selection target in mers-cov and related coronaviruses. *Sci. Rep.* **2015**, *5*, 14480. [CrossRef] [PubMed]

14. Gao, J.; Lu, G.; Qi, J.; Li, Y.; Wu, Y.; Deng, Y.; Geng, H.; Li, H.; Wang, Q.; Xiao, H.; et al. Structure of the fusion core and inhibition of fusion by a heptad repeat peptide derived from the s protein of middle east respiratory syndrome coronavirus. *J. Virol.* **2013**, *87*, 13134–13140. [CrossRef] [PubMed]

15. Raj, V.S.; Mou, H.; Smits, S.L.; Dekkers, D.H.; Muller, M.A.; Dijkman, R.; Muth, D.; Demmers, J.A.; Zaki, A.; Fouchier, R.A.; et al. Dipeptidyl peptidase 4 is a functional receptor for the emerging human coronavirus-emc. *Nature* **2013**, *495*, 251–254. [CrossRef] [PubMed]

16. Lu, L.; Liu, Q.; Zhu, Y.; Chan, K.H.; Qin, L.; Li, Y.; Wang, Q.; Chan, J.F.; Du, L.; Yu, F.; et al. Structure-based discovery of middle east respiratory syndrome coronavirus fusion inhibitor. *Nat. Commun.* **2014**, *5*, 3067. [CrossRef] [PubMed]

17. Xu, Y.; Lou, Z.; Liu, Y.; Pang, H.; Tien, P.; Gao, G.F.; Rao, Z. Crystal structure of severe acute respiratory syndrome coronavirus spike protein fusion core. *J. Boil. Chem.* **2004**, *279*, 49414–49419. [CrossRef]

18. Channappanavar, R.; Lu, L.; Xia, S.; Du, L.; Meyerholz, D.K.; Perlman, S.; Jiang, S. Protective effect of intranasal regimens containing peptidic middle east respiratory syndrome coronavirus fusion inhibitor against mers-cov infection. *J. Infect. Dis.* **2015**, *212*, 1894–1903. [CrossRef]

19. Tao, X.; Garron, T.; Agrawal, A.S.; Algaissi, A.; Peng, B.H.; Wakamiya, M.; Chan, T.S.; Lu, L.; Du, L.; Jiang, S.; et al. Characterization and demonstration of the value of a lethal mouse model of middle east respiratory syndrome coronavirus infection and disease. *J. Virol.* **2016**, *90*, 57–67. [CrossRef]

20. Wang, C.; Xia, S.; Zhang, P.; Zhang, T.; Wang, W.; Tian, Y.; Meng, G.; Jiang, S.; Liu, K. Discovery of hydrocarbon-stapled short alpha-helical peptides as promising middle east respiratory syndrome coronavirus (mers-cov) fusion inhibitors. *J. Med. Chem.* **2018**, *61*, 2018–2026. [CrossRef]

21. Zhao, H.; Zhou, J.; Zhang, K.; Chu, H.; Liu, D.; Poon, V.K.; Chan, C.C.; Leung, H.C.; Fai, N.; Lin, Y.P.; et al. A novel peptide with potent and broad-spectrum antiviral activities against multiple respiratory viruses. *Sci. Rep.* **2016**, *6*, 22008. [CrossRef] [PubMed]

22. Wang, C.; Zhao, L.; Xia, S.; Zhang, T.; Cao, R.; Liang, G.; Li, Y.; Meng, G.; Wang, W.; Shi, W.; et al. De novo design of alpha-helical lipopeptides targeting viral fusion proteins: A promising strategy for relatively broad-spectrum antiviral drug discovery. *J. Med. Chem.* **2018**, *61*, 8734–8745. [CrossRef] [PubMed]

23. Wang, X.; Zou, P.; Wu, F.; Lu, L.; Jiang, S. Development of small-molecule viral inhibitors targeting various stages of the life cycle of emerging and re-emerging viruses. *Front. Med.* **2017**, *11*, 449–461. [CrossRef] [PubMed]

24. Jiang, S.B.; Tao, X.R.; Xia, S.; Garron, T.; Yu, F.; Du, L.Y.; Lu, L.; Tseng, C.T.K. Intranasally administered peptidic viral fusion inhibitor protected hdpp4 transgenic mice from mers-cov infection. *Lancet* **2015**, *386*, S44. [CrossRef]

25. Zumla, A.; Chan, J.F.; Azhar, E.I.; Hui, D.S.; Yuen, K.Y. Coronaviruses—Drug discovery and therapeutic options. *Nat. Rev. Drug Discov.* **2016**, *15*, 327–347. [CrossRef]

26. Liu, Q.; Xia, S.; Sun, Z.; Wang, Q.; Du, L.; Lu, L.; Jiang, S. Testing of middle east respiratory syndrome coronavirus replication inhibitors for the ability to block viral entry. *Antimicrob. Agents Chemother.* **2015**, *59*, 742–744. [CrossRef] [PubMed]

27. Zhao, G.; Du, L.; Ma, C.; Li, Y.; Li, L.; Poon, V.K.; Wang, L.; Yu, F.; Zheng, B.J.; Jiang, S.; et al. A safe and convenient pseudovirus-based inhibition assay to detect neutralizing antibodies and screen for viral entry inhibitors against the novel human coronavirus mers-cov. *Virol. J.* **2013**, *10*, 266. [CrossRef] [PubMed]

28. Bertram, S.; Glowacka, I.; Blazejewska, P.; Soilleux, E.; Allen, P.; Danisch, S.; Steffen, I.; Choi, S.Y.; Park, Y.; Schneider, H.; et al. Tmprss2 and tmprss4 facilitate trypsin-independent spread of influenza virus in caco-2 cells. *J. Virol.* **2010**, *84*, 10016–10025. [CrossRef]

29. Shirogane, Y.; Takeda, M.; Iwasaki, M.; Ishiguro, N.; Takeuchi, H.; Nakatsu, Y.; Tahara, M.; Kikuta, H.; Yanagi, Y. Efficient multiplication of human metapneumovirus in vero cells expressing the transmembrane serine protease tmprss2. *J. Virol.* **2008**, *82*, 8942–8946. [CrossRef]

30. Matsuyama, S.; Nagata, N.; Shirato, K.; Kawase, M.; Takeda, M.; Taguchi, F. Efficient activation of the severe acute respiratory syndrome coronavirus spike protein by the transmembrane protease tmprss2. *J. Virol.* **2010**, *84*, 12658–12664. [CrossRef]

31. Shirato, K.; Kawase, M.; Matsuyama, S. Middle east respiratory syndrome coronavirus infection mediated by the transmembrane serine protease tmprss2. *J. Virol.* **2013**, *87*, 12552–12561. [CrossRef]

32. Zhou, Y.; Vedantham, P.; Lu, K.; Agudelo, J.; Carrion, R., Jr.; Nunneley, J.W.; Barnard, D.; Pohlmann, S.; McKerrow, J.H.; Renslo, A.R.; et al. Protease inhibitors targeting coronavirus and filovirus entry. *Antivir. Res.* **2015**, *116*, 76–84. [CrossRef] [PubMed]

33. Engel, J.C.; Doyle, P.S.; Hsieh, I.; McKerrow, J.H. Cysteine protease inhibitors cure an experimental trypanosoma cruzi infection. *J. Exp. Med.* **1998**, *188*, 725–734. [CrossRef] [PubMed]

34. De Wilde, A.H.; Jochmans, D.; Posthuma, C.C.; Zevenhoven-Dobbe, J.C.; van Nieuwkoop, S.; Bestebroer, T.M.; van den Hoogen, B.G.; Neyts, J.; Snijder, E.J. Screening of an fda-approved compound library identifies four small-molecule inhibitors of middle east respiratory syndrome coronavirus replication in cell culture. *Antimicrob. Agents Chemother.* **2014**, *58*, 4875–4884. [CrossRef] [PubMed]

35. Burkard, C.; Verheije, M.H.; Haagmans, B.L.; van Kuppeveld, F.J.; Rottier, P.J.; Bosch, B.J.; de Haan, C.A. Atp1a1-mediated src signaling inhibits coronavirus entry into host cells. *J. Virol.* **2015**, *89*, 4434–4448. [CrossRef] [PubMed]

36. Kim, J.Y.; Kim, Y.I.; Park, S.J.; Kim, I.K.; Choi, Y.K.; Kim, S.H. Safe, high-throughput screening of natural compounds of mers-cov entry inhibitors using a pseudovirus expressing mers-cov spike protein. *Int. J. Antimicrob. Agents* **2018**, *52*, 730–732. [CrossRef] [PubMed]

37. Millet, J.K.; Whittaker, G.R. Host cell entry of middle east respiratory syndrome coronavirus after two-step, furin-mediated activation of the spike protein. *Proc. Natl. Acad. Sci. USA* **2014**, *111*, 15214–15219. [CrossRef] [PubMed]

38. Chou, C.Y.; Lai, H.Y.; Chen, H.Y.; Cheng, S.C.; Cheng, K.W.; Chou, Y.W. Structural basis for catalysis and ubiquitin recognition by the severe acute respiratory syndrome coronavirus papain-like protease. *Acta Crystallogr. Sect. D Biol. Crystallogr.* **2014**, *70*, 572–581. [CrossRef] [PubMed]

39. Han, Y.S.; Chang, G.G.; Juo, C.G.; Lee, H.J.; Yeh, S.H.; Hsu, J.T.; Chen, X. Papain-like protease 2 (plp2) from severe acute respiratory syndrome coronavirus (sars-cov): Expression, purification, characterization, and inhibition. *Biochemistry* **2005**, *44*, 10349–10359. [CrossRef]

40. Lee, H.; Lei, H.; Santarsiero, B.D.; Gatuz, J.L.; Cao, S.; Rice, A.J.; Patel, K.; Szypulinski, M.Z.; Ojeda, I.; Ghosh, A.K.; et al. Inhibitor recognition specificity of mers-cov papain-like protease may differ from that of sars-cov. *ACS Chem. Biol.* **2015**, *10*, 1456–1465. [CrossRef]

41. Thiel, V.; Ivanov, K.A.; Putics, A.; Hertzig, T.; Schelle, B.; Bayer, S.; Weissbrich, B.; Snijder, E.J.; Rabenau, H.; Doerr, H.W.; et al. Mechanisms and enzymes involved in sars coronavirus genome expression. *J. Gen. Virol.* **2003**, *84*, 2305–2315. [CrossRef] [PubMed]

42. Harcourt, B.H.; Jukneliene, D.; Kanjanahaluethai, A.; Bechill, J.; Severson, K.M.; Smith, C.M.; Rota, P.A.; Baker, S.C. Identification of severe acute respiratory syndrome coronavirus replicase products and characterization of papain-like protease activity. *J. Virol.* **2004**, *78*, 13600–13612. [CrossRef] [PubMed]

43. Mielech, A.M.; Kilianski, A.; Baez-Santos, Y.M.; Mesecar, A.D.; Baker, S.C. Mers-cov papain-like protease has deisgylating and deubiquitinating activities. *Virology* **2014**, *450–451*, 64–70. [CrossRef] [PubMed]

44. Yang, X.; Chen, X.; Bian, G.; Tu, J.; Xing, Y.; Wang, Y.; Chen, Z. Proteolytic processing, deubiquitinase and interferon antagonist activities of middle east respiratory syndrome coronavirus papain-like protease. *J. Gen. Virol.* **2014**, *95*, 614–626. [CrossRef] [PubMed]

45. Galkin, A.; Kulakova, L.; Lim, K.; Chen, C.Z.; Zheng, W.; Turko, I.V.; Herzberg, O. Structural basis for inactivation of giardia lamblia carbamate kinase by disulfiram. *J. Boil. Chem.* **2014**, *289*, 10502–10509. [CrossRef] [PubMed]

46. Paranjpe, A.; Zhang, R.; Ali-Osman, F.; Bobustuc, G.C.; Srivenugopal, K.S. Disulfiram is a direct and potent inhibitor of human o6-methylguanine-DNA methyltransferase (mgmt) in brain tumor cells and mouse brain and markedly increases the alkylating DNA damage. *Carcinogenesis* **2014**, *35*, 692–702. [CrossRef] [PubMed]

47. Lin, M.H.; Moses, D.C.; Hsieh, C.H.; Cheng, S.C.; Chen, Y.H.; Sun, C.Y.; Chou, C.Y. Disulfiram can inhibit mers and sars coronavirus papain-like proteases via different modes. *Antivir. Res* **2018**, *150*, 155–163. [CrossRef]

48. Kumar, V.; Tan, K.P.; Wang, Y.M.; Lin, S.W.; Liang, P.H. Identification, synthesis and evaluation of sars-cov and mers-cov 3c-like protease inhibitors. *Bioorg. Med. Chem.* **2016**, *24*, 3035–3042. [CrossRef]

49. Needle, D.; Lountos, G.T.; Waugh, D.S. Structures of the middle east respiratory syndrome coronavirus 3c-like protease reveal insights into substrate specificity. *Acta Crystallogr. Sect. D Biol. Crystallogr.* **2015**, *71*, 1102–1111. [CrossRef]

50. Hsu, M.F.; Kuo, C.J.; Chang, K.T.; Chang, H.C.; Chou, C.C.; Ko, T.P.; Shr, H.L.; Chang, G.G.; Wang, A.H.; Liang, P.H. Mechanism of the maturation process of sars-cov 3cl protease. *J. Biol. Chem.* **2005**, *280*, 31257–31266. [CrossRef]

51. Hu, T.; Zhang, Y.; Li, L.; Wang, K.; Chen, S.; Chen, J.; Ding, J.; Jiang, H.; Shen, X. Two adjacent mutations on the dimer interface of sars coronavirus 3c-like protease cause different conformational changes in crystal structure. *Virology* **2009**, *388*, 324–334. [CrossRef]

52. Galasiti Kankanamalage, A.C.; Kim, Y.; Damalanka, V.C.; Rathnayake, A.D.; Fehr, A.R.; Mehzabeen, N.; Battaile, K.P.; Lovell, S.; Lushington, G.H.; Perlman, S.; et al. Structure-guided design of potent and permeable inhibitors of mers coronavirus 3cl protease that utilize a piperidine moiety as a novel design element. *Eur. J. Med. Chem.* **2018**, *150*, 334–346. [CrossRef]

53. Kilianski, A.; Mielech, A.M.; Deng, X.; Baker, S.C. Assessing activity and inhibition of middle east respiratory syndrome coronavirus papain-like and 3c-like proteases using luciferase-based biosensors. *J. Virol.* **2013**, *87*, 11955–11962. [CrossRef] [PubMed]

54. Kumar, V.; Shin, J.S.; Shie, J.J.; Ku, K.B.; Kim, C.; Go, Y.Y.; Huang, K.F.; Kim, M.; Liang, P.H. Identification and evaluation of potent middle east respiratory syndrome coronavirus (mers-cov) 3cl(pro) inhibitors. *Antivir. Res.* **2017**, *141*, 101–106. [CrossRef] [PubMed]

55. Kim, Y.; Liu, H.; Galasiti Kankanamalage, A.C.; Weerasekara, S.; Hua, D.H.; Groutas, W.C.; Chang, K.O.; Pedersen, N.C. Reversal of the progression of fatal coronavirus infection in cats by a broad-spectrum coronavirus protease inhibitor. *PLoS Pathog.* **2016**, *12*, e1005531.

56. Ren, Z.; Yan, L.; Zhang, N.; Guo, Y.; Yang, C.; Lou, Z.; Rao, Z. The newly emerged sars-like coronavirus hcov-emc also has an "achilles' heel": Current effective inhibitor targeting a 3c-like protease. *Protein Cell* **2013**, *4*, 248–250. [CrossRef] [PubMed]

57. Muller, C.; Schulte, F.W.; Lange-Grunweller, K.; Obermann, W.; Madhugiri, R.; Pleschka, S.; Ziebuhr, J.; Hartmann, R.K.; Grunweller, A. Broad-spectrum antiviral activity of the eif4a inhibitor silvestrol against corona- and picornaviruses. *Antivir. Res.* **2018**, *150*, 123–129. [CrossRef]

58. Agostini, M.L.; Andres, E.L.; Sims, A.C.; Graham, R.L.; Sheahan, T.P.; Lu, X.; Smith, E.C.; Case, J.B.; Feng, J.Y.; Jordan, R.; et al. Coronavirus susceptibility to the antiviral remdesivir (gs-5734) is mediated by the viral polymerase and the proofreading exoribonuclease. *mBio* **2018**, *9*, e00221-18. [CrossRef]

59. Cong, Y.; Hart, B.J.; Gross, R.; Zhou, H.; Frieman, M.; Bollinger, L.; Wada, J.; Hensley, L.E.; Jahrling, P.B.; Dyall, J.; et al. Mers-cov pathogenesis and antiviral efficacy of licensed drugs in human monocyte-derived antigen-presenting cells. *PLoS ONE* **2018**, *13*, e0194868. [CrossRef]

60. Hart, B.J.; Dyall, J.; Postnikova, E.; Zhou, H.; Kindrachuk, J.; Johnson, R.F.; Olinger, G.G., Jr.; Frieman, M.B.; Holbrook, M.R.; Jahrling, P.B.; et al. Interferon-beta and mycophenolic acid are potent inhibitors of middle east respiratory syndrome coronavirus in cell-based assays. *J. Gen. Virol.* **2014**, *95*, 571–577. [CrossRef]

61. Tao, X.; Mei, F.; Agrawal, A.; Peters, C.J.; Ksiazek, T.G.; Cheng, X.; Tseng, C.T. Blocking of exchange proteins directly activated by camp leads to reduced replication of middle east respiratory syndrome coronavirus. *J. Virol.* **2014**, *88*, 3902–3910. [CrossRef] [PubMed]

62. Warren, T.K.; Wells, J.; Panchal, R.G.; Stuthman, K.S.; Garza, N.L.; Van Tongeren, S.A.; Dong, L.; Retterer, C.J.; Eaton, B.P.; Pegoraro, G.; et al. Protection against filovirus diseases by a novel broad-spectrum nucleoside analogue bcx4430. *Nature* **2014**, *508*, 402–405. [CrossRef] [PubMed]

63. Peters, H.L.; Jochmans, D.; de Wilde, A.H.; Posthuma, C.C.; Snijder, E.J.; Neyts, J.; Seley-Radtke, K.L. Design, synthesis and evaluation of a series of acyclic fleximer nucleoside analogues with anti-coronavirus activity. *Bioorg. Med. Chem. Lett.* **2015**, *25*, 2923–2926. [CrossRef] [PubMed]

64. Shin, J.S.; Jung, E. Saracatinib inhibits middle east respiratory syndrome-coronavirus replication in vitro. *Viruses* **2018**, *10*, 283. [CrossRef] [PubMed]

65. Cheung, N.N.; Lai, K.K.; Dai, J.; Kok, K.H.; Chen, H.; Chan, K.H.; Yuen, K.Y.; Kao, R.Y.T. Broad-spectrum inhibition of common respiratory rna viruses by a pyrimidine synthesis inhibitor with involvement of the host antiviral response. *J. Gen. Virol.* **2017**, *98*, 946–954. [CrossRef] [PubMed]

66. Lin, S.C.; Ho, C.T.; Chuo, W.H.; Li, S.; Wang, T.T.; Lin, C.C. Effective inhibition of mers-cov infection by resveratrol. *BMC Infect. Dis.* **2017**, *17*, 144. [CrossRef] [PubMed]

67. Gupta, P.B.; Onder, T.T.; Jiang, G.; Tao, K.; Kuperwasser, C.; Weinberg, R.A.; Lander, E.S. Identification of selective inhibitors of cancer stem cells by high-throughput screening. *Cell* **2009**, *138*, 645–659. [CrossRef]

68. Huczynski, A. Polyether ionophores-promising bioactive molecules for cancer therapy. *Bioorg. Med. Chem. Lett.* **2012**, *22*, 7002–7010. [CrossRef]

69. McFadden, G. Gleevec casts a pox on poxviruses. *Nat. Med.* **2005**, *11*, 711–712. [CrossRef]

70. Adedeji, A.O.; Singh, K.; Kassim, A.; Coleman, C.M.; Elliott, R.; Weiss, S.R.; Frieman, M.B.; Sarafianos, S.G. Evaluation of ssya10-001 as a replication inhibitor of severe acute respiratory syndrome, mouse hepatitis, and middle east respiratory syndrome coronaviruses. *Antimicrob. Agents Chemother.* **2014**, *58*, 4894–4898. [CrossRef]

71. Lundin, A.; Dijkman, R.; Bergstrom, T.; Kann, N.; Adamiak, B.; Hannoun, C.; Kindler, E.; Jonsdottir, H.R.; Muth, D.; Kint, J.; et al. Targeting membrane-bound viral rna synthesis reveals potent inhibition of diverse coronaviruses including the middle east respiratory syndrome virus. *PLoS Pathog.* **2014**, *10*, e1004166. [CrossRef] [PubMed]

72. Li, H.S.; Kuok, D.I.T.; Cheung, M.C.; Ng, M.M.T.; Ng, K.C.; Hui, K.P.Y.; Peiris, J.S.M.; Chan, M.C.W.; Nicholls, J.M. Effect of interferon alpha and cyclosporine treatment separately and in combination on middle east respiratory syndrome coronavirus (mers-cov) replication in a human in-vitro and ex-vivo culture model. *Antivir. Res.* **2018**, *155*, 89–96. [CrossRef] [PubMed]

73. Clasman, J.R.; Baez-Santos, Y.M.; Mettelman, R.C.; O'Brien, A.; Baker, S.C.; Mesecar, A.D. X-ray structure and enzymatic activity profile of a core papain-like protease of mers coronavirus with utility for structure-based drug design. *Sci. Rep.* **2017**, *7*, 40292. [CrossRef]

74. Josset, L.; Menachery, V.D.; Gralinski, L.E.; Agnihothram, S.; Sova, P.; Carter, V.S.; Yount, B.L.; Graham, R.L.; Baric, R.S.; Katze, M.G. Cell host response to infection with novel human coronavirus emc predicts potential antivirals and important differences with sars coronavirus. *mBio* **2013**, *4*, e00165-13. [CrossRef] [PubMed]

MDPI

St. Alban-Anlage 66

4052 Basel

Switzerland

Tel. +41 61 683 77 34

Fax +41 61 302 89 18

www.mdpi.com

Viruses Editorial Office

E-mail: viruses@mdpi.com

www.mdpi.com/journal/viruses